BASIC
Mathematical
Fourth Edition
Skills
WITH GEOMETRY

INSTRUCTOR'S EDITION

James Streeter
Late Professor of Mathematics
Clackamas Community College

Donald Hutchison
Clackamas Community College

Louis Hoelzle
Bucks County Community College

D1104628

WCB McGraw-Hill

Boston, Massachusetts   Burr Ridge, Illinois   Dubuque, Iowa
Madison, Wisconsin   New York, New York   San Francisco, California   St. Louis, Missouri

WCB/McGraw-Hill

A Division of The **McGraw-Hill** Companies

Basic Mathematical Skills with Geometry

Instructor's Edition for Basic Mathematical Skills with Geometry

This book is printed on acid-free paper.

1 2 3 4 5 6 7 8 9 0 VNH VNH 9 0 0 9 8 7

ISBN 0-07-063266-9 (Student Edition)

ISBN 0-07-063270-7 (Instructor's Edition)

Publisher: Tom Casson
Sponsoring editor: Jack Shira
Marketing manager: Michelle Sala
Project manager: Eva Marie Strock
Production supervisor: Tanya Nigh
Designer: Wanda Kofax
Cover designer: Linear Design Group
Compositor: York Graphic Services, Inc.
Typeface: Times New Roman
Printer: Von Hoffmann Press, Inc.

Library of Congress Cataloging-in-Publication Data

Streeter, James (James A.)
    Basic mathematical skills with geometry / James Streeter, Donald Hutchison, Louis Hoelzle.—4th ed.
        p.   cm.
    Includes index.
    ISBN 0-07-063266-9 (student's ed.   acid-free paper),   —ISBN 0-07-063270-7 (instructor's ed.   acid-free paper)
    1. Arithmetic.   2. Algebra.   3. Geometry.   I. Hutchison, Donald, 1948–   .   II. Hoelzle, Louis F.   III. Title.
QA107.A783   1998
513′. 14—dc21   97–36774
                                                                            CIP

http://www.mhhe.com

ABOUT THE AUTHORS

While a graduate student at the University of Washington, **James Streeter** paid for his education as a math tutor. It was here that he began to formulate the ideas that would eventually become this package. Upon graduation, he taught for 2 years at Centralia Community College. In 1968 he moved on to Clackamas Community College to become the school's first mathematics chair.

At the college, Jim recognized that he faced a very different population than the one he had tutored at the University of Washington. Jim was convinced that to reach the maximum number of these students, he would have to utilize every medium available to him. Jim opened a math lab that included CAI, original slides and tapes (which were eventually published by Harper & Row), and original worksheets and text materials. With the assistance of the people at McGraw-Hill, that package has been refined to include media and supplements that did not even exist when this project began.

Donald Hutchison spent his first 10 years of teaching working with disadvantaged students. He taught in an intercity elementary school and an intercity high school. He also worked for 2 years at Wassaic State School in New York and 2 years at the Portland Habilitation Center. He worked with both physically and mentally disadvantaged students in these two settings.

In 1982, Don was hired by Jim Streeter to teach at Clackamas Community College. In 1989, Don became Chair of the Mathematics Department at the college. It was here at Clackamas that Don discovered two things that, along with his family, form the focus for his life. Jim introduced Don to the joy of writing (with the first edition of *Beginning Algebra*), and Jack Scrivener converted him to a born-again environmentalist.

Don is also active in several professional organizations. He was a member of the ACM committee that undertook the writing of computer curriculum for the 2-year college. From 1989 to 1994 he was the Chair of the Technology in Mathematics Education Committee for AMATYC. He was President of ORMATYC from 1996 to 1998.

Louis Hoelzle has been teaching at Bucks County Community College for 27 years. In 1989, Lou became Chair of the Mathematics Department at Bucks County Community College. He has taught the entire range of courses from Arithmetic to Calculus, giving him an excellent view of the current and future needs of developmental students.

Over the past 34 years, Lou has also taught Physics courses at 4-year colleges, which has enabled him to have the perspective of the practical applications of mathematics. In addition, Lou has extensively reviewed manuscripts and written several solutions manuals for major textbooks. In these materials he has focused on writing for the student.

Lou is also active in professional organizations. He has served on the Placement and Assessment Committee for AMATYC since 1989.

iii

This book is dedicated to the family I grew up with: my parents, Jack and Melinda, and my brothers, Tom, Paul, Lee, and Dean. They formed the foundation of my being.

Don Hutchison

This book is dedicated to my parents, whose love and complete support were always there for me. They molded my character and ideas.

Louis Hoelzle

THIS SERIES IS DEDICATED TO THE MEMORY OF JAMES ARTHUR STREETER, AN ARTISAN WITH WORDS, A GENIUS WITH NUMBERS, AND A VIRTUOSO WITH PICTURES FROM 1940 UNTIL 1989.

CONTENTS

PREFACE

Statement of Philosophy

We believe that the key to learning mathematics, at any level, is active participation. When students are active participants in the learning process, they have the opportunity to construct their own mathematical ideas and make connections to previously studied material. Such participation leads to understanding, success, and confidence. We developed this text with this philosophy in mind. The *Check Yourself* exercises are designed to keep students involved and active with every page of exposition. The calculator references involve students actively in the development of mathematical ideas. The chapter-opening vignettes attract the interest of the students by describing a worker in an occupation to which mathematics is relevant. Many exercise sets have application problems, challenging exercises, writing exercises, and collaborative exercises. Each exercise is designed to awaken interest and insight within students. Not all the exercises will be appropriate for every student, but each one provides another opportunity for both instructor and student. Our hope is that every student who uses this text will be a better mathematical thinker as a result.

Changes from the Third Edition

As we set out to revise *Basic Mathematical Skills with Geometry,* we had to keep in mind that this is a successful text with a very supportive group of adopters. Our goal in this revision was to incorporate new elements to enhance an already proven system. In order to accomplish this goal, we regularly communicated with both current and potential users of the text. We also solicited contributions from professionals with considerable experience in the implementation of collaborative and writing activities in the classroom. We worked hard to incorporate these ideas throughout the text. Every potential change was sent to a set of reviewers. We were very pleased with the support we received from these reviewers. We believe collaborating with so many adept professionals (see the Acknowledgments section) has greatly enhanced this text.

Integration of Applications

Beginning with Chapter 1, in which we introduce the definition of perimeter, every chapter (and virtually every section) has a set of application problems relevant to the material being presented.

Writing in Mathematics

Almost every section includes a set of writing exercises. These exercises encourage students to both research and communicate mathematical ideas. We tried to build a model that helps students understand that being able to solve a problem is useful only if you first understand the problem and then are able to communicate your solution.

Collaborative Projects

A number of collaborative projects are included in this book, for two main reasons. First, they allow students to build group cooperation skills while facing challenging exercises. Second, our students have had fun doing them. We strongly encourage you to assign these projects.

Calculator Sections

In the previous three editions we included a few special sections on calculator usage. In this fourth edition we doubled the number of these sections. The topics include operations on whole numbers, operations on fractions, and exponents, and the sections discuss the use of several types of calculators, including the TI-82 and the TI-83, in carrying out these operations.

Pedagogical Features

This edition has taken seven major directions. The changes are presented below, with examples of each one. Each feature is designed to encourage, facilitate, and motivate problem solving among students. This goal is not just the nature of these books, it is the primary argument for the inclusion of mathematics in virtually every curriculum.

Pretests and Self-Tests

Each chapter begins with a pretest and ends with a self-test. These tests allow each student to perform a self-assessment. Each pretest provides a baseline from which students can measure success when the self-test is taken at the end of the chapter. The self-test also helps prepare students for classroom testing.

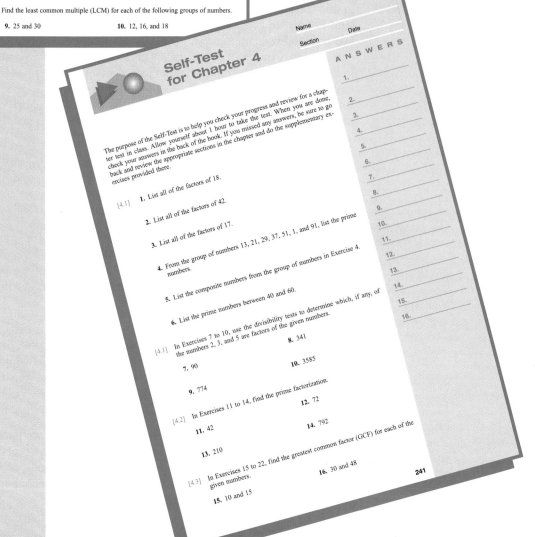

Pretest for Chapter 4

ANSWERS

1. _____
2. _____
3. _____
4. _____
5. _____
6. _____
7. _____
8. _____
9. _____
10. _____

Factors and Multiples

This pretest will point out any difficulties you may be having with the factors and multiples of whole numbers. Do all the problems. Then check your answers with those in the back of the book.

1. List all the factors of 42.

2. For the group of numbers 2, 3, 6, 7, 9, 17, 18, 21, and 23, list the prime and composite numbers.

Using divisibility tests, determine which, if any, of the numbers 2, 3, and 5 are factors of each of the following numbers.

3. 546   4. 5130

Write the prime factorizations for each of the following numbers.

5. 60   6. 350

Find the greatest common factor (GCF) for each of the following groups of numbers.

7. 12 and 32   8. 24, 36, and 42

Find the least common multiple (LCM) for each of the following groups of numbers.

9. 25 and 30   10. 12, 16, and 18

Self-Test for Chapter 4

Name _____ Date _____
Section _____

ANSWERS

1. _____
2. _____
3. _____
4. _____
5. _____
6. _____
7. _____
8. _____
9. _____
10. _____
11. _____
12. _____
13. _____
14. _____
15. _____
16. _____

The purpose of the Self-Test is to help you check your progress and review for a chapter test in class. Allow yourself about 1 hour to take the test. When you are done, check your answers in the back of the book. If you missed any answers, be sure to go back and review the appropriate sections in the chapter and do the supplementary exercises provided there.

[4.1] 1. List all of the factors of 18.

2. List all of the factors of 42.

3. List all of the factors of 17.

4. From the group of numbers 13, 21, 29, 37, 51, 1, and 91, list the prime numbers.

5. List the composite numbers from the group of numbers in Exercise 4.

6. List the prime numbers between 40 and 60.

[4.1] In Exercises 7 to 10, use the divisibility tests to determine which, if any, of the numbers 2, 3, and 5 are factors of the given numbers.

7. 90   8. 341

9. 774   10. 3585

[4.2] In Exercises 11 to 14, find the prime factorization.

11. 42   12. 72

13. 210   14. 792

[4.3] In Exercises 15 to 22, find the greatest common factor (GCF) for each of the given numbers.

15. 10 and 15   16. 30 and 48

241

Check Yourself Exercises

These exercises have been the hallmark of the text; they are designed to actively involve students throughout the learning process. Each example is followed by an exercise that encourages students to solve a problem similar to the one just presented. Answers are provided at the end of the section for immediate feedback.

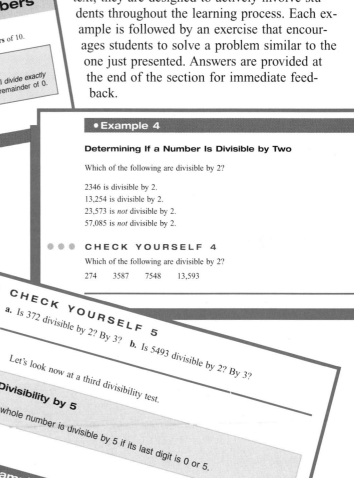

4.1 Prime and Composite Numbers

In Section 2.1 we said that since $2 \times 5 = 10$, we call 2 and 5 **factors** of 10.

4.1 OBJECTIVES
1. Find the factors of a number.
2. Determine whether a number is prime, composite, or neither.
3. Determine whether a number is divisible by 2, 3, or 5.

2 and 5 can also be called divisors of 10. They divide 10 exactly.

This is a complete list of the factors. There are no other whole numbers that divide 18 exactly. Note that the factors of 18, except for 18 itself, are smaller than 18.

Definition of a Factor

A **factor** of a whole number is another whole number that will *divide exactly* into that number. This means that the division will have a remainder of 0.

• Example 1

Finding Factors

List all factors of 18.

$3 \times 6 = 18$ Since $3 \times 6 = 18$, 3 and 6 are factors (or divisors) of 18.

$2 \times 9 = 18$ 2 and 9 are also factors of 18.

$1 \times 18 = 18$ 1 and 18 are factors of 18.

1, 2, 3, 6, 9, and 18 are all the factors of 18.

• • • CHECK YOURSELF 1

List all the factors of 24.

• Example 4

Determining If a Number Is Divisible by Two

Which of the following are divisible by 2?

2346 is divisible by 2.
13,254 is divisible by 2.
23,573 is *not* divisible by 2.
57,085 is *not* divisible by 2.

• • • CHECK YOURSELF 4

Which of the following are divisible by 2?

274   3587   7548   13,593

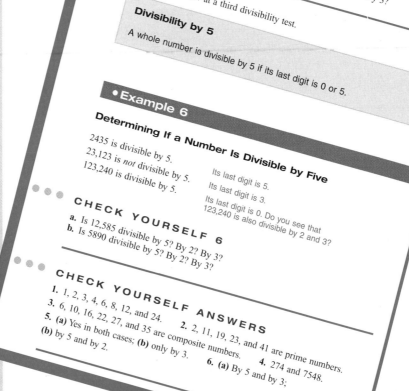

• • • CHECK YOURSELF 5

a. Is 372 divisible by 2? By 3?   **b.** Is 5493 divisible by 2? By 3?

Let's look now at a third divisibility test.

Divisibility by 5

A whole number is divisible by 5 if its last digit is 0 or 5.

• Example 6

Determining If a Number Is Divisible by Five

2435 is divisible by 5.   Its last digit is 5.
23,123 is *not* divisible by 5.   Its last digit is 3.
123,240 is divisible by 5.   Its last digit is 0. Do you see that 123,240 is also divisible by 2 and 3?

• • • CHECK YOURSELF 6

a. Is 12,585 divisible by 5? By 2? By 3?
b. Is 5890 divisible by 5? By 2? By 3?

• • • CHECK YOURSELF ANSWERS

1. 1, 2, 3, 4, 6, 8, 12, and 24.   **2.** 2, 11, 19, 23, and 41 are prime numbers.
3. 6, 10, 16, 22, 27, and 35 are composite numbers.   **4.** 274 and 7548.
5. (a) Yes in both cases; **(b)** only by 3.   **6. (a)** By 5 and by 3;
(b) by 5 and by 2.

Comprehensive Exercise Sets

Complete exercise sets are at the end of each section as well as after the summary at the end of each chapter. These exercises were designed to reinforce basic skills and develop critical thinking and communication abilities. Exercise sets include writing and word problems as well as collaborative and group exercises.

Name

Section   Date

4.1 Exercises

ANSWERS
1.
2.
3.
4.
5.
6.
7.
8.
9.
10.
11.
12.
13.
14.
15.
16.

List the factors of each of the following numbers.

1. 4   2. 6

3. 10   4. 12

5. 15   6. 21

7. 24   8. 32

9. 64   10. 66

11. 11   12. 37

Use the following list of numbers for Exercises 13 and 14.

15, 19, 23, 31, 49, 55, 59, 87, 91, 97, 103, 105

13. Which of the given numbers are prime?

14. Which of the given numbers are composite?

15. List all the prime numbers between 30 and 50.

16. List all the prime numbers between 55 and 75.

209

21.

22.

23.

19. Which of the given numbers are divisible by 5?

20. Which of the given numbers are divisible by 10?

21. Why is the following not a valid divisibility test for 8?

"A number is divisible by 8 if it is divisible by 2 and 4"

Support your answer with an example. Determine a valid divisibility test for 8.

22. Prime numbers that differ by two are called "twin primes." Examples are 3 and 5, 5 and 7, and so on. Find one pair of twin primes between 85 and 105.

23. The Greek mathematician Eratosthenes developed a method to identify prime numbers in a list of numbers. The method is called the "Sieve of Eratosthenes." Research this topic and describe it to your classmates. Then use it to find all the prime numbers less than 100.

ANSWERS
a.
b.
c.
d.
e.
f.

Getting Ready for Section 4.2 [Section 3.3]

Divide, using short division.

a. $3\overline{)72}$ b. $5\overline{)90}$ c. $4\overline{)84}$

d. $2\overline{)384}$ e. $3\overline{)693}$ f. $5\overline{)750}$

Answers

1. 1, 2, and 4   3. 1, 2, 5, and 10   5. 1, 3, 5, and 15   11. 1 and 11
7. 1, 2, 3, 4, 6, 8, 12, and 24   9. 1, 2, 4, 8, 16, 32, and 64
13. 19, 23, 31, 59, 97, 103   15. 31, 37, 41, 43, 47   19. 45, 260, 570, 585, 4530, 8300
17. 72, 158, 260, 378, 570, 585, 4530, 8300
21.   23.   a. 24   b. 18   c. 21   d. 192   e. 231

f. 150

ANSWERS
a.
b.
c.
d.
e.
f.

Getting Ready for Section 4.3 [Section 4.1]

List all factors of the following numbers.

a. 12   b. 20

c. 30   d. 45

e. 17   f. 29

Answers

1. $2 \times 3 \times 3$ 3. $2 \times 3 \times 5$ 5. 3×17 7. $3 \times 3 \times 7$ 9. $2 \times 5 \times 7$
11. $2 \times 3 \times 11$ 13. $2\overline{)130}$ 15. $3 \times 3 \times 5 \times 7$
$5\overline{)65}$
$130 = 2 \times 5 \times 13$
17. $3 \times 3 \times 5 \times 5$ 19. $3\overline{)189}$ 21. $2 \times 2 \times 2 \times 2 \times 3 \times 7$
$3\overline{)63}$
$7\overline{)21}$
$189 = 3 \times 3 \times 3 \times 7$ 27. 5, 6   29.
23. $2 \times 2 \times 2 \times 3 \times 5 \times 7$ 25. 4, 6   c. 1, 2, 3, 5, 6, 10, 15, 30
a. 1, 2, 3, 4, 6, 12   b. 1, 2, 4, 5, 10, 20
d. 1, 3, 5, 9, 15, 45   e. 1, 17   f. 1, 29

Getting Ready Exercises

These exercises draw on problems from previous sections of the text and are designed to help students review concepts that will be applied in the following section. This preview helps students make important connections with upcoming material.

Supplementary Exercises

Each exercise set includes a second set of exercises for which answers are not provided in the text.

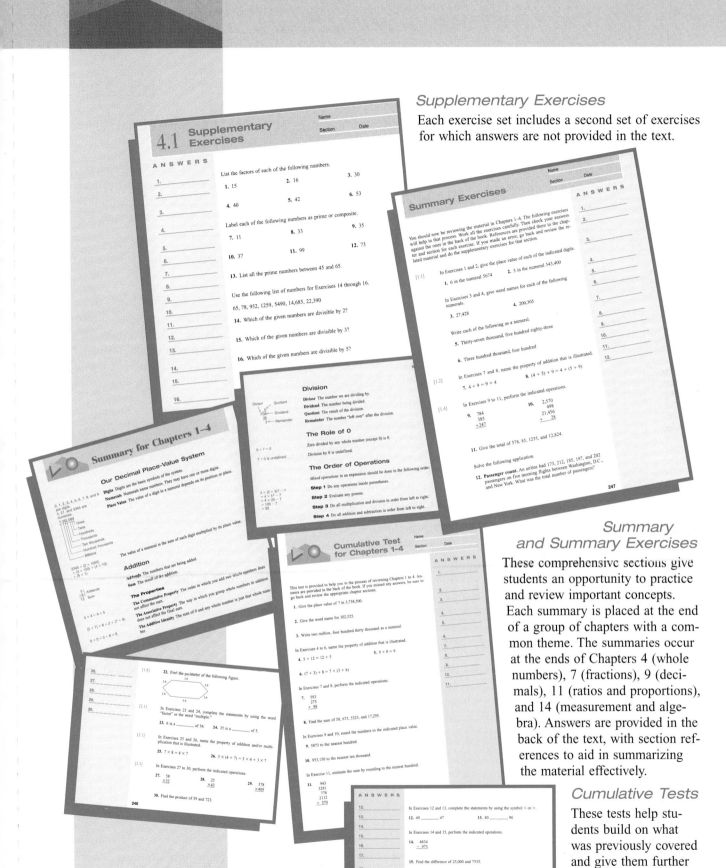

4.1 Supplementary Exercises

Name
Section   Date

ANSWERS

1. _____
2. _____
3. _____
4. _____
5. _____
6. _____
7. _____
8. _____
9. _____
10. _____
11. _____
12. _____
13. _____
14. _____
15. _____
16. _____

List the factors of each of the following numbers.

1. 15   **2.** 16   **3.** 30

4. 40   **5.** 42   **6.** 53

Label each of the following numbers as prime or composite.

7. 11   **8.** 33   **9.** 35

10. 37   **11.** 99   **12.** 73

13. List all the prime numbers between 45 and 65.

Use the following list of numbers for Exercises 14 through 16.

65, 78, 952, 1259, 5490, 14,685, 22,390

14. Which of the given numbers are divisible by 2?

15. Which of the given numbers are divisible by 3?

16. Which of the given numbers are divisible by 5?

Summary Exercises

Name
Section   Date

ANSWERS

1. _____
2. _____
3. _____
4. _____
5. _____
6. _____
7. _____
8. _____
9. _____
10. _____
11. _____
12. _____

You should now be reviewing the material in Chapters 1–4. The following exercises will help in that process. Work all the exercises carefully. Then check your answers against the ones in the back of the book. References are provided there to the chapter and section for each exercise. If you made an error, go back and review the related material and do the supplementary exercises for that section.

[1.1] In Exercises 1 and 2, give the place value of each of the indicated digits.

1. 6 in the numeral 5674   **2.** 5 in the numeral 543,400

In Exercises 3 and 4, give word names for each of the following numerals.

3. 27,428   **4.** 200,305

Write each of the following as a numeral.

5. Thirty-seven thousand, five hundred eighty-three

6. Three hundred thousand, four hundred

[1.2] In Exercises 7 and 8, name the property of addition that is illustrated.

7. $4 + 9 = 9 + 4$ **8.** $(4 + 5) + 9 = 4 + (5 + 9)$

[1.4] In Exercises 9 to 11, perform the indicated operations.

9. 784
385
+247

10. 2,570
498
21,456
+ 28

11. Give the total of 578, 85, 1235, and 12,824.

Solve the following application.

12. Passenger count. An airline had 173, 212, 185, 197, and 202 passengers on five morning flights between Washington, D.C., and New York. What was the total number of passengers?

247

Division

Divisor The number we are dividing by.
Dividend The number being divided.
Quotient The result of the division.
Remainder The number "left over" after the division.

The Role of 0

Zero divided by any whole number (except 0) is 0.

$0 \div 7 = 0$

Division by 0 is undefined.

$7 \div 0$ is undefined.

The Order of Operations

Mixed operations in an expression should be done in the following order:

Step 1 Do any operations inside parentheses.
Step 2 Evaluate any powers.
Step 3 Do all multiplication and division in order from left to right.
Step 4 Do all addition and subtraction in order from left to right.

$4 \times (2 + 3)^2 - 7$
$= 4 \times 5^2 - 7$
$= 4 \times 25 - 7$
$= 100 - 7$
$= 93$

Summary for Chapters 1–4

Our Decimal Place-Value System

Digits Digits are the basic symbols of the system. They may have one or more digits.
Numerals Numerals name numbers.
Place Value The value of a digit in a numeral depends on its position or place.

The value of a numeral is the sum of each digit multiplied by its place value.

Addition

Addends The numbers that are being added.
Sum The result of an addition.

The Properties

The Commutative Property The order in which you add two whole numbers does not affect the sum.
The Associative Property The way in which you group whole numbers in addition does not affect the final sum.
The Additive Identity The sum of 0 and any whole number is just that whole number.

$5 + 4 = 4 + 5$

$(2 + 7) + 8 = 2 + (7 + 8)$

$6 + 0 = 0 + 6 = 6$

Cumulative Test for Chapters 1–4

Name
Section   Date

ANSWERS

1. _____
2. _____
3. _____
4. _____
5. _____
6. _____
7. _____
8. _____
9. _____
10. _____
11. _____

This test is provided to help you in the process of reviewing Chapters 1 to 4. Answers are provided in the back of the book. If you missed any answers, be sure to go back and review the appropriate chapter sections.

1. Give the place value of 7 in 3,738,500.

2. Give the word name for 302,525.

3. Write two million, four hundred thirty thousand as a numeral.

In Exercises 4 to 6, name the property of addition that is illustrated.

4. $5 + 12 = 12 + 5$ **5.** $9 + 0 = 9$

6. $(7 + 3) + 8 = 7 + (3 + 8)$

In Exercises 7 and 8, perform the indicated operations.

7. 593
275
+ 98

8. Find the sum of 58, 673, 5325, and 17,295.

In Exercises 9 and 10, round the numbers to the indicated place value.

9. 5873 to the nearest hundred

10. 953,150 to the nearest ten thousand

In Exercise 11, estimate the sum by rounding to the nearest hundred.

11. 943
3281
778
2112
+ 570

ANSWERS

12. _____
13. _____
14. _____
15. _____
16. _____
17. _____
18. _____
19. _____
20. _____
21. _____
22. _____
23. _____
24. _____

In Exercises 12 and 13, complete the statements by using the symbol < or >.

12. 49 _____ 47   **13.** 80 _____ 90

In Exercises 14 and 15, perform the indicated operations.

14. 4834
– 973

15. Find the difference of 25,000 and 7535.

In Exercises 16 and 17, solve the applications.

16. Attendance. Attendance for five performances of a play was 172, 153, 205, 193, and 182. How many people attended those performances?

17. Balance. Alan bought a Volkswagen with a list price of $8975. He added stereo equipment for $439 and an air conditioner for $615. If he made a down payment of $2450, what balance remained on the car?

In Exercises 18 to 20, name the property of addition and/or multiplication that is illustrated.

18. $3 \times (4 \times 7) = (3 \times 4) \times 7$

[1.8] **22.** Find the perimeter of the following figure.

26. _____
27. _____
28. _____
29. _____
30. _____

[2.1] In Exercises 23 and 24, complete the statements by using the word "factor" or the word "multiple."

23. 6 is a _____ of 36.   **24.** 35 is a _____ of 5.

[2.1] In Exercises 25 and 26, name the property of addition and/or multiplication that is illustrated.

25. $7 \times 8 = 8 \times 7$ **26.** $3 \times (4 + 7) = 3 \times 4 + 3 \times 7$

[2.3] In Exercises 27 to 30, perform the indicated operations.

27. 58
×32

28. 25
×43

29. 378
×409

30. Find the product of 59 and 723.

248

Summary and Summary Exercises

These comprehensive sections give students an opportunity to practice and review important concepts.

Each summary is placed at the end of a group of chapters with a common theme. The summaries occur at the ends of Chapters 4 (whole numbers), 7 (fractions), 9 (decimals), 11 (ratios and proportions), and 14 (measurement and algebra). Answers are provided in the back of the text, with section references to aid in summarizing the material effectively.

Cumulative Tests

These tests help students build on what was previously covered and give them further opportunity for building skills necessary in preparing for midterm and final exams.

SUPPLEMENTS

Supplements

A comprehensive set of ancillary materials for both the student and the instructor is available with this text.

Instructor's Edition

This ancillary includes answers to all exercises and tests. These answers are printed in a second color for ease of use by the instructor and are located on the appropriates pages throughout the text.

Instructor's Solutions Manual

The manual provides worked-out solutions to the odd-numbered exercises in the text.

Instructor's Resource Manual

The resource manual contains multiple-choice placement tests for three levels of testing: (1) a diagnostic pretest for each chapter and three forms of multiple-choice and open-ended chapter tests; (2) two forms of multiple-choice and open-ended cumulative tests; and (3) two forms of multiple-choice and open-ended final tests. Also included is an answer section and appendixes that cover collaborative learning and the implementation of the new standards.

Print and Computerized Testing

The testing materials provide an array of formats that allow the instructor to create tests using both algorithmically generated test questions and those from a standard testbank. This testing system enables the instructor to choose questions either manually or randomly by section, question type, difficulty level, and other criteria. Testing is available for IBM, IBM-compatible, and Macintosh computers. A softcover print version of the testbank provides most questions found in the computerized version.

Streeter Video Series

The video series is completely new to this edition. It gives students additional reinforcement of the topics presented in the book. The videos were developed especially for the Streeter pedagogy, and features are tied directly to the main text's individual chapters and section objectives. The videos feature an effective combination of learning techniques, including personal instruction, state-of-the-art graphics, and real-world applications.

Multimedia Tutorial

This interactive CD-ROM is a self-paced tutorial specifically linked to the text and reinforces topics through unlimited opportunities to review concepts and practice problem solving. It requires virtually no computer training on the part of the students and supports IBM and Macintosh computers.

MathWorks

This DOS-based interactive tutorial software is available and specifically designed to accompany the Streeter pedagogy. The program supports IBM, IBM-compatible, and Macintosh computers as well as a variety of networks. MathWorks can also be used with its companion program, the Instructor's Management System, to track and record the progress of students in the class.

In addition, a number of other technology and Web-based ancillaries are under development; they will support the ever-changing technology needs in developmental mathematics. For further information about these or any supplements, please contact your local McGraw-Hill sales representative.

ACKNOWLEDGMENTS

Acknowledgments

In the process of writing four editions of this text, we learned much more than we could ever have taught. The faculty we work with, the students who do us the honor of signing up for our classes, and the staff at McGraw-Hill have all been part of our education. The first two groups are the most important, but the most difficult to identify. The totality of their contributions is overwhelming. Every student who has sat in our offices struggling to learn this material has helped us write the next edition. Every story that another teacher has told us, every AMATYC session we've attended, and every reviewer comment that we've read has become part of the fabric of this text. A great deal of thanks certainly goes to Zanae Rodrigo, the developmental editor for this book, as well as Norma James, who served as accuracy reviewer. In addition, our thanks goes to the following people for their important contributions to the development of this edition:

Gail Ann Aurand, Cloud County Community College (KA)
Linda Chamblin, Southern State Community College (OH)
Nancy Cholvin, Antelope Valley College (CA)
Katherine Creery, University of Memphis
Albert Deas, Trident Technical College (SC)
Philip Glynn, Naugatuk Valley Community College (CT)
Mary Henderson, Okaloosa–Walton Community College (FL)
Dawn Kindel, Newbury College (MA)
Tom Kremer, University of Wisconsin–Parkside
Nick Lahue, Penn Valley Community College (MO)
Kenneth McClain, University of Memphis
William Merrow, Western Michigan University
Renee Patterson, Cumberland County College (NJ)
Roy Pearson, St. Louis Community College at Florissant Valley
Marilyn Platt, Gaston College (NC)
Kathy Pletsch, Antelope Valley College (CA)
Deborah Puett, Isothermal Community College (NC)
Kathleen Sherman, Napa Valley College
Karen Spriegel, Muskingum Area Technical College (OH)
Angela Stanford, Gulf Coast Community College (FL)
John Thoo, Yuba College (CA)
Michael Turegun, Oklahoma City Community College
Joyce Wellington, Southeastern Community College (NC)
Henry Wyzinski, Indiana University–Northwest
Cora West, Florida Community College–Kent
Stephen Zona, Quinsigamond Community College

But, it is the McGraw-Hill staff who has suffered with us the most, so special thanks to them all.

Donald Hutchison
Louis Hoelzle

TO THE STUDENT

You are about to begin a course in basic mathematics. We made every attempt to provide a text that will help you understand what basic mathematics is about and how to effectively use it. We made no assumptions about your previous experience with mathematics. Your progress through the course will depend on the amount of time and effort you devote to the course and your previous background in math. There are some specific features in this book that will aid you in your studies. Here are some suggestions about how to use this book. (Keep in mind that a review of all the chapter and summary material will further enhance your ability to grasp later topics and to move more effectively through the text.

1. If you are in a lecture class, make sure that you take the time to read the appropriate text section *before* your instructor's lecture on the subject. Then take careful notes on the examples that your instructor presents during class.
2. After class, work through similar examples in the text, making sure that you understand each of the steps shown. Examples are followed in the text by *Check Yourself* exercises. Basic math is best learned by being involved in the process, and that is the purpose of these exercises. Always have a pencil and paper at hand, and work out the problems presented and check your results immediately. If you have difficulty, go back and carefully review the previous examples. Make sure you understand what you are doing and why. The best test of whether you do understand a concept lies in your ability to explain that concept to one of your classmates. Try working together.
3. At the end of each chapter section you will find a set of exercises. Work these carefully in order to check your progress on the section you have just finished. You will find the solutions for the odd-numbered exercises following the problem set. If you have had difficulties with any of the exercises, review the appropriate parts of the chapter section. If your questions are not completely cleared up, by all means do not become discouraged. Ask your instructor or an available tutor for further assistance. A word of caution: Work the exercises on a regular (preferably daily) basis. Again, learning basic math requires becoming involved. As is the case with learning any skill, the main ingredient is practice.
4. When finished with the last section of a chapter, try the *Self-Test* that appears at the end of each chapter. This test will give you an actual practice test to work as you review for in-class testing. Again, answers with section references are provided.
5. When you have completed Chapters 4, 7, 9, 11, and 14, review by using the *Summary.* You will find all the important terms and definitions in this section, along with examples, illustrating all the techniques developed since the previous Summary. Following the summary are *Summary Exercises* for further practice. The exercises are keyed to chapter sections, so you will know where to turn if you are still having problems.
6. Finally, an important element of success in studying basic math is the process of regular review. We provided a series of *Cumulative Tests* throughout the textbook, beginning at the end of Chapter 4. These tests will help you review not only the concepts of the chapter that you have just completed but those of previous chapters. Use these tests in preparation for any midterm or final exams. If it appears that you have forgotten some concepts that are being tested, don't worry. Go back and review the sections where the idea was initially explained, or the appropriate chapter summary. That is the purpose of the cumulative tests. We hope that you will find our suggestions helpful as you work through this material, and we wish you the best of luck in the course.

Donald Hutchison
Louis Hoelzle

CHAPTER 1

ADDITION AND SUBTRACTION OF WHOLE NUMBERS

INTRODUCTION

Numbers were first used for counting people and objects. We still keep track of the number of people in our towns, states, and country. Every 10 years the U.S. government undertakes a complete count of the number of the people in the country. Such a count is called a *census.*

Rosita has worked for the Census Bureau since 1988. She helped organize, conduct, and audit the 1990 census, in which the government counted a total of 248,718,301 people in the United States. (As of July 1, 1996, the Census Bureau established the revised U.S. population at 265,283,783.) Rosita will be working on the census for the year 2000 until the final audit is completed, most likely in the year 2005. ■

© Rhoda Sidney/The Image Works

ANSWERS

1. One hundred seven thousand, nine hundred forty-five

2. Associative property

3. 758

4. 39,662

5. 57,800

6. False

7. 323

8. 4844

9. 93

10. 14 ft

Addition and Subtraction of Whole Numbers

This pretest will point out any difficulties you may be having in adding and subtracting whole numbers. Do all the problems. Then check your answers with those in the back of the book.

1. Write 107,945 in words.

2. The statement $2 + (3 + 5) = (2 + 3) + 5$ illustrates which property of addition?

3. $125 + 431 + 202 =$ **4.** $35,147 + 3673 + 783 + 59 =$

5. Round 57,849 to the nearest hundred.

6. Is the following statement true or false?

$150 > 1500$

7. $596 - 273 =$ **8.** $8473 - 3629 =$

9. Test Scores. Suppose that you need a total of 360 points on four tests during the semester to receive an A for the course. Your scores on the first three tests were 84, 91, and 92. What is the lowest score you can get on the fourth test and still receive the A?

10. Find the perimeter of the following figure:

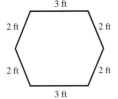

3 ft
2 ft   2 ft
2 ft   2 ft
3 ft

The Decimal Place-Value System

1.1 OBJECTIVES

1. Write numerals in expanded form.
2. Determine the place value of a digit.
3. Write a numeral in words.
4. Write a numeral, given its word name.

Number systems have been developed throughout human history. Starting with simple tally systems used to count and keep track of possessions, more and more complex systems developed. The Egyptians used a set of picturelike symbols called **hieroglyphics** to represent numbers. The Romans and Greeks had their own systems of numeration. We see the Roman system today in the form of Roman numerals. Some examples of these systems are shown in Figure 1.

NUMERALS	EGYPTIAN	GREEK	ROMAN
1	I	I	I
10	∩	△	X
100	ꝑ	H	C

Figure 1

The prefix "deci" means 10. Our word "digit" comes from the Latin word "digitus," which means finger.

Any number, no matter how large, can be represented as a numeral by using the 10 digits of our system.

For convenience, we will not worry about the distinction between the words "number" and "numeral" in the remainder of this book.

Any number system provides a way of naming numbers. The system we use is described as a **decimal place-value system.** This system is based on the number 10 and uses symbols called **digits.** (Other numbers have also been used as bases. The Mayans used 20, and the Babylonians used 60.)

The basic symbols of our system are the digits:

0, 1, 2, 3, 4, 5, 6, 7, 8, 9

These basic symbols, or digits, were first used in India and then adopted by the Arabs. For this reason, our system is called the Hindu-Arabic numeration system.

Numbers are represented by symbols called **numerals.** Numerals may consist of one or more *digits*.

3, 45, 567, and 2359 are numerals, or symbols that *name* numbers. We say that 45 is a two-digit numeral, 567 is a three-digit numeral, and so on.

As we said, our decimal system uses a *place-value* concept based on the number 10. Understanding how this system works will help you see the reasons for the rules and methods of arithmetic that we will be introducing.

●Example 1

Identifying Place Value

Each digit in a numeral has its own place value.

Look at the numeral 438.

8 represents 8 individual items. We call 8 the *ones digit*. Moving to the left, the digit 3 represents 3 groups of 10 objects.

3 is the *tens digit*. Again moving to the left, 4 represents 4 groups of 100.

4 is the *hundreds digit*.

4 hundreds   8 ones

3 tens

In symbols, we can write 438 as

Here the parentheses are used for emphasis. We will discuss parentheses further in the next chapter.

$(4 \times 100) + (3 \times 10) + 8$

This is called the **expanded form** of the numeral.

(4×100) means 4 is multiplied by 100.

(3×10) means 3 is multiplied by 10.

● ● ●   **CHECK YOURSELF 1**

Write 593 in expanded form.

The following place-value diagram shows the place value of digits as we write numerals that represent larger numbers. For the numeral 3,156,024,798, we have

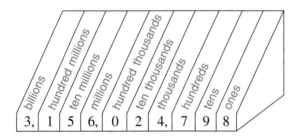

Of course, the naming of place values continues for larger and larger numbers beyond the chart.

For the numeral 3,156,024,798, the place value of 4 is thousands. As we move to the left, each place value is 10 times the value of the previous place. The place value of 2 is ten thousands, the place value of 0 is hundred thousands, and so on.

● **Example 2**

Identifying Place Value

Identify the place value of each digit in the numeral 418,295.

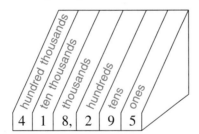

● ● ● **CHECK YOURSELF 2**

Use a place-value diagram to answer the following questions for the numeral 6,831,425,097.

a. What is the place value of 2?
b. What is the place value of 4?
c. What is the place value of 3?
d. What is the place value of 6?

Understanding place value will help you read or write numerals in word form. Look at the numeral

$$7\ 2,\quad 3\ 5\ 8,\quad 6\ 9\ 4$$
Millions   Thousands   Ones

A four-digit numeral, such as 3456, can be written with or without a comma. We have chosen to omit the comma in these materials.

Commas are used to set off groups of three digits in the numeral. The name of each group—millions, thousands, ones, and so on—is then used as we write the numeral in words. To write a word name for a numeral, we work from left to right, writing the numerals in each group, followed by the group name. The following chart summarizes the group names.

Billions Group			Millions Group			Thousands Group			Ones Group		
Hundreds	Tens	Ones	Hundreds	Tens	Ones	Hundreds	Tens	Ones	Hundreds	Tens	Ones

● **Example 3**

Writing Numbers in Words

Note that the commas in the word statements are in the same place as the commas in the number.

27,345 is written in words as twenty-seven *thousand,* three hundred forty-five.
2,305,273 is two *million,* three hundred five *thousand,* two hundred seventy-three.

Note: We do *not* write the name of the ones group. Also, "and" is not used when a number is written in words. It will have a special meaning later.

● ● ● **CHECK YOURSELF 3**

Write the word name for each of the following numerals.

a. 658,942   **b.** 2305

We reverse the process to write numerals for numbers given in word form. Consider the following.

• Example 4

Translating Words into Numbers

Forty-eight thousand, five hundred seventy-nine in numeral form is

48,579

Five hundred three thousand, two hundred thirty-eight in numeral form is

503,238
⌐‾‾‾‾‾‾‾‾ Note the use of 0 as a placeholder
              in writing the numeral.

● ● ● **CHECK YOURSELF 4**

Write twenty-three thousand, seven hundred nine in numeral form.

● ● ● **CHECK YOURSELF ANSWERS**

1. $(5 \times 100) + (9 \times 10) + 3$.   **2. (a)** Ten thousands; **(b)** hundred thousands; **(c)** ten millions; (d) billions.   **3. (a)** Six hundred fifty-eight thousand, nine hundred forty-two; (b) two thousand, three hundred five.   **4.** 23,709.

Write each numeral in expanded form.

1. 456

2. 637

3. 5073

4. 20,721

Give the place values for the indicated digits.

5. 4 in the numeral 416

6. 8 in the numeral 38,615

7. 6 in the numeral 56,489

8. 0 in the numeral 427,083

9. 7 in the numeral 27,243,012

10. 5 in the numeral 3,527,213

11. 2 in the numeral 523,010,000

12. 3 in the numeral 317,008,000

Write the word name for each of the following numerals.

13. 5618

14. 21,812

15. 200,304

16. 103,900

ANSWERS

1. $(4 \times 100) + (5 \times 10) + 6$

2. $(6 \times 100) + (3 \times 10) + 7$

3. $(5 \times 1000) + (7 \times 10) + 3$

4. $(2 \times 10,000) + (7 \times 100) + (2 \times 10) + 1$

5. Hundreds

6. Thousands

7. Thousands

8. Hundreds

9. Millions

10. Hundred thousands

11. Ten millions

12. Hundred millions

13. Five thousand, six hundred eighteen

14. Twenty-one thousand, eight hundred twelve

15. Two hundred thousand, three hundred four

16. One hundred three thousand, nine hundred

17.	253,483
18.	350,359
19.	502,078,000
20.	4,230,000,000
21.	Eight
22.	Twenty-two thousand, two hundred twenty-two
23.	34,215
24.	46,789
25.	Two thousand, five hundred sixty-five
26.	
27.	$2545

Write each of the following as numerals.

17. Two hundred fifty-three thousand, four hundred eighty-three

18. Three hundred fifty thousand, three hundred fifty-nine

19. Five hundred two million, seventy-eight thousand

20. Four billion, two hundred thirty million

Assume that you have alphabetized the word names for every numeral from one to one million.

21. Which number would appear first in the list?

22. Which number would appear last?

Determine the numeral represented by the scrambled place values.

23. 4 thousands
1 tens
3 ten-thousands
5 ones
2 hundreds

24. 7 hundreds
4 ten-thousands
9 ones
8 tens
6 thousands

25. Inci had to write a check for $2565. There is a space on the check to write out the amount of the check in words. What should she write in this space?

26. In addition to personal checks, name two other places where writing amounts in words is necessary.

27. In a rental agreement, the amount of the initial deposit required is Two Thousand, Five Hundred and Forty-Five dollars. Write this amount as a numeral.

28. Several early numeration systems did not use place values. Do some research, and determine at least two of these systems. Describe the system that they used. What were the disadvantages?

29. What are the advantages of a place-value system of numeration?

30. The number 0 was not used initially by the Hindus in our number system (about 250 BC). Go to your library (or "surf the net"), and determine when a symbol for zero was introduced. What do you think is the importance of the role of 0 in a numeration system?

 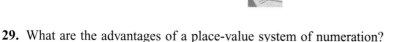
Answers

We will provide the answers (with some worked out in detail) for the odd-numbered exercises at the end of each exercise set. The answers for the even-numbered exercises are provided in the back of the book.

1. $(4 \times 100) + (5 \times 10) + 6$ **3.** $(5 \times 1000) + (7 \times 10) + 3$ **5.** Hundreds

7. Thousands   **9.** Millions   **11.** Ten millions

13. Five thousand, six hundred eighteen

15. Two hundred thousand, three hundred four   **17.** 253,483   **19.** 502,078,000

21. Eight   **23.** 34,215   **25.** Two thousand, five hundred sixty-five

27. $2545   **29.**

Name

Section   Date

A N S W E R S

1. $(6 \times 100) + (3 \times 10) + 8$

2. $(3 \times 1000) + (7 \times 100) + (2 \times 10) + 5$

3. Thousands

4. Ten thousands

5. Millions

6. Ten millions

7. Twenty-five thousand, four hundred eighty-nine

8. Three hundred thousand, two hundred fifty-seven

9. 52,384

10. 503,687,000

Write each numeral in expanded form.

1. 638

2. 3725

Give the place values for the indicated digits.

3. 2 in the numeral 32,785

4. 8 in the numeral 584,123

5. 3 in the numeral 23,000,000

6. 0 in the numeral 1,205,567,293

Give word names for the following numerals.

7. 25,489

8. 300,257

Write each of the following as numerals.

9. Fifty-two thousand, three hundred eighty-four

10. Five hundred three million, six hundred eighty-seven thousand

1.2 The Properties of Addition

1.2 OBJECTIVES

1. Use the language of addition.
2. Add single-digit numbers.
3. Identify the Properties of Addition.

The three dots (. . .) are called **ellipses;** they mean that the set continues the indicated pattern.

The first printed use of the symbol + dates back to 1500.

The point labeled 0 is called the **origin** of the number line.

The *natural* or *counting numbers* are the numbers we use to count objects.

The natural numbers are 1, 2, 3, . . .

When we include the number 0, we then have the set of *whole numbers.*

The whole numbers are 0, 1, 2, 3, . . .

Let's look at the operation of *addition* on the whole numbers.

> **Addition** is the combining of two or more groups of the same kinds of objects.

This concept is extremely important, as we will see in our later work with fractions. We can only combine or add numbers that represent the same kinds of objects.

Each operation of arithmetic has its own special terms and symbols. The addition symbol + is read **plus.** When we write $3 + 4$, 3 and 4 are called the **addends.**

We can use a number line to illustrate the addition process. To construct a number line, we pick a point on the line and label it 0. We then mark off evenly spaced units to the right, naming each point marked off with a successively larger whole number.

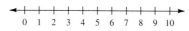

We use an arrowhead to show the direction of increase.

• Example 1

Representing Addition on a Number Line

Again, addition corresponds to combining groups of the same kind of objects.

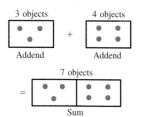

Represent $3 + 4$ on the number line.

To represent an addition, such as $3 + 4$, on the number line, start by moving 3 spaces to the right of the origin. Then move 4 more spaces to the right to arrive at 7. The number 7 is called the *sum* of the addends.

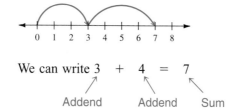

We can write 3   +   4   =   7

Addend   Addend   Sum

● ● ● **CHECK YOURSELF 1**

Represent $5 + 6$ on the number line.

A statement such as $3 + 4 = 7$ is one of the **basic addition facts.** These facts include the sum of every possible pair of digits. Before you can add larger numbers correctly and quickly, you must memorize these basic facts.

Basic Addition Facts

+	0	1	2	3	4	5	6	7	8	9
0	0	1	2	3	4	5	6	7	8	9
1	1	2	3	4	5	6	7	8	9	10
2	2	3	4	5	6	7	8	9	10	11
3	3	4	5	6	7	8	9	10	11	12
4	4	5	6	7	8	9	10	11	12	13
5	5	6	7	8	9	10	11	12	13	14
6	6	7	8	9	10	11	12	13	14	15
7	7	8	9	10	11	12	13	14	15	16
8	8	9	10	11	12	13	14	15	16	17
9	9	10	11	12	13	14	15	16	17	18

To find the sum $5 + 8$, start with the row labeled 5. Move along that row to the column headed 8 to find the sum, 13.

Examining the table of Basic Addition Facts leads us to several important properties of addition on the whole numbers. For instance, we know that the sum $3 + 4$ is 7. What about the sum $4 + 3$? It is also 7. This is an illustration of the fact that addition is a **commutative** operation.

The Commutative Property of Addition

The order in which you add two whole numbers *does not* affect the sum.

•Example 2

Using the Commutative Property

The *order* does not affect the sum.

$8 + 5 = 5 + 8 = 13$

$6 + 9 = 9 + 6 = 15$

● ● ● **CHECK YOURSELF 2**

Show that

$7 + 8 = 8 + 7$

If we wish to add *more* than two numbers, we can group them and then add. In mathematics this grouping is indicated by a set of parentheses (   ). This symbol tells us to perform the operation inside the parentheses first.

• Example 3

Using the Associative Property

We add 3 and 4 as the first step and then add 5.

$$(3 + 4) + 5 = 7 + 5 = 12$$

We also have

Here we add 4 and 5 as the first step and then add 3. Again the final sum is 12.

$$3 + (4 + 5) = 3 + 9 = 12$$

Example 3 suggests the following property of whole numbers.

The Associative Property of Addition

The order in which you add several whole numbers *does not* affect the final sum.

● ● ●   CHECK YOURSELF 3

Find

$$(4 + 8) + 3 \quad \text{and} \quad 4 + (8 + 3)$$

The number 0 has a special property in addition. Looking at the table of Basic Addition Facts, we see that

---

The Additive Identity Property

The sum of 0 and any whole number is just that whole number.

Because of this property, we call 0 the **identity** for the addition operation.

• Example 4

Adding Zero

Find the sum (*a*) $3 + 0$ and (*b*) $0 + 8$.

(*a*) $3 + 0 = 3$

(*b*) $0 + 8 = 8$

● ● ● **CHECK YOURSELF 4**

Find the sum.

a. $4 + 0 =$ **b.** $0 + 7 =$

● ● ● **CHECK YOURSELF ANSWERS**

1.

$5 + 6 = 11.$

2. $7 + 8 = 15$ and $8 + 7 = 15$.

3. $(4 + 8) + 3 = 12 + 3 = 15$; $4 + (8 + 3) = 4 + 11 = 15$.

4. **(a)** 4; **(b)** 7.

Name

Section   Date

A N S W E R S

1. In the statement $5 + 4 = 9$
5 is called the
4 is called the
9 is called the

2. In the statement $7 + 8 = 15$
7 is called the
8 is called the
15 is called the

Add.

3.   4
    + 3

4.   3
    + 4

5.   9
    + 5

6.   5
    + 9

7.   7
    + 3

8.   1
    + 6

9.   4
    + 4

10.   9
     + 7

11.   8
     + 6

12.   6
     + 8

13.   7
     + 4

14.   2
     + 6

15.   6
     + 6

16.   0
     + 9

17.   9
     + 1

18.   5
     + 8

19.   8
     + 7

20.   6
     + 7

21.   5
     + 4

22.   5
     + 7

23.   3
     + 5

1.	addend; addend; sum
2.	addend; addend; sum
3.	7
4.	7
5.	14
6.	14
7.	10
8.	7
9.	8
10.	16
11.	14
12.	14
13.	11
14.	8
15.	12
16.	9
17.	10
18.	13
19.	15
20.	13
21.	9
22.	12
23.	8

24. 12	
25. 10	
26. 12	
27. 7	
28. 16	
29. 15	
30. 11	
31. 12	
32. 15	
33. 12	
34. 15	
35. 12	
36. 15	
37. 12	
38. 15	
39. 3	
40. 7	
41. 8	
42. 5	
43. 18	
44. 15	
45. 18	
46. 23	

24. $\begin{array}{r} 4 \\ +\,8 \\ \hline \end{array}$ **25.** $\begin{array}{r} 5 \\ +\,5 \\ \hline \end{array}$ **26.** $\begin{array}{r} 6 \\ +\,6 \\ \hline \end{array}$

27. $\begin{array}{r} 7 \\ +\,0 \\ \hline \end{array}$ **28.** $\begin{array}{r} 8 \\ +\,8 \\ \hline \end{array}$ **29.** $\begin{array}{r} 6 \\ +\,9 \\ \hline \end{array}$

30. $\begin{array}{r} 4 \\ +\,7 \\ \hline \end{array}$

Do the indicated addition.

31. $5 + 7$ **32.** $7 + 8$

33. $7 + 5$ **34.** $8 + 7$

35. $(2 + 4) + 6$ **36.** $(3 + 7) + 5$

37. $2 + (4 + 6)$ **38.** $3 + (7 + 5)$

39. $3 + 0$ **40.** $0 + 7$

41. $0 + 8$ **42.** $5 + 0$

43. $2 + 7 + 9$ **44.** $3 + 4 + 8$

45. $2 + 3 + 4 + 9$ **46.** $3 + 6 + 9 + 5$

16

Name the property of addition that is illustrated.

47. $5 + 8 = 8 + 5$

48. $2 + (7 + 9) = (2 + 7) + 9$

49. $(4 + 5) + 8 = 4 + (5 + 8)$

50. $9 + 7 = 7 + 9$

51. $3 + (7 + 5) = (3 + 7) + 5$

52. $5 + 0 = 5$

53. $3 + (4 + 0) = 3 + 4$

54. $(3 + 6) + 4 = 3 + (6 + 4)$

55. Adding of Whole Numbers is Commutative. (The order in which you add does not affect the sum.) Can you think of two actions in your daily routine that are commutative. Explain. List two actions that are *not* commutative in your daily routine and explain.

56. Adding of Whole Numbers is Associative. (The way you group whole numbers does not affect the final sum.) If you are following a recipe that lists 10 ingredients that need to be combined, do you think that adding these ingredients is associative? *Be daring!* Find a recipe and combine the ingredients in different orders. Tell the class what happens in each case. (Better yet, bring in the completed product for all to sample.)

Answers

1. 5 is the addend, 4 is the addend, 9 is the sum **3.** 7 **5.** 14 **7.** 10
9. 8 **11.** 14 **13.** 11 **15.** 12 **17.** 10 **19.** 15 **21.** 9
23. 8 **25.** 10 **27.** 7 **29.** 15 **31.** 12
33. 12. The answers to 31 and 33 must be the same since addition is commutative.
35. 12 **37.** 12. The answers to 35 and 37 must be the same since addition is associative. **39.** 3 **41.** 8 **43.** 18 **45.** 18
47. Commutative property of addition **49.** Associative property of addition
51. Associative property of addition **53.** Additive identity property
55.

ANSWERS

47. Commutative property of addition

48. Associative property of addition

49. Associative property of addition

50. Commutative property of addition

51. Associative property of addition

52. Additive identity property

53. Additive identity property

54. Associative property of addition

55.

56.

Name _____

Section _____ Date _____

A N S W E R S

1.	Addend; addend; sum
2.	13
3.	17
4.	17
5.	11
6.	11
7.	8
8.	9
9.	13
10.	13
11.	14
12.	7
13.	12
14.	10
15.	14
16.	9
17.	17
18.	17
19.	16
20.	16
21.	6
22.	7
23.	16
24.	21
25.	Commutative property of addition
26.	Additive identity property
27.	Associative property of addition
28.	Associative property of addition

1. In the statement $9 + 4 = 13$

9 is called the _____

4 is called the _____

13 is called the _____

Add.

2. $\begin{array}{r} 5 \\ + 8 \\ \hline \end{array}$
  3. $\begin{array}{r} 9 \\ + 8 \\ \hline \end{array}$
  4. $\begin{array}{r} 8 \\ + 9 \\ \hline \end{array}$

5. $\begin{array}{r} 2 \\ + 9 \\ \hline \end{array}$
  6. $\begin{array}{r} 6 \\ + 5 \\ \hline \end{array}$
  7. $\begin{array}{r} 8 \\ + 0 \\ \hline \end{array}$

8. $\begin{array}{r} 6 \\ + 3 \\ \hline \end{array}$
  9. $\begin{array}{r} 9 \\ + 4 \\ \hline \end{array}$
  10. $\begin{array}{r} 4 \\ + 9 \\ \hline \end{array}$

11. $\begin{array}{r} 7 \\ + 7 \\ \hline \end{array}$
  12. $\begin{array}{r} 6 \\ + 1 \\ \hline \end{array}$
  13. $\begin{array}{r} 3 \\ + 9 \\ \hline \end{array}$

14. $\begin{array}{r} 8 \\ + 2 \\ \hline \end{array}$
  15. $\begin{array}{r} 6 \\ + 8 \\ \hline \end{array}$
  16. $\begin{array}{r} 0 \\ + 9 \\ \hline \end{array}$

Do the indicated addition.

17. $8 + 9$
  18. $9 + 8$

19. $(3 + 8) + 5$
  20. $3 + (8 + 5)$

21. $6 + 0$
  22. $0 + 7$

23. $3 + 8 + 5$
  24. $2 + 5 + 6 + 8$

Name the property of addition that is illustrated.

25. $6 + 9 = 9 + 6$
  26. $8 + 0 = 8$

27. $(4 + 8) + 3 = 4 + (8 + 3)$
  28. $5 + (7 + 6) = (5 + 7) + 6$

Adding Whole Numbers

© 1998 McGraw-Hill Companies

1.3 OBJECTIVES

1. Add groups of whole numbers without carrying.
2. Solve applications involving simple addition.

Remember that 25 means 2 tens and 5 ones; 34 means 3 tens and 4 ones.

Let's turn now to the process of adding larger numbers. We will apply the following rule.

> We can add the digits of the same place value since they represent the same quantities.

Adding two numbers, such as $25 + 34$, can be done in expanded form. Here we write out the place value for each digit.

$$\begin{aligned} 25 &= 2 \text{ tens} + 5 \text{ ones} \\ + \, 34 &= 3 \text{ tens} + 4 \text{ ones} \qquad \downarrow \quad \text{Add down.} \\ \hline &= 5 \text{ tens} + 9 \text{ ones} \\ &= 59 \end{aligned}$$

In actual practice, we use a more convenient short form to perform the addition.

● Example 1

Adding Two Numbers

Add $25 + 34$.

Step 1 Add first in the ones column.

$$\begin{array}{r} 25 \\ + \, 34 \\ \hline 9 \end{array}$$

In using the short form, be very careful to line up the numbers correctly so that each column contains digits of the same place value.

Step 2 Now add in the tens column.

$$\begin{array}{r} 25 \\ + \, 34 \\ \hline 59 \end{array}$$

● ● ● **CHECK YOURSELF 1**

Add:

$$\begin{array}{r} 46 \\ + \, 32 \\ \hline \end{array}$$

The process is easily extended to even larger numbers. Again, we begin in the ones column.

•Example 2

Adding Two Numbers

Add 352 + 546.

Step 1   Add in the ones column.

```
  352
+ 546
    8
```

Step 2   Add in the tens column.

```
  352
+ 546
   98
```

Step 3   Add in the hundreds column.

```
  352
+ 546
  898
```

● ● ● CHECK YOURSELF 2

Add.

```
  245
+ 632
```

If you want to add more than two numbers, use the idea illustrated in Example 3.

•Example 3

Adding Three Numbers

This uses the associative property. Group the first two numbers and then add the third.

Step 1
```
  531
  142
+  25
    8
```
$1 + 2 = 3$

$3 + 5 = 8$

Think: "In the ones column, $1 + 2 = 3$. Then add the 5 for the sum, 8."

You can complete the addition by using the same idea in the tens and hundreds columns.

Step 2
```
  531
  142
+  25
   98
```

Step 3
```
  531
  142
+  25
  698
```

● ● ● **CHECK YOURSELF 3**

Add.

```
   423
    42
+  332
```

Many problems will require you to "set up" the addition. Let's work through an example.

● Example 4

Adding a Set of Numbers

Add 21, 362, 1403, and 3.
    Start by lining up like place values under each other.

```
    21       You must be very careful to line up the numbers correctly
   362       so that each column contains digits of the same place value.
  1403
+    3
```

Now you can add in each column as before.

```
    21
   362
  1403
+    3
  1789
```

● ● ● **CHECK YOURSELF 4**

Add 301, 24, 4251, and 3.

You have already seen that the word "sum" indicates addition. There are other words that also tell you to use the addition operation.
    The *total* of 12 and 5 is written as

12 + 5 or 17

8 *more than* 10 is written as

10 + 8 or 18

12 *increased by* 3 is written as

12 + 3 or 15

● Example 5

Translating Words That Indicate Addition

Find each of the following.

(*a*) 36 increased by 12.

36 increased by 12 is written as 36 + 12 = 48.

(*b*) The total of 18 and 31.

The total of 18 and 31 is written as 18 + 31 = 49.

● ● ●   **CHECK YOURSELF 5**

Find each of the following.

a. 43 increased by 25   **b.** The total of 22 and 73

You may very well be able to do some of these problems in your head. Get into the habit of writing down *all* your work, rather than just an answer.

Now we consider applications, or word problems, that will use the operation of addition. An organized approach is the key to successful problem solving, and we would suggest the following strategy. First, make sure you understand the problem. Then decide upon the operation, in this case addition, that should be used for the solution. At that point you can do the necessary calculations. Always finish your work by making sure that you have answered the question asked in the problem and that your answer seems reasonable. We can summarize this strategy with the following four basic steps.

Solving Addition Applications

STEP 1   Read the problem carefully to determine the given information and what you are asked to find.
STEP 2   Decide upon the operation (in this case, addition) to be used.
STEP 3   Write down the complete statement necessary to solve the problem and do the calculations.
STEP 4   Check to make sure you have answered the question of the problem and that your answer seems reasonable.

Let's work through some examples, using these steps.

● Example 6

Setting Up a Word Problem

A housing development has 31 finished homes. A builder has already started construction on 22 new homes and plans to start another 26. How many homes will there be in the development when the construction is complete?

Step 1 The given information is the number of existing homes, 31, the number already started, 22, and the number that will be started, 26. We want the total number of homes.

Step 2 Since we want a *sum,* we use addition for the solution.

Step 3 Write

$$
\begin{array}{r}
31 \text{ homes} \\
22 \text{ homes} \\
+\ 26 \text{ homes} \\
\hline
79 \text{ homes}
\end{array}
$$

Be sure to attach the proper units (here it is "homes") to your answer.

Step 4 The answer to the question of the original problem is 79 homes.

● ● ● **CHECK YOURSELF 6**

Ahmed has planted 30 tulips in his front yard, 28 along the side of the house, and 41 in the backyard. How many tulips has he planted?

Let's look at another similar example.

● Example 7

Setting Up a Word Problem

Four sections of algebra were offered in the fall quarter, with enrollments of 33, 24, 20, and 22 students. What was the total number of students taking algebra?

Step 1 The given information is the number of students in each section. We want the total number.

Step 2 Since we wish a total, we use addition.

Step 3 Write $33 + 24 + 20 + 22 = 99$ students.

Remember to attach the proper unit (here "students") to your answer.

Step 4 Our answer is 99 students.

● ● ● **CHECK YOURSELF 7**

Elva Ramos won an election for city council with 3110 votes. Her two opponents had 1022 and 1211 votes. How many votes were cast for that office?

● ● ● **CHECK YOURSELF ANSWERS**

1.
```
    46        46
  + 32      + 32
  ─────     ─────
     8        78
     ↑         ↑
```
Add the   Then add
ones.   the tens.

2. 877

3.
```
    423
     42
  + 332
  ─────
    797
```

4. 4579.   **5. (a)** 68; **(b)** 95.   **6.** 99 tulips.   **7.** 5343 votes.

Name

Section   Date

A N S W E R S

Add.

1.   24
        + 3

2.   13
        + 5

3.   23
        + 56

4.   75
        + 20

5.   332
        + 54

6.   620
        + 67

7.   307
        + 232

8.   349
        + 420

9.   2792
        + 205

10.   5463
         + 435

11.   2345
         + 6053

12.   3271
         + 4715

13.   2531
         + 5354

14.   5003
         + 4205

1. 27

2. 18

3. 79

4. 95

5. 386

6. 687

7. 539

8. 769

9. 2997

10. 5898

11. 8398

12. 7986

13. 7885

14. 9208

15.	64,356
16.	47,728
17.	69
18.	87
19.	3699
20.	2798
21.	467
22.	589
23.	1569
24.	2889
25.	54
26.	29
27.	793
28.	168
29.	159
30.	2475
31.	596
32.	668

15. $\begin{array}{r} 21{,}314 \\ +\ 43{,}042 \end{array}$

16. $\begin{array}{r} 12{,}325 \\ +\ 35{,}403 \end{array}$

17. $\begin{array}{r} 13 \\ 21 \\ +\ 35 \end{array}$

18. $\begin{array}{r} 24 \\ 31 \\ +\ 32 \end{array}$

19. $\begin{array}{r} 3462 \\ 213 \\ +\ \ \ 24 \end{array}$

20. $\begin{array}{r} 2430 \\ 356 \\ +\ \ \ 12 \end{array}$

21. $35 + 432$

22. $527 + 62$

23. $4 + 12 + 340 + 1213$

24. $534 + 2 + 31 + 2322$

Find each of the following.

25. The total of 23 and 31

26. 7 more than 22

27. The sum of 562 and 231

28. 123 increased by 45

29. 34 more than 125

30. The total of 124 and 2351

31. The sum of 23, 122, and 451

32. The total of 112, 24, and 532

Solve each of the following addition applications.

33. **Golf.** A golfer shot a score of 42 on the first nine holes and a score of 46 on the second nine holes. What was her total score for the round?

34. **Bowling.** A bowler scored 201, 153, and 215 in three games. What was the total score for those games?

35. **Test scores.** Angela had a score of 73 on her first mathematics test. On the second test, she increased her score by 23 points. What was her score on the second test?

36. **Driving distance.** Jesse drove 244 miles (mi) from St. Louis to Indianapolis on the first day of a business trip. He then drove 113 mi more to Louisville on the second day of the trip. How far did he travel?

37. **Vacation mileage.** The Burton family drove 325 mi on the first day of a vacation trip and 273 mi on the second day. How far did they drive in those 2 days?

38. **Car purchase.** Susan Compton buys a car with a list price of $8250. She also orders an air conditioner for $445. What will the total cost be?

39. **Bowling.** Marilyn rolled games of 181, 212, and 206 in a three-game bowling series. What was her total score for the series?

40. **Basketball.** A basketball player scored 32, 25, 22, and 20 points in a four-game tournament. How many points did he score in all?

41. **Ticket sales.** For a band concert, an auditorium has 245 $9 seats, 350 $7 seats, and 403 $5 seats. How many tickets can be sold?

42. **Play attendance.** Four performances of a play had attendance figures of 230, 312, 244, and 213. How many people saw the play during this period?

43. **Airline travel.** An airline had 133, 115, 120, and 111 passengers on their four shuttle flights between Los Angeles and San Francisco during 1 day. What was the total number of passengers?

ANSWERS

33. 88
34. 569
35. 96
36. 357 mi
37. 598 mi
38. $8695
39. 599
40. 99
41. 998 tickets
42. 999 people
43. 479 passengers

44. $1999

45. 12, 17

46.

47.

44. Utility costs. A company spent $1321 for rent, $232 for heat, $123 for electricity, $112 for phone service, and $211 for cleaning during 1 month. What was the company's total expense for the month?

45. Omar is opening a restaurant and wants to buy trapezoidal tables that seat seven people each (see figure).

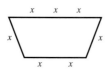

If Omar wants to place the tables in a line touching one another, how many people can be seated at two tables? At three tables?

46. The Romans used a system of numeration that involved symbols called Roman numerals. Describe this system and how the addition process worked. Create a table of basic addition facts for this system using the symbols for one, five, and ten.

47. Investigate two other systems of numeration (other than ours and one the Romans used). Describe the symbols used and process of addition in these systems.

Answers

1. 27   **3.** 79   **5.** 386   **7.** 539   **9.** 2997   **11.** 8398   **13.** 7885

15. 64,356   **17.** 69   **19.** 3699   **21.**

$$\begin{array}{r} 35 \\ + 432 \\ \hline 467 \end{array}$$

23.

$$\begin{array}{r} 4 \\ 12 \\ 340 \\ + 1213 \\ \hline 1569 \end{array}$$

25. 54

27. 793   **29.** 159   **31.** 596   **33.** 88   **35.** 96   **37.** 598 mi

39. 599   **41.** 998 tickets   **43.** 479 passengers   **45.** 12, 17   **47.**

Name

Section   Date

Add:

1. 34
     + 4

2. 40
     + 38

3. 450
     + 27

4. 328
     + 431

5.   240
     + 3157

6. 5315
     + 2463

7. 1253
     + 3644

8. 31,304
     + 52,583

9. 73
     12
   + 13

10. 2450
      134
    + 15

11. 13 + 21 + 34

12. 545 + 31 + 3 + 2020

Find each of the following.

13. 71 increased by 24

14. 15 more than 61

15. A total of 43 and 32

16. 27 increased by 21

Solve each of the following addition applications.

17. Baseball. A baseball team scored 5 runs in each of the first two games and 7 runs in the third game. How many total runs did they score in the three games?

ANSWERS

1. 38
2. 78
3. 477
4. 759
5. 3397
6. 7778
7. 4897
8. 83,887
9. 98
10. 2599
11. 68
12. 2599
13. 95
14. 76
15. 75
16. 48
17. 17 runs

29

18. 1484 mi

19. 585

20. 69

21. $198

22. $488

23. 197 mi

24. 69 points

18. Flying distance. Consuelo flew 1121 miles (mi) from Seattle to Los Angeles and 363 mi from Los Angeles to Phoenix. How many miles did she fly?

19. Test scores. Johanna scored 515 the first time she took the SAT. The next time she took it, she increased her score by 70 points. What was her score the second time?

20. Golf. Chi Chi shot a score of 34 on the first nine holes of golf and a score of 35 on the second nine. What was his total score for the 18 holes?

21. Education costs. Pat spent $115 for tuition and $83 for books and supplies during one term. What were her expenses for tuition, books, and supplies?

22. Purchase costs. Matt buys a television set with a list price of $425. With his payment plan, finance charges are $63. What will be the total cost of the purchase?

23. Driving mileage. A salesman drove 64 mi on Tuesday, 112 mi on Thursday, and 21 mi on Friday. What was his mileage for those 3 days?

24. Basketball. In 4 basketball games, Ryan scores 21, 14, 11, and 23 points. How many points did he score in all four games?

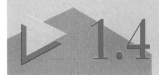

Addition with Carrying

1.4 OBJECTIVES

1. Add any group of whole numbers.
2. Solve applications involving some carrying.

Carrying in addition is also called **regrouping,** or **renaming.** Of course, the name makes no difference as long as you understand the process.

Of course this is true for any size number. The place value thousands is 10 times the place value hundreds, and so on.

In the examples and exercises of the last section, the digits in each column added to 9 or less. Let's look at the situation in which a column has a two-digit sum. This will involve the process of **carrying.** Let's look at the process in expanded form.

• Example 1

Adding in Expanded Form When Carrying Is Needed

$$
\begin{aligned}
67 &= 6 \text{ tens} + 7 \text{ ones} \\
+ 28 &= 2 \text{ tens} + 8 \text{ ones} \\
\hline
& 8 \text{ tens} + 15 \text{ ones}
\end{aligned}
$$

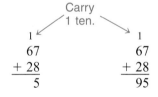

or $\quad 8 \text{ tens} + \overbrace{1 \text{ ten} + 5 \text{ ones}}$

or $\qquad\quad 9 \text{ tens} \qquad + 5 \text{ ones}$

or $\qquad\quad 95$

We have written 15 ones as 1 ten and 5 ones. The 1 ten is then combined with the 8 tens.

The more convenient short form carries the excess units from one column to the next column left. Recall that the place value of the next column left is 10 times the value of the original column. It is this property of our decimal place-value system that makes carrying work. Let's look at the problem again, this time done in the short, or "carrying," form.

Step 1   **Step 2**

Carry
1 ten.

$$
\begin{array}{r}
\overset{1}{}67 \\
+\ 28 \\
\hline
5
\end{array}
\qquad
\begin{array}{r}
\overset{1}{}67 \\
+\ 28 \\
\hline
95
\end{array}
$$

Step 1: The sum of the digits in the ones column is 15, so write 5 and carry 1 to the tens column. **Step 2:** Now add in the tens column, being sure to include the carried 1.

● ● ● CHECK YOURSELF 1

Add. $\quad \begin{array}{r} 58 \\ + 36 \\ \hline \end{array}$

The addition process often requires more than one carrying step, as is shown in Example 2.

● Example 2

Adding in Short Form When Carrying Is Needed

Add 285 and 378.

$$
\begin{array}{r}
1 \quad\longleftarrow \text{ Carry 1 ten.}\\
285\\
+\,378\\
\hline
3
\end{array}
$$

The sum of the digits in the ones column is 13, so write 3 and carry 1 to the tens column.

Carry $\longrightarrow$ 1 hundred.
$$
\begin{array}{r}
1\,1\\
285\\
+\,378\\
\hline
63
\end{array}
$$

Now add in the tens column, being sure to include the carry. We have 16 tens, so write 6 in the tens place and carry 1 to the hundreds column.

$$
\begin{array}{r}
1\,1\\
285\\
+\,378\\
\hline
663
\end{array}
$$

Finally, add in the hundreds column.

● ● ● **CHECK YOURSELF 2**

Add.
$$
\begin{array}{r}
479\\
+\,287\\
\hline
\end{array}
$$

The carrying process is the same if we want to add more than two numbers.

● Example 3

Adding in Short Form With Multiple Carrying Steps

Add 53, 2678, 587, and 27,009.

$$
\begin{array}{r}
1\,1\,2\,2 \quad\longleftarrow \text{ Carries}\\
53\\
2{,}678\\
587\\
+\,27{,}009\\
\hline
30{,}327
\end{array}
$$

Add in the ones column: $3 + 8 + 7 + 9 = 27$. Write 7 in the sum and carry 2 to the tens column.

Now add in the tens column, being sure to include the carry. The sum is 22. Write 2 tens and carry 2 to the hundreds column. Complete the addition by adding in the hundreds column, the thousands column, and the ten thousands column.

● ● ● **CHECK YOURSELF 3**

Add 46, 365, 7254, and 24,006.

● ● ● **CHECK YOURSELF ANSWERS**

1. 94.   **2.** 766.   **3.** 31,671.

1.4 Exercises

Name

Section   Date

ANSWERS

Add.

1. 47
   + 9

2. 64
   + 8

3. 23
   + 48

4. 96
   + 57

5. 31
   27
   + 35

6. 69
   27
   + 58

7. 213
   + 78

8. 392
   + 58

9. 703
   + 287

10. 898
   + 457

11. 589
   306
   + 42

12. 257
   18
   + 504

13. 590
   345
   + 758

14. 358
   271
   + 595

1. 56

2. 72

3. 71

4. 153

5. 93

6. 154

7. 291

8. 450

9. 990

10. 1355

11. 937

12. 779

13. 1693

14. 1224

© 1998 McGraw-Hill Companies

33

15.
$$\begin{array}{r} 2578 \\ + 3455 \\ \hline \end{array}$$

16.
$$\begin{array}{r} 8295 \\ + 4927 \\ \hline \end{array}$$

17.
$$\begin{array}{r} 3490 \\ 548 \\ + 25 \\ \hline \end{array}$$

18.
$$\begin{array}{r} 678 \\ 4533 \\ + 70 \\ \hline \end{array}$$

19.
$$\begin{array}{r} 2289 \\ 38 \\ 578 \\ + 3489 \\ \hline \end{array}$$

20.
$$\begin{array}{r} 3678 \\ 259 \\ 27 \\ + 2356 \\ \hline \end{array}$$

21.
$$\begin{array}{r} 23{,}458 \\ + 32{,}623 \\ \hline \end{array}$$

22.
$$\begin{array}{r} 52{,}591 \\ + 59{,}739 \\ \hline \end{array}$$

23.
$$\begin{array}{r} 26{,}735 \\ 259 \\ 3{,}056 \\ + 35{,}489 \\ \hline \end{array}$$

24.
$$\begin{array}{r} 35{,}607 \\ 2{,}345 \\ 456 \\ + 81{,}247 \\ \hline \end{array}$$

25. Find the sum of 79 and 735.

26. Add 28 and 386.

27. What is the total of 38, 354, and 8?

28. Find the sum of 23, 57, and 236.

29. Add 23, 2845, 5, and 589.

30. Find the sum of 3295, 9, 427, and 56.

31. What is the total of 2195, 348, 640, 59, and 23,785?

32. Add 5637, 78, 690, 28, and 35,589.

The sequences below are called *arithmetic sequences*. Determine the pattern, and write the next four numbers in each sequence.

33. 5, 12, 19, 26, _____, _____, _____, _____

34. 8, 14, 20, 26, _____, _____, _____, _____

35. 7, 13, 19, 25, _____, _____, _____, _____

36. 9, 17, 25, 33, _____, _____, _____, _____

Solving the following applications.

37. Test points. Marsha had grades of 85, 93, 79 and 89 on four tests during a course. What was her total number of points?

38. Basketball. A professional basketball player scores 1814 points in his first year, 1953 in his second, and 1893 in his third year. How many points did he score in his first three years?

39. Purchasing automobiles. Annmarie bought a 1931 Model A for $5200, a 1964 Thunderbird convertible for $7100, and a 1959 Austin Healy Mark I for $7450. How much did she invest in the three cars?

40. Consumer spending. Trinh bought a used Pentium 100 for $2120. In addition, he spent $379 for a printer and $589 for software. How much did he spend?

41. Shipping. Rita's vineyard shipped 4200 pounds (lb) of grapes in August, 5970 lb in September, and 4850 lb in October. How many pounds were shipped?

42. Total distance. A salesman drove 68 miles (mi) on Tuesday, 114 mi on Thursday, and 79 mi on Friday. What was the mileage for those 3 days?

43. Video rentals. The following chart shows Family Video's monthly rentals for the first three months of 1996 by category of film. Complete the totals.

Category of Film	Jan.	Feb.	Mar.	Category Totals
Comedy	4568	3269	2189	10,026
Drama	5612	4129	3879	13,620
Action/Adventure	2654	3178	1984	7816
Musical	897	623	528	2048
Monthly Totals	13,731	11,199	8580	33,510

33. 33, 40, 47, 54

34. 32, 38, 44, 50

35. 31, 37, 43, 49

36. 41, 49, 57, 65

37. 346 points

38. 5660 points

39. $19,750

40. $3088

41. 15,020 lb

42. 261 mi

43. See exercise

44. Business expenses. The following chart shows Regina's Dress Shop's expenses by department for the last three months of the year. Complete the totals.

Department	Oct.	Nov.	Dec.	Department Totals
Office	$31,714	$32,512	$30,826	$95,052
Production	85,146	87,479	81,234	$253,859
Sales	34,568	37,612	33,455	$105,635
Warehouse	16,588	11,368	13,567	$41,523
Monthly Totals	$168,016	$168,971	$159,082	$496,069

Fibonacci numbers occur in the sequence:

1, 1, 2, 3, 5, 8, 13, 21, 34, 55, . . .

This sequence begins with the numbers 1 and 1 again, and each subsequent number is obtained by adding the two preceding numbers.

45. Find the next four numbers in the sequence.

46. You can find more about Fibonacci numbers in an encyclopedia or on the World Wide Web. Do some research and find two examples in nature that exhibit the patterns displayed in the Fibonacci sequence.

Answers

1. 56   **3.** 71   **5.** 93   **7.** 291   **9.** 990   **11.** 937   **13.** 1693

15. 6033   **17.** 4063   **19.**
123
2289
38
578
+ 3489
6394

21. 56,081   **23.** 65,539

25. 814   **27.** 400   **29.** 3462   **31.**
2 32
2,195
348
23,785
640
+   59
27,027

33. 33, 40, 47, 54

35. 31, 37, 43, 49   **37.** 346   **39.** $19,750   **41.** 15,020 lb

43.

Category of Film	Jan.	Feb.	Mar.	Category Totals
Comedy	4568	3269	2189	10,026
Drama	5612	4129	3879	13,620
Action/Adventure	2654	3178	1984	7816
Musical	897	623	528	2048
Monthly Totals	13,731	11,199	8580	33,510

45. 89, 144, 233, 377

Name _____

Section _____ Date _____

A N S W E R S

Add.

1.   58
   + 7

2.   48
   + 76

3.   35
   29
   + 32

4.   987
   + 84

5.   783
   + 529

6.   604
   489
   + 54

7.   704
   583
   + 435

8.   4538
   + 2759

9.   5832
   539
   + 2470

10.   684
   5372
   2358
   + 56

11.   24,078
   + 58,254

12.   48,032
   2,509
   358
   + 53,645

13. Find the sum of 527 and 43.

14. Add 29, 273, and 155.

15. What is the total of 3245, 300, 2891, and 78?

1.	65
2.	124
3.	96
4.	1071
5.	1312
6.	1147
7.	1723
8.	7297
9.	8841
10.	8470
11.	82,332
12.	104,544
13.	570
14.	457
15.	6514

16. Find the sum of 5848, 39, 583, 157, and 29,875.

Solving the following applications.

17. Family budget. The Brackens had the following expenses in 1 month: housing, $350; food, $185; clothing, $60; utilities, $95; and recreation, $45. What were their total expenses for 1 month?

18. Checking accounts. Teresa has $1457 in her checking account, $2794 in her savings account, and $13,450 in a CD. What is the total amount of money Teresa has in all her accounts?

Using Your Calculator to Add

Even though the reason for this book is to help you review the basic skills of arithmetic, many of you will want to be able to use a handheld calculator for some of the problems that are presented. Ideally you should learn to do the basic operations *by hand*. So in each section of this book, start by learning to do the work *without* your calculator. We will then provide these special calculator sections to show how you can use the calculator.

First, to enter a number in your calculator, simply press the digits in order *from left to right*.

● Example 1

Entering a Number on a Calculator

To enter 23,456, press each number, one at a time.

| 2 || 3 || 4 || 5 || 6 | We start with the digit of the largest place value, in this case 2, meaning 2 ten thousands.

Display | 23456 |

From this point on, we will simply say "enter the number" as a single step.

● ● ● **CHECK YOURSELF 1**

Enter 32,196 in your calculator.

● Example 2

Using Your Calculator to Add

To add whole numbers, say, 23 + 45 + 67, follow the indicated steps.

This simply clears the calculator for what follows.

1. Press the clear key. | C |
2. Enter the first number. | 2 || 3 |
3. Press the plus key. | + |
4. Enter the second number. | 4 || 5 |
5. Press the plus key. | + |

The sum of the first two numbers, 68, should now be in the display.

6. Continue until the last number is entered. | 6 || 7 |
7. Press the equals key. | = |

The desired sum should now be in the display.

Display | 135 |

● ● ● **CHECK YOURSELF 2**

Add 39 + 27 + 18 on your calculator.

Let's look at another example.

●Example 3

Using Your Calculator to Add

Add 23 + 3456 + 7 + 985.
Enter:

23 ⊞ 3456 ⊞ 7 ⊞ 985 ⊟

Display |4471|

Enter each of the first three numbers followed by the plus key. Then enter the final number and press the equals key. The sum will be in the display.

As we mentioned, calculators use a variety of patterns in performing operations in arithmetic. If you have a calculator, try the addition of Example 3 now. If you do not get 4471 as the answer, check the operating manual or ask your instructor for assistance.

● ● ● **CHECK YOURSELF 3**

Add 295 + 3 + 4162 + 84.

● ● ● **CHECK YOURSELF ANSWERS**

1. 32196.   **2.** 39 ⊞ 27 ⊞ 18 ⊟ 84.   **3.** 295 ⊞ 3 ⊞ 4162 ⊞ 84 ⊟ 4544.

Calculator Exercises

Add.

1.   23
       + 78

2.   589
       + 257

3.   2173
       + 5899

4.   458
       273
       + 568

5.   2743
       258
       35
       + 5823

6.   29,753
       249
       53
       5,821
       + 4,258

7.   3,295,153
       573,128
       21,257
       2,586,241
       +   5,291

8. 507 + 359 + 259

9. 23 + 5638 + 385 + 7 + 27,345

10. 563 + 9487 + 5 + 35,600 + 39

	ANSWERS
1.	101
2.	846
3.	8072
4.	1299
5.	8859
6.	40,134
7.	6,481,070
8.	1125
9.	33,398
10.	45,694
11.	See exercise

Solve the following applications.

11. Banking. The following table shows the number of customers using three branches of a bank during 1 week. Complete the calculations.

Branch	M	T	W	Th	F	Weekly Totals
Downtown	487	356	429	278	834	2384
Suburban	236	255	254	198	423	1366
Westside	345	278	323	257	563	1766
Daily Totals	1068	889	1006	733	1820	5516
						Grand Total

12. Poker hands. The following table lists the number of possible types of poker hands. What is the total number possible?

Royal flush	4
Straight flush	36
Four of a kind	624
Full house	3,744
Flush	5,108
Straight	10,200
Three of a kind	54,912
Two pairs	123,552
One pair	1,098,240
Nothing	1,302,540

Answers

1. 101   **3.** 8072   **5.** 8859   **7.** 6,481,070   **9.** 33,398

11.

Branch	M	T	W	Th	F	Weekly Totals
Downtown	487	356	429	278	834	2384
Suburban	236	255	254	198	423	1366
Westside	345	278	323	257	563	1766
Daily Totals	1068	889	1006	733	1820	5516
						Grand Total

Rounding, Estimation, and Order

1.5 OBJECTIVES

1. Round a whole number at any place value.
2. Estimate sums by rounding.
3. Estimate distance.
4. Use the symbols < and >.

It is a common practice to express numbers to the nearest hundred, thousand, and so on. For instance, the distance from Los Angeles to New York along one route is 2833 miles (mi). We might say that the distance is 2800 mi. This is called **rounding,** because we have rounded the distance to the nearest hundred miles.

One way to picture this rounding process is with the use of a number line.

• Example 1

Rounding to the Nearest Hundred

To round 2833 to the nearest hundred:

Since 2833 is closer to 2800, we round *down* to 2800.

● ● ● CHECK YOURSELF 1

Round 587 to the nearest hundred.

• Example 2

Rounding to the Nearest Thousand

To round 28,734 to the nearest thousand:

Since 28,734 is closer to 29,000, we round *up* to 29,000.

● ● ● CHECK YOURSELF 2

Locate 1375 and round to the nearest hundred.

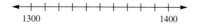

Instead of using a number line, we can apply the following rule.

Rounding Whole Numbers

By a certain *place,* we mean tens, hundreds, thousands, and so on.

STEP 1 To round off a whole number to a certain decimal place, look at the digit to the right of that place.

STEP 2

This is called **rounding up**.

a. If that digit is 5 or more, that digit and all digits to the right become 0. The digit in the place you are rounding to is increased by 1.

This is called **rounding down**.

b. If that digit is less than 5, that digit and all digits to the right become 0. The digit in the place you are rounding to remains the same.

● Example 3

Rounding to the Nearest Ten

Round 587 to the nearest ten:

Tens
↓

5 8 7
↑

The digit to the right of the tens place
↓

587 is between 580 and 590. It is closer to 590, so it makes sense to round up.

5 8 7   is rounded to 590

We shade the tens digit. The digit to the right of the tens place, 7, is 5 or more. So round up.

580   587   590

● ● ● CHECK YOURSELF 3

Round 847 to the nearest ten.

● Example 4

Rounding to the Nearest Hundred

Round 2638 to the nearest hundred:

2638 is closer to 2600 than to 2700. So it makes sense to round down.

↓

2 6 38   is rounded to 2600

We shade the hundreds digit. The digit to the right, 3, is less than 5. So round down.

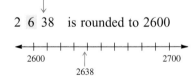

2600   2638   2700

● ● ● **CHECK YOURSELF 4**

Round 3482 to the nearest hundred.

Let's look at some further examples of using the rounding rule.

● **Example 5**

Rounding to the Nearest Hundred

(*a*) Round 2378 to the nearest hundred:

↓

2 3 78 is rounded to 2400   We have shaded the hundreds digit. The digit to the right is 7. Since this is 5 or more, the 7 and all digits to the right become 0. The hundreds digit is increased by 1.

(*b*) Round 53,258 to the nearest thousand:

↓

5 3 ,258 is rounded to 53,000   We have shaded the thousands digit. Since the digit to the right is less than 5, it and all digits to the right become 0, and the thousands digit remains the same.

(*c*) Round 685 to the nearest ten:

↓

6 8 5 is rounded to 690   The digit to the right of the tens place is 5 or more. Round up by our rule.

(*d*) Round 52,813,212 to the nearest million:

↓

5 2 ,813,212 is rounded to 53,000,000

● ● ● **CHECK YOURSELF 5**

a. Round 568 to the nearest ten.
b. Round 5446 to the nearest hundred.

Let's look at a case in which we round up a 9.

• Example 6

Rounding to the Nearest Ten

Suppose we want to round 397 to the nearest ten. We shade the tens digit and look at the next digit to the right.

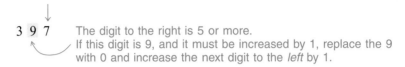

3 9 7   The digit to the right is 5 or more.
          If this digit is 9, and it must be increased by 1, replace the 9
          with 0 and increase the next digit to the *left* by 1.

So 397 is rounded to 400.

CHECK YOURSELF 6

Round 4961 to the nearest hundred.

An estimate is basically a good guess. If your answer is close to your estimate, then your answer is reasonable.

Whether you are doing an addition problem by hand or using a calculator, rounding numbers gives you a handy way of deciding if the answer seems reasonable. The process is called **estimating.** Let's illustrate with an example.

• Example 7

Estimating a Sum

Begin by rounding to the nearest hundred

456	500
235	200
976	1000
+ 344	+ 300
2011	2000	← Estimate

By rounding to the nearest hundred and adding quickly, we get an estimate or guess of 2000. Since this is close to the sum calculated, 2011, our answer seems reasonable.

CHECK YOURSELF 7

Estimate the sum by rounding each addend to the nearest hundred then add.

287 + 526 + 311 + 378

Estimation is a wonderful tool to use while you're shopping. Every time you go to the store, you should try to estimate the total bill by rounding the price of each item. If you do this regularly, both your addition skills and your rounding skills will improve. The same holds true when you eat in a restaurant. It is always a good idea to know approximately how much you are spending.

• Example 8

Estimating a Sum in a Word Problem

Samantha has taken the family out to dinner, and she's now ready to pay the bill. The dinner check has no total, only the individual entries, as below:

Soup	$2.95
Soup	$2.95
Salad	1.95
Salad	1.95
Salad	1.95
Lasagna	7.25
Spaghetti	4.95
Ravioli	5.95

What is the approximate cost of the dinner?

Rounding each entry to the nearest whole dollar, we can estimate the total by finding the sum

$$3 + 3 + 2 + 2 + 2 + 7 + 5 + 6 = \$30$$

●●● **CHECK YOURSELF 8**

Jason is doing the weekly food shopping at FoodWay. So far his basket has items that cost $3.99, $7.98, $2.95, $1.15, $2.99, and $1.95. Approximate the total cost of these items.

Earlier in this section, we used the number line to illustrate the idea of rounding numbers. The number line also gives us an excellent way to picture the concept of **order** for whole numbers, which means that numbers become larger as we move from left to right on the line.

For instance, we know that 3 is less than 5. On the number line

3 is less than or smaller than 5.

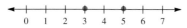

we see that 3 lies *to the left* of 5.

We also know that 4 is greater than 2. On the number line

4 is greater than or larger than 2.

we see that 4 lies *to the right* of 2.

Two symbols are used to indicate these relationships.

> For whole numbers *a* and *b*, we can write
>
> **1.** $a < b$ (read "*a* is less than *b*") when *a* is *to the left* of *b* on the number line.
>
> **2.** $a > b$ (read "*a* is greater than *b*") when *a* is *to the right* of *b* on the number line.

Example 9 illustrates the use of this notation.

● Example 9

Indicating Order with $<$ or $>$

Use the symbols $<$ or $>$ to complete each statement.

(*a*) 7 _____ 10
(*b*) 25 _____ 20
(*c*) 200 _____ 300
(*d*) 8 _____ 0

Solutions

(*a*) $7 < 10$ 7 lies to the left of 10 on the number line.
(*b*) $25 > 20$ 25 lies to the right of 20 on the number line.
(*c*) $200 < 300$
(*d*) $8 > 0$

● ● ●   CHECK YOURSELF 9

Use one of the symbols $<$ or $>$ to complete each of the following statements.

a. 35 ____ 25   **b.** 0 ____ 4   **c.** 12 ____ 18   **d.** 1000 ____ 100

● ● ●   CHECK YOURSELF ANSWERS

1. 600.   **2.** Round 1375 *up* to 1400.

3. 850.   **4.** 3500.   **5. (a)** 570; **(b)** 5400.   **6.** 5000.   **7.** 1500.
8. \$21.   **9. (a)** $35 > 25$; **(b)** $0 < 4$; **(c)** $12 < 18$; **(d)** $1000 > 100$.

A N S W E R S

Round each of the following numbers to the indicated place.

1. 38, the nearest ten

2. 72, the nearest ten

3. 253, the nearest ten

4. 578, the nearest ten

5. 696, the nearest ten

6. 683, the nearest hundred

7. 3482, the nearest hundred

8. 6741, the nearest hundred

9. 5962, the nearest hundred

10. 4352, the nearest thousand

11. 4927, the nearest thousand

12. 39,621, the nearest thousand

13. 23,429, the nearest thousand

14. 38,589, the nearest thousand

15. 787,000, the nearest ten thousand

16. 582,000, the nearest hundred thousand

17. 21,800,000, the nearest million

18. 931,000, the nearest ten thousand

Estimate each of the following sums by rounding to the indicated place. Then do the addition and use your estimate to see if your actual sum seems reasonable.

Round to the nearest ten.

19.
```
  58
  27
+ 33
```

20.
```
  92
  37
  85
+ 64
```

21.
```
  87
  53
  41
  93
+ 62
```

22.
```
  78
  67
  53
  42
+ 86
```

1.	40
2.	70
3.	250
4.	580
5.	700
6.	700
7.	3500
8.	6700
9.	6000
10.	4000
11.	5000
12.	40,000
13.	23,000
14.	39,000
15.	790,000
16.	600,000
17.	22,000,000
18.	930,000
19.	Estimate: 120, actual sum: 118
20.	Estimate: 280, actual sum: 278
21.	Estimate: 330, actual sum: 336
22.	Estimate: 330, actual sum: 326

23. Estimate: 2100,
actual sum: 2122

24. Estimate: 4600,
actual sum: 4614

25. Estimate: 4700,
actual sum: 4730

26. Estimate: 7500,
actual sum: 7503

27. Estimate: 11,000,
actual sum: 11,380

28. Estimate: 10,000,
actual sum: 9925

29. Estimate: 22,000,
actual sum: 22,187

30. Estimate: 35,000,
actual sum: 35,255

31. <

32. <

33. >

34. >

35. <

36. >

37. $29

38. $27

Round to the nearest hundred.

23.
```
   379
  1215
+  528
```

24.
```
   967
  2365
   544
+  738
```

25.
```
  1378
   519
   792
+ 2041
```

26.
```
  3145
   889
   259
   692
+ 2518
```

Round to the nearest thousand.

27.
```
  2238
  3925
+ 5217
```

28.
```
  3678
  4215
+ 2032
```

29.
```
  9137
  2315
  7643
+ 3092
```

30.
```
  11,548
   3,874
  14,435
+  5,398
```

Use the symbol < or > to complete each statement.

31. 4 _____ 8

32. 0 _____ 5

33. 500 _____ 400

34. 20 _____ 15

35. 100 _____ 1000

36. 3000 _____ 2000

Solve the following applications.

37. Lunch bills. Ed and Sharon go to lunch. The lunch check has no total but only lists individual items:

Soup $1.95	Soup $1.95
Salad $1.80	Salad $1.80
Salmon $8.95	Flounder $6.95
Pecan pie $3.25	Vanilla ice cream $2.25

Estimate the total amount of the lunch check.

38. Consumer spending. Mary will purchase several items at the stationery store. Thus far, the items she has collected cost $2.99, $6.97, $3.90, $2.15, $9.95, and $1.10. Approximate the total cost of these items.

39. Clothes shopping. Mrs. Murphy went shopping for clothes. She bought a sweater for $32.95, a scarf for $9.99, boots for $68.29, a coat for $125.90, and socks for $18.15. Estimate the total amount of Mrs. Murphy's purchases.

40. Food shopping. Maritza went to the local supermarket and purchased the following items: milk, $2.89; butter, $1.75; bread, $1.10; orange juice, $1.25; cereal, $3.95; and coffee, $3.80. Approximate the total cost of these items.

41. A bag contains 60 marbles. The number of blue marbles, rounded to the nearest 10, is 40, and the number of green marbles in the bag, rounded to the nearest 10, is 20. How many blue marbles are in the bag? (List all answers that satisfy the conditions of the problem.)

42. Describe some situations in which estimating and rounding would not produce a result that would be suitable or acceptable. Review the instructions for filing your federal income tax. What rounding rules are used in the preparation of your tax returns? Do the same rules apply to the filing of your state tax returns? If not, what are these rules?

Answers

1. 40   **3.** 250   **5.** 700   **7.** 3500   **9.** 6000   **11.** 5000   **13.** 23,000

15. 790,000   **17.** 22,000,000   **19.** Estimate: 120, actual sum: 118

21. Estimate: 330, actual sum: 336   **23.** Estimate: 2100, actual sum: 2122

25. Estimate: 4700, actual sum: 4730   **27.** Estimate: 11,000, actual sum: 11,380

29. Estimate: 22,000, actual sum: 22,187   **31.** $4 < 8$ **33.** $500 > 400$

35. $100 < 1000$ **37.** $29   **39.** $255   **41.** 36, 37, 38, 39, 40, 41, 42, 43, 44

39. $255

40. $15

41. 36, 37, 38, 39, 40, 41, 42, 43, 44

42.

$$1.5$$ **Supplementary Exercises**

Name

Section   Date

Round each of the following numbers to the indicated place.

1. 59, the nearest ten

2. 73, the nearest ten

3. 583, the nearest ten

4. 768, the nearest ten

5. 896, the nearest hundred

6. 2981, the nearest hundred

ANSWERS

1. 60

2. 70

3. 580

4. 770

5. 900

6. 3000

7. 44,000

8. 81,000

9. 800,000

10. Estimate: 190
Sum: 192

11. Estimate: 300
Sum: 298

12. Estimate: 4000
Sum: 4015

13. Estimate: 5000
Sum: 4975

14. Estimate: 21,000
Sum: 21,092

15. Estimate: 30,000
Sum: 30,075

16. >

17. >

18. <

19. $25

20. $64

7. 43,587, the nearest thousand

8. 81,243, the nearest thousand

9. 791,000, the nearest hundred thousand

Estimate each of the following sums by rounding to the indicated place. Then do the actual addition, and use your estimate to see if your actual sum seems reasonable.

Round to the nearest ten.

10.
```
   57
   73
 + 62
```

11.
```
   93
   68
   75
 + 62
```

Round to the nearest hundred.

12.
```
    687
   2320
    573
 +  435
```

13.
```
    480
    515
    693
   2560
 +  727
```

Round to the nearest thousand.

14.
```
    2,788
    7,215
 + 11,089
```

15.
```
    8,319
   12,835
    5,902
 +  3,019
```

Use the symbol < or > to complete each statement.

16. 18 _____ 12   **17.** 10 _____ 0   **18.** 500 _____ 600

Solve the following applications.

19. Consumer spending. Amir bought several items at the hardware store: hammer, $8.95; screwdriver, $3.15; pliers, $6.90; wire cutters, $4.25; and sandpaper, $1.89. Estimate the total cost of Amir's bill.

20. Clothes shopping. Olga went to the dress shop and purchased items that cost $3.75, $8.90, $19.95, $23.15, and $7.80. Approximate the total cost of her purchases.

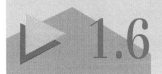

1.6 Subtraction of Whole Numbers

1.6 OBJECTIVES

1. Use the language of subtraction.
2. Subtract whole numbers without borrowing.
3. Solve applications of simple subtraction.

By *opposite* we mean that subtracting a number "undoes" an addition of that same number. Start with 1. Add 5 and then subtract 5. Where are you?

We are now ready to consider a second operation of arithmetic—subtraction. In Section 1.2, we described addition as the process of combining two or more groups of the same kinds of objects. Subtraction can be thought of as the *opposite operation* to addition. Every arithmetic operation has its own notation. The symbol for subtraction, −, is called a **minus sign.**

When we write 8 − 5, we wish to subtract 5 from 8. We call 5 the **subtrahend.** This is the number being subtracted. And 8 is the **minuend.** This is the number we are subtracting from. The **difference** is the result of the subtraction.

To find the *difference* of two numbers, we look for a number which, when added to the number being subtracted, will give the number that we started with. For example,

$$8 - 5 = 3 \qquad \text{since} \qquad 3 + 5 = 8$$

This special relationship between addition and subtraction provides a method of checking subtraction.

> The sum of the difference and the subtrahend must be equal to the minuend.

• Example 1

Subtracting a Single-Digit Number

$$12 - 5 = 7$$

Check:

$$7 + 5 = 12$$

Difference   Subtrahend   Minuend

Our check works because 12 − 5 asks for the number that must be added to 5 to get 12.

• • •   CHECK YOURSELF 1

Subtract, and check your work.

$$13 - 9 =$$

The procedure for subtracting larger whole numbers is similar to the procedure for addition. We subtract digits of the same place value.

Let's look at an example in the expanded form.

•Example 2

Subtracting a Two-Digit Number in Expanded Form

$$87 = 8 \text{ tens} + 7 \text{ ones}$$
$$\underline{-53 = 5 \text{ tens} + 3 \text{ ones}}$$
$$3 \text{ tens} + 4 \text{ ones}$$

We subtract the digits of the same place value, first ones and then tens.

or   34

In the short form we write

$$\begin{array}{r} 87 \\ -\ 53 \\ \hline 34 \end{array}$$

Again we subtract in the ones column and then the tens column. In practice we will always use this short form.

To check the subtraction:

$$34 + 53 = 87$$

● ● ● **CHECK YOURSELF 2**

Subtract, and check your work.

$$\begin{array}{r} 79 \\ -\ 36 \\ \hline \end{array}$$

The subtraction process is easily extended to larger numbers.

•Example 3

Subtracting a Larger Number

Step 1   **Step 2**   **Step 3**

$$\begin{array}{r} 789 \\ -\ 246 \\ \hline 3 \end{array} \qquad \begin{array}{r} 789 \\ -\ 246 \\ \hline 43 \end{array} \qquad \begin{array}{r} 789 \\ -\ 246 \\ \hline 543 \end{array}$$

We subtract in the ones column, then in the tens column, and finally in the hundreds column.

To check:
$$\begin{array}{r} 789 \\ -\ 246 \\ \hline 543 \end{array}$$
Add $543 + 246 = 789$

The sum of the difference and the subtrahend must be the minuend.

● ● ● **CHECK YOURSELF 3**

Subtract, and check your work.

$$\begin{array}{r} 3468 \\ -\ 2248 \\ \hline \end{array}$$

You know that the word *difference* indicates subtraction. There are other words that also tell you to use the subtraction operation. For instance, 5 *less than* 12 is written as

$12 - 5$ or 7

20 *decreased* by 8 is written as

$20 - 8$ or 12

● Example 4

Translating Words That Indicate Subtraction

Find each of the following.

(*a*) 4 less than 11

4 less than 11 is written $11 - 4 = 7$.

(*b*) 27 decreased by 6

27 decreased by 6 is written $27 - 6 = 21$.

● ● ● CHECK YOURSELF 4

Find each of the following.

a. 6 less than 19   **b.** 18 decreased by 3

Now we consider subtraction word problems. The strategy is the same one presented in Section 1.3 for addition word problems. It is summarized with the following four basic steps.

Solving Subtraction Applications

STEP 1 Read the problem carefully to determine the given information and what you are asked to find.
STEP 2 Decide upon the operation (in this case, subtraction) to be used.
STEP 3 Write down the complete statement necessary to solve the problem and do the calculations.
STEP 4 Check to make sure you have answered the question of the problem and that your answer seems reasonable.

Let's work an example using these steps.

• Example 5

Setting Up a Subtraction Word Problem

Tory has $37 in his wallet. He is thinking about buying a $24 pair of pants and a $10 shirt. If he buys them both, how much money will he have?

First we must add the cost of the pants and the shirt

$24 + $10 = $34

Now, that amount must be subtracted from the $37.

$37 − $34 = $3

He will have $3 left.

● ● ● **CHECK YOURSELF 5**

Sonya has $97 left in her checking account. If she writes checks for $12, $32, and $21, how much will she have in the account?

● ● ● **CHECK YOURSELF ANSWERS**

1.   13 − 9 = 4   **2.**   79
   Check: 4 + 9 = 13.   − 36
                              ─────
                               43
                          Check: 43 + 36 = 79.

3. 1220.   **4. (a)** 13; **(b)** 15.   **5.** $32.

1. In the statement $9 - 6 = 3$
9 is called the
6 is called the
3 is called the
Write the related addition statement.

2. In the statement $7 - 5 = 2$
5 is called the
2 is called the
7 is called the
Write the related addition statement.

Do the indicated subtraction, and check your results by addition.

3. $\begin{array}{r} 86 \\ -\ 23 \\ \hline \end{array}$

4. $\begin{array}{r} 97 \\ -\ 54 \\ \hline \end{array}$

5. $97 - 45$

6. $86 - 32$

7. Subtract 57 from 98.

8. Subtract 35 from 87.

9. $\begin{array}{r} 347 \\ -\ 201 \\ \hline \end{array}$

10. $\begin{array}{r} 575 \\ -\ 302 \\ \hline \end{array}$

11. $\begin{array}{r} 689 \\ -\ 245 \\ \hline \end{array}$

12. $\begin{array}{r} 598 \\ -\ 278 \\ \hline \end{array}$

13. $\begin{array}{r} 3446 \\ -\ 2326 \\ \hline \end{array}$

14. $\begin{array}{r} 5896 \\ -\ 3862 \\ \hline \end{array}$

15. $\begin{array}{r} 8540 \\ -\ 2320 \\ \hline \end{array}$

16. $\begin{array}{r} 5830 \\ -\ 3220 \\ \hline \end{array}$

17. $\begin{array}{r} 23{,}689 \\ -\ 2{,}523 \\ \hline \end{array}$

18.	59,786	**19.**	47,235	**20.**	59,342
	$-$ 3,214		$-$ 23,025		$-$ 27,140

21. 25 less than 76   **22.** 58 decreased by 23   **23.** The difference of 97 and 43

24. 125 less than 265   **25.** 298 decreased by 47   **26.** The difference of 167 and 57

Solve the following applications.

27. Test scores. Danielle's score on a math test was 87 while Tony's score was 23 points less than Danielle's. What was Tony's score on the test?

28. New pay. Monica's monthly pay of $879 was decreased by $175 for withholding. What amount of pay did she receive?

29. Number problem. The difference between two numbers is 134. If the larger number is 655, what is the smaller number?

30. Family budget. In Jason's monthly budget, he set aside $375 for housing, and $165 less than that for food. How much did he budget for food?

31. Consumer purchases. Brenda has $228 in cash and wants to buy a television set that costs $449. How much more money does she need?

32. Voting. A bond election for schools has the following results: yes, 3457 votes; no, 3125 votes. By how much of a margin did the bond pass?

33. Distance traveled. At the beginning of a trip your odometer reads 21,342 miles (mi), and at the end it reads 22,578 mi. How far did you drive?

34. Consumer purchases. Alan owes $598 after buying a stereo. If he makes a payment of $175, how much does he still owe?

35. Write an English version of the following equation. (Make sure you use a complete sentence.) Then exchange your sentence with other students and see if their interpretations result in the same equation you used.

    (a) $69 - 23 = 46$ **(b)** $17 + 13 = 30$

36. Evaluate the following two expressions:

    (1) $8 - (4 - 2)$ **(2)** $(8 - 4) - 2$

    (a) Do you obtain the same answer? What conclusion can you draw about the associative property of subtraction?

    (b) Is there any action in your daily life that involves "removing" or "reducing" several items? Is this process associative; that is, does the order in which you remove items affect the end result?

37. Think of any whole number.
Add 5.
Subtract 3.
Subtract two less than the original number.
What number do you end up with?
Check with other people. Does everyone have the same answer? Can you explain the results?

Answer lines (right margin):

32. 332 votes

33. 1236 mi

34. $423

35.

36.

37.

Answers

1. 9 is the minuend, 6 is the subtrahend, and 3 is the difference. $3 + 6 = 9$ **3.** 63

5. 52   **7.** 98   Check: $41 + 57 = 98$.   **9.** 146   **11.** 444   **13.** 1120

$$\begin{array}{r} 98 \\ -57 \\ \hline 41 \end{array}$$

15. 6220   **17.** 21,166   **19.** 24,210   **21.** 51   **23.** 54   **25.** 251

27. 64   **29.** 521   **31.** $221   **33.** 1236 mi   **35.**   **37.**

Name

Section   Date

1. 8 minuend, 5 subtrahend, 3 difference; $3 + 5 = 8$

2. 54

3. 35

4. 42

5. 541

6. 630

7. 1245

8. 1280

9. 50,130

10. 23,033

11. $8233

12. $1545

13. $341

14. 63

1. In the statement $8 - 5 = 3$
   5 is called the
   3 is called the
   8 is called the
   Write the related addition statement.

Do the indicated subtraction, and check your results by addition.

2.   97
       $- 43$

3. $88 - 53$

4. Subtract 36 from 78.

5.   779
       $- 238$

6.   835
       $- 205$

7.   2587
       $- 1342$

8.   3489
       $- 2209$

9.   52,486
       $- 2,356$

10.   43,495
        $- 20,462$

11. Consumer affairs. Miryam bought a new Toyota with a list price of $8958 and was given a discount of $725. What did she pay for the car?

12. Car purchases. The down payment on a new car is $2895. Curt has $1350 in his savings account now. How much more does he need for the down payment?

13. Banking. Arlene wrote checks of $54, $37, and $143. Her balance before that was $575. What is her new balance?

14. Education. Kuan had grades of 82, 73, and 81 on three tests during a course. What score must he make on a fourth test in order to have a total of 299 points for the course?

Subtraction with Borrowing

1.7 OBJECTIVES

1. Use borrowing in subtracting whole numbers.
2. Estimate differences by rounding.
3. Solve applications of subtraction.

Difficulties can arise in subtraction if one or more of the digits of the subtrahend are larger than the corresponding digits in the minuend. We will solve this problem by using a process called **borrowing.**

First, we'll look at an example in expanded form.

• Example 1

Subtracting When Borrowing Is Needed

$$52 = 5 \text{ tens} + 2 \text{ ones}$$
$$-27 = 2 \text{ tens} + 7 \text{ ones}$$

Do you see that we cannot subtract in the ones column?

Regrouping, we borrow 1 ten in the minuend and write that ten as 10 ones:

$$5 \text{ tens} + 2 \text{ ones}$$

becomes $\qquad 4 \text{ tens} + 1 \text{ ten} + 2 \text{ ones}$

or $\qquad 4 \text{ tens} + 10 \text{ ones} + 2 \text{ ones}$

or $\qquad 4 \text{ tens} + 12 \text{ ones}$

We now have

$$52 = 4 \text{ tens} + 12 \text{ ones}$$
$$-27 = 2 \text{ tens} + 7 \text{ ones}$$
$$2 \text{ tens} + 5 \text{ ones}$$

or $\qquad 25$

We can now subtract as before.

In practice, we will use a more convenient short form for the subtraction.

$$
\begin{array}{r}
52 \\
-\ 27 \\
\end{array}
\qquad
\begin{array}{r}
4\,1 \\
\not 5 2 \\
-\ 27 \\
\hline
25 \\
\end{array}
$$

We indicate the fact that we have borrowed 1 ten by putting a slash through the 5 and then writing 4 tens. Add 10 ones to the original 2 ones to get 12 ones. We can then subtract.

Check: $25 + 27 = 52$

● ● ● CHECK YOURSELF 1

Subtract, and check your work.

$$
\begin{array}{r}
64 \\
-\ 38 \\
\end{array}
$$

The subtraction process may involve more than one borrowing step.

● Example 2

Subtracting in Short Form When Borrowing Is Needed

Step 1

$$
\begin{array}{r}
^{1\,1}\\
62\!\!\!/4\\
-\,346\\
\hline
8
\end{array}
$$

We borrow 1 ten to write 14 ones.
Subtracting, we have 8 ones.
Now we cannot subtract in the tens
column. We must borrow again.

Step 2

$$
\begin{array}{r}
^{5\,^{1}1\,1}\\
62\!\!\!/4\\
-\,346\\
\hline
278
\end{array}
$$

Borrow 1 hundred. This is written as 10 tens
and combined with the existing 1 ten.
We then have 11 tens and can continue.

● ● ● **CHECK YOURSELF 2**

Subtract.

$$
\begin{array}{r}
536\\
-\,258\\
\hline
\end{array}
$$

Let's work through another subtraction example that will require a number of borrowing steps. Here, zero appears as a digit in the minuend.

● Example 3

Subtracting When Borrowing Is Needed

Step 1

$$
\begin{array}{r}
^{4\,1}\\
405\!\!\!/3\\
-\,2365\\
\hline
8
\end{array}
$$

In this first step we borrow 1 ten. This is written
as 10 ones and combined with the original 3
ones. We can then subtract in the ones column.

Here we borrow 1 thousand; this is written as 10 hundreds.

Step 2

$$
\begin{array}{r}
^{3\,14\,1}\\
405\!\!\!/3\\
-\,2365\\
\hline
8
\end{array}
$$

We must borrow again to subtract in the tens
column. There are no hundreds, and
so we move to the thousands column.

We now borrow 1 hundred; this is written as 10 tens and combined with the remaining 4 tens.

Step 3

$$
\begin{array}{r}
^{\quad 9}\\
^{3\,1\,14\,1}\\
40\!\!\!/5\!\!\!/3\\
-\,2365\\
\hline
8
\end{array}
$$

The minuend is now renamed as 3 thousands,
9 hundreds, 14 tens, and 13 ones.

Step 4

$$
\begin{array}{r}
^{\quad 9}\\
^{3\,1\,14\,1}\\
40\!\!\!/5\!\!\!/3\\
-\,2365\\
\hline
1688
\end{array}
$$

The subtraction can now be completed.

To check our subtraction: $1688 + 2365 = 4053$

● ● ●   **CHECK YOURSELF 3**

Subtract, and check your work.

5024
− 1656

Estimation is also a useful skill when used with subtraction, as Example 4 illustrates.

● Example 4

Estimating a Difference

Estimate the difference.

48,378
− 34,429

Round the minuend to 48,000 (the nearest thousand) and the subtrahend to 34,000. Our estimate of the difference is then 48,000 − 34,000 = 14,000.
   We will leave it to you to perform the subtraction and find the actual difference. Does your answer seem reasonable given our estimate?

● ● ●   **CHECK YOURSELF 4**

Estimate the difference below by rounding to the nearest thousand. Then perform the actual subtraction to see if your answer seems reasonable, given your estimate.

53,928
− 38,199

You will need to use both addition and subtraction to solve some problems, as Example 5 illustrates.

● Example 5

Solving a Subtraction Application

Bernard wants to buy a new piece of stereo equipment. He has $142 and can trade in his old amplifier for $135. How much more does he need if the new equipment costs $449?

Solution First we must add to find out how much money Bernard has available. Then we subtract to find out how much more money he needs.

$142 + $135 = $277   The money available to Bernard

$449 − $277 = $172   The money Bernard still needs

● ● ● **CHECK YOURSELF 5**

Martina spent $239 in airfare, $174 for lodging, and $108 for food on a business trip. Her company allowed her $375 for the expenses. How much of these expenses will she have to pay herself?

● ● ● **CHECK YOURSELF ANSWERS**

1.
$$
\begin{array}{r}
^{5\,1}\\
\cancel{6}4\\
-\ 38\\
\hline
26
\end{array}
$$
To check:
26 + 38 = 64

2.
$$
\begin{array}{r}
4\,^{1}2\,1\\
\cancel{5}\cancel{3}6\\
-\ 258\\
\hline
278
\end{array}
$$

3. 3368.   4. Estimate: 16,000; difference: 15,729.

5.
$$
\begin{array}{r}
\$239\\
174\\
+\ 108\\
\hline
\$521
\end{array}
$$
←— Total expenses

$$
\begin{array}{r}
\$521\\
-\ 375\\
\hline
\$146
\end{array}
$$
←— Total expenses
←— Amount allowed

Name

Section   Date

Subtract, and check your results by addition.

1.   64
        − 27

2.   73
        − 36

3.   50
        − 36

4.   40
        − 23

5.   372
        − 58

6.   500
        − 65

7.   534
        − 263

8.   867
        − 483

9.   627
        − 358

10.   642
         − 367

11.   280
         − 185

12.   370
         − 193

13.   603
         − 259

14.   705
         − 368

15.   2358
         − 562

16.   3547
         − 673

17.   3537
         − 2675

18.   4693
         − 2736

1. 37

2. 37

3. 14

4. 17

5. 314

6. 435

7. 271

8. 384

9. 269

10. 275

11. 95

12. 177

13. 344

14. 337

15. 1796

16. 2874

17. 862

18. 1957

19.   6423
        − 3678

20.   5352
        − 2577

21.   6034
        − 2569

22.   5206
        − 1748

23.   4000
        − 2345

24.   6000
        − 4349

25.   33,486
        − 14,047

26.   53,487
        − 25,649

27.   29,400
        − 17,900

28.   53,500
        − 28,700

29.   59,000
        − 23,458

30.   41,000
        − 27,645

Solve the following applications.

31. Checking account balance. To keep track of a checking account, you must subtract the amount of each check from the current balance. Complete the following statement.

Beginning balance	$351
Check #1	29
Balance	322
Check #2	139
Balance	183
Check #3	75
Ending balance	108

32. Expense accounts. Complete the following record of a monthly expense account.

Monthly income	$1620
House payment	343
Balance	1277
Car payment	183
Balance	1094
Food	312
Balance	782
Clothing	89
Amount remaining	693

Estimate each of the following differences by rounding the minuend and the subtrahend. Use your estimate to determine which of the answers are correct and which are incorrect.

ANSWERS

33. Correct

34. Incorrect

35. Incorrect

36. Correct

37. 204 ft

38. 174 students

39. $264

40. $28

41. 40 cal over

42. 43 points

33. 5846
    − 1938
    ———
     3908

34. 7983
    − 3579
    ———
     3404

35. 29,857
    − 2,098
    ———
     26,759

36. 53,174
    − 30,098
    ———
     23,076

Solve the following applications.

37. Construction. The Sears Tower in Chicago is 1454 feet (ft) tall. The Empire State Building is 1250 ft tall. How much taller is the Sears Tower than the Empire State Building?

38. Education. A college's enrollment was 2479 students in the fall of 1995 and 2653 students in the fall of 1996. What was the increase in enrollment?

39. Net pay. In 1 week, Margaret earned $278 in regular pay and $53 for overtime work, and $49 was deducted from her paycheck for income taxes and $18 for social security. What was her take-home pay?

40. Savings. Fritz opened a checking account and made deposits of $85 and $272. He wrote checks during the month for $35, $27, $89, and $178. What was his balance at the end of the month?

41. Dieting. Sandra is trying to limit herself to 1500 calories per day (cal/day). Her breakfast was 270 cal, her lunch was 450 cal, and her dinner was 820 cal. By how much was she *under* or *over* her diet?

42. Recreation. A professional basketball team scored 98, 136, and 113 points in three games. If its opponents scored 102, 109, and 93 points, by how much did the team outscore its opponents?

43. 190 points

44. 7595 mi

45. See exercise

46. See exercise

47. 15

48. 34

49. See exercise

50. See exercise

51. See exercise

52. See exercise

43. Education. A course outline states that you must have 540 points on five tests during the term to receive an A for the course. Your scores on the first four tests have been 95, 84, 82, and 89. How many points must you score on the 200-point final to receive an A?

44. Travel. Carmen's frequent-flyer program requires 30,000 miles (mi) for a free flight. During 1987 she accumulated 13,850 mi. In 1988 she took three more flights of 2800, 1475, and 4280 mi. How much further must she fly for her free trip?

Determine the pattern for the following arithmetic sequences and write the next four numbers in each sequence.

45. 53, 47, 41, 35, __29__, __23__, __17__, __11__

46. 158, 141, 124, 107, __90__, __73__, __56__, __39__

The figures below are called **magic squares.** Let's see why.

47.

4	9	2
3	5	7
8	1	6

Add each row.
Add each column.
Add each diagonal.
Compare the sums.

48.

1	15	14	4
12	6	7	9
8	10	11	5
13	3	2	16

Add each row.
Add each column.
Add each diagonal.
Compare the sums.

Complete the magic squares.

49.

6	7	2
1	5	9
8	3	4

50.

4	3	8
9	5	1
2	7	6

51.

16	3	2	13
5	10	11	8
9	6	7	12
4	15	14	1

52.

7	12	1	14
2	13	8	11
16	3	10	5
9	6	15	4

68

53. Sam has lost track of his checking account transactions. He knows he started with $50 and has deposited $120, $85, and $120. He also knows he has withdrawn $200 and $55. He just can't remember the order in which he did all this activity.

(a) What is Sam's balance after all these transactions?

(b) Does the order of the transactions make any difference from the math point of view?

(c) Does the order of transactions make any difference from the banking point of view?

Explain your answers.

54. Based on the past three censuses, determine the population of Arizona, California, Oregon, and Pennsylvania in each census year.

(a) Find the total change in each state's population over this time period.

(b) Which state shows the most change over the past three censuses?

(c) Write a brief essay describing the changes and any trends you see in this data. List any implications that they might have for future planning.

Answers

1. 37   **3.** 14   **5.** 314   **7.**
$$\begin{array}{r} {}^{4\,1}\!\!534 \\ -\ 263 \\ \hline 271 \end{array}$$
9. 269   **11.** 95   **13.** 344

15.
$$\begin{array}{r} {}^{1\,2\,1}\!2358 \\ -\ 562 \\ \hline 1796 \end{array}$$
17. 862   **19.**
$$\begin{array}{r} {}^{5\,3\,1\,1}\!6423 \\ -\ 3678 \\ \hline 2745 \end{array}$$
21.
$$\begin{array}{r} {}^{9}_{5\,1\,2\,1}\!6034 \\ -\ 2569 \\ \hline \end{array}$$
23. 1655

25. 19,439   **27.** 11,500   **29.** 35,542   **31.** End balance: $108

33. Correct   **35.** Incorrect   **37.** 204 ft

39.

Total pay	Deductions	Take-home pay
$278	$49	$331
+ 53	+ 18	− 67
$331	$67	$264

41. Number of calories:
270
450
+ 820
1540   40 cal over

43. 190 points   **45.** 29, 23, 17, 11   **47.** 15

49.

6	7	2
1	5	9
8	3	4

51.

16	3	2	13
5	10	11	8
9	6	7	12
4	15	14	1

53.

Name

Section   Date

A N S W E R S

1. 37	
2. 14	
3. 646	
4. 157	
5. 272	
6. 153	
7. 654	
8. 2797	
9. 3093	
10. 2784	
11. 679	
12. 1518	
13. 16,515	
14. 27,888	
15. 14,015	
16. See exercise	
17. Correct	
18. Incorrect	

Subtract, and check your results by addition.

1.   75
      $- 38$

2.   90
      $- 76$

3.   700
      $- 54$

4.   835
      $- 678$

5.   540
      $- 268$

6.   607
      $- 454$

7.   900
      $- 246$

8.   3592
      $- 795$

9.   6585
      $- 3492$

10.   5372
      $- 2588$

11.   2058
      $- 1379$

12.   5000
      $- 3482$

13.   42,024
      $- 25,509$

14.   65,482
      $- 37,594$

15.   32,000
      $- 17,985$

Solve the following application.

16. Checking account balance. Complete the following checking account statement.

Beginning balance	$1268
Check #1	349
Balance	919
Check #2	57
Balance	862
Check #3	459
Ending balance	403

Estimate each of the following differences by rounding. Use your estimate to determine which of the answers are correct and which are incorrect.

17.   6732
      $- 3843$
      2889

18.   73,524
      $- 42,678$
      20,846

Using Your Calculator to Subtract

Particularly after working with borrowing in subtraction, you may be tempted to use your calculator for problems besides the ones in these special calculator sections.

Remember, the point is to brush up on your arithmetic skills by hand, *along with* learning to use the calculator in a variety of situations.

Now that you have reviewed the process of subtracting by hand, let's look at the use of the calculator in performing that operation.

To compute $56 - 29$, follow the indicated calculator steps.

1. Press the clear key. $\boxed{C}$
2. Enter the first number.   56
3. Press the minus key. $\boxed{-}$
4. Enter the second number.   29   The difference, 27, will now be
5. Press the equals key. $\boxed{=}$ in the display.

Display $\boxed{27}$

The calculator can be very helpful in a problem that involves both addition and subtraction operations.

• Example 1

Using Your Calculator to Subtract

Find

$$23 - 13 + 56 - 29$$

Enter the numbers and the operation signs exactly as they appear in the expression.

$23 \; \boxed{-} \; 13 \; \boxed{+} \; 56 \; \boxed{-} \; 29 \; \boxed{=}$

Display $\boxed{37}$

An alternative approach would be to add 23 and 56 first and then subtract 13 and 29. The result is the same in either case.

● ● ●   **CHECK YOURSELF 1**

Find

$$58 - 12 + 93 - 67$$

● ● ●   **CHECK YOURSELF ANSWER**

$58 \; \boxed{-} \; 12 \; \boxed{+} \; 93 \; \boxed{-} \; 67 \; \boxed{=} \; 72.$

Calculator Exercises

A N S W E R S

1. 41

2. 187

3. 1843

4. 7685

5. 143,958

6. 178,750

7. 977

8. 402

9. See exercise

10. 61,712 mi^2

11. 90,501 mi^2

Do the indicated operations.

1. 89
    $-$ 48

2. 576
    $-$ 389

3. 5830
    $-$ 3987

4. 15,280
    $-$ 7,595

5. $193,243 - 49,285$

6. $257,500 - 78,750$

7. Subtract 235 from the sum of 534 and 678.

8. Subtract 476 from the sum of 306 and 572.

Solve the following applications.

9. **Gas sales.** Readings from the Fast Service Station's storage tanks were taken at the beginning and the end of a month. How much of each type of gas was sold? What was the total sold?

	Regular	Unleaded	Super Unleaded	Total
Beginning reading	73,255	82,349	81,258	
End reading	28,387	19,653	8,654	
Gallons used	44,868	62,696	72,604	180,168

The land areas, in square miles (mi^2), of three Pacific coast states are California, 158,693 mi^2; Oregon, 96,981 mi^2; Washington, 68,192 mi^2.

10. How much larger is California than Oregon?

11. How much larger is California than Washington?

Answers

1. 41   **3.** 1843   **5.** 143,958   **7.** 977   **9.** Regular, 44,868 gal; unleaded, 62,696 gal; super unleaded, 72,604 gal; total, 180,168 gal   **11.** 90,501 mi^2

1.8   Measuring Perimeter

1.8   OBJECTIVES

1. Find a perimeter.
2. Solve applications that involve perimeter.

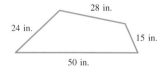

One application of addition is in finding the *perimeter* of a figure.

Perimeter

The **perimeter** is the distance around a closed figure.

If the figure has straight sides, the perimeter is the sum of the lengths of its sides.

● Example 1

Finding the Perimeter

We wish to fence in the field shown in Figure 1. How much fencing, in feet (ft), will be needed?

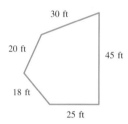

Figure 1

Solution   The fencing needed is the perimeter of (or the distance around) the field. We must add the lengths of the five sides.

Make sure to include the unit with the number in the answer.

$$20 + 30 + 45 + 25 + 18 = 138 \text{ ft}$$

So the perimeter is 138 ft.

● ● ●   **CHECK YOURSELF 1**

What is the perimeter of the region shown?

28 in.

24 in.

15 in.

50 in.

A **rectangle** is a figure, like a sheet of paper, with four equal corners. The perimeter of a rectangle is found by adding the distances of the four sides.

• Example 2

Finding the Perimeter of a Rectangle

Find the perimeter, in inches (in.) of the rectangle pictured below.

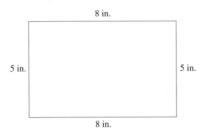

The perimeter is the sum of the distances 8 in., 5 in., 8 in., and 5 in.

$$8 + 5 + 8 + 5 = 26$$

The perimeter of the rectangle is 26 in.

● ● ● **CHECK YOURSELF 2**

Find the perimeter of the rectangle pictured below.

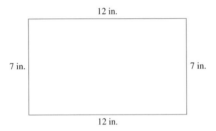

In general, we can find the perimeter of a rectangle by using a *formula*. A **formula** is a set of symbols that describe a general solution to a problem.

Let's look at a picture of a rectangle.

The perimeter can be found by adding the distances, so

Perimeter = length + width + length + width

To make this formula a little more readable, we abbreviate each of the words, using just the first letter.

Formula for the Perimeter of a Rectangle

$$P = L + W + L + W \tag{1}$$

There is one other version of this formula that we could use. Since we're adding the length (L) twice, we could write that as $2 \times L$. Since we're adding the width (W) twice, we could write that as $2 \times W$. This gives us another version of the formula.

Formula for the Perimeter of a Rectangle

$$P = 2 \times L + 2 \times W \tag{2}$$

In words, we say that the perimeter of a rectangle is twice its length plus twice its width.

Example 3 uses formula (1).

• Example 3

Finding the Perimeter of a Rectangle

A rectangle has length 11 in. and width 8 in. What is its perimeter?
Start by drawing a picture of the problem.

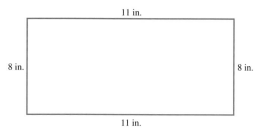

Now use formula (1)

$P = 11$ in. $+ 8$ in. $+ 11$ in. $+ 8$ in.

$\quad = 38$ in.

The perimeter is 38 in.

● ● ● **CHECK YOURSELF 3**

A bedroom is 9 ft by 12 ft. What is its perimeter?

● ● ● **CHECK YOURSELF ANSWERS**

1. 117 in. (or 9 ft 9 in.)   **2.** 38 in.   **3.** 42 ft.

NumberCross 1

The following puzzle will give you a chance to practice some of your addition skills. The answer is upside down at the bottom of the page.

ACROSS

1. 23 + 22
3. 103 + 42
6. 29 + 58 + 19
8. 3 + 3 + 4
9. 1480 + 1624
11. 568 + 730
13. 25 + 25
14. 131 + 132
16. The total of 121, 146, 119, and 132
17. The perimeter of a 4 × 6 rug

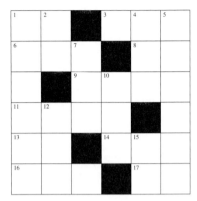

DOWN

1. The sum of 224, 155 and 186,000
2. 20 + 30
4. 210 + 200
5. 5000,000 + 4730
7. 130 + 509
10. 90 + 92
12. 100 + 101
15. The perimeter of a 15 × 16 room

SOLUTION TO NUMBERCROSS1

1.8 Exercises

A N S W E R S

Find the perimeter of each figure.

1.
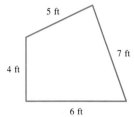
5 ft
4 ft
6 ft
7 ft

2.
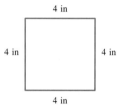
4 in
4 in
4 in
4 in

3.
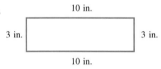
6 yd
8 yd
7 yd

4.
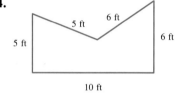
5 ft
6 ft
5 ft
6 ft
10 ft

5.
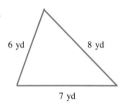
10 in.
3 in.
3 in.
10 in.

6.
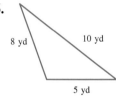
8 yd
10 yd
5 yd

7.
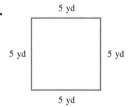
5 yd
5 yd
5 yd
5 yd

8.
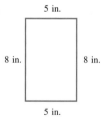
5 in.
8 in.
8 in.
5 in.

9.
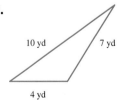
10 yd
7 yd
4 yd

10.
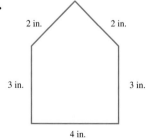
2 in.
2 in.
3 in.
3 in.
4 in.

1. 22 ft

2. 16 in.

3. 21 yd

4. 32 ft

5. 26 in.

6. 23 yd

7. 20 yd

8. 26 in.

9. 21 yd

10. 14 in.

11. Window size. A rectangular picture window is 4 feet (ft) by 5 ft. Meg wants to put a trim molding around the window. How many feet of molding should she buy?

12. Fencing material. You are fencing in a backyard that measures 30 ft by 20 ft. How much fencing should you buy?

Answers

1. 22 ft   **3.** 21 yd   **5.** 26 in.   **7.** 20 yd   **9.** 21 yd   **11.** 18 ft

1.8 Supplementary Exercises

Name _____

Section _____ Date _____

Find the perimeter of each figure.

1.

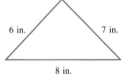

2.

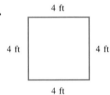

3.

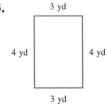

4.

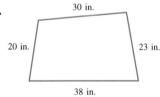

5.

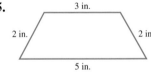

6.

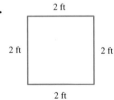

Solve the following applications.

7. Quantity of molding. A rectangular room measures 16 ft by 18 ft. If Jose wants to install baseboard molding around the room, how much molding should he buy?

8. Framing. If you want to frame a photograph that measures 11 in. by 14 in., how much material should you buy?

© 1998 McGraw-Hill Companies

Self-Test for Chapter 1

Name

Section   Date

The purpose of the Self-Test is to help you check your progress and review for a chapter test in class. Allow yourself about 1 hour to take the test. When you are done, check your answers in the back of the book. If you missed any answers, be sure to go back and review the appropriate sections in the chapter and do the supplementary exercises provided there.

[1.1]   **1.** What is the place value of 3 in 238,157?

2. Write 23,543 in words.

3. Write four hundred eight million, five hundred twenty thousand as a numeral.

[1.2]   In Exercises 4 to 6, name the property of addition that is illustrated.

4. $6 + 7 = 7 + 6$

5. $8 + (3 + 7) = (8 + 3) + 7$

6. $5 + 0 = 5$

[1.3]   In Exercises 7 and 8, add.

7. $23 + 542 =$

8. $14 + 223 + 4321 =$

[1.4]   In Exercises 9 to 14, perform the indicated operations.

9. $\begin{array}{r} 1369 \\ + 5804 \\ \hline \end{array}$

10. $\begin{array}{r} 489 \\ 562 \\ 613 \\ + 254 \\ \hline \end{array}$

11. $\begin{array}{r} 357 \\ 28 \\ + 2346 \\ \hline \end{array}$

12. $\begin{array}{r} 13 \\ 2,543 \\ + 10,547 \\ \hline \end{array}$

13. Find the sum of 2459, 53, and 23,467.

14. What is the total of 392, 95, 9237, and 11,972?

Solving the following application.

15. The attendance for the games of a playoff series in basketball was 12,438, 14,325, 14,581, and 14,634. What was the total attendance for the series?

ANSWERS

1. Ten thousands

2. Twenty-three thousand, five hundred forty-three

3. 408,520,000

4. Commutative property of addition

5. Associative property of addition

6. Additive identity property

7. 565

8. 4558

9. 7173

10. 1918

11. 2731

12. 13,103

13. 25,979

14. 21,696

15. 55,968

[1.5]   In Exercises 16 and 17, complete the following table by rounding the numbers to the indicated place.

		Nearest 10	Nearest 100	Nearest 1000
(Example)	6743	6740	6700	7000
	8546	**16.** 8550	8500	9000
	2973	**17.** 2970	3000	3000

In Exercises 18 and 19, use estimation to determine which of the problems is correct.

18. $\begin{array}{r} 293 \\ 521 \\ + 789 \\ \hline 1503 \end{array}$

19. $\begin{array}{r} 23{,}875 \\ + 14{,}230 \\ \hline 38{,}105 \end{array}$

In Exercises 20 and 21, use the symbol $<$ or $>$ to complete the statement.

20. 12 _____ 14

21. 500 _____ 400

[1.6]   In Exercises 22 and 23, subtract.

22. $289 - 54$

23. $53{,}294 - 41{,}074$

[1.7]   In Exercises 24 to 28, perform the indicated operations.

24. $503 - 74$

25. $5731 - 2492$

26. $32{,}345 - 1575$

27. $55{,}342 - 14{,}787$

28. The maximum load for a light plane with full gas tanks is 500 pounds (lb). Mr. Whitney weighs 215 lb, his wife 135 lb, and their daughter 78 lb. How much luggage can they take on a trip without exceeding the load limit?

[1.8]   In Exercises 31 and 32, find the perimeter of the figures shown.

29.

30.
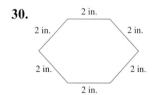

2 MULTIPLICATION OF WHOLE NUMBERS

INTRODUCTION

Photographers have to be both artists and scientists. Conceiving the picture and taking it at the right moment are artistic achievements. Scientific feats include developing and printing the pictures.

Gary studied chemistry and physics at New York University as part of his preparation for becoming a photographer. He especially enjoys nature photography. He is particularly proud of his photograph of a butterfly emerging from a cocoon that recently won an award at an international art show. ▪

© Inga Spence/The Picture Cube

Pretest
for Chapter 2

A N S W E R S

1. factors, product

2. 1, 2, 3, 6, 11, 22, 33, 66

3. Associative property

4. 21,080

5. 19,992

6. 56,100

7. (a) 26; (b) 56

8. $151

9. 192

10. (a) 10 yd^2; (b) 12 ft^3

Multiplication of Whole Numbers

This pretest will point out any difficulties you may be having in multiplying whole numbers. Do all the problems. Then check your answers with those in the back of the book.

1. If $5 \times 11 = 55$, we call 5 and 11 the _____ of 55.

55 is the _____ of 5 and 11.

2. List all the factors of 66.

3. The statement $3 \times (5 \times 7) = (3 \times 5) \times 7$ is an illustration of which property of multiplication?

4. $4216 \times 5 =$ **5.** $392 \times 51 =$

6. $187 \times 300 =$ **7.** (a) $5 + 3 \times 7 =$
                                                        (b) $7 \times (3 + 5) =$

8. Installment Buying. A refrigerator is advertised as follows: "Pay $50 down and $30 a month for 24 months." If the cash price of the refrigerator is $619, how much extra will you pay if you buy on the installment plan?

9. $3 \times 4^3 =$

10. (a) Find the area of the following figure:

(b) Find the volume of the following figure:

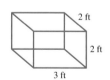

2.1 The Properties of Multiplication

2.1 OBJECTIVES

1. Use the language of multiplication.
2. Multiply single-digit whole numbers.
3. Identify the properties of multiplication.

The use of the symbol × dates back to the 1600s.

Our work in this chapter deals with multiplication, another of the basic operations of arithmetic. Multiplication is closely related to addition. In fact, we can think of multiplication as a shorthand method for repeated addition. The symbol × is used to indicate multiplication.

3×4 can be interpreted as 3 rows of 4 objects. By counting we see that $3 \times 4 = 12$. Similarly, 4 rows of 3 means $4 \times 3 = 12$.

```
        4                      3
   ┌ ★ ★ ★ ★             ┌ ★ ★ ★
 3 │ ★ ★ ★ ★           4 │ ★ ★ ★
   │ ★ ★ ★ ★             │ ★ ★ ★
                         │ ★ ★ ★
```

● Example 1

Multiplying Single-Digit Numbers

3×5 means 5 multiplied by 3. It is read 3 *times* 5. To find 3×5, we can add 5 three times.

$$3 \times 5 = 5 + 5 + 5 = 15$$

In a multiplication problem such as $3 \times 5 = 15$, we call 3 and 5 the **factors**. The answer, 15, is the **product** of the factors, 3 and 5.

$$3 \times 5 = 15$$

Factor   Factor   Product

● ● ● CHECK YOURSELF 1

Name the factors and the product in the following statement.

$$2 \times 9 = 18$$

> There are other ways of indicating multiplication. Later we will use a raised dot or parentheses. For example, $3 \cdot 5$ is the same as 3×5. The dot was introduced so that the multiplication symbol × would not be confused with the letter *x*. Parentheses can also be used to indicate multiplication. For example, $(3)(5)$ is the same as 3×5.

In any multiplication statement, we call the numbers being multiplied *factors*. Sometimes we want to know all the pairs of whole numbers that can be multiplied to give us a certain product.

• Example 2

Finding Factors

A whole number greater than 1 will always have itself and 1 as factors.

Since $4 \times 6 = 24$, we know that 4 and 6 are factors of 24. And 24 is the product of 4 and 6. Also, because 1×24, 2×12, and 3×8 are all 24, we say that 1, 2, 3, 8, 12, and 24 are also factors of 24.

● ● ● CHECK YOURSELF 2

List all the pairs of numbers that can be multiplied to get a product of 50.

Let's find another list of factors.

• Example 3

Finding Factors

Since 1×18, 2×9, and 3×6 are all 18, a list of the factors of 18 is 1, 2, 3, 6, 9, and 18. Be sure that you include the number itself, and 1, when you list the factors of a whole number.

● ● ● CHECK YOURSELF 3

List all the factors of 50.

The product of two numbers is also called a **multiple** of each of the numbers. We know that $5 \times 6 = 30$. We can say that 30 is a multiple of both 5 and 6.

• Example 4

Finding the Multiples of a Number

Find the multiples of 5.

$1 \times 5 = 5$ $4 \times 5 = 20$
$2 \times 5 = 10$ $5 \times 5 = 25$
$3 \times 5 = 15$ $6 \times 5 = 30$

Recall that the three dots mean that the list goes on indefinitely.

The numbers 5, 10, 15, 20, 25, 30, . . . are the multiples of 5. One easy way to list these multiples is to think of "counting by fives."

● ● ● CHECK YOURSELF 4

Find the first five multiples of 10.

•Example 5

Finding the Multiples of a Number

List some multiples of 6.
The numbers 6, 12, 18, 24, 30, . . . are multiples of 6.

● ● ● CHECK YOURSELF 5

List some multiples of 8.

Statements such as $3 \times 5 = 15$ and $4 \times 6 = 24$ are called the **basic multiplication facts**. If you have difficulty with multiplication, it may be that you do not know some of these facts. The Basic Multiplication Facts Table will help you review before you go on.

To use the table to find the product of 7×6: Find the row labeled 7, then move to the right in this row until you are in the column labeled 6 at the top. We see that 7×6 is 42.

Basic Multiplication Facts Table

×	0	1	2	3	4	5	6	7	8	9
0	0	0	0	0	0	0	0	0	0	0
1	0	1	2	3	4	5	6	7	8	9
2	0	2	4	6	8	10	12	14	16	18
3	0	3	6	9	12	15	18	21	24	27
4	0	4	8	12	16	20	24	28	32	36
5	0	5	10	15	20	25	30	35	40	45
6	0	6	12	18	24	30	36	42	48	54
7	0	7	14	21	28	35	42	49	56	63
8	0	8	16	24	32	40	48	56	64	72
9	0	9	18	27	36	45	54	63	72	81

From the Basic Multiplication Facts Table, we can discover some important properties of multiplication. For instance, did you see that the order in which we multiply two numbers does not affect the product?

•Example 6

Using the Commutative Property

Looking at the table, we see

$4 \times 6 = 6 \times 4 = 24$

This illustrates the **commutative property** of multiplication.

> **The Commutative Property of Multiplication**
>
> Multiplication, like addition, is a *commutative* operation. The order in which you multiply two whole numbers does not affect the product.

● ● ● **CHECK YOURSELF 6**

Show that $5 \times 8 = 8 \times 5$.

The next example will lead us to another property of multiplication.

• Example 7

Using the Associative Property

$(2 \times 3) \times 4 = 6 \times 4 = 24$ We do the multiplication in the parentheses first, $2 \times 3 = 6$. Then multiply 6×4.

Also,

$2 \times (3 \times 4) = 2 \times 12 = 24$ Here we multiply 3×4 as the first step. Then multiply 2×12.

We see that

$(2 \times 3) \times 4 = 2 \times (3 \times 4)$

The product is the same no matter which way we *group* the factors. This is called the **associative property** of multiplication.

> **The Associative Property of Multiplication**
>
> Multiplication is an *associative* operation. The way in which you group numbers in multiplication does not affect the final product.

● ● ● **CHECK YOURSELF 7**

Find the products.

a. $(5 \times 3) \times 6$ **b.** $5 \times (3 \times 6)$

Examining the Basic Multiplication Facts Table, we can also see that the two numbers 1 and 0 have special properties in multiplication.

We call 1 the **multiplicative identity** because of this property.

> ### The Multiplicative Identity Property
>
> Multiplying any whole number by 1 simply gives that number as a product.

• Example 8

Multiplying by One

$5 \times 1 = 5$
$1 \times 8 = 8$

● ● ● **CHECK YOURSELF 8**

Find the product.

a. $7 \times 1 =$ **b.** $1 \times 9 =$

> ### The Multiplication Property of Zero
>
> Multiplying any whole number by 0 gives the product 0.

• Example 9

Multiplying by Zero

$3 \times 0 = 0$
$0 \times 5 = 0$

● ● ● **CHECK YOURSELF 9**

Find the product.

a. $5 \times 0 =$ **b.** $0 \times 9 =$

The next property involves *both* multiplication and addition.

• Example 10

Using the Distributive Property

$2 \times (3 + 4) = 2 \times 7 = 14$ We have added $3 + 4$ and then multiplied.

Also,

$$2 \times (3 + 4) = (2 \times 3) + (2 \times 4)$$ We have multiplied 2×3 and 2×4 as the first step.
$$= 6 + 8$$
$$= 14$$ The result is the same.

We see that $2 \times (3 + 4) = (2 \times 3) + (2 \times 4)$. This is an example of the **distributive property of multiplication over addition** because we distributed the multiplication (in this case by 2) over the sum.

The Distributive Property of Multiplication over Addition

To multiply a factor by a sum of numbers, multiply the factor by each number inside the parentheses. Then add the products.

● ● ● CHECK YOURSELF 10

Show that

$$3 \times (5 + 2) = (3 \times 5) + (3 \times 2)$$

● Example 11

Using the Distributive Property

Evaluate $5 \times (6 + 3)$ in two ways.

(a) $5 \times (6 + 3) = 5 \times 9 = 45$ Add $6 + 3$. Then multiply.
(b) $5 \times (6 + 3) = (5 \times 6) + (5 \times 3)$ Multiply by 5 and then add the products.
$$= \quad 30 \quad + \quad 15$$
$$= 45$$

● ● ● CHECK YOURSELF 11

Evaluate $3 \times (4 + 6)$ in two ways.

The distributive law will be very important later in this chapter when we discuss multiplication involving larger numbers.

● ● ● CHECK YOURSELF ANSWERS

1. 2 and 9 are the factors, 18 is the product.   **2.** 1×50, 2×25, 5×10.
3. 1, 2, 5, 10, 25, 50.   **4.** 10, 20, 30, 40, 50.   **5.** 8, 16, 24, 32, . . .
6. $5 \times 8 = 40$ and $8 \times 5 = 40$.   **7.** **(a)** $15 \times 6 = 90$; **(b)** $5 \times 18 = 90$.
8. **(a)** 7; **(b)** 9.   **9.** **(a)** 0; **(b)** 0.   **10.** $21 = 15 + 6$.   **11.** $3 \times (4 + 6) = 3 \times$
$10 = 30$ or $3 \times (4 + 6) = (3 \times 4) + (3 \times 6) = 12 + 18 = 30$.

1. Find 3×7 and 7×3 by repeated addition.

2. Find 4×5 and 5×4 by repeated addition.

3. If $6 \times 7 = 42$, we call 6 and 7 _____ of 42. And 42 is the _____ of 6 and 7.

4. If $5 \times 8 = 40$, we call 5 and 8 _____ of 40. And 40 is the _____ of 5 and 8.

5. List all the factors of 30.

6. List all the factors of 36.

7. The numbers 3, 6, 9, 12, 15, . . . are the _____ of 3.

8. The numbers 4, 8, 12, 16, 20, . . . are the _____ of 4.

Multiply.

9. $\begin{array}{r} 5 \\ \times 3 \\ \hline \end{array}$	**10.** $\begin{array}{r} 7 \\ \times 4 \\ \hline \end{array}$	**11.** $\begin{array}{r} 8 \\ \times 1 \\ \hline \end{array}$	**12.** $\begin{array}{r} 9 \\ \times 5 \\ \hline \end{array}$
13. $\begin{array}{r} 6 \\ \times 0 \\ \hline \end{array}$	**14.** $\begin{array}{r} 6 \\ \times 6 \\ \hline \end{array}$	**15.** $\begin{array}{r} 2 \\ \times 9 \\ \hline \end{array}$	**16.** $\begin{array}{r} 1 \\ \times 7 \\ \hline \end{array}$
17. $\begin{array}{r} 5 \\ \times 6 \\ \hline \end{array}$	**18.** $\begin{array}{r} 0 \\ \times 5 \\ \hline \end{array}$	**19.** $\begin{array}{r} 4 \\ \times 9 \\ \hline \end{array}$	**20.** $\begin{array}{r} 8 \\ \times 4 \\ \hline \end{array}$
21. $\begin{array}{r} 3 \\ \times 8 \\ \hline \end{array}$	**22.** $\begin{array}{r} 8 \\ \times 7 \\ \hline \end{array}$	**23.** $\begin{array}{r} 5 \\ \times 7 \\ \hline \end{array}$	**24.** $\begin{array}{r} 7 \\ \times 5 \\ \hline \end{array}$

1. 21

2. 20

3. Factors, product

4. Factors, product

5. 1, 2, 3, 5, 6, 10, 15, 30

6. 1, 2, 3, 4, 6, 9, 12, 18, 36

7. Multiples

8. Multiples

9. 15

10. 28

11. 8

12. 45

13. 0

14. 36

15. 18

16. 7

17. 30

18. 0

19. 36

20. 32

21. 24

22. 56

23. 35

24. 35

25. $\begin{array}{r} 6 \\ \times 9 \\ \hline \end{array}$ **26.** $\begin{array}{r} 7 \\ \times 9 \\ \hline \end{array}$ **27.** $\begin{array}{r} 8 \\ \times 8 \\ \hline \end{array}$ **28.** $\begin{array}{r} 6 \\ \times 7 \\ \hline \end{array}$

29. $\begin{array}{r} 9 \\ \times 8 \\ \hline \end{array}$ **30.** $\begin{array}{r} 8 \\ \times 6 \\ \hline \end{array}$ **31.** $\begin{array}{r} 5 \\ \times 8 \\ \hline \end{array}$ **32.** $\begin{array}{r} 9 \\ \times 9 \\ \hline \end{array}$

33. $\begin{array}{r} 4 \\ \times 6 \\ \hline \end{array}$ **34.** $\begin{array}{r} 7 \\ \times 7 \\ \hline \end{array}$ **35.** $\begin{array}{r} 3 \\ \times 9 \\ \hline \end{array}$ **36.** $\begin{array}{r} 5 \\ \times 5 \\ \hline \end{array}$

Exercises 9 to 36 refer to the Basic Multiplication Facts Table. You should have been able to work quickly through these exercises. If you made errors or had to refer to the table, you should go back and review the table now!

As we indicated early in this section, both parentheses and a raised dot can be used to represent multiplication. Find the following products.

37. $5 \cdot 9$ **38.** $3(7)$

39. $(4)(8)$ **40.** $6(0)$

41. $(9)(6)$ **42.** $8 \cdot 7$

43. $(3)(2)(4)$ **44.** $7 \cdot 0 \cdot 5$

Name the property of addition and/or multiplication that is illustrated.

45. $5 \times 8 = 8 \times 5$ **46.** $8 \times 1 = 8$

47. $5 \times 0 = 0$ **48.** $7 \times 6 = 6 \times 7$

49. $2 \times (3 \times 5) = (2 \times 3) \times 5$ **50.** $0 \times 8 = 0$

90

51. $9 \times 3 = 3 \times 9$

52. $2 \times (5 + 7) = (2 \times 5) + (2 \times 7)$

53. $1 \times 5 = 5$

54. $4 \times (3 \times 5) = (4 \times 3) \times 5$

55. $0 \times 9 = 0$

56. $1 \times 7 = 7$

57. $5 \times (2 \times 3) = (5 \times 2) \times 3$

58. $9 \times 8 = 8 \times 9$

59. $3 \times (2 + 8) = (3 \times 2) + (3 \times 8)$

60. $4 \times 0 = 0$

61. We have seen that addition and multiplication are commutative operations. Decide which of the following activities are commutative.

(a) Taking a shower and eating breakfast

(b) Getting dressed and taking a shower

(c) Putting on your shoes and your socks

(d) Brushing your teeth and combing your hair

(e) Putting your key in the ignition and starting your car

● Getting Ready for Section 2.2 [Section 1.2]

Name the property of addition that is illustrated.

a. $3 + 5 = 5 + 3$

b. $(2 + 7) + 4 = 2 + (7 + 4)$

c. $5 + 0 = 5$

d. $3 + (8 + 9) = (3 + 8) + 9$

e. $8 + 9 = 9 + 8$

f. $0 + 9 = 9$

Answers

Solutions for the even-numbered exercises are provided in the back of the book.

1. 21   **3.** Factors, product   **5.** 1, 2, 3, 5, 6, 10, 15, and 30. The order in which you list these factors makes no difference. Did you remember 1 and 30?
7. Multiples   **9.** 15   **11.** 8   **13.** 0   **15.** 18   **17.** 30   **19.** 36
21. 24   **23.** 35   **25.** 54   **27.** 64   **29.** 72   **31.** 40   **33.** 24
35. 27   **37.** 45   **39.** 32   **41.** 54   **43.** 24   **45.** Commutative property of multiplication   **47.** Multiplication property of zero   **49.** Associative property of multiplication   **51.** Commutative property of multiplication
53. Multiplicative identity property   **55.** Multiplication property of zero
57. Associative property of multiplication   **59.** Distributive property of multiplication over addition   **61.**   **a.** Commutative property

b. Associative property   **c.** Additive identity   **d.** Associative property
e. Commutative property   **f.** Additive identity property

51. Commutative property of multiplication

52. Distributive property of multiplication

53. Multiplicative identity property

54. Associative property of multiplication

55. Multiplication property of zero

56. Multiplicative identity property

57. Associative property of multiplication

58. Commutative property of multiplication

59. Distributive property of multiplication over addition

60. Multiplication property of zero

61.

a. Commutative property of addition

b. Associative property of addition

c. Additive identity property

d. Associative property of addition

e. Commutative property of addition

f. Additive identity property

Supplementary Exercises

Name _____

Section _____ Date _____

ANSWERS

1. Find 3×6 and 6×3 by repeated addition.

2. If $6 \times 8 = 48$, we call 6 and 8 _____ of 48. And 48 is the _____ of 6 and 8.

3. List all the factors of 42.

4. The numbers 6, 12, 18, 24, 30, . . . are the _____ of 6.

Multiply.

5. $\begin{array}{r} 4 \\ \times 5 \\ \hline \end{array}$ **6.** $\begin{array}{r} 7 \\ \times 3 \\ \hline \end{array}$ **7.** $\begin{array}{r} 9 \\ \times 1 \\ \hline \end{array}$ **8.** $\begin{array}{r} 9 \\ \times 6 \\ \hline \end{array}$

9. $\begin{array}{r} 0 \\ \times 6 \\ \hline \end{array}$ **10.** $\begin{array}{r} 7 \\ \times 7 \\ \hline \end{array}$ **11.** $\begin{array}{r} 3 \\ \times 8 \\ \hline \end{array}$ **12.** $\begin{array}{r} 1 \\ \times 5 \\ \hline \end{array}$

13. $\begin{array}{r} 4 \\ \times 9 \\ \hline \end{array}$ **14.** $\begin{array}{r} 6 \\ \times 8 \\ \hline \end{array}$ **15.** $\begin{array}{r} 8 \\ \times 0 \\ \hline \end{array}$ **16.** $\begin{array}{r} 8 \\ \times 9 \\ \hline \end{array}$

17. $\begin{array}{r} 9 \\ \times 5 \\ \hline \end{array}$ **18.** $\begin{array}{r} 5 \\ \times 5 \\ \hline \end{array}$

Name the property of addition and/or multiplication that is illustrated.

19. $9 \times 6 = 6 \times 9$

20. $7 \times 1 = 7$

21. $5 \times (2 \times 7) = (5 \times 2) \times 7$

22. $(2 \times 8) \times 3 = 2 \times (8 \times 3)$

23. $8 \times 0 = 0$

24. $4 \times (3 + 8) = (4 \times 3) + (4 \times 8)$

25. $0 \times 3 = 0$

26. $3 \times (5 + 6) = (3 \times 5) + (3 \times 6)$

Multiplying by a One-Digit Number

2.2 OBJECTIVES

1. Multiply by a single-digit whole number.
2. Solve an application of multiplication by a single digit.

Here's where we use the distributive property to multiply over the addition.

Let's look at the multiplication of larger numbers. To multiply 2×34, recall that 34 means 3 tens and 4 ones. So, in the expanded form,

$$2 \times 34 = 2 \times (3 \text{ tens} + 4 \text{ ones})$$
$$= 2 \times 3 \text{ tens} + 2 \times 4 \text{ ones}$$
$$= 6 \text{ tens} + 8 \text{ ones}$$
$$= 68$$

We multiply each digit of 34 by 2, keeping track of the place value. A more convenient short form will allow us to do the same thing.

• Example 1

Multiplying by a Single-Digit Number

To multiply 2×34, write

Step 1
$$\begin{array}{r} 34 \\ \times\ 2 \\ \hline \end{array}$$
Line up the ones digits.

Step 2
$$\begin{array}{r} 34 \\ \times\ 2 \\ \hline 8 \end{array}$$
Multiply 2×4. Write the product, 8, in the ones place of the product.

Step 3
$$\begin{array}{r} 34 \\ \times\ 2 \\ \hline 68 \end{array}$$
Now multiply 2×3. Write the product in the tens place.

● ● ● **CHECK YOURSELF 1**

Multiply.

$$\begin{array}{r} 21 \\ \times\ 3 \\ \hline \end{array}$$

Carrying must often be used to multiply larger numbers. Let's see how carrying works in multiplication by looking at an example in the expanded form.

● Example 2

Multiplying by a Single-Digit Number

$3 \times 25 = 3 \times (2 \text{ tens} + 5 \text{ ones})$

$\quad\quad = 3 \times 2 \text{ tens} + 3 \times 5 \text{ ones}$ We use the distributive property again.

$\quad\quad = 6 \text{ tens} \quad + 15 \text{ ones}$ Write the 15 ones as 1 ten and 5 ones.

$\quad\quad = 6 \text{ tens} + 1 \text{ ten} + 5 \text{ ones}$

$\quad\quad = 7 \text{ tens} + 5 \text{ ones}$ Carry 10 ones or 1 ten to the tens place.

$\quad\quad = 75$

Here is the same multiplication problem using the short form.

Step 1
```
  1 ← Carry
  25
× 3
───
  5
```
Multiplying 3×5 gives us 15 ones. Write 5 ones and carry 1 ten.

Step 2
```
  1
  25
× 3
───
 75
```
Now multiply 3×2 tens and add the carry to get 7, the tens digit of the product.

● ● ● **CHECK YOURSELF 2**

Multiply.

```
  34
× 6
```

When a multiplication problem involves larger whole numbers, you may have to carry more than once while doing the multiplication.

● Example 3

Multiplying by a Single-Digit Number

Multiply 438×4.

Step 1
```
   3 ← Carry
  438
×   4
────
    2
```
We multiply 4×8 to get 32. Write 2 ones and carry 3 tens.

Step 2

$$\overset{13 \longleftarrow \text{Carry}}{\underset{52}{\begin{array}{r} 438 \\ \times \quad 4 \\ \hline \end{array}}}$$

Now multiply 4×3 tens. The product is 12 tens, and we add the carry of 3 tens. The result is 15 tens. Write 5 as the tens digit of the product and carry 1 to the hundreds place.

Step 3

$$\overset{13 \longleftarrow \text{Carry}}{\underset{1752}{\begin{array}{r} 438 \\ \times \quad 4 \\ \hline \end{array}}}$$

Multiply again and add the carry. We have 17 hundreds, and the multiplication is complete.

● ● ● **CHECK YOURSELF 3**

Multiply.

$$\begin{array}{r} 527 \\ \times \quad 5 \\ \hline \end{array}$$

Our next example uses multiplication to solve an application.

Application problems are those that apply what we have learned to a real-world problem.

● Example 4

Solving an Application Involving Multiplication

A car rental agency orders a fleet of 7 new subcompact cars at a cost of $7258 per automobile. What will the company pay for the entire order?

Step 1 We know the number of cars and the price per car. We want to find the total cost.

Step 2 Multiplication is the best approach to the solution.

Step 3 Write

$7 \times \$7258 = \$50,806$ We could, of course, *add* $7258, the cost, 7 times, but multiplication is certainly more efficient.

Step 4 The total cost of the order is $50,806

● ● ● **CHECK YOURSELF 4**

Tires sell for $47 apiece. What is the total cost for five tires?

Remember that it is best to write down the complete statement necessary for the solution of any application.

Let's review our discussion of applications, or word problems, from the last chapter. Read the problem carefully. Reread it if necessary. Pick out the important facts and write them down. Know what you are being asked to find; then decide which operation or operations must be used for the solution.

As you will see, the process of solving applications is the same no matter which operation is required for the solution. In fact, the four-step procedure we suggested in Section 1.4 can be effectively applied here.

Solving Applications

STEP 1   Read the problem carefully to determine the given information and what you are asked to find.

STEP 2   Decide upon the operation or operations to be used.

STEP 3   Write down the complete statement necessary to solve the problem, and do the calculations.

STEP 4   Check to make sure you have answered the question of the problem and that your answer seems reasonable.

Let's apply the procedure to another example.

• Example 5

Solving an Application of Multiplication

Peter earns $4 for every case of oranges he sells. How much has he earned when he sells 14 cases of oranges?

To find Peter's total earnings, multiply the number of cases (14) by his earnings per case ($4).

$$14 \times 4 = \$56$$

Peter has earned $56 selling oranges.

● ● ●   **CHECK YOURSELF 5**

Dominique's income from her summer job is $245 per week. How much will she make in 9 weeks?

● ● ●   **CHECK YOURSELF ANSWERS**

1. 63.   **2.** 204.   **3.** 2635.   **4.** $235.   **5.** $2205.

Multiply.

1.   23
   $\times\ 2$

2.   32
   $\times\ 3$

3.   48
   $\times\ 4$

4.   53
   $\times\ 5$

5.   508
   $\times\ 6$

6.   903
   $\times\ 9$

7.   523
   $\times\ 8$

8.   635
   $\times\ 7$

9.   2035
   $\times\ \ \ 9$

10.   5018
   $\times\ \ \ 7$

11.   5478
   $\times\ \ \ 7$

12.   6893
   $\times\ \ \ 9$

13.   26,555
   $\times\ \ \ \ \ 7$

14.   30,524
   $\times\ \ \ \ \ 6$

15.   20,108
   $\times\ \ \ \ \ 7$

16.   31,015
   $\times\ \ \ \ \ 5$

17. 245×8

18. 9×306

19. 3249×5

20. 7×1258

21. Find the product of 304 and 7.

22. Find the product of 409 and 4.

23. Find the product of 8 and 5679.

24. Find the product of 23,452 and 5.

Solve the following applications.

25. Counting. There are 18 boys in a Cub Scout pack. Each boy brings 6 packs of baseball cards to a meeting. How many packs of baseball cards are there at the meeting?

26. Counting. Each of the packs of baseball cards in the previous exercise had 9 cards in it. How many baseball cards were there?

27. Recycling. Jon collects 225 aluminum cans a day for 7 days to take to the recycling plant. How many cans does he collect?

	ANSWERS
1.	46
2.	96
3.	192
4.	265
5.	3048
6.	8127
7.	4184
8.	4445
9.	18,315
10.	35,126
11.	38,346
12.	62,037
13.	185,885
14.	183,144
15.	140,756
16.	155,075
17.	1960
18.	2754
19.	16,245
20.	8806
21.	2128
22.	1636
23.	45,432
24.	117,260
25.	108 packs
26.	972 cards
27.	1575 cans

28. 825 cal

29. $750

30. 45 pages

31. $130

32. 84 seats

33.

34.

28. **Recreation.** A tennis player burns an average of 275 calories per hour (cal/h). How many calories will be burned in playing a 3-h match?

29. **Savings.** Jan saves $125 per month in a payroll savings plan. What amount will she save in 6 months?

30. **Computers.** A new laser printer can print 9 pages per minute. How many pages will it print in 5 minutes?

31. **Payroll deduction.** Kay has $5 taken from each paycheck as a contribution to United Charities. If she receives 26 paychecks a year, what is her annual contribution?

32. **Number of seats.** A small seminar room contains 12 rows of seats. Each row contains 7 seats. How many seats are in the room?

33. Most maps contain legends that allow you to convert the distance between two points on the map to actual miles. For instance, if a map uses a legend that equates 1 in. to 5 miles (mi) and the distance between two towns is 4 in. on the map, then the towns are actually 20 mi apart.

    (a) Obtain a map of your state and determine the shortest distance between any two major cities.

    (b) Could you actually travel the route you measured in part **(a)**?

    (c) Plan a trip between the two cities you selected in part **(a)** over established roads. Determine the distance that you actually traveled using this route.

34. Mario wanted to surprise his wife and two children by making cream of wheat for himself and them one cold winter morning. Unfortunately, the dog had eaten all the recipe card except for the following small part. Mario could only read this:

	1 serving
Milk	1 cup
Cream of Wheat	3 tbs

    Mario did make cream of wheat for everyone. How much of each ingredient did he use? Check a box of cream of wheat and see if Mario left anything out of his mixture.

Getting Ready for Section 2.3
[Section 1.4]

Add.

a.
```
    56
  +420
```

b.
```
    72
  +840
```

c.
```
      28
     840
   +9100
```

a. 476

b. 912

c. 9968

d. 11,084

e. 34,671

f. 76,732

d.
```
      34
     850
 +10,200
```

e.
```
     521
   2,450
 +31,700
```

f.
```
     782
   3,150
 +72,800
```

Answers

1. 46   **3.** 192   **5.** 3048   **7.** 4184   **9.** 18,315   **11.**
```
   3 5 5
   5478
 ×     7
  38,346
```

13. 185,885   **15.** 140,756   **17.** 1960   **19.**
```
   1 2 4
   3249
 ×     5
  16,245
```
   21. 2128

23. 45,432   **25.** 108 packs   **27.** 1575 cans   **29.** $750   **31.** $130

33.   **a.** 476   **b.** 912   **c.** 9968   **d.** 11,084   **e.** 34,671

f. 76,732

Name

Section   Date

A N S W E R S

1. 392

2. 2640

3. 2863

4. 10,240

5. 38,584

6. 59,283

7. 139,626

8. 248,346

9. 2187

10. 18,011

11. 28,656

12. 67,435

13. $4365

14. 1080 eggs

15. 105 points

16. 224 students

Multiply.

1.   56
    $\times\ 7$

2.   528
    $\times\ 5$

3.   409
    $\times\ 7$

4.   2048
    $\times\ 5$

5.   4823
    $\times\ 8$

6.   6587
    $\times\ 9$

7.   23,271
    $\times\ \ \ \ 6$

8.   35,478
    $\times\ \ \ \ \ 7$

9. 243×9

10. 7×2573

11. 3582×8

12. Find the product of 5 and 13,487.

Solve the following applications.

13. Living expenses. A student averages $485 per month for living expenses. What will she spend during the 9-month school year?

14. Eating habits. The average person in this country eats 270 eggs in 1 year. How many eggs will a family of four eat in 1 year?

15. Sports. Damien scores 7 points per game during a 15-game season. How many total points did he score?

16. Education. The local elementary school has 8 classrooms with 28 students in each class. How many students are in the school?

2.3 Multiplying by Numbers with More Than One Digit

2.3 OBJECTIVES

1. Multiply any two whole numbers.
2. Solve an application of multiplication of two numbers.

To multiply by numbers with more than one digit, we must multiply each digit of the first factor by each digit of the second. To do this, we form a series of partial products and then add them to arrive at the final product.

• Example 1

Multiplying by a Two-Digit Number

Multiply 32×13.

Step 1
$$\begin{array}{r} 32 \\ \times 13 \\ \hline 96 \end{array}$$
Always start the multiplication with the ones digit. The first partial product is 3×32, or 96.

Step 2
$$\begin{array}{r} 32 \\ \times 13 \\ \hline 96 \\ 320 \end{array}$$
The second partial product is 1 ten $\times$ 32, or 320.

Step 3
$$\begin{array}{r} 32 \\ \times 13 \\ \hline 96 \\ 320 \\ \hline 416 \end{array}$$
Add. As the final step of the process, we add the partial products to arrive at the final product, 416.

● ● ● **CHECK YOURSELF 1**

Multiply.

$$\begin{array}{r} 31 \\ \times 23 \end{array}$$

It may be necessary to carry in forming one or both of the partial products.

• Example 2

Multiplying by a Two-Digit Number

Multiply 56×47.

Step 1

$$\begin{array}{r} 4 \\ 56 \\ \times 47 \\ \hline 392 \end{array}$$

The first partial product is 7×56, or 392. Note that we had to carry 4 to the tens column.

Step 2

$$\begin{array}{r} 2 \\ 4 \\ 56 \\ \times 47 \\ \hline 392 \\ 224 \end{array}$$

The second partial product is 4 tens $\times$ 56, or 224 tens. We must carry 2 during the process.

Note that we shift the product *one place to the left* to indicate the multiplication by 4 *tens*. This means that the digit 4 is placed below the tens digit of 47. This is more efficient than writing the unnecessary 0 as we did in Example 1.

Step 3

$$\begin{array}{r} 2 \\ 4 \\ 56 \\ \times 47 \\ \hline 392 \\ 224 \\ \hline 2632 \end{array}$$

We add the partial products for our final result.

● ● ● **CHECK YOURSELF 2**

Multiply.

$$\begin{array}{r} 38 \\ \times 76 \end{array}$$

Let's work through another multiplication example.

● Example 3

Multiplying by a Two-Digit Number

Multiply 56×673.

Step 1

$$\begin{array}{r} 673 \\ \times \ 56 \end{array}$$

Write the product as shown.

Step 2

$$\begin{array}{r} 41 \\ 673 \\ \times \ 56 \\ \hline 4038 \end{array}$$

Multiply 6×673 for the first partial product. We must carry to the tens and hundreds columns.

Note: This places 5 below 3, the *tens* digit of the first partial product.

Step 3

$$\begin{array}{r} 31 \\ 41 \\ 673 \\ \times \ 56 \\ \hline 4038 \\ 3365 \end{array}$$

Multiply 5 tens $\times$ 673 for the second partial product.

Again we shift the product one place left to show the multiplication by 5 tens.

Step 4
$$
\begin{array}{r}
31 \\
41 \\
673 \\
\times 56 \\
\hline
4038 \\
3365 \\
\hline
37,688
\end{array}
$$
We now add the partial products for our result.

In practice these steps are combined, and your work should look like that shown in step 4.

● ● ● CHECK YOURSELF 3

Multiply.

74×538

The three partial products are formed when we multiply by the ones, tens, and then the hundreds digits.

If multiplication involves two three-digit numbers, another step is necessary. In this case we form three partial products. This will ensure that each digit of the first factor is multiplied by each digit of the second.

● Example 4

Multiplying Two Three-Digit Numbers

Multiply.

$$
\begin{array}{r}
22 \\
33 \\
22 \\
278 \\
\times 343 \\
\hline
834 \\
1112 \\
834 \\
\hline
95,354
\end{array}
$$

In forming the third partial product, we must multiply by 3 hundreds. To indicate this, we shift that product *two* places left. Again the unnecessary zeros are omitted.

● ● ● CHECK YOURSELF 4

Multiply.

$$
\begin{array}{r}
352 \\
\times 249 \\
\end{array}
$$

Let's look at an example of multiplying by a number involving 0 as a digit. There are several ways to arrange the work, as our example shows.

●Example 5

Multiplying Larger Numbers

Multiply 573×205.

Method 1

$$
\begin{array}{r}
1 \\
3\,1 \\
573 \\
\times 205 \\
\hline
2865 \\
000 \\
1146 \\
\hline
1117,465
\end{array}
$$

We can write the second partial product as 000 to indicate the multiplication by 0 in the tens place.

Let's look at a second approach to the problem.

Method 2

$$
\begin{array}{r}
1 \\
3\,1 \\
573 \\
\times 205 \\
\hline
2865 \\
11460 \\
\hline
1117,465
\end{array}
$$

We can write a single 0 as our second step. If we place the third partial product on the same line, that product will be shifted *two* places left, indicating that we are multiplying by 2 hundreds.

Since this second method is more compact, it is usually used.

● ● ● CHECK YOURSELF 5

Multiply.

$$
\begin{array}{r}
489 \\
\times 304 \\
\hline
\end{array}
$$

Applications of multiplication occur in many everyday situations. Anytime a number is repeated over and over, it makes more sense to multiply than to count or add. Let's look at a common example.

●Example 6

Solving a Multiplication Application

A theater has 12 seats in each row. If there are 11 rows, how many seats are in the theater?

We could draw a picture and count the seats, but that would be a long solution to a short problem. How about addition? We know that each row has 12 seats, so we could add $12 + 12 + 12 + \ldots$ a total of 11 times. Clearly this problem is most easily solved by multiplication. The 12 seats in each row are to be multiplied by the 11 rows.

$11 \times 12 = 132$

● ● ● **CHECK YOURSELF 6**

A large band is marching in 6 columns. There are 24 rows of band members. How many members are in the band? (No picture is necessary!)

Some applications require more than one multiplication step for their solution. .

● Example 7

Solving a Multiplication Application

Jamal owns three rental units. Each unit rents for $685 per month, and all three are occupied for an entire year. What is Jamal's income for the year from the three units?

Solution We multiply $685 by 12 to find the *yearly income* for *each* unit. Then multiply by 3 to find the combined income for the three units.

Note: You could have written $685 $\times$ 12 $\times$ 3 = $24,660 for the solution.

$685 $\times$ 12 = $8220   Jamal will receive 12 monthly payments during the year for each unit.

$8220 $\times$ 3 = $24,660   There were three units.

● ● ● **CHECK YOURSELF 7**

Ken pays $35 per month for parking. How much will he have paid in 3 years?

To solve an application of mathematics, you may also have to combine multiplication with other operations, as our final examples illustrate.

● Example 8

Suzanne buys a used car on the following terms. She pays $1500 down and agrees to make payments of $180 per month for 24 months. What is the total amount she pays in monthly payments? What will be her total cost for the car?

Two operations are necessary. We first multiply to find the total of the monthly payments. Then we add the down payment to find the total amount Suzanne paid for the car.

Solution There are three pieces of information that you need: the amount of the down payment, the monthly payment, and the number of months that the payment must be made. Write

Total monthly payments = 24 $\times$ 180   Suzanne pays 24 payments of $180 each.

= $4320

Cost = down payment + total monthly payments

= $1500 + $4320

= $5820   Total cost

● ● ● **CHECK YOURSELF 8**

Jovita put $350 down on a new stereo system. She is also paying $25 a month for 30 months. What will be her total cost for the system?

● Example 9

Solving an Application of Multiplication

A farmer harvested 58 bushels of wheat per acre from 27 acres in July. He then harvested 69 bushels per acre from 39 acres in August. What was the total harvest?

Solution We must find the harvest for each month, then add to find the total.

July:	27×58 bushels =	1566 bushels
August:	39×69 bushels =	2691 bushels
Total harvest:		4257 bushels

● ● ● **CHECK YOURSELF 9**

Larry buys a $699 dishwasher but decides to finance it by paying $115 down and making $30 monthly payments for 24 months. How much would he have saved by paying cash?

● ● ● **CHECK YOURSELF ANSWERS**

1. 713.　**2.**　$5 \leftarrow$ Carry (tens)　**3.** 39,812.　**4.** 87,648.
　　　　　　　$4 \leftarrow$ Carry (ones)
　　　　　　　38
　　　　　　$\times 76$
　　　　　　288
　　　　　　266
　　　　　2888.

5. 148,656.　**6.** 144 band members.　**7.** $1260.　**8.** $1100.　**9.** $136.

Multiply.

1. 47
   ×38

2. 58
   ×49

3. 98
   ×57

4. 75
   ×68

5. 235
   × 49

6. 327
   × 59

7. 2364
   × 67

8. 4075
   × 84

9. 315
   ×243

10. 124
    ×225

11. 345
    ×267

12. 639
    ×358

13. 547
    ×203

14. 668
    ×305

15. 2458
    × 135

16. 3219
    × 207

17. 1208
    × 305

18. 2407
    × 521

19. 2534
    ×3106

20. 3158
    ×2034

21. What is the product of 21 and 551?

22. What is the product of 112 and 168?

23. What is the product of 135 and 507?

24. Find the product of 2409 and 68.

1. 1786

2. 2842

3. 5586

4. 5100

5. 11,515

6. 19,293

7. 158,388

8. 342,300

9. 76,545

10. 27,900

11. 92,115

12. 228,762

13. 111,041

14. 203,740

15. 331,830

16. 666,333

17. 368,440

18. 1,254,047

19. 7,870,604

20. 6,451,794

21. 11,571

22. 18,816

23. 68,445

24. 163,812

Solve the following applications.

25. Gas mileage. The gas tank of a Honda holds 12 gallons (gal). If the car gets 42 miles per gallon (mi/gal) for highway driving, how far will it travel on one tankful of gas?

26. Counting. Sharon shot 18 rolls of 24-exposure film on a trip to England. How many pictures did she take?

27. Transportation. A convoy company can transport 8 new cars on one of its trucks. If 34 truck shipments were made in one week, how many cars were shipped?

28. Printing. A computer printer can print 40 mailing labels per minute. How many labels can be printed in 1 hour (h)?

29. Parking. A rectangular parking lot has 14 rows of parking spaces and each row contains 24 spaces. How many cars can be parked in the lot?

30. Sales. A college purchased 28 computers for a new laboratory at a cost of $879 per computer. What was the total cost of the order?

31. Copying. A ream of paper is 500 sheets. If 29 reams of paper were used in a copy machine during 1 week, how many copies were made?

32. Science. If sound waves travel at a rate of 1088 feet per second (ft/s) and you hear thunder 23 seconds (s) after seeing a lightning flash, how far away did the lightning flash?

33. Salary. Margaret has a job that pays $356 per week. What will she earn in 1 year? (Use 52 weeks per year.)

34. Medicine. A hospital has 46 bottles of aspirin and each bottle contains 275 tablets. How many tablets does the hospital have in all?

35. Assembly cost. A company pays its shop workers $15 for every assembly. How much would be paid to a person who had 104 assemblies?

36. Politics. A petition to get Tom on the ballot for treasurer of student council has 28 signatures on each of 43 pages. How many signatures were collected?

37. Manufacturing. The manufacturer of woodburning stoves can make 15 stoves in one day. How many stoves can be made in 28 days?

38. Quantity of papers. Each bundle of newspapers contains 25 papers. If 43 bundles are delivered to Jose's house, how many papers are delivered?

39. Total cost. Erin agrees to buy a boat by paying $2500 down and $139 a month for 36 months. What is the total cost of the boat?

40. Farming. Celeste harvested 34 bushels of corn per acre from 32 acres in June and 43 bushels of corn per acre from 36 acres in July. How many bushels of corn did Celeste harvest?

41. Consider the lists of the multiples of 3, 5, and 15. What is the relationship between the lists?

42. The associative property of addition and multiplication indicates that the result of the operation is the same regardless of where the grouping symbol is placed. This is not always the case in the use of the English language. Many phrases can have different meanings based on how the words are grouped. In each of the following, explain why the associative property would not hold.

   (a) Cat fearing dog   **(b)** Hard test question

   (c) Defective parts department   **(d)** Man eating animal

Write some phrases where the associative property is satisfied.

● Getting Ready for Section 2.4 [Section 1.5]

Round each of the numbers as indicated.

a. 78 to the nearest ten   **b.** 533 to the nearest hundred

c. 92 to the nearest ten   **d.** 689 to the nearest hundred

e. 2541 to the nearest hundred   **f.** 23,745 to the nearest thousand

Answers

1. 1786   **3.** 5586   **5.** 11,515   **7.** 2364   **9.** 76,545   **11.** 92,115

$$\begin{array}{r} 2364 \\ \times\ 67 \\ \hline 16548 \\ 14184 \\ \hline 158,388 \end{array}$$

13. $\begin{array}{r} 547 \\ \times 203 \\ \hline 1641 \\ 1\,0940 \\ \hline 111,041 \end{array}$ **15.** 331,830   **17.** 368,440   **19.** 7,870,604   **21.** 11,571

23. 68,445   **25.** 504 mi   **27.** 272 cars   **29.** 336 cars

31. 14,500 copies   **33.** $18,512   **35.** $1560   **37.** 420 stoves

39. $7504   **41.**   **a.** 80   **b.** 500   **c.** 90   **d.** 700

   e. 2500   **f.** 24,000

ANSWERS

38. 1075 papers

39. $7504

40. 2636 bushels

41.

42.

a. 80

b. 500

c. 90

d. 700

e. 2500

f. 24,000

Name

Section   Date

A N S W E R S

1. 1620

2. 4464

3. 27,589

4. 199,325

5. 159,036

6. 199,691

7. 528,078

8. 320,252

9. 765,351

10. 5,193,728

11. 20,532

12. 93,480

13. 768 seats

14. 756 pictures

15. $6642

16. 8760 h

17. $6816

18. 8280 pretzels

19. 1541 papers

20. 23,535 cars

Multiply.

1. $\begin{array}{r} 45 \\ \times 36 \end{array}$

2. $\begin{array}{r} 93 \\ \times 48 \end{array}$

3. $\begin{array}{r} 587 \\ \times\ 47 \end{array}$

4. $\begin{array}{r} 2345 \\ \times\ \ 85 \end{array}$

5. $\begin{array}{r} 457 \\ \times 348 \end{array}$

6. $\begin{array}{r} 397 \\ \times 503 \end{array}$

7. $\begin{array}{r} 933 \\ \times 566 \end{array}$

8. $\begin{array}{r} 1357 \\ \times\ 236 \end{array}$

9. $\begin{array}{r} 2493 \\ \times\ 307 \end{array}$

10. $\begin{array}{r} 2536 \\ \times 2048 \end{array}$

11. Find the product of 708 and 29.

12. What is the product of 456 and 205?

Solve the following applications.

13. **Seating capacity.** An auditorium has 32 rows of seats. If there are 24 seats in each row, how many seats are there in the auditorium?

14. **Quantity.** Paul shot 21 rolls of 36-exposure film on a trip to Europe. How many pictures did he take?

15. **Cost.** A school purchases 18 television sets for $369 each. What is the cost of the order?

16. **Time.** How many hours (h) are there in a 365-day year? (Use 24 h/day.)

17. **Salary.** Shannon worked as a medical transcriptionist. She earned $426 per week for 16 weeks. What was the total amount she earned?

18. **Baking.** Barry's Pretzels makes 276 soft pretzels per day. How many pretzels did Barry's make in September?

19. **Paper deliveries.** Tom runs a paper delivery service and employs 23 carriers. Each carrier needs 67 papers. How many papers should Tom order?

20. **Car sales.** Maria Ruiz owns 15 car dealerships. Each dealership has an inventory of 1569 cars. How many cars does Maria have at all of her dealerships?

2.4 Multiplying by Numbers Ending in Zero

2.4 OBJECTIVES

1. Multiply by whole numbers ending in zero.
2. Use rounding to estimate a product.
3. Solve an application of multiplication by a number ending in 0.

There are some shortcuts that will let you simplify your work when you are multiplying by a number that ends in 0. Let's see what we can discover by looking at some examples.

● Example 1

Multiplying by Ten

First we'll multiply by 10.

$$\begin{array}{r} 67 \\ \times 10 \\ \hline 670 \end{array}$$ $10 \times 67 = 670$

Next we'll multiply by 100.

$$\begin{array}{r} 537 \\ \times 100 \\ \hline 53,700 \end{array}$$ $100 \times 537 = 53,700$

Finally, we'll multiply by 1000.

$$\begin{array}{r} 489 \\ \times 1000 \\ \hline 489,000 \end{array}$$ $1000 \times 489 = 489,000$

●●● CHECK YOURSELF 1

Multiply

$$\begin{array}{r} 257 \\ \times 100 \end{array}$$

We'll talk about powers of 10 in more detail later in Section 2.6.

Do you see a pattern? Rather than writing out the multiplication, there is an easier way!

We call the numbers 10, 100, 1000, and so on, **powers of 10**.

Multiplying by Powers of 10

When a whole number is multiplied by a power of 10, the product is just that number followed by as many zeros as there are in the power of 10.

● Example 2

Multiplying by One Hundred

There are 2 zeros in 100.

$$45 \times 100 = 4500$$

2 zeros

There are 3 zeros in 1000.

$$1000 \times 67 = 67,000$$

3 zeros

10,000 has 4 zeros.

$$543 \times 10,000 = 5,430,000$$

4 zeros

This method works because multiplying by a power of 10 is the same as multiplying by 10 a certain number of times, each time adding a 0 to the product.

● ● ● **CHECK YOURSELF 2**

Multiply

a. 100×58 **b.** 395×1000

Let's see how we can use this to simplify any multiplication problem in which one of the factors ends in a zero or zeros.

● Example 3

Multiplying by Numbers That End in Zero

Multiply 30×56.

Solution We could write

$$\begin{array}{r} 56 \\ \times 30 \end{array}$$

We have lined up the ones digits as before.

This is better:

$$\begin{array}{r} 56 \\ \times\ 30 \end{array}$$

Shift 30 to the right so that the 0 is *to the right* of the digits above.

$$\begin{array}{r} {\scriptstyle 1} \\ 56 \\ \times\ 30 \\ \hline 1680 \end{array}$$

Now bring down the 0 and multiply 3×56 as before to write the product.

● ● ● CHECK YOURSELF 3

Multiply.

50 × 49

●Example 4

Multiplying by Numbers That End in Zero

Multiply 400 × 678.

Solution Write

```
    678
×   400
```
Shift 400 so that the two zeros are
to the right of the digits above.

```
   3 3
    678
×   400
271,200
```
Bring down the two zeros, then multiply
4 × 678 to find the product.

There is no mystery about why this works. We know that 400 is 4 × 100. In this method, we are multiplying 678 by 4 and then by 100, adding two zeros to the product by our earlier rule.

● ● ● CHECK YOURSELF 4

Multiply.

300 × 574

Your work in this section, together with our earlier rounding techniques, provides a convenient means of using estimation to check the reasonableness of our results in multiplication, as Example 5 illustrates.

●Example 5

Estimating a Product by Rounding

Estimate the product below by rounding each factor to the nearest hundred.

```
           Rounded
 512  →       500
×289  →      ×300
           150,000
```
You might want to now find the *actual* product and use our estimate to see if your result seems reasonable.

● ● ●   **CHECK YOURSELF 5**

Estimate the product by rounding each factor to the nearest hundred.

$$689 \\ \times 425$$

Rounding the factors can be a very useful way of estimating the solution to an application problem.

• Example 6

Estimating the Solution to a Multiplication Application

Bart is thinking of running an ad in the local newspaper for an entire year. The ad costs $19.95 per week. Approximate the annual cost of the ad.

Rounding the charge to $20 and rounding the number of weeks in a year to 50, we get

$$50 \times 20 = 1000$$

The ad would cost approximately $1000.

● ● ●   **CHECK YOURSELF 6**

Phyllis is debating whether to join the health club for $450 per year or just pay $7 per visit. If she goes about once a week, approximately how much would she spend at $7 per visit?

● ● ●   **CHECK YOURSELF ANSWERS**

1. 25,700.   **2. (a)** 5800; **(b)** 395,000.   **3.** 2450.   **4.** 172,200.

                       ↑   ↑

           Two zeros   Three zeros

5. 700 × 400, or 280,000.   **6.** $350 (exactly $364).

Multiply.

1. 53 × 10

2. 10 × 67

3. 89 × 100

4. 1000 × 73

5. 1000 × 567

6. 3456 × 100

7. 236 × 10,000

8. 10,000 × 42

9.   43
     ×70

10.   58
      ×40

11.   562
      ×400

12.   907
      ×900

13.   345
      ×230

14.   362
      ×310

15.   157
      ×3200

16.   253
      ×5300

17. 367 × 20

18. 30 × 563

19. 249 × 300

20. 700 × 612

21. 238 × 4000

22. 134 × 2500

23. 5000 × 408

24. 3200 × 265

A N S W E R S

1. 530

2. 670

3. 8900

4. 73,000

5. 567,000

6. 345,600

7. 2,360,000

8. 420,000

9. 3010

10. 2320

11. 224,800

12. 816,300

13. 79,350

14. 112,220

15. 502,400

16. 1,340,900

17. 7340

18. 16,890

19. 74,700

20. 428,400

21. 952,000

22. 335,000

23. 2,040,000

24. 848,000

Estimate each of the following products by rounding each factor to the nearest ten.

25.   36
    ×23

26.   27
    ×34

27.   93
    ×48

28.   74
    ×57

Estimate each of the following products by rounding each factor to the nearest hundred.

29.   212
    ×278

30.   179
    ×431

31.   391
    ×531

32.   729
    ×481

Solve the following applications.

33. Recreation. A movie theater has its seats arranged so that there are 42 seats per row. The theater has 48 rows. Estimate the number of seats in the theater.

34. Education. There are 52 mathematics classes with 28 students in each class. Estimate the total number of students in the mathematics classes.

35. Manufacturing. A company can manufacture 45 sleds per day. Approximately how many can this company make in 128 days?

36. Recreation. The attendance at a basketball game was 2345. The cost of admission was $12 per person. Estimate the total gate receipts for the game.

37. Calculate the product 378×215 in two ways.

Method 1: Round each factor to the nearest hundred and then multiply.

Method 2: Multiply first and then round the product to the nearest hundred.

(a) Compare your answers, and comment on the difference between the two results.

(b) List the advantages and disadvantages of each method.

(c) Describe situations in which each method is the preferred approach.

38. There are many different ways of rounding. One way used in computer applications is called **truncating.**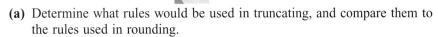

   (a) Determine what rules would be used in truncating, and compare them to the rules used in rounding.

   (b) Round 7473 to the nearest hundred using truncating and rounding.

   (c) State some possible problems that could occur in truncating.

39. Maria has been asked to estimate the number of pieces of paper in five large piles. She does not want to count every piece. Devise a plan to help her estimate the total number of pieces of paper.

38. _____

39. _____

a. 16 _____

b. 36 _____

c. 40 _____

d. 16 _____

e. 36 _____

f. 40 _____

Getting Ready for Section 2.5 [Section 1.2]

Evaluate by adding inside the parentheses and then multiplying.

a. $2 \times (3 + 5)$ **b.** $3 \times (5 + 7)$ **c.** $4 \times (2 + 8)$

Evaluate by distributing the multiplication over the sum.

d. $2 \times (3 + 5)$ **e.** $3 \times (5 + 7)$ **f.** $4 \times (2 + 8)$

Answers

1. 530   **3.** 8900; 100 has _two_ zeros   **5.** 567,000   **7.** 2,360,000

9. 3010   **11.** $\begin{array}{r} 562 \\ \times\ 400 \\ \hline 224{,}800 \end{array}$ **13.** 79,350   **15.** 502,400   **17.** 7340

19. 74,700   **21.** $\begin{array}{r} 238 \\ \times\ 4000 \\ \hline 952{,}000 \end{array}$ **23.** 2,040,000   **25.** 800   **27.** 4500

29. 60,000   **31.** 200,000   **33.** 2000 seats   **35.** 6500 sleds   **37.**

39.   **a.** 16   **b.** 36   **c.** 40   **d.** 16   **e.** 36   **f.** 40

Name

Section   Date

1. 480

2. 5700

3. 63,000

4. 2,750,000

5. 6510

6. 469,600

7. 152,960

8. 955,800

9. 1260

10. 34,200

11. 215,200

12. 259,200

13. 800

14. 3600

15. 60,000

16. 150,000

17. 400 mi

18. 2500 eggs

Multiply.

1. 48×10

2. 100×57

3. 1000×63

4. $275 \times 10,000$

5. 93
$\times 70$

6. 587
$\times 800$

7. 478
$\times 320$

8. 354
$\times 2700$

9. 30×42

10. 57×600

11. 400×538

12. 2400×108

Estimate by rounding to the nearest ten.

13. 37
$\times 22$

14. 86
$\times 42$

Estimate by rounding to the nearest hundred.

15. 195
$\times 321$

16. 529
$\times 278$

Solve the following applications.

17. Travel. Estimate how far a car can go without running out of gas if the car's tank holds 23 gallons (gal) and the car gets an average of 18 miles (mi) per gallon.

18. Farming. The hens at a chicken farm lay 53 eggs per day. Approximate the number of eggs laid in 48 days.

2.5 The Order of Operations

2.5 OBJECTIVES

1. Know the order of operations.
2. Evaluate an expression by using the order of operations.

CAUTION

If multiplication is combined with addition or subtraction, you must know which operation to do first in finding the expression's value. We can easily illustrate this problem. How should we simplify the following statement?

$$3 \times 4 + 5 = ?$$

Both multiplication and addition are involved in this expression, and we must decide which to do first to find the answer.

1. Multiplying first gives us

$$12 + 5 = 17$$

2. Adding first gives us

$$3 \times 9 = 27$$

The answers differ depending on which operation is done first!

Only one of these results can be correct, which is why mathematicians developed a rule to tell us the order in which the operations should be performed. The rules are as follows.

The Order of Operations

If multiplication, addition, and subtraction are involved in the same expression, do the operations in the following order:

STEP 1 Do all multiplication in order from left to right.
STEP 2 Do all addition or subtraction in order from left to right.

● Example 1

Using the Order of Operations

By this rule, we see that strategy (1) in the introduction was correct.

(a) $3 \times 4 + 5 = 12 + 5 = 17$ Multiply *first,* then add or subtract.
(b) $5 + 3 \times 6 = 5 + 18 = 23$
(c) $16 - 2 \times 3 = 16 - 6 = 10$
(d) $7 \times 8 - 20 = 56 - 20 = 36$
(e) $5 \times 6 + 4 \times 3 = 30 + 12 = 42$

● ● ● CHECK YOURSELF 1

Evaluate.

a. $8 + 3 \times 5$ **b.** $15 \times 5 - 3$ **c.** $4 \times 3 + 2 \times 6$

We now want to extend our rule for the order of operations. Let's see what happens when parentheses are involved in an expression.

• Example 2

Using the Order of Operations

Evaluate $3 \times (4 + 5)$.

Do you see the importance of understanding parentheses?

$3 \times (4 + 5) \neq 3 \times 4 + 5$

The notation $\neq$ means n*ot equal to*. Check the statement above for yourself.

$3 \times (4 + 5) = 3 \times 9 = 27$ We saw this expression earlier. In this case, add $4 + 5$ as the first step.

The following explains the rules concerning parentheses.

The Order Of Operations

If an expression contains an operation in parentheses, do that operation first. Then follow the order of our previous rule.

● ● ● **CHECK YOURSELF 2**

Evaluate.

$(2 + 5) \times 4$

• Example 3

Using the Order of Operations

(*a*) Evaluate $(12 - 5) \times 6$.

$(12 - 5) \times 6 = 7 \times 6 = 42$ Perform the subtraction in the parentheses as the first step.

(*b*) Evaluate $(8 + 3) \times 4 + 3$.

$$(8 + 3) \times 4 + 3 = \underbrace{11 \times 4}_{} + 3$$
$$= \quad 44 \quad + 3 = 47$$

Add $8 + 3$ to get 11 as the first step. Then multiply 11×4. As the final step, we add.

● ● ● **CHECK YOURSELF 3**

Evaluate.

$5 \times (8 - 3)$

● ● ● **CHECK YOURSELF ANSWERS**

1. (a) $8 + 3 \times 5 = 8 + 15 = 23$; **(b)** $15 \times 5 - 3 = 75 - 3 = 72$; **(c)** $4 \times 3 + 2 \times 6 = 12 + 12 = 24$.   **2.** 28.   **3.** 25.

Evaluate.

1. $4 \times 5 + 7$ **2.** $5 \times 2 + 6$

3. $3 + 6 \times 4$ **4.** $7 - 3 \times 2$

5. $8 \times 5 - 20$ **6.** $7 + 4 \times 8$

7. $48 - 8 \times 5$ **8.** $3 \times 5 + 8$

9. $9 + 6 \times 8$ **10.** $8 \times 6 - 48$

11. $20 \times 6 - 5$ **12.** $8 \times 7 + 2$

13. $20 \times (6 - 5)$ **14.** $8 \times (7 + 2)$

15. $9 \times (4 + 3)$ **16.** $7 \times (9 - 5)$

17. $4 \times 5 + 7 \times 3$ **18.** $8 \times 5 - 7 \times 4$

19. $9 \times 8 - 12 \times 6$ **20.** $4 \times 5 + 2 \times 10$

21. $5 \times (3 + 4)$ **22.** $7 \times (6 + 8)$

23. $5 \times 3 + 5 \times 4$ **24.** $7 \times 6 + 7 \times 8$

Solve the following applications.

25. Total cost. You buy a stereo for $125 down and agree to pay $25 per month for 12 months on the balance that is owed. What is your total cost?

26. Income. Clyde owns two rental units. During 1 year the first unit was rented for 9 months at $245 per month. The other was occupied for 11 months at $285 per month. What was the income from the two units for the year?

121

1.	27
2.	16
3.	27
4.	1
5.	20
6.	39
7.	8
8.	23
9.	57
10.	0
11.	115
12.	58
13.	20
14.	72
15.	63
16.	28
17.	41
18.	12
19.	0
20.	40
21.	35
22.	98
23.	35
24.	98
25.	$425
26.	$5340

27. $6720

28. 172 people

a. 29

b. 23

c. 38

d. 26

e. 82

f. 19

27. Total cost. Curly buys a $31,000 car but decides to finance it by paying $13,000 down and making $412 monthly payments for 60 months. How much would he have saved by paying cash?

28. Capacity. A restaurant has 16 booths that will seat 4 people each, 10 tables that will seat 6 people each and 6 tables that will seat 8 people each. How many people can be seated in the restaurant if every space is full?

 Getting Ready for Section 2.6
[Sections 1.2 and 1.6]

Find each of the following.

a. 12 more than 17

b. 9 less than 32

c. 23 increased by 15

d. The difference of 49 and 23

e. The sum of 53 and 29

f. 58 decreased by 39

Answers

1. 27   **3.** $3 + 6 \times 4 = 3 + 24 = 27$ **5.** 20   **7.** 8   **9.** 57
11. $20 \times 6 - 5 = 120 - 5 = 115$ **13.** $20 \times (6 - 5) = 20 \times 1 = 20$ **15.** 63
17. $4 \times 5 + 7 \times 3 = 20 + 21 = 41$ **19.** 0   **21.** 35   **23.** 35   **25.** $425
27. $6720   **a.** 29   **b.** 23   **c.** 38   **d.** 26   **e.** 82   **f.** 19

2.5 Supplementary Exercises

Name

Section   Date

1. 29

2. 13

3. 30

4. 18

5. 18

6. 28

7. 10

8. 18

9. 26

10. 0

11. 54

12. 54

13. $17,375

14. $875

Evaluate

1. $4 \times 6 + 5$ **2.** $7 + 3 \times 2$ **3.** $6 + 8 \times 3$ **4.** $36 - 6 \times 3$

5. $5 \times 4 - 2$ **6.** $6 \times 5 - 2$ **7.** $5 \times (4 - 2)$ **8.** $6 \times (5 - 2)$

9. $3 \times 2 + 4 \times 5$ **10.** $8 \times 3 - 4 \times 6$ **11.** $6 \times (4 + 5)$ **12.** $6 \times 4 + 6 \times 5$

13. Salary. Sarah received a year-end bonus of $475 in addition to her weekly salary of $325. What was her total income for the year?

14. Utility budget. The local electric company has put Nick on a budget plan that requires that he pay $305 per month for 12 months. If his actual annual bill is $2785, how much does the company owe Nick?

Using Your Calculator to Multiply

As we pointed out earlier, there are many differences among calculators. If any of the examples we consider in this section do not come out the same on the model you are using, check with your instructor.

Multiplication, like addition and subtraction, is easy to do on your calculator. (We hope you haven't been using one in this chapter so far!).

To multiply 23×37 on the calculator:

1. Press the clear key. $\boxed{C}$
2. Enter the first factor. 23
3. Press the times key. $\boxed{\times}$
4. Enter the second factor. 37
5. Press the equals key. $\boxed{=}$ Your display will now show the desired product, 851.

Display $\boxed{851}$

● Example 1

Using a Calculator to Multiply Three or More Numbers

To multiply $3 \times 5 \times 7$, use this sequence:

$3 \boxed{\times} 5 \boxed{\times} 7 \boxed{=}$ When you press the times key the second time, note that the product of 3 and 5 is in the display.

Display $\boxed{105}$

● ● ● CHECK YOURSELF 1

Multiply $9 \times 11 \times 15$ with your calculator.

● Example 2

Using a Calculator to Evaluate an Expression

To evaluate $3 \times 4 + 5$, use this sequence:

$3 \boxed{\times} 4 \boxed{+} 5 \boxed{=}$

Display $\boxed{17}$

● ● ● CHECK YOURSELF 2

Evaluate $5 \times 6 + 11$ with your calculator.

© 1998 McGraw-Hill Companies

So far, our calculator has done the multiplication and then the addition. This means it operates according to our rule of the previous section for the order of operations.

Let's change the order of the addition and multiplication in an expression and see what the calculator does.

•Example 3

Using a Calculator to Evaluate an Expression

To evaluate $6 + 3 \times 5$, use this sequence:

6 ⊞ 3 ⊠ 5 ⊟

Display 21

Again the calculator follows the order of operations. Try this with your calculator.

● ● ● CHECK YOURSELF 3

Evaluate $9 + 3 \times 7$ with your calculator.

Most calculators have parentheses keys that will allow you to evaluate more complicated expressions easily.

•Example 4

Using a Calculator to Evaluate an Expression

To evaluate $3 \times (4 + 5)$, use this sequence:

3 ⊠ ⬛(4 ⊞ 5 ⬛) ⊟

Display 27

Now the calculator does the addition in the parentheses as the first step. Then it does the multiplication.

● ● ● CHECK YOURSELF 4

Evaluate $4 \times (2 + 9)$ with your calculator.

● ● ● CHECK YOURSELF ANSWERS

1. 1485.   **2.** 41.   **3.** 30.   **4.** 44.

Calculator Exercises

ANSWERS

Multiply.

1.   57
   $\times 89$

2.   98
   $\times 25$

3.   256
   $\times 508$

4.   285
   $\times 820$

5.   23,456
   $\times\ 2,358$

6.   18,569
   $\times\ 3,286$

7. $12 \times 15 \times 8 =$

8. $32 \times 5 \times 18 =$

9. $78 \times 145 \times 36 =$

10. $358 \times 39 \times 928 =$

11. $24 \times 35 \times 48 \times 36 =$

12. $37 \times 15 \times 42 \times 29 =$

Evaluate each of these expressions by hand. Then use your calculator to verify your results.

13. $4 \times 5 - 7 =$

14. $3 \times 7 + 8 =$

15. $9 + 3 \times 7 =$

16. $6 \times 0 + 3 =$

17. $4 + 5 \times 0 =$

18. $23 - 4 \times 5 =$

19. $5 \times (4 + 7) =$

20. $8 \times (6 + 5) =$

21. $5 \times 4 + 5 \times 7 =$

22. $8 \times 6 + 8 \times 5 =$

ANSWERS	
1.	5073
2.	2450
3.	130,048
4.	233,700
5.	55,309,248
6.	61,017,734
7.	1440
8.	2880
9.	407,160
10.	12,956,736
11.	1,451,520
12.	675,990
13.	13
14.	29
15.	30
16.	3
17.	4
18.	3
19.	55
20.	88
21.	55
22.	88

Solve the following applications.

23. Sales. A car dealer kept the following record of a month's sales. Complete the table.

Model	Number Sold	Profit per Sale	Monthly Profit
Subcompact	38	$528	$20,064
Compact	33	647	21,351
Standard	19	912	17,328
		Monthly Total Profit	$58,743

24. Salary. You take a job paying $1 the first day. On each following day your pay doubles. That is, on day 2 your pay is $2, on day 3 the pay is $4, and so on. Complete the table.

Day	Daily Pay	Total Pay
1	$1	$1
2	2	3
3	4	7
4	8	15
5	16	31
6	32	63
7	64	127
8	128	255
9	256	511
10	512	1023

Answers

1. 5073   **3.** 130,048   **5.** 55,309,248   **7.** 1440   **9.** 407,160
11. 1,451,520   **13.** 13   **15.** 30   **17.** 4   **19.** 55   **21.** 55
23. Monthly profit:

$20,064
21,351
17,328
$58,743

Powers of Whole Numbers

2.6 OBJECTIVES

1. Use exponent notation.
2. Evaluate expressions that contain powers of whole numbers.

Recall that

$3 + 3 + 3 + 3 = 4 \times 3$

Repeated addition was written as multiplication.

René Descartes, a French philosopher and mathematician, is generally credited with first introducing our modern exponent notation in about 1637.

Earlier we described multiplication as a shorthand for repeated addition. There is also a shorthand for repeated multiplication. It uses **powers of a whole number**.

● Example 1

Writing Repeated Multiplication as a Power

$3 \times 3 \times 3 \times 3$ can be written as 3^4. This is read as "3 to the fourth power."

In this case, repeated multiplication is written as the power of a number.

In this example, 3 is the **base** of the expression, and the raised number, 4, is the **exponent**, or **power**.

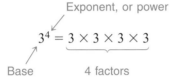

Exponent, or power

$$3^4 = \underbrace{3 \times 3 \times 3 \times 3}_{\text{4 factors}}$$

Base   4 factors

We count the factors and make this the power (or exponent) of the base.

CHECK YOURSELF 1

Write $2 \times 2 \times 2 \times 2 \times 2 \times 2$ as a power of 2.

Exponents

The *exponent* tells us the number of times the base is to be used as a factor.

● Example 2

Evaluating a Number Raised to a Power

2^5 is read "2 to the fifth power."

$$2^5 = \underbrace{2 \times 2 \times 2 \times 2 \times 2}_{\text{5 times}} = 32$$

Here 2 is the base, and 5 is the exponent.

2^5 tells us to use 2 as a factor 5 times. The result is 32.

© 1998 McGraw-Hill Companies

●●● **CHECK YOURSELF 2**

How would you read 6^4?

● Example 3

Evaluating a Number Raised to a Power

Evalute 5^3 and 8^2.

CAUTION

Be careful: 5^3 is *entirely different* from 5×3.

$5^3 = 125$ while $5 \times 3 = 15$.

$5^3 = 5 \times 5 \times 5 = 125$ Use 3 factors of 5.

5^3 is read "5 to the third power" or "5 cubed."

$8^2 = 8 \times 8 = 64$ Use 2 factors of 8.

And 8^2 is read "8 to the second power" or "8 squared."

●●● **CHECK YOURSELF 3**

Evaluate.

a. 6^2 **b.** 2^4

We need two special definitions for powers of whole numbers.

> ### Raising a Number to the First Power
>
> A whole number raised to the first power is just that number.
>
> For any whole number a, $a^1 = a$.

This definition may look a bit strange. Don't concern yourself at this point. Later, in algebra, you will see that we must define the 0 exponent this way to be consistent with the laws for exponents.

> ### Raising a Number to the Zero Power
>
> A whole number, other than 0, raised to the zero power is 1.
>
> For any whole number a, where $a \neq 0$, $a^0 = 1$.

● Example 4

Evaluating Numbers Raised to the Power of Zero or One

a. $8^0 = 1$ **b.** $4^0 = 1$ **c.** $5^1 = 5$ **d.** $3^1 = 3$

● ● ● ● **CHECK YOURSELF 4**

Evaluate.

a. 7^0 **b.** 7^1

We talked about *powers of 10* earlier when we multiplied by numbers that end in 0. Since the powers of 10 have a special importance, let's list some of them.

$10^0 = 1$

$10^1 = 10$

$10^2 = 10 \times 10 = 100$

$10^3 = 10 \times 10 \times 10 = 1000$

$10^4 = 10 \times 10 \times 10 \times 10 = 10,000$

$10^5 = 10 \times 10 \times 10 \times 10 \times 10 = 100,000$

Do you see why the powers of 10 are so important?

Note that 10^3 is just a 1 followed by *three zeros.*

10^5 is a 1 followed by *five zeros.*

Archimedes (about 250 B.C.) reportedly estimated the number of grains of sand in the universe to be 10^{63}. This would be a 1 followed by 63 zeros!

> **Powers of 10**
>
> The powers of 10 correspond to the place values of our number system, ones, tens, hundreds, thousands, and so on.

This is what we meant earlier when we said that our number system was based on the number 10.

To deal with powers of numbers in an expression, we must again extend our rule from Section 2.5 for the order of operations.

> **The Order of Operations**
>
> Mixed operations in an expression should be done in the following order:
>
> **STEP 1** Do any operations inside parentheses.
> **STEP 2** Evaluate any powers.
> **STEP 3** Do all multiplication in order from left to right.
> **STEP 4** Do all addition or subtraction in order from left to right.

● **Example 5**

Evaluating an Expression

Evalute $4^3 + 5$.

$4^3 + 5 = 64 + 5 = 69$

● ● ● **CHECK YOURSELF 5**

Evaluate.

$2 + 3^3$

● Example 6

Evaluating an Expression

Evalute 4×2^3.

$4 \times 2^3 = 4 \times 8 = 32$ Find 2^3 as the first step and then multiply by 4.

● ● ● **CHECK YOURSELF 6**

Evaluate.

3×3^2

● Example 7

Evaluating an Expression

Evaluate.

$(2 + 3)^2 + 4 \times 3 = 5^2 + 4 \times 3$ Perform the operation in parentheses first, $2 + 3 = 5$. Then evaluate the power, $5^2 = 25$.

$\qquad\qquad\quad = 25 + 4 \times 3$ Multiply.

$\qquad\qquad\quad = 25 + 12$ Add.

$\qquad\qquad\quad = 37$

● ● ● **CHECK YOURSELF 7**

Evaluate.

$4 + (8 - 5)^2$

● ● ● **CHECK YOURSELF ANSWERS**

1. 2^6.   **2.** 6 to the fourth power.   **3. (a)** $6^2 = 6 \times 6 = 36$;
(b) $2^4 = 2 \times 2 \times 2 \times 2 = 16$.   **4. (a)** 1; **(b)** 7.   **5.** $2 + 3^3 = 2 + 27 = 29$.
6. $3 \times 3^2 = 3 \times 9 = 27$.   **7.** $4 + (8 - 5)^2 = 4 + 3^2 = 4 + 9 = 13$.

Name _____

Section _____ Date _____

Evaluate.

1. 3^2 **2.** 2^3

3. 2^4 **4.** 5^2

5. 8^3 **6.** 3^5

7. 1^5 **8.** 4^4

9. 5^1 **10.** 6^0

11. 9^0 **12.** 7^1

13. 10^3 **14.** 10^2

15. 10^6 **16.** 10^7

17. 2×4^3 **18.** $(2 \times 4)^3$

19. 2×3^2 **20.** $(2 \times 3)^2$

21. $5 + 2^2$ **22.** $(5 + 2)^2$

23. $(3 \times 2)^4$ **24.** 3×2^4

1. 9
2. 8
3. 16
4. 25
5. 512
6. 243
7. 1
8. 256
9. 5
10. 1
11. 1
12. 7
13. 1000
14. 100
15. 1,000,000
16. 10,000,000
17. 128
18. 512
19. 18
20. 36
21. 9
22. 49
23. 1296
24. 48

25. 72

26. 144

27. 5

28. 28

29. 105

30. 21

31. 51

32. 69

33. Yes

34. No

35. Yes

36. Yes

37. No

38. Yes

39.

40.

25. 2×6^2 **26.** $(2 \times 6)^2$

27. $14 - 3^2$ **28.** $12 + 4^2$

29. $(3 + 2)^3 - 20$ **30.** $5 + (9 - 5)^2$

31. $(7 - 4)^4 - 30$ **32.** $(5 + 2)^2 + 20$

Numbers such as 3, 4, and 5 are called **Pythagorean triples,** after the Greek mathematician Pythagoras (sixth century BC), because

$$3^2 + 4^2 = 5^2$$

Which of the following sets of numbers are Pythagorean triples?

33. 6, 8, 10 **34.** 6, 11, 12

35. 5, 12, 13 **36.** 7, 24, 25

37. 8, 16, 18 **38.** 8, 15, 17

39. Is $(a + b)^P = a^P + b^P$?

Try a few numbers and decide if you think this is true for all whole numbers, for some whole numbers, or never true. Write an explanation of your findings, and give examples.

40. Is $(a \cdot b)^P = a^P \cdot b^P$?

Try a few numbers and decide if you think this is true for all whole numbers, for some whole numbers, or never true. Write an explanation of your findings, and give examples.

Getting Ready for Section 2.7 [Section 2.5]

Evaluate the following

a. $18 \times 4 + 18 \times 8$ **b.** $2 \times 15 + 2 \times 27$ **c.** $2 \times (15 + 27)$

d. $3 \times (16 + 31)$ **e.** $5 \times 16 + 3 \times 31$ **f.** $8 \times 6 + 12 \times 6$

a. 216

b. 84

c. 84

d. 141

e. 173

f. 120

Answers

1. 9   **3.** 16   **5.** $8^3 = 8 \times 8 \times 8 = 512$ **7.** 1   **9.** 5   **11.** 1
13. 1000   **15.** 1,000,000   **17.** $2 \times 4^3 = 2 \times 64 = 128$ **19.** 18   **21.** 9
23. 1296   **25.** 72   **27.** 5   **29.** 105   **31.** 51   **33.** Yes   **35.** Yes
37. No   **39.**   **a.** 216   **b.** 84   **c.** 84   **d.** 141   **e.** 173
f. 120

NumberCross 2

ACROSS

  1. 6×551
  5. 7×2^3
  6. 27×27
  7. 19×50
10. 3×67
12. 6×5^2
13. 9×8
15. $2^4 \times 303$

DOWN

  1. 5×7
  2. $3^2 \times 41$
  3. 67×100
  4. $2 \times (7^2 + 100)$
  8. 4×1301
  9. $10^2 + 10 + 1$
11. 2×87
14. $5^2 + 3$

SOLUTION TO NUMBERCROSS2

Supplementary Exercises

Name

Section   Date

A N S W E R S

1. 16

2. 125

3. 81

4. 81

5. 64

6. 32

7. 6

8. 1

9. 10,000

10. 100,000,000

11. 24

12. 216

13. 25

14. 11

15. 20

16. 6

Evaluate.

1. 4^2

2. 5^3

3. 3^4

4. 9^2

5. 4^3

6. 2^5

7. 6^1

8. 7^0

9. 10^4

10. 10^8

11. 3×2^3

12. $(3 \times 2)^3$

13. $(2 + 3)^2$

14. $2 + 3^2$

15. $(7 - 3)^2 + 4$

16. $(2 + 4)^2 - 30$

Measuring Area and Volume

2.7 OBJECTIVES

1. Find the area of a rectangular figure.
2. Apply area formulas.
3. Find the volume of a rectangular solid.
4. Apply volume formulas.

The unit inch (in.) can be treated as though it were a number. So in. × in. can be written in.^{2} It is read "square inches."

Let's look now at the idea of **area.** Area is a measure that we give to a surface. It is measured in terms of **square units.** The area is the number of square units that are needed to cover the surface.

One standard unit of area measure is the **square inch,** written in.^{2}. This is the measure of the surface contained in a square with sides of 1 in. See Figure 1.

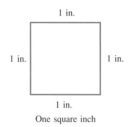

One square inch

Figure 1

Other units of area measure are the square foot (ft^2), the square yard (yd^2), the square centimeter (cm^2), and the square meter (m^2).

Finding the area of a figure means finding the number of square units it contains. One simple case is a rectangle.

Figure 2 shows a rectangle. The length of the rectangle is 4 inches (in.), and the width is 3 in. The area of the rectangle is measured in terms of square inches. We can simply count to find the area, 12 square inches (in.^{2}). However, since each of the four vertical strips contains 3 in.^{2}, we can multiply:

$$\text{Area} = 4 \text{ in.} \times 3 \text{ in.} = 12 \text{ in.}^2$$

The length and width must be in terms of the same unit.

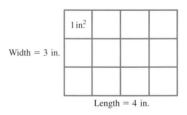

Width = 3 in.

Length = 4 in.

Figure 2

Note that area is always measured in *square units*, and the length and width must be in terms of the same unit of measure.

A **formula** is a mathematical rule or statement written with algebraic symbols.

Formula for the Area of a Rectangle

In general, we can write the formula for the **area of a rectangle**: If the length of a rectangle is L units and the width is W units, then the formula for the area, A, of the rectangle can be written as

$$A = L \times W \text{ (square units)} \tag{1}$$

• Example 1

Find the Area of a Rectangle

A room has dimensions 12 by 15 feet (ft). Find its area.

Solution Use Formula (1), with $L = 15$ ft and $W = 12$ ft.

$$A = L \times W$$
$$= 15 \times 12 = 180 \text{ ft}^2$$

The area of the room is 180 ft^2.

CHECK YOURSELF 1

A desktop has dimensions 50 in. by 25 in. What is the area of its surface?

We can also write a convenient formula for the area of a square. If the sides of the square have length S, we can write

S^2 is read "S squared."

Formula for the Area of a Square

$$A = S \times S = S^2 \qquad (2)$$

• Example 2

Finding Area

You wish to cover a square table with a plastic laminate that costs 60¢ a square foot. If each side of the table measures 3 ft, what will it cost to cover the table?

Solution We first must find the area of the table. Use Formula (2), with $S = 3$ ft.

$$A = S^2$$
$$= (3 \text{ ft})^2 = 3 \text{ ft} \times 3 \text{ ft} = 9 \text{ ft}^2$$

Now, multiply by the cost per square foot.

$$\text{Cost} = 9 \times 60¢ = \$5.40$$

CHECK YOURSELF 2

You wish to carpet a room that is a square, 4 yd by 4 yd, with carpet that costs \$12 per square yard. What will be the total cost of the carpeting?

Sometimes the total area of an oddly shaped figure is found by adding the smaller areas. The next example shows how this is done.

● Example 3

Finding the Area of an Oddly Shaped Figure

Find the area of Figure 3.

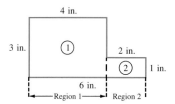

Figure 3

The area of the figure is found by adding the areas of regions 1 and 2. Region 1 is a 4 in. by 3 in. rectangle; the area of region 1 = 4 in. × 3 in. = 12 in.². Region 2 is a 2 in. by 1 in. rectangle; the area of region 2 = 2 in. × 1 in. = 2 in.²

The total area is the sum of the two areas:

Total area = 12 in.² + 2 in.² = 14 in.²

● ● ● CHECK YOURSELF 3

Find the area of Figure 4.

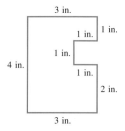

Figure 4

Our next measurement deals with finding **volumes.** The volume of a **solid** is the measure of the space contained in the solid.

Definition of a Solid

A *solid* is a three-dimensional figure. It has length, width, and height.

Volume is measured in **cubic units.** Examples include cubic inches (in.³), cubic feet (ft³), and cubic centimeters (cm³). A cubic inch, for instance, is the measure of the space contained in a cube that is 1 in. on each edge. See Figure 5.

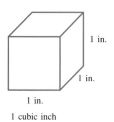

1 cubic inch

Figure 5

In finding the volume of a figure, we want to know how many cubic units are contained in that figure. Let's start with a simple example, a **rectangular solid.** A rectangular solid is a very familiar figure. A box, a crate, and most rooms are rectangular solids. Say that the dimensions of the solid are 5 in. by 3 in. by 2 in. as pictured in Figure 6. If we divide the solid into units of 1 in., we have two layers, each containing 3 units by 5 units, or 15 in.³ Since there are two layers, the volume is 30 in.³

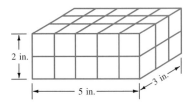

Figure 6

In general, we can see that the volume of a rectangular solid is the product of its length, width, and height.

Formula for the Volume of a Rectangular Solid

$V = L \times W \times H$ (3)

● Example 4

Finding Volume

A crate has dimensions 3 ft by 4 ft by 2 ft. Find its volume.

We are not particularly worried about which is the length, which is the width, and which is the height, since the order in which we multiply won't change the result.

Solution Use Formula (3), with $L = 4$ ft, $W = 3$ ft, and $H = 2$ ft.

$$V = L \cdot W \cdot H$$
$$= 4 \text{ ft} \times 3 \text{ ft} \times 2 \text{ ft}$$
$$= 24 \text{ ft}^3$$

● ● ● **CHECK YOURSELF 4**

A room is 15 ft long, 10 ft wide, and 8 ft high. What is its volume?

● ● ● **CHECK YOURSELF ANSWERS**

1. 1250 in.²   **2.** $192.   **3.** 11 in.²   **4.** 1200 ft³.

A N S W E R S

Find the area of each figure.

1.

6 yd
6 yd

2.

2 in.
9 in.

3.

3 in.
6 in.

4.

4 ft
4 ft

5.
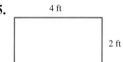
4 ft
2 ft

6.

5 in.
6 in.

7.
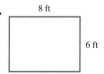
8 ft
6 ft

8.

3 in.
3 in.

9.
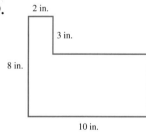
2 in.
3 in.
8 in.
10 in.

10.
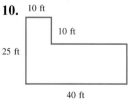
10 ft
10 ft
25 ft
40 ft

11.
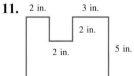
2 in.   3 in.
2 in.
2 in.
5 in.

12.
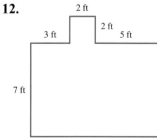
2 ft
2 ft
3 ft   5 ft
7 ft

1. 36 yd²

2. 18 in.²

3. 18 in.²

4. 16 ft²

5. 8 ft²

6. 30 in.²

7. 48 ft²

8. 9 in.²

9. 56 in.²

10. 700 ft²

11. 31 in.²

12. 74 ft²

13.

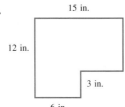

15 in.

12 in.

3 in.

6 in.

14.

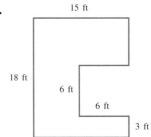

15 ft

18 ft

6 ft

6 ft

3 ft

Find the volume of each solid shown.

15.

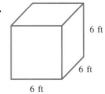

6 ft

6 ft

6 ft

16.

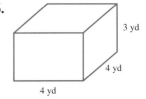

3 yd

4 yd

4 yd

17.

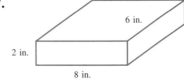

6 in.

2 in.

8 in.

18.

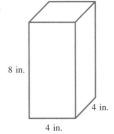

8 in.

4 in.

4 in.

19.

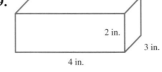

2 in.

3 in.

4 in.

20.

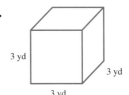

3 yd

3 yd

3 yd

21.

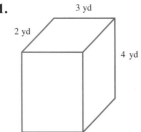

3 yd

2 yd

4 yd

22.

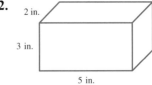

2 in.

3 in.

5 in.

Solve each of the following applications.

23. **Tile costs.** You wish to cover a bathroom floor with 1-square-foot (1 ft^2) tiles that cost $2 each. If the bathroom is rectangular, 5 ft by 8 ft, how much will the tile cost?

24. **Roofing.** A rectangular shed roof is 30 ft long and 20 ft wide. Roofing is sold in squares of 100 ft^2. How many squares will be needed to roof the shed?

25. **House repairs.** A plate glass window measures 5 ft by 7 ft. If glass costs $8 per square foot, how much will it cost to replace the window?

26. **Paint costs.** In a hallway, Bill is painting 2 walls that are 10 ft high by 22 ft long. The instructions on the paint can say that it will cover 400 ft^2 per gallon (gal). Will 1 gal be enough for the job?

27. **Tile costs.** Tile for a kitchen counter will cost $7 per square foot to install. If the counter measures 12 ft by 3 ft, what will the tile cost?

28. **Carpet costs.** You wish to cover a floor 4 yards (yd) by 5 yd with a carpet costing $13 per square yard (yd^2). What will the carpeting cost?

29. **Frame costs.** A mountain cabin has a rectangular front that measures 30 ft long and 20 ft high. If the front is to be glass that costs $12 per square foot, what will the glass cost?

30. **Posters.** You are making posters 12 ft by 15 ft. How many square feet of material will you need for four posters?

31. **Shipping.** A shipping container is 5 by 3 by 2 ft. What is its volume?

32. **Size of a cord.** A cord of wood is 4 by 4 by 8 ft. What is its volume?

33. **Storage.** The inside dimensions of a meat market's cooler are 9 by 9 by 6 ft. What is the capacity of the cooler in cubic feet?

34. **Storage.** A storage bin is 18 ft long, 6 ft wide, and 3 ft high. What is its volume in cubic feet?

Answers

1. 36 yd^2 **3.** 18 in.^{2} **5.** 8 ft^2 **7.** 48 ft^2 **9.** 56 in.^{2} **11.** 31 in.^{2}
13. 153 in.^{2} **15.** 216 ft^3 **17.** 96 in.^{3} **19.** 24 in.^{3} **21.** 24 yd^3
23. $80   **25.** $280   **27.** $252   **29.** $7200   **31.** 30 ft^3 **33.** 486 ft^3

141

23. $80

24. 6 squares

25. $280

26. No, the area is 440 ft^2.

27. $252

28. $260

29. $7200

30. 720 ft^2

31. 30 ft^3

32. 128 ft^3

33. 486 ft^3

34. 324 ft^3

Name

Section   Date

A N S W E R S

1. 1200 ft²

2. 15 in.²

3. 16 ft²

4. 12 in.²

5. 14 in.²

6. 14 in.²

7. 72 in.³

8. 125 in.³

9. $429

10. $320

11. 108 ft²

12. 240 ft³

Find the area of each figure.

1.
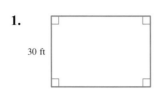

2.
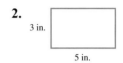

3.
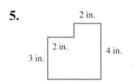

4.

5.

6.
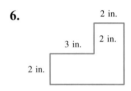

Find the volume of each solid shown.

7.

8.
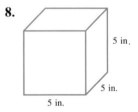

Solve each of the following applications.

9. **Carpeting.** You are carpeting a room that measures 13 ft by 11 ft. If the carpeting costs $3 per square ft, how much will it cost to carpet the room?

10. **Cost.** A shopping center designs a rectangular sign, 10 ft across its base and 8 ft high. If the material for the sign costs $4 a square foot, what will the material cost?

11. **Shipping.** A shipping crate is 9 ft long, 3 ft wide, and 4 ft tall. What is its volume?

12. **Capacity.** A rectangular holding tank is 10 ft long, 8 ft wide, and 3 ft deep. What is its volume?

The purpose of the Self-Test is to help you check your progress and review for a chapter test in class. Allow yourself about 1 hour to take the test. When you are done, check your answers in the back of the book. If you missed any answers, be sure to go back and review the appropriate sections in the chapter and do the supplementary exercises provided there.

[2.1]   **1.** List all the factors of 48.   **2.** List the first five multiples of 6.

In Exercises 3 to 7, name the property that is illustrated.

3. $7 \times 9 = 9 \times 7$ **4.** $5 \times 1 = 5$

5. $8 \times 0 = 0$ **6.** $3 \times (2 \times 7) = (3 \times 2) \times 7$

7. $4 \times (3 + 6) = (4 \times 3) + (4 \times 6)$

[2.2]   In Exercises 8 to 11, do the indicated operations.

8.   58
       $\times\ 3$ **9.** Find the product of 273 and 7.

[2.3]   **10.**   89
              $\times 56$ **11.**   538
                              $\times 103$

12. A college lecture hall has 14 rows of seats with 18 seats per row. How many people will the room seat?

13. A truck rental firm has ordered 25 new vans at a cost of $12,350 per van. What will be the total cost of the order?

14. A school must order new sports uniforms. There are 25 different teams, and each team must have 36 uniforms. How many uniforms must the school order?

15. Each of 16 classrooms in a school contains 250 books in its reference library. What is the total number of books in the reference libraries?

16. Find the product of 4568 and 537.

ANSWERS

1. 1, 2, 3, 4, 6, 8, 12, 16, 24, 48

2. 6, 12, 18, 24, 30

3. Commutative property of multiplication

4. Multiplicative identity property

5. Multiplication property of 0

6. Associative property of multiplication

7. Distributive property of multiplication over addition

8. 174

9. 1911

10. 4984

11. 55,414

12. 252 people

13. $308,750

14. 900 uniforms

15. 4000 books

16. 2,453,016

143

17. 53,000

18. 226,800

19. 321,840

20. 150,000

21. 920,000

22. 17

23. 22

24. 1

25. 75

26. 75

27. $2224

28. 3769 mi

29. 625

30. 1

31. 108

32. 9

33. 12 in.²

34. 36 ft²

35. 24 ft³

36. 81 yd³

37. $1200

[2.4]   **17.**   53
            $\times 1000$

18.   567
            $\times 400$

19.   894
            $\times 360$

In Exercises 20 and 21, estimate the products by rounding each factor to the nearest hundred.

20.   325
          $\times 468$

21.   2345
          $\times\ 389$

[2.5]   In Exercises 22 to 26, evaluate the expressions.

22. $5 + 6 \times 2$ **23.** $(5 + 6) \times 2$ **24.** $2 \times 8 - 3 \times 5$

25. $5 \times (7 + 8)$ **26.** $5 \times 7 + 5 \times 8$

27. Gretchen has invested in the stocks of two companies. Her dividend is $59 per month on the stock of the first company and $128 per month on the stock of the second. What is her yearly income from the two stocks?

28. Darcie drove 467 miles (mi) for each of 3 days and 592 mi every day for 4 days. How far did she drive?

[2.6]   In Exercises 29 to 32, evaluate the expressions.

29. 5^4 **30.** 8^0 **31.** 3×6^2 **32.** $(4 + 3)^2 - 40$

[2.7]   Find the area of the given figure.

33.
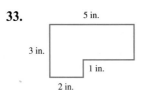

34.

Find the volume of the given figure.

35.
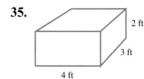

36.

37. A store designs a rectangular sign 20 feet (ft) long by 12 ft high. If the material for the sign costs $5 per square foot (ft²), what will the material cost?

144

CHAPTER 3

DIVISION OF WHOLE NUMBERS

INTRODUCTION

The United States was the first industrialized nation to pass laws regulating levels of air pollutants. Congress passed the Clean Air Act in 1970. Since then, the level of many of the major air pollutants in the larger U.S. cities has dropped. Congress was encouraged enough to pass a tougher version of the Clean Air Act in 1990.

Jason works for the Environmental Protection Agency (EPA) in Michigan. He works with manufacturers that are trying to develop cars that run on alternative fuels such as alcohol and electricity. He hopes to see an electric car with a range of 200 miles (mi) and a top speed of 90 km/h (55 mi/h) by the year 2001.

© Michael Rosenfeld/Tony Stone Images

Pretest for Chapter 3

Division of Whole Numbers

ANSWERS

1. Divisor: 8; dividend: 61; quotient: 7; remainder: 5
2. 7
3. Undefined
4. 1204 r3
5. 270 r1
6. 81 r131
7. 1041 r3
8. 7
9. $47
10. 277 mi

This pretest will point out any difficulties you may be having in dividing whole numbers. Do all the problems. Then check your answers with those in the back of the book.

1. In the division problem shown, identify the divisor, the dividend, the quotient, and the remainder.

$$\begin{array}{r} 7 \\ 8\overline{)61} \\ \underline{56} \\ 5 \end{array}$$

2. $7 \div 1 =$

3. $5 \div 0 =$

Divide by long division.

4. $7\overline{)8431}$

5. $34\overline{)9181}$

6. $267\overline{)21{,}758}$

Divide by short division.

7. $6\overline{)6249}$

Evaluate.

8. $(3 + 5^2) \div 4 =$

9. Installment Buying. Marcia purchases furniture costing $614. She agrees to pay $50 down and the balance in 12 equal monthly payments. Find the amount of each monthly payment.

10. Driving Distance. Five company-owned cars are driven 243, 181, 357, 193, and 411 mi in 1 week. What is the mean number of miles driven during the week per car?

3.1 The Language of Division

3.1 OBJECTIVES

1. Use the language of division.
2. Divide by using repeated subtraction.

Let's look at the fourth of the basic arithmetic operations—division. **Division** asks *how many times* one number is contained in another. Just as multiplication was thought of as repeated addition, division can be carried out by repeated subtraction steps.

• Example 1

Dividing by Using Subtraction

How many times is 7 contained in 35?

To answer the question, subtract 7 repeatedly until the difference is smaller than 7.

$$
\begin{array}{ccccc}
35 & 28 & 21 & 14 & 7 \\
-7 & -7 & -7 & -7 & -7 \\
\hline
28 & 21 & 14 & 7 & 0
\end{array}
$$

We can see that there are exactly 5 sevens in 35. This means that 35 *divided* by 7 is 5.

CHECK YOURSELF 1

How many times is 6 contained in 30?

The symbol $\overline{)}$ was used first in the 1500s. The division sign ÷ was introduced about a hundred years later.

The division sign ÷ and the symbol $\overline{)}$ are two common ways to indicate division.

• Example 2

Writing a Division Statement

A third way to indicate division is the fraction bar:

$$\frac{35}{7} = 5$$

This will be considered later in the book.

35 divided by 7 can be written as

$$35 \div 7 = 5 \qquad \text{or} \qquad 7\overline{)35}^{\,5}$$

The number being divided (35 in our example) is called the **dividend.** The number we are dividing by (7 in the example) is the **divisor.** The result of the division, here 5, is the **quotient.**

CHECK YOURSELF 2

Write the statement "30 divided by 6 is 5," using the symbol $\overline{)}$.

© 1998 McGraw-Hill Companies

● Example 3

Dividing by Using Subtraction

By repeated subtraction we can easily show that 8 is contained 4 times in 32.

$$
\begin{array}{cccc}
32 & 24 & 16 & 8 \\
-\ 8 & -\ 8 & -\ 8 & -8 \\
\hline
24 & 16 & 8 & 0
\end{array}
$$

As a division statement we can write

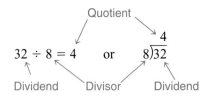

Quotient

$$32 \div 8 = 4 \qquad \text{or} \qquad 8\overline{)32}$$

Dividend   Divisor   Dividend

In either notation, we see that the divisor is 8. The dividend is 32, and the answer, 4, is the quotient.

● ● ● CHECK YOURSELF 3

In $40 \div 5 = 8$, identify the dividend, divisor, and quotient.

Because of the special relationship between multiplication and division, we say that they are **inverse** operations.

Division is closely related to multiplication. In fact, for every division statement there is a related multiplication operation.

● Example 4

Comparing Multiplication and Division

$32 \div 8 = 4$ because $32 = 8 \times 4$.
$42 \div 6 = 7$ because $42 = 6 \times 7$.

$$7\overline{)56}^{\ 8} \quad \text{because} \quad 56 = 7 \times 8.$$

● ● ● CHECK YOURSELF 4

Rewrite the statement $36 \div 4 = 9$ as a multiplication statement.

● Example 5

Checking Division by Using Multiplication

For a division problem to check, the *product* of the divisor and the quotient *must equal the dividend*.

(*a*) $7\overline{)21}^{\ 3}$ Check: $7 \times 3 = 21$

(*b*) $48 \div 6 = 8$ Check: $6 \times 8 = 48$

⦿⦿⦿ **CHECK YOURSELF 5**

Complete the division statements, and check your results.

a. $9\overline{)45}$

b. $28 \div 7 =$

Since $36 \div 9 = 4$, we say that 36 is *exactly divisible* by 9.

In our examples so far, the product of the divisor and the quotient has been equal to the dividend. This means that the dividend is *exactly divisible* by the divisor. That is not always the case. Let's look at another example using repeated subtraction.

● Example 6

Dividing by Using Subtraction, Leaving a Remainder

How many times is 5 contained in 23?

Note that the remainder must be smaller than the divisor or we could subtract again.

$$
\begin{array}{cccc}
23 & 18 & 13 & 8 \\
-\ 5 & -\ 5 & -\ 5 & -5 \\
\hline
18 & 13 & 8 & 3
\end{array}
$$

We see that 5 is contained 4 times in 23, but 3 is "left over."

23 is not exactly divisible by 5. The "left over" 3 is called the **remainder** in the division. To check the division operation when a remainder is involved, we have the following rule:

Dividend = divisor × quotient + remainder

⦿⦿⦿ **CHECK YOURSELF 6**

How many times is 7 contained in 38?

● Example 7

Checking Division by a Single-Digit Number

Using the work of the previous example, we can write

Another way to write the result is

$\dfrac{4\ \text{r}3}{5\overline{)23}}$ The "r" stands for remainder.

$\dfrac{4}{5\overline{)23}}$ with remainder 3

To apply our previous rule, we have

Note that the rule for the order of operations must be used.

$$\text{Dividend} \longrightarrow 23 = 5 \times 4 + 3 \longleftarrow \text{Remainder}$$

with *Divisor* and *Quotient* labeling 5 and 4.

$$23 = 20 + 3$$

$$23 = 23 \qquad \text{The division checks.}$$

● ● ● CHECK YOURSELF 7

Evaluate $7\overline{)38}$.

You have now looked at the four operations of arithmetic: addition, subtraction, multiplication, and division. So, in dealing with applications, or word problems, it becomes even more important that you read the problem carefully and then decide which operation or operations must be used for the solution.

Let's look at an example.

● Example 8

Solving a Division Application

Six people estimate that the car expenses for a short trip will be $54. What will be the amount of each person's share?

Division is the necessary operation for the solution.

Solution The needed information from the problem is the total expense, $54, and the number of people sharing that expense, here six.

To solve the problem, think, "If the total expense is $54, we must divide that expense into six equal parts." Write

$$\$54 \div 6 = \$9$$

Each person's share is $9.

● ● ● CHECK YOURSELF 8

There are 56 students going on a field trip. If each van holds 8 students, how many vans will be needed for the field trip?

● ● ● CHECK YOURSELF ANSWERS

1. 5. **2.** $6\overline{)30}$ with 5 above. **3.** 40 is the dividend; 5 is the divisor, 8 is the quotient.
4. $4 \times 9 = 36$. **5. (a)** 5; **(b)** 4. **6.** 5. **7.** 5 r3. **8.** 7 vans.

Name _____

Section _____   Date _____

ANSWERS

1. If $48 \div 8 = 6$, 8 is the _____, 48 is the _____, and 6 is the _____.

2. In the statement $5\overline{)45}^{\,9}$, 9 is the _____, 5 is the _____, and 45 is the _____.

3. Find $36 \div 9$ by repeated subtraction.

4. Find $40 \div 8$ by repeated subtraction.

5. If $30 \div 6 = 5$, write the related multiplication statement.

6. If $35 \div 5 = 7$, write the related multiplication statement.

7. Check: $4\overline{)35}^{\,8\text{ r}3}$

8. Check: $7\overline{)46}^{\,6\text{ r}4}$

Do the indicated division, and check your results.

9. $35 \div 7$	**10.** $6\overline{)36}$	**11.** $5\overline{)40}$
12. $54 \div 9$	**13.** $21 \div 3$	**14.** $6\overline{)42}$
15. $7\overline{)63}$	**16.** $4\overline{)32}$	**17.** $56 \div 8$

1.	Divisor, dividend, quotient
2.	Quotient, divisor, dividend
3.	4
4.	5
5.	$30 = 6 \times 5$
6.	$35 = 5 \times 7$
7.	$4 \times 8 + 3 = 35$
8.	$7 \times 6 + 4 = 46$
9.	5
10.	6
11.	8
12.	6
13.	7
14.	7
15.	9
16.	8
17.	7

18. 7

19. 8

20. 7

21. 5 r1

22. 5 r3

23. 3 r3

24. 5 r5

25. 5 r4

26. 6 r2

27. 8 r3

28. 4 r4

29. 7 r2

30. 8 r3

31. 7 r1

32. 9 r2

33. 4 classrooms

34. $9

35. 7 boxes

36. 8 cars

37. 9 bars

38. 6 printers

18. $63 \div 9$ **19.** $7\overline{)56}$ **20.** $49 \div 7$

21. $16 \div 3$ **22.** $5\overline{)28}$ **23.** $27 \div 8$

24. $40 \div 7$ **25.** $7\overline{)39}$ **26.** $8\overline{)50}$

27. $5\overline{)43}$ **28.** $40 \div 9$ **29.** $9\overline{)65}$

30. $6\overline{)51}$ **31.** $57 \div 8$ **32.** $74 \div 8$

Solve the following applications.

33. Education. There are 60 pupils in a school. There are 15 students in each classroom. How many classrooms are there?

34. Cost. A customer pays $81 for 9 calculators. How much does each calculator cost?

35. Counting. Ramon bought 56 bags of candy. There were 8 bags in each box. How many boxes were there?

36. Capacity. There are 32 students who are taking a field trip. If each car can hold 4 students, how many cars will be needed for the field trip?

37. Packaging. There are 63 candy bars in 7 boxes. How many candy bars are in each box?

38. Business. A total of 54 printers were shipped to 9 stores. How many printers were shipped to each store?

39. You are going to recarpet your living room. You have budgeted $1000 for the carpet and installation.

    (a) Determine how much carpet you will need to do the job. Draw a sketch to support your measurements.

    (b) What is the highest price per square yard you can pay and still stay within budget?

    (c) Go to a local store and determine the total cost of doing the job for three different grades of carpet. Be sure to include padding, labor costs, and any other expenses.

    (d) What considerations (other than cost) would affect your decision as to what type of carpet you would install?

    (e) Write a brief paragraph indicating your final decision, and give supporting reasons.

40. Division is the inverse operation of multiplication. Many daily activities have inverses. For each of the following activities, state the inverse activity:

    (a) Spending money   **(b)** Going to sleep
    (c) Turning down the volume on your CD player   **(d)** Getting dressed

39.

40.

a. 8

b. 9

c. 0

d. 0

e. 10

f. 0

● Getting Ready for Section 3.2 [Section 2.1]

Find each of the following products.

a. 8×1 **b.** 1×9

c. 0×7 **d.** 5×0

e. 10×1 **f.** 12×0

Answers

1. Divisor, dividend, quotient   **3.** 4   **5.** $30 = 6 \times 5$
7. $4 \times 8 + 3 = 32 + 3 = 35$ **9.** 5   **11.** 8   **13.** 7   **15.** 9   **17.** 7
19. 8   **21.** 5 r1   **23.** 3 r3   **25.** 5 r4   **27.** 8 r3   **29.** 7 r2
31. 7 r1   **33.** 4 classrooms   **35.** 7 boxes   **37.** 9 bars   **39.**
a. 8   **b.** 9   **c.** 0   **d.** 0   **e.** 10   **f.** 0

Name _____

Section _____ Date _____

1. Divisor, dividend, quotient

2. 7

3. $42 = 7 \times 6$

4. $8 \times 4 + 7 = 39$

5. 5

6. 6

7. 7

8. 9

9. 8

10. 9

11. 7 r4

12. 6 r3

13. 6 r1

14. 4 r1

15. 8 r3

16. 9 r2

17. $5

18. 4 mi

1. If $36 \div 9 = 4$, then 9 is the _____, 36 is the _____, and 4 is the _____.

2. Find $35 \div 5$ by repeated subtraction.

3. If $42 \div 7 = 6$, write the related multiplication statement.

4. Check: $8\overline{)39}$ with quotient $4\,r7$

Do the indicated division, and check your results.

5. $45 \div 9$

6. $8\overline{)48}$

7. $4\overline{)28}$

8. $72 \div 8$

9. $7\overline{)56}$

10. $54 \div 6$

11. $5\overline{)39}$

12. $51 \div 8$

13. $43 \div 7$

14. $37 \div 9$

15. $75 \div 9$

16. $7\overline{)65}$

Solve the following applications.

17. **Business.** The bill for 9 light bulbs is $45. Find the cost of one lightbulb.

18. **Exercise.** Joan walks a total of 28 miles (mi) in a week. If she walks every day, how many miles does she walk daily?

3.2 Division with Zero and One

3.2 OBJECTIVES

1. Perform division by one.
2. Understand division by zero.
3. Divide into zero.

The relationship between division and multiplication allows us to illustrate some special division results. First let's look at the role that 1, the multiplicative identity, plays in division.

Division and the Multiplicative Identity

1. Any whole number (except 0) divided by itself is 1.
2. Any whole number divided by 1 is just that number.

Since the related multiplication statement is

$a = a \times 1$

Since the related multiplication statement is

$a = 1 \times a$

• Example 1

Dividing when the Multiplicative Identity Is Involved

(*a*) $8 \div 8 = 1$ because $8 \times 1 = 8$
(*b*) $6 \div 1 = 6$ because $6 = 1 \times 6$.

● ● ● **CHECK YOURSELF 1**

a. $5 \div 5 =$ **b.** $7 \div 1 =$

We must be careful when 0 is involved in a division problem. Again there are two special cases.

Division and Zero

1. 0 divided by any whole number (except 0) is 0.
2. Division by 0 is undefined.

The first case involving zero occurs when we are dividing into zero.

• Example 2

Dividing into Zero

Since $0 = a \times 0$

$0 \div 5 = 0$ because $0 = 5 \times 0$.

155

● ● ● **CHECK YOURSELF 2**

a. $0 \div 7 =$ **b.** $0 \div 12 =$

Our second case illustrates what happens when 0 is the *divisor*. Here we have a special problem.

● Example 3

Dividing by Zero

$8 \div 0 = ?$ This means that $8 = 0 \times ?$

Can 0 times some number ever be 8? From our multiplication facts, the answer is *no!* There is no answer to this problem, so we say that $8 \div 0$ is undefined.

● ● ● **CHECK YOURSELF 3**

Decide whether each problem results in 0 or is undefined.

a. $9 \div 0$ **b.** $0 \div 9$ **c.** $0 \div 15$ **d.** $15 \div 0$

Let's look at a special case of division involving zero.

$0 \div 0 = ?$

Suppose that we write

In this case, both $0 = 0 \times 5$ and $0 = 0 \times 9$ are *true* statements. For this reason, we say that $0 \div 0$ is **indeterminate.**

$0 \div 0 = 5$ Because $0 = 0 \times 5$

or

$0 \div 0 = 9$ Because $0 = 0 \times 9$

Do you see the problem? We now have two answers to the same question. In fact, any number would work as a quotient. There is *no* specific answer.

● ● ● **CHECK YOURSELF ANSWERS**

1. (a) 1; **(b)** 7.   **2. (a)** 0; **(b)** 0.   **3. (a)** Undefined; **(b)** 0; **(c)** 0;
(d) undefined.

Name

Section   Date

Divide if possible.

1. $5\overline{)5}$

2. $9 \div 1$

3. $0 \div 5$

4. $7\overline{)7}$

5. $1\overline{)1}$

6. $1\overline{)10}$

7. $1\overline{)6}$

8. $5 \div 0$

9. $9 \div 9$

10. $1\overline{)8}$

11. $4 \div 0$

12. $12 \div 0$

13. $8\overline{)8}$

14. $0 \div 7$

15. $8 \div 1$

16. $3 \div 0$

17. $0 \div 6$

18. $6 \div 6$

19. $10 \div 1$

20. $0 \div 8$

21. If you have no money in your pocket and want to divide it equally among your four friends, how much does each person get? Use this situation to explain division of zero by a nonzero number.

22. Explain the difference between division by zero and division of zero by a natural number.

1. 1

2. 9

3. 0

4. 1

5. 1

6. 10

7. 6

8. Undefined

9. 1

10. 8

11. Undefined

12. Undefined

13. 1

14. 0

15. 8

16. Undefined

17. 0

18. 1

19. 10

20. 0

21.

22.

Getting Ready for Section 3.3
[Section 1.7]

Subtract.

a. $\begin{array}{r} 45 \\ -40 \\ \hline \end{array}$

b. $\begin{array}{r} 53 \\ -48 \\ \hline \end{array}$

c. $\begin{array}{r} 73 \\ -64 \\ \hline \end{array}$

d. $\begin{array}{r} 61 \\ -56 \\ \hline \end{array}$

e. $\begin{array}{r} 41 \\ -35 \\ \hline \end{array}$

f. $\begin{array}{r} 62 \\ -54 \\ \hline \end{array}$

Answers

1. 1   **3.** 0   **5.** 1   **7.** 6   **9.** 1   **11.** Undefined   **13.** 1   **15.** 8
17. 0   **19.** 10   **21.**   **a.** 5   **b.** 5   **c.** 9   **d.** 5   **e.** 6
f. 8

3.2 **Supplementary Exercises**

Name _____

Section _____ Date _____

Divide if possible.

1. $1\overline{)5}$

2. $8 \div 1$

3. $7 \div 0$

4. $6\overline{)6}$

5. $0 \div 4$

6. $9 \div 0$

7. $8 \div 8$

8. $7 \div 7$

9. $7 \div 1$

10. $0 \div 5$

Division by Single-Digit Divisors

3.3 OBJECTIVES

1. Divide by single-digit numbers.
2. Solve applications of single-digit division.

With larger numbers, repeated subtraction is just too time-consuming to be practical.

It is easy to divide when small whole numbers are involved, since much of the work can be done mentally. In working with larger numbers, we turn to a process called **long division.** This is a shorthand method for performing the steps of repeated subtraction.

To start, let's look at an example in which we subtract multiples of the divisor.

• Example 1

Dividing by a Single-Digit Number

Divide 176 by 8.

Solution   Since 20 eights are 160, we know that there are at least 20 eights in 176.

Step 1   Write

$$
\begin{array}{r}
20 \\
8)\overline{176} \\
160 \\
\hline
16
\end{array}
$$

20 eights $\longrightarrow$ 160

Subtracting 160 is just a shortcut for subtracting eight 20 times.

After subtracting the 20 eights, or 160, we are left with 16. There are 2 eights in 16, and so we continue.

Step 2

$$
\left.\begin{array}{r}
2 \\
20
\end{array}\right\} 22
$$

$$
\begin{array}{r}
8)\overline{176} \\
160 \\
\hline
16 \\
16 \\
\hline
0
\end{array}
$$

2 eights $\longrightarrow$ 16

Adding 20 and 2 gives us the quotient, 22.

Subtracting the 2 eights, we have a 0 remainder. So

$$176 \div 8 = 22$$

● ● ●   **CHECK YOURSELF 1**

Verify the results of Example 1, using multiplication.

The next step is to simplify this repeated-subtraction process one step further. The result will be the long-division method.

• Example 2

Dividing by a Single-Digit Number

Divide 358 by 6.

Solution   The dividend is 358. We look at the first digit, 3. We cannot divide 6 into 3, and so we look at the *first two digits,* 35. There are 5 sixes in 35, and so we write 5 above the tens digit of the dividend.

$$
\begin{array}{r}
5 \\
6\overline{)358}
\end{array}
$$
When we place 5 as the tens digit, we really mean 5 tens, or 50.

Now multiply 5×6, place the product below 35, and subtract.

$$
\begin{array}{r}
5 \\
6\overline{)358} \\
\underline{30} \\
5
\end{array}
$$
We have actually subtracted 50 sixes (300) from 358.

To continue the division, we bring down 8, the ones digit of the dividend.

$$
\begin{array}{r}
5 \\
6\overline{)358} \\
\underline{30} \\
58
\end{array}
$$

Now divide 6 into 58. There are 9 sixes in 58, and so 9 is the ones digit of the quotient. Multiply 9×6 and subtract to complete the process.

Because the 4 is smaller than the divisor, we have a remainder of 4.

$$
\begin{array}{r}
59 \\
6\overline{)358} \\
\underline{30}\downarrow \\
58 \\
\underline{54} \\
4
\end{array}
$$
We now have:
$358 \div 6 = 59$ r4

Verify that this is true and that the division checks.

To check: $358 = 6 \times 59 + 4$

● ● ● CHECK YOURSELF 2

Divide $7\overline{)453}$.

Here is an example in which a special problem comes up during the division.

• Example 3

Dividing by a Single-Digit Divisor

Divide 3278 by 8.

Solution We start by dividing 8 into the first two digits of the dividend. Multiply 4×8 and subtract the product.

$$
\begin{array}{r}
4 \\
8\overline{)3278} \\
\underline{32} \\
0
\end{array}
$$

We continue by bringing down 7, the tens digit of the dividend. Do you see the problem? We cannot divide 8 into 7, so we place a 0 in the quotient.

$$
\begin{array}{r}
40 \\
8\overline{)3278} \\
\underline{32}\downarrow \\
7
\end{array}
$$

To continue, we bring down 8, the last digit of the dividend, and complete the process

$$
\begin{array}{r}
409 \\
8\overline{)3278} \\
\underline{32}\downarrow \\
78 \\
\underline{72} \\
6
\end{array}
$$

The quotient is 409 r6.

Check: $3278 = 8 \times 409 + 6$

● ● ● **CHECK YOURSELF 3**

Divide 5641 by 7.

Short division is a way of simplifying the long-division process. You can use it whenever you have a single-digit divisor. Look at the following example comparing the two approaches.

● Example 4

Dividing by a Single-Digit Number

Long Division   **Short Division**

$$
\begin{array}{r}
132 \\
3\overline{)396} \\
\underline{3} \\
9 \\
\underline{9} \\
6 \\
\underline{6} \\
0
\end{array}
\qquad
\begin{array}{r}
132 \\
3\overline{)396}
\end{array}
$$

Step 1 3 divides into 3 one time. Write 1 in the quotient as shown.

Step 2 3 divides into 9 three times. Write 3 in the quotient.

Step 3 3 divides into 6 two times. Write 2 as the final digit of the quotient.

● ● ● **CHECK YOURSELF 4**

Divide 2)$\overline{486}$.

When you use short division, just carry out the steps mentally without writing down all the steps of the long-division process.

● Example 5

Using Short Division to Divide by a Single-Digit Number

(*a*) Divide 375 by 5

Step 1 $5)\overline{375}$ with quotient 7

5 will not divide into 3, so look at the first two digits. 37 divided by 5 is 7 with a remainder of 2. Write 7 in the tens place of the quotient.

Step 2 $5)\overline{37^{2}5}$ with quotient 75

Insert the remainder, 2, from the last step before the ones digit of the dividend. Then divide 5 into 25 for the last digit of the quotient.

(*b*) Divide 327 by 6.

Step 1 $6)\overline{32^{2}7}$ with quotient 5

To start the short-division process, think, "32 divided by 6 is 5 with remainder 2." Insert 2 before the ones digit of the dividend.

Step 2 $6)\overline{32^{2}7}$ with quotient 54 r3

Now divide 6 into 27. The last digit of the quotient is 4, and we have a remainder of 3.

● ● ● **CHECK YOURSELF 5**

Divide.

a. 4)$\overline{256}$ **b.** 8)$\overline{425}$ **c.** 5)$\overline{873}$

● ● ● **CHECK YOURSELF ANSWERS**

1. $8 \times 22 = 176$.

2.
$$
\begin{array}{r}
64 \\
7)\overline{453} \\
\underline{42} \\
33 \\
\underline{28} \\
5
\end{array}
$$
We have:
$453 \div 7 = 64$ r5.

3.
$$
\begin{array}{r}
805 \\
7)\overline{5641} \\
\underline{56} \\
41 \\
\underline{35} \\
6
\end{array}
$$
We have:
$5641 \div 7 = 805$ r6.

4. 243.

5. (a) 64; **(b)** 53 r1; **(c)** 174 r3.

Name

Section   Date

Divide using long division, and check your work.

1. 5)83

2. 9)78

3. 3)162

4. 4)232

5. 8)293

6. 7)346

7. 8)3136

8. 5)4938

9. 8)5438

10. 9)3527

11. 8)22,153

12. 5)43,287

13. 7)82,013

14. 6)73,108

Divide, using short division.

15. 4)85

16. 5)78

17. 3)88

18. 6)79

19. 4)848

20. 2)486

ANSWERS

1. 16 r3
2. 8 r6
3. 54
4. 58
5. 36 r5
6. 49 r3
7. 392
8. 987 r3
9. 679 r6
10. 391 r8
11. 2769 r1
12. 8657 r2
13. 11,716 r1
14. 12,184 r4
15. 21 r1
16. 15 r3
17. 29 r1
18. 13 r1
19. 212
20. 243

21.	48 r6
22.	318
23.	104 r1
24.	131 r1
25.	472 r4
26.	488 r5
27.	1087 r3
28.	406 r3
29.	1830 r1
30.	1274 r5
31.	4473 r2
32.	8573 r1
33.	4589 r5
34.	3128 r3
35.	2
36.	6
37.	451
38.	58

21. $7\overline{)342}$ **22.** $3\overline{)954}$ **23.** $6\overline{)625}$

24. $7\overline{)918}$ **25.** $5\overline{)2364}$ **26.** $7\overline{)3421}$

27. $4\overline{)4351}$ **28.** $8\overline{)3251}$ **29.** $4\overline{)7321}$

30. $7\overline{)8923}$ **31.** $3\overline{)13,421}$ **32.** $4\overline{)34,293}$

33. $7\overline{)32,128}$ **34.** $9\overline{)28,155}$

35. Find the remainder when 2708 is divided by 6.

36. Find the remainder when 412 is divided by 7.

37. Find the quotient when 2708 is divided by 6.

38. Find the quotient when 412 is divided by 7.

Solve the following applications.

39. **Recreation.** Joaquin is putting pictures in an album. He can fit 8 pictures on each page. If he has 77 pictures, how many will be left over after he has filled the last 8-picture page?

40. **Counting.** Kathy is separating a deck of 52 cards into 6 equal piles. How many cards will be left over?

41. **Recreation.** Ticket receipts for a play were $552. If the tickets were $4 each, how many tickets were purchased?

42. **Construction.** Construction of a fence section requires 8 boards. If you have 256 boards available, how many sections can you build?

43. Division is not associative. For example, $8 \div 4 \div 2$ will produce different results if 8 is divided by 4 and then divided by 2 or if 8 is divided by the result of $4 \div 2$. In the following, place parentheses in the proper place so that the expression is true.

   (a) $16 \div 8 \div 2 = 4$ **(b)** $16 \div 8 \div 2 = 1$

   (c) $125 \div 25 \div 5 = 1$ **(d)** $124 \div 25 \div 5 = 25$

   (e) Is there any situation in which the order of how the operation of division is performed produces the same result? Give an example.

⬤ **Getting Ready for Section 3.4**
[Section 1.5]

Round each number to the indicated place.

a. 87 to the nearest ten   **b.** 134 to the nearest ten

c. 758 to the nearest hundred   **d.** 1225 to the nearest hundred

Answers

1. 16 r3   **3.** 54   **5.** 36 r5   **7.** 392   **9.** 679 r6   **11.** 2769 r1
13. 11,716 r1   **15.** 21 r1   **17.** 29 r1   **19.** 212   **21.** 48 r6   **23.** 104 r1
25. 472 r4   **27.** 1087 r3   **29.** 1830 r1   **31.** 4473 r2   **33.** 4589 r5
35. 2   **37.** 451   **39.** 5 pictures   **41.** 138 tickets   **43.**
a. 90   **b.** 130   **c.** 800   **d.** 1200

Name

Section   Date

ANSWERS

1.	23 r1
2.	218
3.	591 r8
4.	1225 r4
5.	506 r3
6.	2926 r2
7.	14
8.	132
9.	109 r2
10.	121 r3
11.	507 r3
12.	1137 r1
13.	1089 r1
14.	5037 r4
15.	4138 r1
16.	228 mi
17.	$833
18.	32 boxes; 4 left over
19.	15 cookies; 4 for Lou

Divide using long division, and check your work.

1. $4\overline{)93}$ **2.** $3\overline{)654}$ **3.** $9\overline{)5327}$

4. $6\overline{)7354}$ **5.** $7\overline{)3545}$ **6.** $8\overline{)23,410}$

Divide, using short division.

7. $3\overline{)42}$ **8.** $3\overline{)396}$ **9.** $4\overline{)438}$

10. $8\overline{)971}$ **11.** $8\overline{)4059}$ **12.** $3\overline{)3412}$

13. $4\overline{)4357}$ **14.** $7\overline{)35,263}$ **15.** $9\overline{)37,243}$

Solve the following applications.

16. Leisure. Elsie plans to travel 1824 miles (mi) on her vacation. If she plans to be away for 8 days, how many miles will she travel each day?

17. Financial grants. A college awards $5831 in grants to 7 faculty members. How much does each faculty member receive?

18. Baking. Sarah has baked 196 cupcakes. She wants to put them into boxes that contain 6 cupcakes each. How many boxes will Sarah need? How many cupcakes will be left over?

19. Distribution. Lou had 139 cookies to distribute equally among 9 boys. How many cookies did each boy receive? How many cookies were left over for Lou?

3.4 Division by Multiple-Digit Divisors

© 1998 McGraw-Hill Companies

3.4 OBJECTIVES

1. Divide with two-digit divisors.
2. Divide with three-digit divisors.

Long division becomes a bit more complicated when we have a two-digit divisor. It is now a matter of trial and error. We round the divisor and dividend to form a *trial divisor and a trial dividend*. We then estimate the proper quotient and must determine whether our estimate was correct.

• Example 1

Dividing by a Two-Digit Number

Divide

$$38\overline{)293}$$

Round the divisor and dividend to the nearest ten. So 38 is rounded to 40, and 293 is rounded to 290. The trial divisor is then 40, and the trial dividend is 290.

Now look at the nonzero digits in the trial divisor and dividend. They are 4 and 29. We know that there are 7 fours in 29, and so 7 is our first estimate of the quotient. Now let's see if 7 works.

Think: $4\overline{)29}^{\,7}$

```
        7  ←—— Your estimate
38)293
    266       Multiply 7 × 38. The product, 266, is less
     27       than 293, and so we can subtract.
```

The remainder, 27, is less than the divisor, 38, and so the process is complete.

$$293 \div 38 = 7 \text{ r}27$$

Check: $293 = 38 \times 7 + 27$ You should verify that this statement is true.

● ● ● CHECK YOURSELF 1

Divide.

$$57\overline{)482}$$

Since this process is based on estimation, we can't expect our first guess to always be right.

• Example 2

Dividing by a Two-Digit Number

Divide

$$46\overline{)342}$$

Round to the nearest ten to form the trial divisor, 50, and the trial dividend, 340.

Using the nonzero digits of our trial divisor and dividend, we think, "How many fives are in 34?" There are 6, and this is our first estimate.

$$\begin{array}{r} 6 \\ 46\overline{)342} \end{array}$$

Now multiply 6×46 and subtract.

$$\begin{array}{r} 6 \\ 46\overline{)342} \\ \underline{276} \\ 66 \end{array}$$ Do you see the problem? The remainder, 66, is larger than the divisor, 46.

Since the remainder was larger than the divisor after our first guess, we must use a *larger quotient*. We will now try 7 as our quotient.

$$\begin{array}{r} 7 \\ 46\overline{)342} \\ \underline{322} \\ 20 \end{array}$$ We can complete the problem.

$342 \div 46 = 7 \text{ r}20$

Check: $342 = 46 \times 7 + 20$

● ● ● **CHECK YOURSELF 2**

Divide.

$$57\overline{)463}$$

Let's look at a second example in which our estimate of the quotient must be changed.

• Example 3

Dividing by a Two-Digit Number

Divide

$$54\overline{)428}$$

Think: $5\overline{)43}$ with 8 above

Rounding to the nearest ten, we have a trial divisor of 50 and a trial dividend of 430.

Looking at the nonzero digits, how many fives are in 43? There are 8. This is our first estimate.

$$\begin{array}{r} 8 \\ 54\overline{)428} \\ \underline{432} \longleftarrow \text{Too large} \end{array}$$

We multiply 8×54. Do you see what's wrong? The product, 432, is too large. We can't subtract. Our estimate of the quotient must be adjusted *downward*.

We adjust the quotient downward to 7. We can now complete the division.

$$\begin{array}{r} 7 \\ 54\overline{)428} \\ \underline{378} \\ 50 \end{array}$$

We have

$$428 \div 54 = 7 \text{ r}50$$

Check: $428 = 54 \times 7 + 50$

● ● ● **CHECK YOURSELF 3**

Divide.

$$63\overline{)557}$$

<hr>

● Example 4

Dividing by a Two-Digit Number

Divide.

$$43\overline{)948}$$

There is a difference between this and our previous examples. Look at the first two digits of the dividend. Our divisor, 43, will divide into 94.

Rounding to the nearest ten, we have a trial divisor of 40 and a trial dividend of 90. (We use only the first two digits of the dividend.)

Our first estimate of the quotient is 2.

$$\begin{array}{r} 2 \\ 43\overline{)948} \\ \underline{86} \\ 8 \end{array}$$

We bring down 8, the next digit of the dividend, and repeat the process of estimating the quotient.

$$
\begin{array}{r}
22 \\
43\overline{)948} \\
\underline{86} \\
88 \\
\underline{86} \\
2
\end{array}
$$

$948 \div 43 = 22 \text{ r}2$

Check: $948 = 43 \times 22 + 2$

● ● ● **CHECK YOURSELF 4**

Divide.

$37\overline{)859}$

We saw earlier that we have to be careful when a 0 appears as a digit in the quotient. Let's look at an example in which this happens with a two-digit divisor.

● Example 5

Dividing with Large Dividends

Again our divisor, 32, will divide into 98, the first two digits of the dividend.

Divide

$32\overline{)9871}$

Rounding to the nearest ten, we have a trial divisor of 30 and a trial dividend of 100. Think, "How many threes are in 10?" There are 3, and this is our first estimate of the quotient.

$$
\begin{array}{r}
3 \\
32\overline{)9871} \\
\underline{96} \\
2
\end{array}
$$
Everything seems fine so far!

Bring down 7, the next digit of the quotient.

$$
\begin{array}{r}
30 \\
32\overline{)9871} \\
\underline{96\downarrow} \\
27
\end{array}
$$
Now do you see the difficulty? We cannot divide 32 into 27, and so we place 0 in the tens place of the quotient to indicate this fact.

We continue by bringing down 1, the last digit of the dividend.

$$
\begin{array}{r}
30 \\
32\overline{)9871} \\
\underline{96\downarrow} \\
271
\end{array}
$$

Another problem develops here. We round 32 to 30 for our trial divisor, and we round 271 to 270, which is the trial dividend at this point. Our estimate of the last digit of the quotient must be 9.

$$
\begin{array}{r}
309 \\
32\overline{)9871} \\
\underline{96} \\
271 \\
\underline{288} \leftarrow \text{Too large}
\end{array}
$$

We can't subtract. The trial quotient must be adjusted downward to 8. We can now complete the division.

$$
\begin{array}{r}
308 \\
32\overline{)9871} \\
\underline{96} \\
271 \\
\underline{256} \\
15
\end{array}
$$

$9871 \div 32 = 308 \text{ r}15$

Check: $9871 = 32 \times 308 + 15$

● ● ●   **CHECK YOURSELF 5**

Divide.

$43\overline{)8857}$

We can also simplify long-division problems involving three-digit divisors by using trial divisors and dividends. In this case we round those divisors and dividends to the nearest hundred.

● Example 6

Dividing by a Three-Digit Number

Divide

$205\overline{)6585}$

Round 205 to the nearest hundred for a trial divisor of 200. To find the trial dividend, round 658 to the nearest hundred, 700.

Using the nonzero digits, think "How many twos are in 7?" There are 3, and this is our estimate.

Think: $2\overline{)7}^{\,3}$

$$
\begin{array}{r}
3 \\
205\overline{)6585} \\
\underline{615} \\
43
\end{array}
$$
We multiply and subtract. Our estimate, 3, was correct.

Continue the process by bringing down 5, the last digit of the dividend.

$$
\begin{array}{r}
3 \\
205\overline{)6585} \\
\underline{615} \\
435
\end{array}
$$

The trial divisor is still 200, but the trial dividend is now 400. Our estimate for the last digit of the quotient is 2.

$$
\begin{array}{r}
32 \\
205\overline{)6585} \\
\underline{615} \\
435 \\
\underline{410} \\
25
\end{array}
$$
We multiply and subtract to complete the process.

$6585 \div 205 = 32$ r25

Check: $6585 = 205 \times 32 + 25$

● ● ● **CHECK YOURSELF 6**

Divide.

$321\overline{)7582}$

● ● ● **CHECK YOURSELF ANSWERS**

1. 8 r26.

2. The trial divisor is 60; the trial dividend is 460.

Let's try 7 as a quotient.   Try a larger quotient, 8.

$$
\begin{array}{r}
7 \\
57\overline{)463} \\
\underline{399} \\
64
\end{array}
\longleftarrow \text{Too large}
\qquad
\begin{array}{r}
8 \\
57\overline{)463} \\
\underline{456} \\
7
\end{array}
$$

3. 8 r53.   **4.** $859 \div 37 = 23$ r8.   **5.** 205 r42.   **6.** 23 r199.

Divide and check your results.

1. 5 r55

2. 21 r2

3. 6 r51

4. 7 r36

5. 18 r28

6. 6 r48

7. 23 r5

8. 14 r1

9. 52 r27

10. 65 r35

11. 257 r10

12. 345 r20

13. 189 r14

14. 164 r37

1. $58\overline{)345}$

2. $39\overline{)821}$

3. $63\overline{)429}$

4. $49\overline{)379}$

5. $48\overline{)892}$

6. $54\overline{)372}$

7. $23\overline{)534}$

8. $67\overline{)939}$

9. $45\overline{)2367}$

10. $53\overline{)3480}$

11. $34\overline{)8748}$

12. $27\overline{)9335}$

13. $42\overline{)7952}$

14. $53\overline{)8729}$

15. $28\overline{)8547}$ 16. $38\overline{)7892}$

17. $763\overline{)3871}$ 18. $871\overline{)4321}$

19. $326\overline{)7564}$ 20. $229\overline{)8312}$

21. $432\overline{)8770}$ 22. $375\overline{)7610}$

23. $454\overline{)32,751}$ 24. $527\overline{)27,563}$

25. $103\overline{)21,185}$ 26. $205\overline{)61,825}$

27. $234\overline{)125,000}$ 28. $179\overline{)126,500}$

Solve the following applications.

29. **Sports.** A basketball player scored 476 points in 17 games. How many points did she score per game?

30. **Travel.** You drive 559 miles (mi) in your new Geo, using 13 gallons (gal) of gas. What is your gas mileage (miles per gallon)?

31. **Education.** There were 522 students enrolled in 18 equal-size sections of English composition. How many students were enrolled in each class?

32. **Luncheon cost.** The bill for a luncheon was $560. If 35 people attended the luncheon, what was the cost per person?

33. **Business.** The homeowners along a street must share the $2030 cost of new street lighting. If there are 14 homes, what amount will each owner pay?

34. **Cost.** A bookstore ordered 325 copies of a textbook at a cost of $7800. What was the cost to the store for an individual textbook?

35. **Telephone calls.** The records of an office show that 1702 calls were made in 1 day. If there are 37 phones in the office, how many calls were placed per phone?

36. **Television costs.** A television dealer purchased 23 sets, each the same model, for $5267. What was the cost of each set?

37. **Computers.** A computer printer can print 340 lines per minute (min). How long will it take to complete a report of 10,880 lines?

38. **Distance.** A train traveled 1364 mi in 22 h. What was the speed of the train? *Hint:* Speed is the distance traveled divided by the time.

39. **Bonuses.** A company distributes $16,488 in year-end bonuses. If each of the 36 employees receives the same amount, what bonus will each receive?

40. Travel. A total of 17,949 cars use a toll bridge in 1 month (31 days). How many cars use the bridge in a single day?

41. Payments. You purchase a new car for $9852, make a down payment of $1500, and agree to pay off the balance in equal monthly payments for 36 months. What is the amount of each monthly payment?

42. Interest. Brad borrows $1000 and is charged $92 interest. If he arranges to pay off the loan and the interest charge in monthly payments over 1 year, what will his monthly payments be?

43. Painting. A gallon of paint should cover 450 square feet (ft^2). How many gallons must be purchased to paint a wall of a warehouse which is 135 ft long and 20 ft high?

44. Investment. Angelo has $14,310 to invest in the stock market. If Nevergood stock is selling at $135 for each share, how many shares of stock can Angelo purchase?

45. Division is not commutative. For example, $15 \div 5 \ne 5 \div 15$. What must be true of the numbers a and b if $a \div b = b \div a$? Besides the fact that $b = 0$?

46. Your class goes to a local amusement park. A ride can carry 15 passengers in each cycle.

(a) If a new cycle starts every 5 min, how many cycles does the ride make every hour?

(b) How many passengers can ride every hour?

(c) How long would it take all the students in your class to complete the ride?

Getting Ready for Section 3.5
[Section 3.1]

Divide.

a. $5\overline{)35}$ **b.** $7\overline{)63}$ **c.** $8\overline{)65}$

d. $9\overline{)84}$ **e.** $6\overline{)40}$ **f.** $4\overline{)39}$

Answers

1. 5 r55   **3.** 6 r51   **5.** 18 r28   **7.** 23 r5

9. $45\overline{)2367}$ ^(52 r27) Check: 2367 = 45 × 52 + 27   **11.** 257 r10   **13.** 189 r14

15. 305 r7   **17.** 5 r56   **19.** $326\overline{)7564}$ ^(23 r66) Check: 7564 = 326 × 23 + 66

21. 20 r130   **23.** 72 r63   **25.** $103\overline{)21,185}$ ^(205 r70) Check: 21,185 = 103 × 205 + 70

27. 534 r44   **29.** 28 points   **31.** 29 students   **33.** $145   **35.** 46 calls

37. 32 min   **39.** $458   **41.** $232   **43.** 6 gal   **45.**

a. 7   **b.** 9   **c.** 8 r1   **d.** 9 r3   **e.** 6 r4   **f.** 9 r3

A N S W E R S

a. 7

b. 9

c. 8 r1

d. 9 r3

e. 6 r4

f. 9 r3

Name

Section   Date

A N S W E R S

Divide, and check your results.

1. $47\overline{)392}$ 2. $54\overline{)429}$ 3. $36\overline{)837}$

4. $28\overline{)692}$ 5. $63\overline{)3542}$ 6. $45\overline{)2791}$

7. $52\overline{)7621}$ 8. $31\overline{)9427}$ 9. $563\overline{)2935}$

10. $218\overline{)7473}$ 11. $329\overline{)8527}$ 12. $434\overline{)35,202}$

13. $228\overline{)69,750}$ 14. $325\overline{)275,000}$ 15. $357\overline{)147,100}$

Solve the following applications.

16. **Cost.** A bookstore ordered 225 copies of a textbook at a cost of $4050. What was the cost per book?

17. **Production.** A machine can produce 137 items per hour. How long will it take to produce 2192 items?

18. **Donations.** A charity collected $5888 in equal donations from 256 people. How much did each person donate?

19. **Monthly payment.** Juwau buys a color television set for $595 and agrees to pay for the set with monthly payments for 1 year. If the interest charge is $89, find the amount of each monthly payment.

20. **Recreation.** The receipts from the performance of a show were $15,300. If 255 people were in attendance, how much did each ticket cost?

Estimating the Results of Division

3.5 OBJECTIVES

1. Estimate the quotient of a long-division problem.
2. Solve applications for long division.

In Section 3.4, we saw that long division can be extended to include any divisor. Because of the availability of the hand-held calculator, it is rarely necessary that people find the exact answer. On the other hand, it is frequently important that one be able to either estimate the result of long division, or confirm that a given answer (particularly from a calculator) is reasonable. As a result, the emphasis in this section will be to improve your estimation skills in division.

Let's divide a three-digit number by a two-digit number. Generally, we will round the divisor to the nearest ten and the dividend to the nearest hundred.

• Example 1

Estimating a Result of Division

Estimate the result when 486 is divided by 97.

The divisor is rounded to the nearest 10 and the dividend is rounded to the nearest 100.

We will estimate $97)\overline{486}$. We will round the 97 up to 100. Having increased the divisor, we will almost always get a better result if we also increase the dividend. We will round the 486 up to 500.

$$\begin{array}{r} 5 \\ 100)\overline{500} \end{array}$$

Our estimate for the quotient is 5.

CHECK YOURSELF 1

Estimate the result when 379 is divided by 95.

Even if we don't get an exact answer when estimating a quotient, we don't report a remainder from estimation.

• Example 2

Estimating a Result of Division

Estimate the result when 1568 is divided by 58.

Here, we use the short-division method to estimate the quotient.

We'll round 58 up to 60 and 1568 up to 1600.

$$\begin{array}{r} 26 \\ 60)\overline{1600} \end{array}$$

Our estimate for the quotient is 26.

CHECK YOURSELF 2

Estimate the result when 1753 is divided by 77.

• Example 3

Estimating the Result of a Division Application

The Ramirez family took a trip of 2394 miles (mi) in their new car, using 68 gallons (gal) of gas. Estimate their gas mileage (mi/gal).

Our estimate will be based on dividing 2400 by 70.

$$
\begin{array}{r}
34 \\
70\overline{)2400}
\end{array}
$$

They got approximately 34 mi/gal.

CHECK YOURSELF 3

Troy flew a light plane on a trip of 2844 mi that took 21 hours (h). What was the average speed in miles per hour (mi/h)?

As before, we may have to combine operations to solve an application of the mathematics you have learned.

• Example 4

Estimating the Result of a Division Application

Charles purchases a new car for $8574. Interest charges will be $978. He agrees to make payments for four years. Approximately what should his payments be?

First, we find the amount that Charles owes:

$8574 + $978 = $9552

Now, to find the monthly payment, we divide that amount by 48 (months). To estimate the payment, we'll divide $9600 by 50.

$$
\begin{array}{r}
192 \\
50\overline{)9600}
\end{array}
$$

The payments will be approximately $192 per month.

CHECK YOURSELF 4

One bag of fertilizer will cover 310 square feet (ft²). How many bags should be purchased to cover an area that is 70 ft by 20 ft?

CHECK YOURSELF ANSWERS

1. 4.   **2.** 22 (or 23).   **3.** 140 mi/h.   **4.** 5 bags.

Estimate the result in the following division problems.

1. 478 divided by 53

2. 963 divided by 37

3. 567 divided by 23

4. 327 divided by 46

5. 890 divided by 38

6. 458 divided by 18

7. 4967 divided by 64

8. 3871 divided by 39

9. 8971 divided by 91

10. 3921 divided by 58

11. 5789 divided by 126

12. 8236 divided by 178

13. 3812 divided by 269

14. 5245 divided by 215

A N S W E R S

1. 10

2. 25

3. 30

4. 6

5. 23

6. 25

7. 83

8. 100

9. 100

10. 67

11. 60

12. 40

13. 13

14. 25

15. 15 mi/gal

16. 10 houses

17. $2700

18. 20 boxes

19. 85

20. 15 shirts

a. 12

b. 10

c. 20

d. 30

e. 23

f. 2

Solve the following applications.

15. Gas mileage. Ross drove 279 miles (mi) on 18 gallons of gas. Estimate his mileage. (*Hint:* Find the number of miles per gallon.)

16. Construction. A contractor can build a house in 27 days. Estimate how many houses can be built in 265 days.

17. Inheritances. Twelve people are to share equally in an estate totaling $26,875. Estimate how much money each person will receive.

18. Business. There is $365 left in the budget to purchase pens. If each box of pens costs $18, estimate the number of boxes of pens that can be ordered.

19. Monthly payments. Tara purchased a used car for $1850 by paying $275 down and the rest in equal monthly payments over a period of 18 months. Estimate the amount of her monthly payments.

20. Consumer purchases. Art has $275 to spend on shirts. If the cost of a shirt is $23, estimate the number of shirts that Art can buy.

 **Getting Ready for Section 3.6
[Sections 2.5 and 2.6]**

Do the indicated operations.

a. $2 \times 3 + 6$ **b.** $4 \times 5 - 10$ **c.** $2 + 3 \times 6$

d. $(2 + 3) \times 6$ **e.** $2 \times 3^2 + 5$ **f.** $20 - 3^2 \times 2$

Answers

1. 10   **3.** 30   **5.** 23   **7.** 83   **9.** 100   **11.** 60   **13.** 13
15. 15 mi/gal   **17.** $2700   **19.** 85   **a.** 12   **b.** 10   **c.** 20   **d.** 30
e. 23   **f.** 2

NumberCross3

ACROSS

1. 48/4
3. 1296/8
6. 2025/5
8. 4 × 5
9. 11 × 11
12. 15/3 × 111
14. 144/(2 × 6)
16. 1404/6
18. 2500/5
19. 3 × 5

DOWN

1. (12 + 16)/2
2. 67 × 3
4. 744/12
5. 2600/13
7. 6300/12
10. 304/2
11. 5 × (161/7)
13. 9027/17
15. 400/20
17. 9 × 5

SOLUTION TO NUMBERCROSS3

Name

Section   Date

A N S W E R S

1.	20
2.	10
3.	20
4.	4
5.	40
6.	20
7.	18
8.	50
9.	140
10.	12
11.	200 shares
12.	$170

Estimate the result in the following division problems.

1. 637 divided by 28   **2.** 788 divided by 83   **3.** 976 divided by 48

4. 213 divided by 52   **5.** 3678 divided by 123   **6.** 6128 divided by 331

7. 8976 divided by 489   **8.** 2356 divided by 43   **9.** 6756 divided by 54

10. 5893 divided by 527

Solve the following applications.

11. Stock. Tim has $7800 with which to buy shares of stock in the Do-Good company. If each share costs $38, estimate the number of shares Tim can purchase.

12. Monthly payments. Samantha has purchased a new washer and dryer combination for $2350 by paying $325 down and agreeing to pay the rest in equal monthly payments over a period of 12 months. Estimate the amount of the monthly payment she will have to make.

3.6 The Order of Operations

3.6 OBJECTIVES

1. Know the order of operations
2. Evaluate expressions that contain division.

Now that we have introduced division, we can write our rule for the order of operations in its final form.

The Order of Operations

Mixed operations in an expression should be done in the following order:

STEP 1 Do any operations inside parentheses.
STEP 2 Evaluate any powers.
STEP 3 Do all multiplication and division in order from left to right.
STEP 4 Do all addition and subtraction in order from left to right.

• Example 1

Using the Order of Operations

(*a*) Evaluate $2 \times 3 + 8 - 4$.

$$2 \times 3 + 8 - 4 \qquad \text{Do the multiplication as the first step.}$$
$$= \quad 6 \; + 8 - 4$$
$$= \qquad 14 \quad - 4 \qquad \text{Then add and subtract, working from left to right.}$$
$$= \qquad 10$$

So $2 \times 3 + 8 - 4 = 10$.

(*b*) Evaluate $36 \div 3^2 - 4$.

$$36 \div 3^2 - 4$$
$$= 36 \div 9 \; - 4 \qquad \text{First, } 3^2 \text{ is 9. Now divide 36 by 9.}$$
$$= \quad 4 \quad - 4 \qquad \text{Subtract.}$$
$$= \quad 0$$

So $36 \div 3^2 - 4 = 0$.

● ● ● **CHECK YOURSELF 1**

Evaluate.

a. $3 \times 5 + 2 - 7$ **b.** $24 \div 2^3 + 7$

If a problem contains both multiplication and division, we work left to right. Operations in parentheses are always evaluated first.

• Example 2

Using the Order of Operations

(*a*) Evaluate $20 \div 2 \times 5$.

$$
\underbrace{20 \div 2}_{} \times 5
$$
$$
= \quad 10 \quad \times 5
$$
$$
= \quad 50
$$

Since the multiplication and division appear next to each other, work in order from left to right. Try it the other way and see what happens!

So $20 \div 2 \times 5 = 50$.

(*b*) Evaluate $(5 + 13) \div 6$.

$$
\underbrace{(5 + 13)}_{} \div 6
$$
$$
= \quad 18 \quad \div 6
$$
$$
= \quad 3
$$

Do the addition in the parentheses as the first step.

So $(5 + 13) \div 6 = 3$.

● ● ● **CHECK YOURSELF 2**

Evaluate.

a. $36 \div 4 \times 2$ **b.** $(8 + 22) \div 5$

● ● ● **CHECK YOURSELF ANSWERS**

1. (a) 10; **(b)** 10.   **2. (a)** 18; **(b)** 6.

A N S W E R S

Do the indicated operations.

1. $8 \div 4 + 2$

2. $3 \times 5 + 2$

3. $24 - 6 \div 3$

4. $3 + 9 \div 3$

5. $(24 - 6) \div 3$

6. $(3 + 9) \div 3$

7. $12 + 3 \div 3$

8. $6 \times 12 \div 3$

9. $18 \div 6 \times 3$

10. $30 \div 5 \times 2$

11. $30 \div 6 - 12 \div 3$

12. $5 + 8 \div 4 - 3$

13. $4^2 \div 2$

14. 2×4^3

15. $5^2 \times 3$

16. $6^2 \div 3$

17. 3×3^3

18. $2^5 \times 3$

19. $(3^3 + 3) \div 10$

20. $(2^4 + 4) \div 5$

21. $15 \div (5 - 3 + 1)$

22. $20 \div (3 + 4 - 2)$

23. $27 \div (2^2 + 5)$

24. $48 \div (2^3 + 4)$

#	Answer
1.	4
2.	17
3.	22
4.	6
5.	6
6.	4
7.	13
8.	24
9.	9
10.	12
11.	1
12.	4
13.	8
14.	128
15.	75
16.	12
17.	81
18.	96
19.	3
20.	4
21.	5
22.	4
23.	3
24.	4

Getting Ready for Section 3.7

Find each of the following.

a. The quotient of 75 and 5

b. The sum of 20 and 30

c. The product of 8 and 9

d. 48 divided by 8

e. The difference of 45 and 30

f. 20 more than 30

Answers

1. 4   **3.** $24 - 6 \div 3 = 24 - 2 = 22$ **5.** 6   **7.** 13

9. $18 \div 6 \times 3 = 3 \times 3 = 9$ **11.** 1   **13.** $4^2 \div 2 = 16 \div 2 = 8$ **15.** 75

17. 81   **19.** 3   **21.** 5   **23.** $27 \div (2^2 + 5) = 27 \div (4 + 5) = 27 \div 9 = 3$

a. 15   **b.** 50   **c.** 72   **d.** 6   **e.** 15   **f.** 50

3.6 Supplementary Exercises

Name

Section   Date

Do the indicated operations.

1. $9 + 6 \div 3$

2. $(9 + 6) \div 3$

3. $36 \div 9 \times 2$

4. $48 \div 6 \times 2$

5. $12 \div 3 - 15 \div 5$

6. $10 - 18 \div 3 + 4$

7. $6^2 \div 3$

8. 3×4^2

9. 5×2^4

10. $64 \div 4^2$

11. $(2^3 + 7) \div 5$

12. $40 \div (3^2 + 1)$

Using Your Calculator to Divide

Of course, division is easily done by using your calculator. However, as we will see, some special things come up when we use a calculator to divide. First let's outline the steps of division as it is done on a calculator.

Divide $35\overline{)2380}$.

1. Enter the dividend.   2380
2. Press the divide key. $\boxed{\div}$
3. Enter the divisor.   35
4. Press the equals key. $\boxed{=}$ The desired quotient is now in your display.

Display $\boxed{68}$

We mentioned some of the difficulties related to division with 0 earlier. Let's experiment on the calculator.

●Example 1

Using Your Calculator to Divide

To find $0 \div 5$, we use this sequence:

$0 \boxed{\div} 5 \boxed{=}$

Display $\boxed{0}$

There is no problem with this. Zero divided by any whole number other than 0 is just 0.

● ● ● **CHECK YOURSELF 1**

What is the result when you use your calculator to perform the following operation?

$0 \div 17$

We've seen that there is difficulty dividing zero by another number, but what happens when we try to divide by zero? More importantly to this section, how does the calculator handle division by zero? Example 2 illustrates this concept.

● Example 2

Using Your Calculator to Divide

To find $5 \div 0$, we use this sequence:

$5 \boxed{\div} 0 \boxed{=}$

Display $\boxed{\text{Error}}$

You may find that you must "clear" your calculator after trying this.

If we try this sequence, the calculator gives us an error! Do you see why? Division by 0 is not allowed. Try this on your calculator to see how this error is indicated.

● ● ● **CHECK YOURSELF 2**

What is the result when you use your calculator to perform the following operation?

$17 \div 0$

Another special problem comes up when a remainder is involved in a division problem.

● Example 3

Using Your Calculator to Divide

In a previous section, we divided 293 by 38 and got 7 with remainder 27.

We will say more about this later. For now, just be aware that the calculator will not give you a remainder in the form we have been using in this chapter.

$293 \boxed{\div} 38 \boxed{=} \boxed{7.7105263}$

Quotient

Remainder

7 is the *whole-number part* of the quotient as before.

0.7105263 is the *decimal form* of the *remainder as a fraction.*

● ● ● **CHECK YOURSELF 3**

What is the result when you use your calculator to perform the following operation?

$458 \div 36$

The calculator can also help you combine division with other operations.

• Example 4

Using Your Calculator to Divide

To find $18 \div 2 + 3$, use this sequence:

18 $\boxed{\div}$ 2 $\boxed{+}$ 3 $\boxed{=}$

Display $\boxed{12}$ Do you see that the calculator has done the division as the first step according to our rules for the order of operations?

● ● ●   **CHECK YOURSELF 4**

Use your calcualtor to compute.

$15 \div 5 + 7$

• Example 5

Using Your Calculator to Divide

Again, the calculator has followed our rules for the order of operations, working from *left to right* to do the division first and then the multiplication.

To find $6 \div 3 \times 2$, use this sequence:

6 $\boxed{\div}$ 3 $\boxed{\times}$ 2 $\boxed{=}$

Display $\boxed{4}$

● ● ●   **CHECK YOURSELF 5**

Use your calculator to compute.

$18 \div 6 \times 5$

● ● ●   **CHECK YOURSELF ANSWERS**

1. 0.   **2.** Error message.   **3.** 12.72222.   **4.** 10.   **5.** 15.

Calculator Exercises

Name

Section   Date

A N S W E R S

Use your calculator to perform the indicated operations.

1. 132

2. 78

3. 458

4. 647

5. 4523

6. 2456

7. 9

8. 16

9. 16

10. 16

11. 100

12. 25

13. 128

14. 10

15. 7

16. 3

1. $5940 \div 45$

2. $2808 \div 36$

3. $36{,}182 \div 79$

4. $36{,}232 \div 56$

5. $583{,}467 \div 129$

6. $464{,}184 \div 189$

7. $6 + 9 \div 3$

8. $18 - 6 \div 3$

9. $24 \div 6 \times 4$

10. $32 \div 8 \times 4$

11. $4368 \div 56 + 726 \div 33$

12. $1176 \div 42 - 1572 \div 524$

13. $3 \times 8^3 \div 12$

14. $5 \times 6^2 \div 18$

15. $(18 + 87) \div 15$

16. $(89 - 14) \div 25$

Answers

1. 132   **3.** 458   **5.** 4523   **7.** 9   **9.** 16   **11.** 100   **13.** 128
15. 7

3.7 Finding the Average of a Group of Numbers

3.7 OBJECTIVES

1. Calculate and interpret the mean.
2. Find and interpret the median.

A very useful concept is the **average** of a group of numbers. An average is a number that is typical of a larger group of numbers. In mathematics we have several different kinds of averages that we can use to represent a larger group of numbers. The first of these is the **mean.**

Finding the Mean

To find the mean for a group of numbers, follow these two steps:

STEP 1   Add all of the numbers in the group.
STEP 2   Divide that sum by the number of items in the group.

• Example 1

Finding the Mean

Find the mean of the group of numbers 12, 19, 15, and 14.

Step 1   Add all of the numbers.

$12 + 19 + 15 + 14 = 60$

Step 2   Divide that sum by the number of items.

$60 \div 4 = 15$ There are 4 items in this group.

The mean of this group of numbers is 15.

● ● ●   **CHECK YOURSELF 1**

Find the mean of the group of numbers 17, 24, 19, and 20.

Let's apply the mean to a word problem.

•Example 2

Finding the Mean

The ticket prices (in dollars) for the nine concerts held at the Civic Arena this school year were

23, 21, 20, 49, 20, 25, 22, 24, 48

What was the mean price for these tickets?

Step 1 Add all the numbers.

$23 + 21 + 20 + 49 + 20 + 25 + 22 + 24 + 48 = 252$

Divide by 9 because there are 9 ticket prices.

Step 2 Divide by 9.

$252 \div 9 = 28$

The mean ticket price was $28.

● ● ● **CHECK YOURSELF 2**

The costs (in dollars) of the six textbooks that Aaron needs for the fall quarter are

65, 59, 47, 67, 56, 60

Find the mean cost of these books.

Although the mean is probably the most common way to find an average for a group of numbers, it is not always the easiest to find. Another kind of average is called the **median.**

Finding the Median

The *median* is the number for which there are as many instances that are above that number as there are instances below it. To find the median, follow these steps:

STEP 1 Rewrite the numbers in order from smallest to largest.
STEP 2 Count from both ends to find the number in the middle.
STEP 3 If there are two numbers in the middle, add them together and divide the result by 2.

• Example 3

Finding the Median

Find the median for the following groups of numbers.

(*a*) 35, 18, 27, 38, 19, 63, 22

Step 1 Rewrite the numbers in order from smallest to largest.

18, 19, 22, 27, 35, 38, 63

Step 2 Count from both ends to find the number in the middle.
    Counting from both ends, we find that 27 is the median. There are 3 numbers above 27 and three numbers below it.

(*b*) 29, 88, 74, 62, 81, 37

Step 1 Rewrite the numbers in order from smallest to largest.

29, 37, 62, 74, 81, 88

Step 2 Count from both ends to find the number in the middle.
    Counting from both ends, we find that there are two numbers in the middle, 62 and 74. We go on to step 3.

Step 3 If there are two numbers in the middle, add them together and divide the result by 2.

$(62 + 74) \div 2 = 138 \div 2 = 69$

● ● ● CHECK YOURSELF 3

Find the median for each group of numbers:

a. 8, 6, 19, 4, 21, 5, 27
b. 43, 29, 13, 37, 29, 53

There are also times in which the median is a better representative of a group of numbers than the mean is. Example 4 illustrates such a case.

•Example 4

Comparing the Mean and the Median

The following numbers represent the hourly wage of seven employees of a local chip manufacturing plant.

8, 7, 10, 12, 28, 9, 10

(*a*) Find the mean hourly wage.

Step 1   Add all of the numbers in the group.

$8 + 7 + 10 + 12 + 28 + 9 + 10 = 84$

Step 2   Divide that sum by the number of items in the group.

$84 \div 7 = 12$

The mean wage is $12 an hour.

(*b*) Find the median wage for the seven workers.

Step 1   Rewrite the numbers in order from smallest to largest.

7, 8, 9, 10, 10, 12, 18

Step 2   Count from both ends to find the number in the middle.
   The middle number is 10. There are three numbers above it and three numbers below it. The median salary is $10 per hour. Which salary do you think is more typical of the workers? Why?

● ● ●   **CHECK YOURSELF 4**

Following are Jessica's phone bills for each month of 1996:

26, 35, 31, 24, 15, 17, 41, 27, 17, 22, 26, 43

a. Find the mean amount of her phone bills.
b. Find the median amount of her phone bills.

● ● ●   **CHECK YOURSELF ANSWERS**

   1. 20.   **2.** $59.   **3.** **(a)** 8; **(b)** 33.   **4.** **(a)** $27; **(b)** $26.

Find the mean for each set of numbers.

1. 6, 9, 10, 8, 12

2. 13, 15, 17, 17, 18

3. 13, 15, 17, 19, 24, 26

4. 41, 43, 56, 67, 69, 72

5. 12, 14, 15, 16, 16, 16, 17, 22, 25, 27

6. 21, 25, 27, 32, 36, 37, 43, 44, 44, 51

7. 5, 8, 9, 11, 12

8. 7, 18, 11, 7, 12

9. 9, 8, 11, 14, 8

10. 21, 23, 25, 27, 22, 20

Find the median for each set of numbers.

11. 2, 3, 5, 6, 10

12. 12, 13, 15, 17, 18

13. 23, 24, 27, 31, 36, 38, 41

14. 1, 4, 9, 16, 25, 36, 49

15. 46, 13, 47, 25, 68, 51, 71

16. 26, 71, 33, 69, 71, 25, 75

1.	9
2.	16
3.	19
4.	58
5.	18
6.	36
7.	9
8.	11
9.	10
10.	23
11.	5
12.	15
13.	31
14.	16
15.	47
16.	69

Solve the following applications.

17. Temperature. High temperatures of 86°, 91°, 92°, 103°, and 98° were recorded for the first 5 days of July. What was the mean high temperature?

18. Travel. A salesperson drove 238, 159, 87, 163, and 198 miles (mi) on a 5-day trip. What was the mean number of miles driven per day?

19. Mileage rating. Highway mileage ratings for seven new diesel cars were 43, 29, 51, 36, 33, 42, and 32 miles per gallon (mi/gal). What was the mean rating?

20. Enrollments. The enrollments in the four elementary schools of a district are 278, 153, 215, and 198 students. What is the mean enrollment?

21. Test scores. To get an A in history, you must have a mean of 90 on five tests. Your scores thus far are 83, 93, 88, and 91. How many points must you have on the final test to receive an A? (*Hint:* First find the total number of points you need to get an A.)

22. Test scores. To pass biology, you must have a mean of 70 on six quizzes. So far your scores have been 65, 78, 72, 66, and 71. How many points must you have on the final quiz to pass biology?

23. Test scores. Louis had scores of 87, 82, 93, 89, and 84 on five tests. Cheryl had scores of 92, 83, 89, 94, and 87 on the same five tests. Who had the higher mean score? By how much?

24. Heating bills. The Sheehan family had heating bills of $85, $78, $64, and $57 in the first 4 months of 1995. The bills for the same months of 1996 were $82, $86, $68, and $56. In which year was the mean monthly bill higher? By how much?

Monthly energy use, in kilowatthours (kwh), by appliance type for four typical U.S. families is shown below.

	Wong Family	McCarthy Family	Abramowitz Family	Gregg Family
Electric range	97	115	80	96
Electric heat	1200	1086	1103	975
Water heater	407	386	368	423
Refrigerator	127	154	98	121
Lights	75	99	108	94
Air conditioner	123	117	96	120
Color TV	39	45	21	47

25. Heating. What is the mean number of kilowatthours used each month by the four families for heating their homes?

26. Heating. What is the mean number of kilowatthours used each month by the four families for hot water?

27. Heating. What is the mean number of kilowatthours used per appliance by the McCarthy family?

28. Heating. What is the mean number of kilowatthours used per appliance by the Gregg family?

Use your calculator for the following applications.

29. Utility bills. Fred kept the following records of his utility bills for 12 months: $53, $51, $43, $37, $32, $29, $34, $41, $58, $55, $49, and $58. What was the mean monthly bill?

30. Test scores. The following scores were recorded on a 200-point final examination: 193, 185, 163, 186, 192, 135, 158, 174, 188, 172, 168, 183, 195, 165, 183. What was the mean of the scores?

31. In a certain math class, you take four tests and the final, which counts as two tests. Your grade is the average of the six tests. At the end of the course, you compute both the mean and the median.

(a) You want to convince the teacher to use the mean to compute your average. Write a note to your teacher explaining why this is a better choice. Choose numbers that make a convincing argument.

(b) You want to convince the teacher to use the median to compute your average. Write a note to your teacher explaining why this is a better choice. Choose numbers that make a convincing argument.

Answers

1. 9   **3.** 19   **5.** 18   **7.** 9   **9.** 10   **11.** 5   **13.** 31   **15.** 47
17. 94°   **19.** 38 mi/gal   **21.** 95 points   **23.** Louis's average score was 87, Cheryl's was 89. Cheryl's average score was 2 points higher than Louis's.
25. 1091 kWh   **27.** 286 kWh   **29.** $45   **31.**

ANSWERS

25. 1091 kWh

26. 396 kWh

27. 286 kWh

28. 268 kWh

29. $45

30. 176

31.

Name _____

Section _____ Date _____

ANSWERS

Find the mean and median for each set of numbers.

1. 8, 10, 9, 9, 9, 10, 8

2. 12, 13, 7, 14, 4, 11, 9

3. 19, 17, 8, 13, 10, 11

4. 8, 14, 13, 9, 11

Solve the following applications.

5. Costs. Sarah purchased five record albums. The total cost was $48.65. What was the mean cost of the albums?

6. Costs. Rebecca had $100 to do her Christmas shopping with. She bought eight gifts and had $28.50 left over. How much did she spend for each gift?

7. Selling price. Henry sold eight pieces of pottery at the Saturday Market. They sold for $15, $15, $15, $20, $20, $25, $25, and $35. Find the mean selling price.

8. Shift time. Jennifer worked six shifts in the math lab this week. The shifts lasted 4, 3, 6.5, 3.5, 2.5, and 7.5 hours (h). What was the mean length of her shifts?

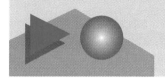

Self-Test for Chapter 3

A N S W E R S

1. Divisor, dividend
    quotient, remainder

2. 7

3. 6

4. 8 r5

5. 9 r4

6. 8

7. 1

8. 0

9. Undefined

10. 123

11. 492 r6

12. 3041 r2

13. 6 r23

14. 14 r41

15. 76 r7

16. 24 r191

17. 22 r21

18. 209 r145

19. $223

The purpose of the Self-Test is to help you check your progress and review for a chapter test in class. Allow yourself about 1 hour to take the test. When you are done, check your answers in the back of the book. If you missed any answers, be sure to go back and review the appropriate sections in the chapter and do the supplementary exercises provided there.

[3.1]   **1.** In the following division problem:

$$\begin{array}{r} 5 \\ 8\overline{)43} \\ 40 \\ \hline 3 \end{array}$$

8 is the _____

43 is the _____

5 is the _____

3 is the _____

Do the indicated division and check your result.

2. $56 \div 8$ **3.** $9\overline{)54}$ **4.** $6\overline{)53}$ **5.** $85 \div 9$

[3.2]   In Exercises 6 to 9, divide if possible.

6. $8 \div 1$ **7.** $5\overline{)5}$ **8.** $0 \div 6$ **9.** $3 \div 0$

[3.3]   In Exercises 10 to 18, divide, using long division.

10. $6\overline{)738}$ **11.** $8\overline{)3942}$ **12.** $9\overline{)27,371}$

[3.4]   **13.** $38\overline{)251}$ **14.** $53\overline{)783}$ **15.** $28\overline{)2135}$

16. $281\overline{)6935}$ **17.** $571\overline{)12,583}$ **18.** $293\overline{)61,382}$

In Exercises 19 to 22, solve each application.

19. Trip expenses. Eight people estimate that the total expenses for a trip they are planning to take together will be $1784. If each person pays an equal amount, what will be each person's share?

20. Mileage. On a trip Ray drove 2345 miles (mi) while using 67 gallons (gal) of gas. What was his mileage on the trip?

21. Weekly pay. Sharon's income was $12,896 last year. What was her weekly pay? (Use 52 weeks per year.)

22. Interest payments. Tom buys a television set for $722. He pays $50 down and agrees to pay the balance in 24 monthly payments. What will be the amount of each monthly payment?

[3.5]   Estimate the result of the following division problems.

23. 569 divided by 118

24. 1649 divided by 83

[3.6]   In Exercises 25 to 30, evaluate each of the expressions.

25. $12 \div 6 + 3 =$

26. $4 + 12 \div 4 =$

27. $3^3 \div 9 =$

28. $28 \div 7 \times 4 =$

29. $5 \times 8 \div 2 =$

30. $36 \div (3^2 + 3) =$

[3.7]   **31.** Find the mean of the numbers 12, 19, 15, 20, 11, and 13.

32. Find the median of the numbers 8, 9, 15, 3, 1.

33. Bus riders. A bus carried 234 passengers on the first day of a newly scheduled route. The next 4 days there were 197, 172, 203, and 214 passengers. What was the mean number of riders per day?

34. Test scores. To earn an A in biology, you must have a mean of 90 on four tests. Your scores thus far are 87, 89, and 91. How many points must you have on the final test to earn the A?

FACTORS AND MULTIPLES

INTRODUCTION

The United States, along with the rest of the world, is trying to be prepared for a steadily increasing population. One of the needs anticipated for a growing population is more housing. The solution is not as simple as just building more houses and apartment buildings. As the population increases, we also need more farm land for food, more access room for highways and other forms of transit, and more recreation space.

Jennifer has worked as a land development architect for 15 years. She helps other developers come up with plans that meet the needs of both the builders and the communities. Her award-winning designs minimize the effect on the environment and maximize the usefulness of the space. ⟶ ■

© Michael Newman/PhotoEdit

Factors and Multiples

This pretest will point out any difficulties you may be having with the factors and multiples of whole numbers. Do all the problems. Then check your answers with those in the back of the book.

1. List all the factors of 42.

2. For the group of numbers 2, 3, 6, 7, 9, 17, 18, 21, and 23, list the prime and composite numbers.

Using divisibility tests, determine which, if any, of the numbers 2, 3, and 5 are factors of each of the following numbers.

3. 546

4. 5130

Write the prime factorizations for each of the following numbers.

5. 60

6. 350

Find the greatest common factor (GCF) for each of the following groups of numbers.

7. 12 and 32

8. 24, 36, and 42

Find the least common multiple (LCM) for each of the following groups of numbers.

9. 25 and 30

10. 12, 16, and 18

Prime and Composite Numbers

4.1 OBJECTIVES

1. Find the factors of a number.
2. Determine whether a number is prime, composite, or neither.
3. Determine whether a number is divisible by 2, 3, or 5.

2 and 5 can also be called *divisors* of 10. They divide 10 exactly.

This is a complete list of the factors. There are no other whole numbers that divide 18 exactly. Note that the factors of 18, except for 18 itself, are *smaller* than 18.

In Section 2.1 we said that since $2 \times 5 = 10$, we call 2 and 5 **factors** of 10.

Definition of a Factor

A **factor** of a whole number is another whole number that will *divide exactly* into that number. This means that the division will have a remainder of 0.

• Example 1

Finding Factors

List all factors of 18.

$3 \times 6 = 18$ Since $3 \times 6 = 18$, 3 and 6 are factors (or divisors) of 18.

$2 \times 9 = 18$ 2 and 9 are also factors of 18.

$1 \times 18 = 18$ 1 and 18 are factors of 18.

1, 2, 3, 6, 9, and 18 are all the factors of 18.

● ● ●   **CHECK YOURSELF 1**

List all the factors of 24.

A whole number greater than 1 will always have itself and 1 as factors. Sometimes these will be the *only* factors. For instance, 1 and 3 are the only factors of 3.

How large can a prime number be? There is no largest prime number. To date, the largest *known* prime is $2^{756839} - 1$. This is a number with 227,832 digits, if you are curious. Of course, a computer had to be used to verify that a number of this size is prime. By the time you read this, someone may very well have found an even larger prime number.

Listing factors leads us to an important classification of whole numbers. Any whole number larger than 1 will be either a *prime* or a *composite* number. Let's look at the following definitions.

Definition of a Prime Number

A **prime number** is any whole number greater than 1 that has only 1 and itself as factors.

As examples, 2, 3, 5, and 7 are prime numbers. Their only factors are 1 and themselves.

To check whether a number is prime, one approach is simply to divide the smaller primes, 2, 3, 5, 7, and so on, into the given number. If no factors other than 1 and the given number are found, the number is prime.

• Example 2

Identifying Prime Numbers

Which of the following numbers are prime?

17 is a prime number.   1 and 17 are the only factors.

29 is a prime number.   1 and 29 are the only factors.

33 is *not* prime.   1, 3, 11, and 33 are all factors of 33.

Note: For two-digit numbers, if the number is *not* a prime, it will have one or more of the numbers 2, 3, 5, or 7 as factors.

● ● ●   **CHECK YOURSELF 2**

Which of the following numbers are prime numbers?

2, 6, 9, 11, 15, 19, 23, 35, 41

We can now define a second class of whole numbers.

This definition tells us that a composite number *does* have factors other than 1 and itself.

Definition of a Composite Number

A **composite number** is any whole number greater than 1 that is not prime.

• Example 3

Identifying Composite Numbers

Which of the following numbers are composite?

18 is a composite number.   1, 2, 3, 6, 9, and 18 are all factors of 18.

23 is not a composite number.   1 and 23 are the only factors. This means that 23 is a *prime number*.

25 is a composite number.   1, 5, and 25 are factors.

38 is a composite number.   1, 2, 19, and 38 are factors.

● ● ●   **CHECK YOURSELF 3**

Which of the following numbers are composite numbers?

2, 6, 10, 13, 16, 17, 22, 27, 31, 35

By the definitions of prime and composite numbers:

The whole numbers 0 and 1 are neither prime nor composite.

This is simply a matter of the way in which prime and composite numbers are defined in mathematics. The numbers 0 and 1 are the *only* two whole numbers that can't be classified as one or the other.

For our work in this and the following sections, it is very useful to be able to tell whether a given number is divisible by 2, 3, or 5. The tests that follow will give you some tools to check divisibility without actually having to divide.

Tests for divisibility by other numbers are also available. However, we have limited this section to those tests involving 2, 3, and 5 because they are very easy to use and occur frequently in our work.

This, of course, means that the number is *even*.

Divisibility by 2

A whole number is divisible by 2 if its last digit is 0, 2, 4, 6, or 8.

● Example 4

Determining If a Number Is Divisible by Two

Which of the following are divisible by 2?

2346 is divisible by 2.
13,254 is divisible by 2.
23,573 is *not* divisible by 2.
57,085 is *not* divisible by 2.

● ● ● CHECK YOURSELF 4

Which of the following are divisible by 2?

274   3587   7548   13,593

Divisibility by 3

A whole number is divisible by 3 if the sum of its digits is divisible by 3.

● Example 5

Determining If a Number Is Divisible by Three

Which of the following are divisible by 3?

345 is divisible by 3.   The sum of the digits, 3 + 4 + 5, is 12, and 12 is divisible by 3.

1243 is *not* divisible by 3.   The sum of the digits, 1 + 2 + 4 + 3, is 10, and 10 is not divisible by 3.

25,368 is divisible by 3.   The sum of the digits, 2 + 5 + 3 + 6 + 8, is 24, and 24 is divisible by 3. Note that 25,368 is also divisible by 2.

● ● ● **CHECK YOURSELF 5**

a. Is 372 divisible by 2? By 3?   **b.** Is 5493 divisible by 2? By 3?

Let's look now at a third divisibility test.

Divisibility by 5

A whole number is divisible by 5 if its last digit is 0 or 5.

● Example 6

Determining If a Number Is Divisible by Five

2435 is divisible by 5.	Its last digit is 5.
23,123 is *not* divisible by 5.	Its last digit is 3.
123,240 is divisible by 5.	Its last digit is 0. Do you see that 123,240 is also divisible by 2 and 3?

● ● ● **CHECK YOURSELF 6**

a. Is 12,585 divisible by 5? By 2? By 3?
b. Is 5890 divisible by 5? By 2? By 3?

● ● ● **CHECK YOURSELF ANSWERS**

1. 1, 2, 3, 4, 6, 8, 12, and 24.   **2.** 2, 11, 19, 23, and 41 are prime numbers.
3. 6, 10, 16, 22, 27, and 35 are composite numbers.   **4.** 274 and 7548.
5. (a) Yes in both cases; **(b)** only by 3.   **6. (a)** By 5 and by 3;
(b) by 5 and by 2.

List the factors of each of the following numbers.

1. 4   **2.** 6

3. 10   **4.** 12

5. 15   **6.** 21

7. 24   **8.** 32

9. 64   **10.** 66

11. 11   **12.** 37

Use the following list of numbers for Exercises 13 and 14.

15, 19, 23, 31, 49, 55, 59, 87, 91, 97, 103, 105

13. Which of the given numbers are prime?

14. Which of the given numbers are composite?

15. List all the prime numbers between 30 and 50.

16. List all the prime numbers between 55 and 75.

A N S W E R S

1. 1, 2, 4

2. 1, 2, 3, 6

3. 1, 2, 5, 10

4. 1, 2, 3, 4, 6, 12

5. 1, 3, 5, 15

6. 1, 3, 7, 21

7. 1, 2, 3, 4, 6, 8, 12, 24

8. 1, 2, 4, 8, 16, 32

9. 1, 2, 4, 8, 16, 32, 64

10. 1, 2, 3, 6, 11, 22, 33, 66

11. 1, 11

12. 1, 37

13. 19, 23, 31, 59, 97, 103

14. 15, 49, 55, 87, 91, 105

15. 31, 37, 41, 43, 47

16. 59, 61, 67, 71, 73

209

Use the following list of numbers for Exercises 17 through 20.

45, 72, 158, 260, 378, 569, 570, 585, 3541, 4530, 8300

17. Which of the given numbers are divisible by 2?

18. Which of the given numbers are divisible by 3?

19. Which of the given numbers are divisible by 5?

20. Which of the given numbers are divisible by 10?

21. Why is the following not a valid divisibility test for 8?

"A number is divisible by 8 if it is divisible by 2 and 4"

Support your answer with an example. Determine a valid divisibility test for 8.

22. Prime numbers that differ by two are called "twin primes." Examples are 3 and 5, 5 and 7, and so on. Find one pair of twin primes between 85 and 105.

23. The Greek mathematician Eratosthenes developed a method to identify prime numbers in a list of numbers. The method is called the "Sieve of Eratosthenes." Research this topic and describe it to your classmates. Then use it to find all the prime numbers less than 100.

Getting Ready for Section 4.2
[Section 3.3]

Divide, using short division.

a. $3\overline{)72}$ **b.** $5\overline{)90}$ **c.** $4\overline{)84}$

d. $2\overline{)384}$ **e.** $3\overline{)693}$ **f.** $5\overline{)750}$

Answers

1. 1, 2, and 4 **3.** 1, 2, 5, and 10 **5.** 1, 3, 5, and 15

7. 1, 2, 3, 4, 6, 8, 12, and 24 **9.** 1, 2, 4, 8, 16, 32, and 64 **11.** 1 and 11

13. 19, 23, 31, 59, 97, 103 **15.** 31, 37, 41, 43, 47

17. 72, 158, 260, 378, 570, 585, 4530, 8300 **19.** 45, 260, 570, 585, 4530, 8300

21. [image] **23.** [image] **a.** 24 **b.** 18 **c.** 21 **d.** 192 **e.** 231

f. 150

ANSWERS

a. 24

b. 18

c. 21

d. 192

e. 231

f. 150

A N S W E R S

1. 1, 3, 5, 15

2. 1, 2, 4, 8, 16

3. 1, 2, 3, 5, 6, 10, 15, 30

4. 1, 2, 4, 5, 8, 10, 20, 40

5. 1, 2, 3, 6, 7, 14, 21, 42

6. 1, 53

7. Prime

8. Composite

9. Composite

10. Prime

11. Composite

12. Prime

13. 47, 53, 59, 61

14. 78, 952, 5490, 22,390

15. 78, 5490, 14,685

16. 65, 5490, 14,685, 22,390

List the factors of each of the following numbers.

1. 15   **2.** 16   **3.** 30

4. 40   **5.** 42   **6.** 53

Label each of the following numbers as prime or composite.

7. 11   **8.** 33   **9.** 35

10. 37   **11.** 99   **12.** 73

13. List all the prime numbers between 45 and 65.

Use the following list of numbers for Exercises 14 through 16.

65, 78, 952, 1259, 5490, 14,685, 22,390

14. Which of the given numbers are divisible by 2?

15. Which of the given numbers are divisible by 3?

16. Which of the given numbers are divisible by 5?

4.2 Writing Composite Numbers as a Product of Prime Factors

4.2 OBJECTIVE

Find the prime factorization of a number.

To **factor a number** means to write the number as a product of its whole-number factors.

● Example 1

Factoring a Composite Number

Factor the number 10.

$10 = 2 \times 5$ The order in which you write the factors does not matter, so $10 = 5 \times 2$ would also be correct.

Of course, $10 = 10 \times 1$ is also a correct statement. However, in this section we are interested in factors other than 1 and the given number.

Factor the number 21.

$21 = 3 \times 7$

● ● ● CHECK YOURSELF 1

Factor 35.

___

In writing composite numbers as a product of factors, there will be a number of different possible factorizations.

● Example 2

Factoring a Composite Number

Find three pairs of factors of 72.

There have to be at least two different factorizations, since a composite number has factors other than 1 and itself.

$$72 = 8 \times 9 \quad (1)$$
$$= 6 \times 12 \quad (2)$$
$$= 3 \times 24 \quad (3)$$

● ● ● CHECK YOURSELF 2

Find three pairs of factors of 42.

We now want to write composite numbers as a product of their **prime factors.** Look again at the first factored line of Example 2. The process of factoring can be continued until all the factors are prime numbers.

• Example 3

Factoring a Composite Number

This is often called a **factor tree.**

$$72 = \quad 8 \quad \times \quad 9$$
$$= \quad 2 \times 4 \quad \times 3 \times 3 \qquad \text{4 is still not prime, and so we continue by factoring 4.}$$
$$= 2 \times 2 \times 2 \times 3 \times 3 \qquad \text{72 is now written as a product of prime factors.}$$

Finding the prime factorization of a number will be important in our later work in adding fractions.

When we write 72 as $2 \times 2 \times 2 \times 3 \times 3$, no further factorization is possible. This is called the *prime factorization* of 72.

Now, what if we start with the second factored line from the same example, $72 = 6 \times 12$?

• Example 3 (Continued)

Factoring a Composite Number

$$72 = \quad 6 \quad \times \quad 12 \qquad \text{Continue to factor 6 and 12.}$$
$$= 2 \times 3 \times \quad 3 \times 4 \qquad \text{Continue again to factor 4. Other choices for the factors of 12 are possible. As we shall see, the end result will be the same.}$$
$$= 2 \times 3 \times 3 \times 2 \times 2$$

No matter which pair of factors you start with, you will find the same prime factorization. In this case, there are three factors of 2 and two factors of 3. Since multiplication is commutative, the order in which we write the factors does not matter.

● ● ●   **CHECK YOURSELF 3**

We could also write

$$72 = 2 \times 36$$

Continue the factorization.

The Fundamental Theorem of Arithmetic

There is exactly one prime factorization for any composite number.

● Example 4

Factoring a Composite Number

$$108 = \quad 9 \quad \times \quad 12$$

Continue to factor 9 and 12.

$$= 3 \times 3 \times \quad 3 \times 4$$

$$= 3 \times 3 \times 3 \times 2 \times 2$$

This is the prime factorization for 108.

It makes no difference which factors of 108 you start with. Try the following exercise.

● ● ● **CHECK YOURSELF 4**

$108 = 36 \times 3$

Continue the factorization.

The method of the previous examples and exercises will always work. However, an easier method for factoring composite numbers exists. This method is particularly useful when numbers get large, in which case factoring with a number tree becomes unwieldy.

Note: The prime factorization is then the product of all the prime divisors and the final quotient.

Factoring by Division

To find the prime factorization of a number, divide the number by a series of primes until the final quotient is a prime number.

● Example 5

Finding Prime Factors

To write 60 as a product of prime factors, divide 2 into 60 for a quotient of 30. Continue to divide by 2 again for the quotient 15. Since 2 won't divide evenly into 15, we try 3. Since the quotient 5 is prime, we are done.

Do you see how the divisibility tests are used here? 60 is divisible by 2, 30 is divisible by 2, and 15 is divisible by 3.

$$2\overline{)60} = 30 \quad \rightarrow \quad 2\overline{)30} = 15 \quad \rightarrow \quad 3\overline{)15} = 5 \quad \text{Prime}$$

Our factors are the prime divisors and the final quotient. We have

$$60 = 2 \times 2 \times 3 \times 5$$

● ● ● ● **CHECK YOURSELF 5**

Complete the process to find the prime factorization of 90.

$$\begin{array}{c} \overset{45}{2)\overline{90}} \longrightarrow \overset{?}{?)\overline{45}} \end{array}$$

Remember to continue until the final quotient is prime.

    Writing composite numbers in their completely factored form can be simplified if we use a format called **continued division.** This method is based on the short-division process that you studied in Chapter 3.

● Example 6

Finding Prime Factors Using Continued Division

Use the continued-division method to divide 60 by a series of prime numbers.

In each short division, we write the quotient *below* rather than above the dividend. This is just a convenience for the next division.

$$\text{Primes} \begin{cases} 2)\overline{60} \\ 2)\overline{30} \\ 3)\overline{15} \\ \quad 5 \end{cases}$$

    Stop when the final quotient is prime.

To write the factorization of 60, we list each divisor used and the final prime quotient. In our example, we have

$$60 = 2 \times 2 \times 3 \times 5$$

● ● ● ● **CHECK YOURSELF 6**

Find the prime factorization of 234.

    Here is a similar example.

● Example 7

Finding Prime Factors Using Continued Division

There is no particular reason for the order of the division. We simply find an easy prime factor, here 2, as our first step, and then keep dividing until the quotient is prime. Note again how useful the divisibility tests are.

To find the prime factorization of 126:

$$\begin{array}{r} 2)\overline{126} \\ 3)\overline{63} \\ 3)\overline{21} \\ 7 \end{array} \longleftarrow \text{Prime}$$

We can write

$$126 = 2 \times 3 \times 3 \times 7$$

The original number, 126, is the product of the series of divisors and the final prime quotient.

● ● ● ● CHECK YOURSELF 7

Find the prime factorization of 150.

Let's look at one more complete example of finding the prime factorization of a composite number.

● Example 8

Finding Prime Factors Using Continued Division

To find the prime factorization of 400:

$$
\begin{array}{r}
2\,)\overline{400} \\
2\,)\overline{200} \\
2\,)\overline{100} \\
2\,)\overline{50} \\
5\,)\overline{25} \\
5
\end{array}
$$

In each of the examples presented, we have removed the *smallest* available prime factor first, since that is often the easiest way to begin. However, it makes no difference in what order the factors are removed. The result will be the same, for there is only one prime factorization for any whole number.

So

$$400 = 2 \times 2 \times 2 \times 2 \times 5 \times 5$$

You can, of course, use the exponent notation to write

$$400 = 2^4 \times 5^2$$

● ● ● ● CHECK YOURSELF 8

Find the prime factorization of 780.

● ● ● **CHECK YOURSELF ANSWERS**

1. 5×7.

2. 2×21, 3×14, 6×7.

3. $2 \times 2 \times 2 \times 3 \times 3$.

4. $108 = \quad 36 \quad \times 3$ You should have two factors of 2 and three factors of 3. Compare with Example 4.

$= \quad 6 \quad \times \quad 6 \quad \times 3$

$= 2 \times 3 \times 2 \times 3 \times 3$

5.
$$\underset{2\overline{)90}}{45} \longrightarrow \underset{3\overline{)45}}{15} \longrightarrow \underset{3\overline{)15}}{5}$$

$90 = 2 \times 3 \times 3 \times 5$

6. $2 \times 3 \times 3 \times 13$.

7. $2 \times 3 \times 5 \times 5$.

8. $2 \times 2 \times 3 \times 5 \times 13$.

Find the prime factorization of each number.

1. 18   **2.** 22

3. 30   **4.** 35

5. 51   **6.** 42

7. 63   **8.** 94

9. 70   **10.** 90

11. 66   **12.** 100

13. 130   **14.** 88

15. 315   **16.** 400

17. 225   **18.** 132

19. 189   **20.** 330

A N S W E R S

1. $2 \times 3 \times 3$

2. 2×11

3. $2 \times 3 \times 5$

4. 5×7

5. 3×17

6. $2 \times 3 \times 7$

7. $3 \times 3 \times 7$

8. 2×47

9. $2 \times 5 \times 7$

10. $2 \times 3 \times 3 \times 5$

11. $2 \times 3 \times 11$

12. $2 \times 2 \times 5 \times 5$

13. $2 \times 5 \times 13$

14. $2 \times 2 \times 2 \times 11$

15. $3 \times 3 \times 5 \times 7$

16. $2 \times 2 \times 2 \times 2 \times 5 \times 5$

17. $3 \times 3 \times 5 \times 5$

18. $2 \times 2 \times 3 \times 11$

19. $3 \times 3 \times 3 \times 7$

20. $2 \times 3 \times 5 \times 11$

21. 336

22. 500

23. 840

24. 1170

In later mathematics courses, you often will want to find factors of a number with a given sum or difference. The following problems use this technique.

25. Find two factors of 24 with a sum of 10.

26. Find two factors of 15 with a difference of 2.

27. Find two factors of 30 with a difference of 1.

28. Find two factors of 28 with a sum of 11.

29. A natural number is said to be perfect if it is equal to the sum of its divisors.

(a) Show that 28 is a perfect number.

(b) Identify another perfect number less than 28.

30. Find the smallest natural number that is divisible by all of the following: 2, 3, 4, 6, 8, 9.

Getting Ready for Section 4.3
[Section 4.1]

List all factors of the following numbers.

a. 12

b. 20

c. 30

d. 45

e. 17

f. 29

Answers

1. $2 \times 3 \times 3$ **3.** $2 \times 3 \times 5$ **5.** 3×17 **7.** $3 \times 3 \times 7$ **9.** $2 \times 5 \times 7$

11. $2 \times 3 \times 11$ **13.** $2\overline{)130}$ **15.** $3 \times 3 \times 5 \times 7$
$ 5\overline{)65}$
$ 13$

$ 130 = 2 \times 5 \times 13$

17. $3 \times 3 \times 5 \times 5$ **19.** $3\overline{)189}$ **21.** $2 \times 2 \times 2 \times 2 \times 3 \times 7$
$ 3\overline{)63}$
$ 3\overline{)21}$
$ 7$

$ 189 = 3 \times 3 \times 3 \times 7$

23. $2 \times 2 \times 2 \times 3 \times 5 \times 7$ **25.** 4, 6   **27.** 5, 6   **29.**

a. 1, 2, 3, 4, 6, 12   **b.** 1, 2, 4, 5, 10, 20   **c.** 1, 2, 3, 5, 6, 10, 15, 30
d. 1, 3, 5, 9, 15, 45   **e.** 1, 17   **f.** 1, 29

Name

Section   Date

A N S W E R S

1. $2 \times 2 \times 3$

2. 2×13

3. $2 \times 5 \times 5$

4. $2 \times 2 \times 13$

5. $2 \times 3 \times 13$

6. $2 \times 5 \times 11$

7. $2 \times 2 \times 2 \times 5 \times 5$

8. $3 \times 5 \times 7$

9. $2 \times 7 \times 11$

10. $2 \times 2 \times 3 \times 3 \times 7$

11. $2 \times 2 \times 3 \times 5 \times 5$

12. $2 \times 2 \times 3 \times 3 \times 5 \times 7$

Find the prime factorization of each number.

1. 12

2. 26

3. 50

4. 52

5. 78

6. 110

7. 200

8. 105

9. 154

10. 252

11. 300

12. 1260

4.3 Finding the Greatest Common Factor (GCF)

4.3 OBJECTIVES

1. Find the greatest common factor of two numbers.
2. Find the greatest common factor of a group of numbers.

Again the factors of 20, other than 20 itself, are less than 20.

We know that a factor or a divisor of a whole number divides that number exactly.
The factors or divisors of 20 are

1, 2, 4, 5, 10, 20

Each of these numbers divides 20 exactly, that is, with no remainder.
Our work in this section involves common factors or divisors. A **common factor** or **divisor** for two numbers is any factor that divides both the numbers exactly.

● Example 1

Finding Common Factors

Look at the numbers 20 and 30. Is there a common factor for the two numbers?
First, we list the factors. Then we circle the ones that appear in both lists.

Factors

20: ①, ②, 4, ⑤, ⑩, 20
30: ①, ②, 3, ⑤, 6, ⑩, 15, 30

We see that 1, 2, 5, and 10 are common factors of 20 and 30. Each of these numbers divides both 20 and 30 exactly.
Our later work with fractions will require that we find the greatest common factor (GCF) of a group of numbers.

Definition of Greatest Common Factor

The **greatest common factor** (GCF) of a group of numbers is the *largest* number that will divide each of the given numbers exactly.

● Example 1 (Continued)

Finding Common Factors

In the first part of Example 1, the common factors of the numbers 20 and 30 were listed as

1, 2, 5, 10   Common factors of 20 and 30

The greatest common factor of the two numbers is then 10, because 10 is the *largest* of the four common factors.

● ● ● **CHECK YOURSELF 1**

List the factors of 30 and 36, and then find the greatest common factor.

The method of Example 1 will also work in finding the greatest common factor of a group of more than two numbers.

● Example 2

Finding the Greatest Common Factor (GCF) by Listing Factors

Find the GCF of 24, 30, and 36. We list the factors of each of the three numbers.

Looking at the three lists, we see that 1, 2, 3, and 6 are common factors.

24: 1 , 2 , 3 , 4, 6 , 8, 12, 24

30: 1 , 2 , 3 , 5, 6 , 10, 15, 30

36: 1 , 2 , 3 , 4, 6 , 9, 12, 18, 36

6 is the greatest common factor of 24, 30, and 36.

● ● ● **CHECK YOURSELF 2**

Find the greatest common factor (GCF) of 16, 24, and 32.

The process shown in Example 2 is very time-consuming where larger numbers are involved. A better approach to the problem of finding the GCF of a group of numbers uses the prime factorization of each number. Let's outline the process.

Finding the Greatest Common Factor

STEP 1 Write the prime factorization for each of the numbers in the group.
STEP 2 Locate the prime factors that are *common* to all the numbers.
STEP 3 The greatest common factor (GCF) will be the *product* of all the common prime factors.

If there are no common prime factors, the GCF is 1.

● Example 3

Finding the Greatest Common Factor (GCF)

Find the GCF of 20 and 30.

Step 1 Write the prime factorization of 20 and 30.

$20 = 2 \times 2 \times 5$

$30 = 2 \times 3 \times 5$

Step 2   Find the prime factors common to each number.

$20 = ②\times 2 \times ⑤$

$30 = ②\times 3 \times ⑤$

2 and 5 are the common prime factors.

Step 3   Form the product of the common prime factors.

$2 \times 5 = 10$

10 is the greatest common factor.

● ● ●   CHECK YOURSELF 3

Find the GCF of 30 and 36.

To find the greatest common factor of a group of more than two numbers, we use the same process.

● Example 4

Finding the Greatest Common Factor (GCF)

Find the GCF of 24, 30, and 36.

$24 = ②\times \ 2 \ \times \ 2 \ \times ③$

$30 = ②\times ③\times \ 5$

$36 = ②\times \ 2 \ \times ③\times \ 3$

2 and 3 are the prime factors common to *all three numbers.*

$2 \times 3 = 6$ is the GCF.

● ● ●   CHECK YOURSELF 4

Find the GCF of 15, 30, and 45.

● Example 5

Finding the Greatest Common Factor (GCF)

Find the greatest common factor of 36, 84, and 120.

Note that 2 appears as a factor twice in all three numbers.

$$36 = \textcircled{2} \times \textcircled{2} \times \textcircled{3} \times 3$$

$$84 = \textcircled{2} \times \textcircled{2} \times \textcircled{3} \times 7$$

$$120 = \textcircled{2} \times \textcircled{2} \times 2 \times \textcircled{3} \times 5$$

Because 2 appears *twice* as a common factor of the three numbers, we multiply by two 2's and one 3 to find the GCF.

$12 = 2 \times 2 \times 3$ is the GCF of 36, 84, and 120.

● ● ● **CHECK YOURSELF 5**

Find the greatest common factor of 16, 24, and 32.

• Example 6

If two numbers, such as 15 and 28, have no common factor other than 1, they are called **relatively prime.**

Finding the Greatest Common Factor (GCF)

Find the greatest common factor of 15 and 28.

$15 = 3 \times 5$
$28 = 2 \times 2 \times 7$

There are no common prime factors listed. But remember that 1 is a factor of every whole number.

The greatest common factor of 15 and 28 is 1.

● ● ● **CHECK YOURSELF 6**

Find the greatest common factor of 30 and 49.

● ● ● **CHECK YOURSELF ANSWERS**

1. 30:   1, 2, 3, 5, $\textcircled{6}$, 10, 15, 30
   36:   1, 2, 3, 4, $\textcircled{6}$, 9, 12, 18, 36

   6 is the greatest common factor.

2. 16: $\textcircled{1}$, $\textcircled{2}$, $\textcircled{4}$, $\textcircled{8}$, 16
   24:   $\textcircled{1}$, $\textcircled{2}$, 3, $\textcircled{4}$, 6, $\textcircled{8}$, 12, 24
   32:   $\textcircled{1}$, $\textcircled{2}$, $\textcircled{4}$, $\textcircled{8}$, 16, 32

   The GCF is 8.

3. $30 = \textcircled{2} \times \textcircled{3} \times 5$
   $36 = \textcircled{2} \times 2 \times \textcircled{3} \times 3$

   The GCF is $2 \times 3 = 6$.

4. 15.   **5.** 8.   **6.** GCF is 1; 30 and 49 are relatively prime.

4.3 Exercises

Name

Section   Date

Find the greatest common factor (GCF) for each of the following groups of numbers.

1. 4 and 6

2. 6 and 9

3. 10 and 15

4. 12 and 14

5. 21 and 24

6. 22 and 33

7. 20 and 21

8. 28 and 42

9. 18 and 24

10. 35 and 36

11. 18 and 54

12. 12 and 48

13. 36 and 48

14. 36 and 54

15. 84 and 105

16. 70 and 105

A N S W E R S

1. 2

2. 3

3. 5

4. 2

5. 3

6. 11

7. 1

8. 14

9. 6

10. 1

11. 18

12. 12

13. 12

14. 18

15. 21

16. 35

17. 15

18. 18

19. 12

20. 15

21. 35

22. 16

23. 25

24. 36

25.

26.

17. 45, 60, and 75

18. 36, 54, and 180

19. 12, 36, and 60

20. 15, 45, and 90

21. 105, 140, and 175

22. 32, 80, and 112

23. 25, 75, and 150

24. 36, 72, and 144

25. Tom and Dick both work the night shift at the steel mill. Tom has every sixth night off, and Dick has every eighth night off. If they both have August 1 off, when will they both be off together again?

26. Mercury, venus, and earth revolve around the sun once every 3, 7, and 12 months, respectively. If the three planets are now in the same straight line, what is the smallest number of months that must pass before they line up again?

228

A N S W E R S

a. 7

b. 14

c. 21

d. 28

e. 35

f. 42

● Getting Ready for Section 4.4
[Section 2.2]

Find each of the following products.

a. 7×1 **b.** 7×2

c. 7×3 **d.** 7×4

e. 7×5 **f.** 7×6

Answers

1. 2   **3.** 5   **5.** 3   **7.** 1   **9.** 6   **11.** 18   **13.** 12   **15.** 21
17. 15   **19.** 12   **21.** 35   **23.** 25   **25.**   **a.** 7   **b.** 14
c. 21   **d.** 28   **e.** 35   **f.** 42

A N S W E R S

Find the greatest common factor (GCF) for each of the following groups of numbers.

1. 16 and 20

2. 15 and 20

3. 24 and 42

4. 15 and 16

5. 21 and 25

6. 30 and 40

7. 36 and 48

8. 40 and 56

9. 30, 45, and 60

10. 36, 72, and 90

11. 24, 36, and 60

12. 70, 105, and 140

Finding the Least Common Multiple (LCM)

4.4 OBJECTIVES

1. Find the least common multiple of two numbers.
2. Find the least common multiple of a group of numbers.

Another idea that will be important in our work with fractions is the concept of **multiples.** Every whole number has an associated group of multiples.

Definition of Multiples

The *multiples* of a number are the product of that number with the natural numbers 1, 2, 3, 4, 5, ...

● Example 1

Listing Multiples

List the multiples of 3.
    The multiples of 3 are

$3 \times 1, 3 \times 2, 3 \times 3, 3 \times 4, \ldots$

or

Note that the multiples, except for 3 itself, are *larger* than 3.

3, 6, 9, 12, . . .   The three dots indicate that the list will go on without stopping.

An easy way of listing the multiples of 3 is to think of *counting by threes.*

● ● ● CHECK YOURSELF 1

List the first seven multiples of 3.

● Example 2

Listing Multiples

List the multiples of 5.
    The multiples of 5 are

5, 10, 15, 20, 25, 30, 35, 40, 45, 50, . . .

● ● ● **CHECK YOURSELF 2**

List the first six multiples of 4.

Sometimes we need to find common multiples of two or more numbers.

Definition of Common Multiples

If a number is a multiple of each of a group of numbers, it is called a *common multiple* of the numbers; that is, it is a number that is evenly divisible by all the numbers in the group.

● Example 3

Finding Common Multiples

15, 30, 45, and 60 are multiples of *both* 3 and 5.

Find four common multiples of 3 and 5.

Some common multiples of 3 and 5 are

15, 30, 45, 60

These numbers will occur in the lists of both our previous examples. Of course, there will be many others if you extend the lists.

● ● ● **CHECK YOURSELF 3**

List the first six multiples of 6. Then look at your list from Check Yourself 2 and list some common multiples of 4 and 6.

For our later work, we will use the *least common multiple* of a group of numbers.

Definition of Least Common Multiple

The **least common multiple** (LCM) of a group of numbers is the *smallest* number that is a multiple of each number in the group.

It is possible to simply list the multiples of each number and then find the LCM by inspection.

● Example 4

Finding the Least Common Multiple (LCM)

Find the least common multiple of 6 and 8.

Multiples

48 is also a common multiple of 6 and 8, but we are looking for the *smallest* such number.

6:    6, 12, 18, ⟨24⟩, 30, 36, 42, 48, . . .

8:    8, 16, ⟨24⟩, 32, 40, 48, . . .

We see that 24 is the smallest number common to both lists. So 24 is the LCM of 6 and 8.

● ● ● **CHECK YOURSELF 4**

Find the least common multiple of 20 and 30 by listing the multiples of each number.

The technique of the last example will work for any group of numbers. However, it becomes tedious for larger numbers. Let's outline a different approach.

For instance, if a number appears 3 times in the factorization of a number, it must be included at least 3 times in forming the least common multiple.

Finding the Least Common Multiple

STEP 1 Write the prime factorization for each of the numbers in the group.
STEP 2 Find all the prime factors that appear in any one of the prime factorizations.
STEP 3 Form the product of those prime factors, using each factor the greatest number of times it occurs in any one factorization.

● Example 5

Finding the Least Common Multiple (LCM)

Let's try this method on the numbers of the last example. To find the LCM of 6 and 8:

Step 1 We write the prime factorizations.

This first step is exactly the same as the first step in finding the GCF.

$6 = 2 \times 3$
$8 = 2 \times 2 \times 2$

Step 2 The prime factors that appear are 2 and 3.

Step 3 Look at the factor 2. Now 2 appears 3 times in the factorization of 8. Look at the factor 3. Here 3 appears only once in any factorization.

$2 \times 2 \times 2 \times 3 = 24$ 2 is included 3 times, and
                                          3 is used once as a factor.

So 24 is the LCM of 6 and 8.

● ● ● **CHECK YOURSELF 5**

Find the LCM of 12 and 18.

Some students prefer a slightly different method of lining up the factors to help in remembering the process of finding the LCM of a group of numbers.

• Example 6

Finding the Least Common Multiple (LCM)

To find the LCM of 10 and 18, factor:

Line up the *like* factors vertically.

$$
\begin{array}{l}
10 = 2 \qquad\quad \times 5 \\
18 = \underline{2 \times 3 \times 3} \\
\; 2 \times 3 \times 3 \times 5 \qquad \text{Bring down the factors.}
\end{array}
$$

2 and 5 appear, at most, one time in any one factorization. And 3 appears 2 times in one factorization.

$2 \times 3 \times 3 \times 5 = 90$

So 90 is the LCM of 10 and 18.

● ● ● **CHECK YOURSELF 6**

Use the method of Example 6 to find the LCM of 24 and 36.

The procedure is the same for a group of more than two numbers.

• Example 7

Finding the Least Common Multiple (LCM)

To find the LCM of 12, 18, and 20, we factor:

The different factors that appear are 2, 3, and 5.

$$
\begin{array}{l}
12 = 2 \times 2 \times 3 \\
18 = 2 \qquad\;\; \times 3 \times 3 \\
20 = \underline{2 \times 2 \qquad\qquad \times 5} \\
\; 2 \times 2 \times 3 \times 3 \times 5
\end{array}
$$

2 and 3 appear twice in one factorization, and 5 appears just once.

$2 \times 2 \times 3 \times 3 \times 5 = 180$

So 180 is the LCM of 12, 18, and 20.

● ● ● **CHECK YOURSELF 7**

Find the LCM of 3, 4, and 6.

Let's work through another example of finding the least common multiple of three numbers, again using the vertical format.

● Example 8

Finding the Least Common Multiple (LCM)

To find the LCM of 15, 20, and 25, factor:

Line up the like factors vertically.

$$
\begin{array}{rl}
15 = & 3 \times 5 \\
20 = 2 \times 2 & \times 5 \\
25 = & \underline{5 \times 5} \\
& 2 \times 2 \times 3 \times 5 \times 5 \\
= & 300 \leftarrow \text{The LCM}
\end{array}
$$

Bring down the factors and multiply.

● ● ● **CHECK YOURSELF 8**

Find the LCM of 6, 8, and 20.

If the numbers have no prime factors in common, the LCM of the group of numbers is just the product of the numbers.

● Example 9

Finding the Least Common Multiple (LCM)

To find the LCM of 10 and 21, factor:

$$
\begin{array}{rl}
10 = 2 & \times 5 \\
21 = & \underline{3 \quad \times 7} \\
& 2 \times 3 \times 5 \times 7
\end{array}
$$

Note that 10 and 21 have *no* prime factors in common, and so the LCM, 210, is just the product of 10 and 21.

No factor appears more than once, so the least common multiple is

$2 \times 3 \times 5 \times 7 = 210$

● ● ● **CHECK YOURSELF 9**

Find the LCM of 9 and 20.

● ● ● **CHECK YOURSELF ANSWERS**

1. 3, 6, 9, 12, 15, 18, 21.

2. The first six multiples of 4 are 4, 8, 12, 16, 20, and 24.

3. 6, 12, 18, 24, 30, 36. Some common multiples of 4 and 6 are 12, 24, and 36.

4. The multiples of 20 are 20, 40, 60, 80, 100, 120, . . . ; the multiples of 30 are 30, 60, 90, 120, 150, . . . ; the least common multiple of 20 and 30 is 60, the smallest number common to both lists.

5. $12 = 2 \times 2 \times 3$; $18 = 2 \times 3 \times 3$. The LCM is $2 \times 2 \times 3 \times 3$, or 36.

6. $2 \times 2 \times 2 \times 3 \times 3 = 72$.   **7.** 12.   **8.** 120.   **9.** 180.

Find the least common multiple (LCM) for each of the following groups of numbers. Use whichever method you wish.

1. 2 and 3

2. 3 and 5

3. 4 and 6

4. 6 and 9

5. 10 and 20

6. 12 and 36

7. 9 and 12

8. 20 and 30

9. 12 and 16

10. 10 and 15

11. 12 and 15

12. 12 and 21

13. 18 and 36

14. 25 and 50

15. 25 and 40

16. 10 and 14

17. 30 and 40

18. 18 and 24

A N S W E R S

1. 6
2. 15
3. 12
4. 18
5. 20
6. 36
7. 36
8. 60
9. 48
10. 30
11. 60
12. 84
13. 36
14. 50
15. 200
16. 70
17. 120
18. 72

19. 8 and 15

20. 20 and 21

21. 30 and 150

22. 36 and 72

23. 8 and 48

24. 15 and 60

25. 2, 3, and 5

26. 3, 4, and 7

27. 3, 5, and 6

28. 2, 8, and 10

29. 18, 21, and 28

30. 8, 15, and 20

31. 20, 30, and 45

32. 12, 20, and 35

33. A company uses two types of boxes 8 cm and 10 cm long. They are packed in larger cartons to be shipped. What is the shortest length container that will accommodate boxes of either size without any room left over? (Each container can contain only boxes of one size—no mixing allowed.)

34. There is an alternate approach to finding the least common multiple of two numbers. The LCM of two numbers can be found by dividing the product of the two numbers by the greatest common factor (GCF) of those two numbers. For example, the GCF of 24 and 36 is 12. If we use the above formula, we obtain

$$\text{LCM of 24 and 36} = \frac{24 \cdot 36}{12} = 72$$

(a) Use the above formula to find the LCM of 150 and 480.

(b) Verify the result by finding the LCM using the method of prime factorization.

(c) The above approach can be extended so that it can be used to find the LCM of three numbers. Describe this extension.

(d) Use the results of part **(c)** to find the LCM of 48, 315, and 450.

Answers

1. 6   **3.** 12   **5.** 20   **7.** 36   **9.** 48

11. $12 = 2 \times 2 \times 3$; $15 = 3 \times 5$; the LCM is $2 \times 2 \times 3 \times 5 = 60$ **13.** 36

15. 200   **17.** 120   **19.** 120   **21.** 150   **23.** 48   **25.** 30   **27.** 30

29. $18 = 2 \times 3 \times 3$; $21 = 3 \times 7$; $28 = 2 \times 2 \times 7$; the LCM is $2 \times 2 \times 3 \times 3 \times 7 = 252$

31. 180

33.

NumberCross 4

ACROSS

2. The LCM of 11 and 13
4. The GCF of 120 and 300
7. The GCF of 13 and 52
8. The GCF of 360 and 540

DOWN

1. The LCM of 8, 14, and 21
3. The LCM of 16 and 12
5. The LCM of 2, 5, and 13
6. The GCF of 54 and 90

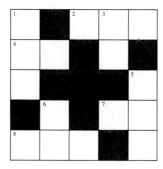

SOLUTION TO NUMBERCROSS4

Name

Section   Date

1. 12

2. 16

3. 18

4. 60

5. 25

6. 42

7. 24

8. 100

9. 72

10. 36

11. 63

12. 88

13. 72

14. 120

15. 84

16. 198

Find the least common multiple (LCM) for each of the following groups of numbers. Use whichever method you wish.

1. 4 and 12   **2.** 8 and 16

3. 6 and 9   **4.** 15 and 20

5. 5 and 25   **6.** 14 and 21

7. 8 and 12   **8.** 20 and 25

9. 18 and 24   **10.** 12 and 18

11. 7 and 9   **12.** 8 and 11

13. 9, 12, and 24   **14.** 8, 12, and 15

15. 14, 21, and 28   **16.** 6, 18, and 33

Name _____

Section _____ Date _____

The purpose of the Self-Test is to help you check your progress and review for a chapter test in class. Allow yourself about 1 hour to take the test. When you are done, check your answers in the back of the book. If you missed any answers, be sure to go back and review the appropriate sections in the chapter and do the supplementary exercises provided there.

[4.1]   **1.** List all of the factors of 18.

2. List all of the factors of 42.

3. List all of the factors of 17.

4. From the group of numbers 13, 21, 29, 37, 51, 1, and 91, list the prime numbers.

5. List the composite numbers from the group of numbers in Exercise 4.

6. List the prime numbers between 40 and 60.

[4.1]   In Exercises 7 to 10, use the divisibility tests to determine which, if any, of the numbers 2, 3, and 5 are factors of the given numbers.

7. 90   **8.** 341

9. 774   **10.** 3585

[4.2]   In Exercises 11 to 14, find the prime factorization.

11. 42   **12.** 72

13. 210   **14.** 792

[4.3]   In Exercises 15 to 22, find the greatest common factor (GCF) for each of the given numbers.

15. 10 and 15   **16.** 30 and 48

ANSWERS

1. 1, 2, 3, 6, 9, 18
2. 1, 2, 3, 6, 7, 14, 21, 42
3. 1, 17
4. 13, 29, 37
5. 21, 51, 91
6. 41, 43, 47, 53, 59
7. 2, 3, and 5
8. None
9. 2 and 3
10. 3 and 5
11. $2 \times 3 \times 7$
12. $2^3 \times 3^2$
13. $2 \times 3 \times 5 \times 7$
14. $2^3 \times 3^2 \times 11$
15. 5
16. 6

ANSWERS

17. 8

18. 1

19. 10

20. 4

21. 14

22. 22

23. 35

24. 18

25. 60

26. 90

27. 20

28. 24

29. 168

30. 180

17. 8 and 24

18. 15 and 16

19. 10, 20, and 50

20. 8, 36, and 60

21. 28, 42, and 70

22. 66, 110, and 154

[4.4] In Exercises 23 to 30, find the least common multiple (LCM) for each of the given numbers.

23. 5 and 7

24. 6 and 9

25. 12 and 60

26. 15 and 18

27. 4, 5, and 10

28. 6, 8, and 12

29. 8, 14, and 21

30. 36, 20, and 30

Summary for Chapters 1–4

Our Decimal Place-Value System

0, 1, 2, 3, 4, 5, 6, 7, 8, and 9 are digits.
5, 27, and 5345 are numerals.

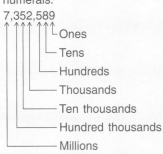

Digits Digits are the basic symbols of the system.

Numerals Numerals name numbers. They may have one or more digits.

Place Value The value of a digit in a numeral depends on its position or place.

$2345 = (2 \times 1000)$
$+ (3 \times 100) + (4 \times 10)$
$+ (5 \times 1)$

The value of a numeral is the sum of each digit multiplied by its place value.

Addition

$$\begin{array}{r} 5 \\ +8 \\ \hline 13 \end{array} \right\} \text{Addends} \\ \text{Sum}$$

Addends The numbers that are being added.

Sum The result of the addition.

The Properties

$5 + 4 = 4 + 5$

The Commutative Property The order in which you add two whole numbers does not affect the sum.

$(2 + 7) + 8 = 2 + (7 + 8)$

The Associative Property The way in which you group whole numbers in addition does not affect the final sum.

$6 + 0 = 0 + 6 = 6$

The Additive Identity The sum of 0 and any whole number is just that whole number.

Rounding Whole Numbers

Step 1 To round a whole number to a certain decimal place, look at the digit to the *right* of that place.

To the nearest hundred, 43,578 is rounded to 43,600.
To the nearest thousand, 273,212 is rounded to 273,000.

Step 2 (*a*) If that digit is 5 or more, that digit and all digits to the right become 0. The digit in the place you are rounding to is increased by 1.

(*b*) If that digit is less than 5, that digit and all digits to the right become 0. The digit in the place you are rounding to remains the same.

Measuring Perimeter

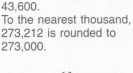

The perimeter is the total distance around the outside edge of a shape. The perimeter of a rectangle is $P = 2L + 2W$.

Order on the Whole Numbers

$8 < 12$

$15 > 10$

For the numbers a and b, we can write

1. $a < b$ (read "a is less than b") when a is *to the left* of b on the number line.
2. $a > b$ (read "a is greater than b") when a is *to the right* of b on the number line.

Subtraction

$$\begin{array}{r} 15 \\ -9 \\ \hline 6 \end{array} \begin{array}{l} \leftarrow \text{Minuend} \\ \leftarrow \text{Subtrahend} \\ \leftarrow \text{Difference} \end{array}$$

Minuend The number we are subtracting from.

Subtrahend The number that is being subtracted.

Difference The result of the subtraction.

Multiplication

$\underbrace{7 \times 9}_{\text{Factors}} = 63 \leftarrow$ Product

Factors The numbers being multiplied.

Product The result of the multiplication.

The Properties

$7 \times 9 = 9 \times 7$

The Commutative Property Multiplication, like addition, is a *commutative* operation. The order in which you multiply two whole numbers does not affect the product.

$(3 \times 5) \times 6 = 3 \times (5 \times 6)$

The Associative Property Multiplication is an *associative* operation. The way in which you group numbers in multiplication does not affect the final product.

$7 \times 1 = 1 \times 7 = 7$

The Multiplicative Identity Multiplying any whole number by 1 simply gives that number as a product.

$9 \times 0 = 0 \times 9 = 0$

The Multiplication Property of 0 Multiplying any whole number by 0 gives the product 0.

$2 \times (3 + 7) = (2 \times 3) + (2 \times 7)$

The Distributive Property To multiply a factor by a sum of numbers, multiply the factor by each number inside the parentheses. Then add the products.

Finding the Area of a Rectangle

6 ft

2 ft

$A = L \times W = 6 \text{ ft} \times 2 \text{ ft} = 12 \text{ ft}^2$

The area of a rectangle is found using the formula $A = L \times W$.

Using Exponents

Exponent

$5^3 = 5 \times 5 \times 5 = 125$

Base   Three factors

This is read "5 to the third power" or "5 cubed."

Base The number that is raised to a power.

Exponent The exponent is written to the right and above the base. The exponent tells the number of times the base is to be used as a factor.

Division

Divisor ⟋ Quotient

5
7)38 ← Dividend
35
3 ← Remainder

Divisor The number we are dividing by.

Dividend The number being divided.

Quotient The result of the division.

Remainder The number "left over" after the division.

The Role of 0

$0 \div 7 = 0$

$7 \div 0$ is undefined.

Zero divided by any whole number (except 0) is 0.

Division by 0 is undefined.

The Order of Operations

Mixed operations in an expression should be done in the following order:

$4 \times (2+3)^2 - 7$
$= 4 \times 5^2 - 7$
$= 4 \times 25 - 7$
$= 100 - 7$
$= 93$

Step 1 Do any operations inside parentheses.

Step 2 Evaluate any powers.

Step 3 Do all multiplication and division in order from left to right.

Step 4 Do all addition and subtraction in order from left to right.

Averages

Finding the Mean

To find the *mean* for a group of numbers follow these two steps:

Given the numbers 4, 8, 17, 23

$4 + 8 + 17 + 23 = 52$

$\bar{x} = \dfrac{52}{4} = 13$

Step 1 Add all the numbers in the group.

Step 2 Divide that sum by the number of items in the group.

Finding the Median

The *median* is the number for which there are as many instances that are above that number as there are instances below it. To find the median follow these steps:

Given the numbers 9, 2, 5, 13, 7, 3

Rewrite them

2, 3, 5, 7, 9, 13

The middle numbers are 5 and 7

$\dfrac{5+7}{2} = \dfrac{12}{2} = 6$

$\bar{x} = 6$

Step 1 Rewrite the numbers in order from smallest to largest.

Step 2 Count from both ends to find the number in the middle.

Step 3 If there are two numbers in the middle, add them together and divide the result by 2.

Prime and Composite Numbers

7, 13, 29, and 73 are prime numbers.

8, 15, 42, and 65 are composite numbers.

Prime Number Any whole number greater than 1 that has only 1 and itself as factors.

Composite Number Any whole number greater than 1 that is not prime.

Zero and 1 0 and 1 are not classified as prime or composite numbers.

Divisibility Tests

932 is divisible by 2; 1347 is not.

546 is divisible by 3; 2357 is not.

865 is divisible by 5; 23,456 is not.

By 2 A whole number is divisible by 2 if its last digit is 0, 2, 4, 6, or 8.

By 3 A whole number is divisible by 3 if the sum of its digits is divisible by 3.

By 5 A whole number is divisible by 5 if its last digit is 0 or 5.

Prime Factorization

$$2)\overline{630}$$
$$3)\overline{315}$$
$$3)\overline{105}$$
$$5)\overline{35}$$
$$7$$

So $630 = 2 \times 3 \times 3 \times 5 \times 7$.

To find the prime factorization of a number, divide the number by a series of primes until the final quotient is a prime number. The prime factors include each prime divisor and the final quotient.

The Greatest Common Factor

Greatest Common Factor (GCF) The GCF is the *largest* number that is a factor of each of a group of numbers.

To Find the GCF

To find the GCF of 24, 30, and 36:

$24 = ② \times 2 \times 2 \times ③$
$30 = ② \times ③ \times 5$
$36 = ② \times 2 \times ③ \times 3$
The GCF is $2 \times 3 = 6$ (230)

Step 1 Write the prime factorization for each of the numbers in the group.

Step 2 Locate the prime factors that are common to all the numbers.

Step 3 The greatest common factor (GCF) will be the product of all of the common prime factors. If there are no common prime factors, the GCF is 1.

The Least Common Multiple

Least Common Multiple (LCM) The LCM is the *smallest* number that is a multiple of each of a group of numbers.

To Find the LCM

To find the LCM of 12, 15, and 18:

$12 = 2 \times 2 \times 3$
$15 = 3 \times 5$
$18 = 2 \times 3 \times 3$
_____
$ 2 \times 2 \times 3 \times 3 \times 5$

The LCM is $2 \times 2 \times 3 \times 3 \times 5$, or 180.

Step 1 Write the prime factorization for each of the numbers in the group.

Step 2 Find all the prime factors that appear in any one of the prime factorizations.

Step 3 Form the product of those prime factors, using each factor the greatest number of times it occurs in any one factorization.

Summary Exercises

You should now be reviewing the material in Chapters 1–4. The following exercises will help in that process. Work all the exercises carefully. Then check your answers against the ones in the back of the book. References are provided there to the chapter and section for each exercise. If you made an error, go back and review the related material and do the supplementary exercises for that section.

[1.1]   In Exercises 1 and 2, give the place value of each of the indicated digits.

1. 6 in the numeral 5674   **2.** 5 in the numeral 543,400

In Exercises 3 and 4, give word names for each of the following numerals.

3. 27,428   **4.** 200,305

Write each of the following as a numeral.

5. Thirty-seven thousand, five hundred eighty-three

6. Three hundred thousand, four hundred

[1.2]   In Exercises 7 and 8, name the property of addition that is illustrated.

7. $4 + 9 = 9 + 4$ **8.** $(4 + 5) + 9 = 4 + (5 + 9)$

[1.4]   In Exercises 9 to 11, perform the indicated operations.

9.   784   **10.**   2,570
         385   498
        +247   21,456
                                        +   28

11. Give the total of 578, 85, 1235, and 12,824.

Solve the following application.

12. Passenger count. An airline had 173, 212, 185, 197, and 202 passengers on five morning flights between Washington, D.C., and New York. What was the total number of passengers?

	ANSWERS
1.	Hundreds
2.	Hundred thousands
3.	Twenty-seven thousand, four hundred twenty-eight
4.	Two hundred thousand, three hundred five
5.	37,583
6.	300,400
7.	Commutative property of addition
8.	Associative property of addition
9.	1389
10.	24,552
11.	14,722
12.	969 people

13. 7000

14. 16,000

15. 550,000

16. <

17. >

18. 4478

19. 18,800

20. 1763

21. $536

22. 18 ft

23. Factor

24. Multiple

25. Commutative property of multiplication

26. Distributive property of multiplication over addition

27. 1856

28. 1075

29. 154,602

30. 42,657

[1.5] In Exercises 13 to 15, round the numbers to the indicated place.

13. 6975 to the nearest hundred   **14.** 15,897 to the nearest thousand

15. 548,239 to the nearest ten thousand

[1.5] In Exercises 16 and 17, complete the statements by using the symbol < or >.

16. 60 _____ 70   **17.** 38 _____ 35

[1.7] In Exercises 18 to 20, perform the indicated operations.

18. 5325   **19.** 38,400
      − 847   −19,600

20. Find the difference of 7342 and 5579.

21. Credit card payments. Chuck owes $795 on a credit card after a trip. He makes payments of $75, $125, and $90. Interest of $31 is charged. How much remains to be paid on the account?

[1.8] **22.** Find the perimeter of the following figure.

[2.1] In Exercises 23 and 24, complete the statements by using the word "factor" or the word "multiple."

23. 6 is a _____ of 36.   **24.** 35 is a _____ of 5.

[2.1] In Exercises 25 and 26, name the property of addition and/or multiplication that is illustrated.

25. $7 \times 8 = 8 \times 7$ **26.** $3 \times (4 + 7) = 3 \times 4 + 3 \times 7$

[2.3] In Exercises 27 to 30, perform the indicated operations.

27.   58   **28.**   25   **29.**   378
         ×32   ×43   ×409

30. Find the product of 59 and 723.

Solve the following application.

31. Costs. You wish to carpet a room that is 5 yards by 7 yards. The carpet costs $18 per square yard. What will be the total cost of the materials?

[2.4]   In Exercise 32, perform the indicated operation.

32.   129
         ×240

[2.4]   Estimate the product by rounding each factor to the nearest hundred.

33.   1217
         × 494

[2.5]   In Exercises 34 to 36, evaluate the expressions.

34. $4 + 8 \times 3$ **35.** $(4 + 8) \times 3$ **36.** $4 \times 3 + 8 \times 3$

37. Rental income. Mr. Parks owns five apartments that rent for $425, $455, $485, $495, and $505 per month, respectively. What is his rental income for 1 year if all are occupied?

[2.6]   In Exercises 38 and 39, evaluate the expressions.

38. 5×2^3 **39.** $(5 \times 2)^3$

[2.7]   **40.** Find the area of the given figure.   **41.** Find the volume of the given figure.

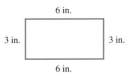

[3.2]   In Exercises 42 and 43, divide if possible.

42. $0 \div 8$ **43.** $5 \div 0$

[3.3]–[3.4]   In Exercises 44 and 45, divide, using long division.

44. $8\overline{)2469}$ **45.** $39\overline{)2157}$

31.	$630
32.	30,960
33.	600,000
34.	28
35.	36
36.	36
37.	$28,380
38.	40
39.	1000
40.	18 in.²
41.	144 in.³
42.	0
43.	Undefined
44.	308 r5
45.	55 r12

46.	28 mi/gal
47.	10
48.	21
49.	4
50.	88
51.	1, 2, 4, 13, 26, 52
52.	1, 41
53.	Prime: 2, 5, 7, 11, 17, 23, 43 Composite: 14, 21, 27, 39
54.	2 and 5
55.	None
56.	$2^4 \times 3$
57.	$2^2 \times 3 \times 5 \times 7$
58.	5
59.	1
60.	120

Solve the following application.

46. Mileage. Terry's odometer read 25,235 miles (mi) at the beginning of a trip and 26,215 mi at the end. If she used 35 gallons (gal) of gas for the trip, what was her mileage (mi/gal)?

[3.5]   Estimate the following.

47. 356 divided by 37   **48.** 2125 divided by 123

[3.6]   In Exercise 49, evaluate the expression.

49. $48 \div (2^3 + 4)$

Solve the following application.

[3.7]   **50. Test scores.** Sally had test scores of 82, 85, 93, 95, 79, and 94 in her biology course. What was her mean test score?

[4.1]   In Exercises 51 and 52, list all the factors of the given numbers.

51. 52   **52.** 41

[4.1]   In Exercise 53, use the group of numbers 2, 5, 7, 11, 14, 17, 21, 23, 27, 39, and 43.

53. List the prime numbers; then list the composite numbers.

[4.1]   In Exercises 54 and 55, use the divisibility tests to determine which, if any, of the numbers 2, 3, and 5 are factors of the following numbers.

54. 2350   **55.** 33,451

[4.2]   In Exercises 56 and 57, find the prime factorization for the given numbers.

56. 48   **57.** 420

[4.3]   In Exercises 58 and 59, find the greatest common factor (GCF).

58. 15 and 20   **59.** 30 and 31

[4.4]   Find the least common multiple (LCM) for the given numbers.

60. 8, 12, and 20

Name

Section   Date

A N S W E R S

1.	Hundred thousands
2.	Three hundred two thousand, five hundred twenty-five
3.	2,430,000
4.	Commutative property of addition
5.	Additive identity
6.	Associative property of addition
7.	966
8.	23,351
9.	5900
10.	950,000
11.	7700

This test is provided to help you in the process of reviewing Chapters 1 to 4. Answers are provided in the back of the book. If you missed any answers, be sure to go back and review the appropriate chapter sections.

1. Give the place value of 7 in 3,738,500.

2. Give the word name for 302,525.

3. Write two million, four hundred thirty thousand as a numeral.

In Exercises 4 to 6, name the property of addition that is illustrated.

4. $5 + 12 = 12 + 5$ **5.** $9 + 0 = 9$

6. $(7 + 3) + 8 = 7 + (3 + 8)$

In Exercises 7 and 8, perform the indicated operations.

7.   593
          275
       $+ \ 98$

8. Find the sum of 58, 673, 5325, and 17,295.

In Exercises 9 and 10, round the numbers to the indicated place value.

9. 5873 to the nearest hundred

10. 953,150 to the nearest ten thousand

In Exercise 11, estimate the sum by rounding to the nearest hundred.

11.   943
         3281
          778
         2112
       $+ \ 570$

12. >

13. <

14. 3861

15. 17,465

16. 905

17. $7579

18. Associative property of multiplication

19. Multiplicative identity

20. Distributive property of multiplication over addition

21. 378,214

22. 686,000

23. 7695

24. 600,000

In Exercises 12 and 13, complete the statements by using the symbol $<$ or $>$.

12. 49 _____ 47   **13.** 80 _____ 90

In Exercises 14 and 15, perform the indicated operations.

14.   4834
      $-$ 973

15. Find the difference of 25,000 and 7535.

In Exercises 16 and 17, solve the applications.

16. Attendance. Attendance for five performances of a play was 172, 153, 205, 193, and 182. How many people attended those performances?

17. Balance. Alan bought a Volkswagen with a list price of $8975. He added stereo equipment for $439 and an air conditioner for $615. If he made a down payment of $2450, what balance remained on the car?

In Exercises 18 to 20, name the property of addition and/or multiplication that is illustrated.

18. $3 \times (4 \times 7) = (3 \times 4) \times 7$

19. $7 \times 1 = 7$

20. $5 \times (2 + 4) = 5 \times 2 + 5 \times 4$

In Exercises 21 to 23, perform the indicated operations.

21.   538   **22.**   1372
      $\times 703$ $\times\ 500$

23. Find the product of 27 and 285.

In Exercise 24, estimate the product by rounding each factor to the nearest hundred.

24.   1475
      $\times\ 418$

In Exercise 25, solve the application.

25. **Carpet cost.** A classroom is 8 yards (yd) wide by 9 yd long. If the room is to be recarpeted with material costing $14 per square yard, find the cost of the carpeting.

In Exercises 26 and 27, divide, using long division.

26. $48\overline{)3259}$

27. $458\overline{)47,350}$

In Exercises 28 to 31, evaluate the expressions.

28. $3 + 5 \times 7$

29. $(3 + 5) \times 7$

30. 4×3^2

31. $2 + 8 \times 3 \div 4$

In Exercises 32 and 33, solve the applications.

32. **Monthly payments.** William bought a washer-dryer combination that, with interest charges, cost $841. He paid $145 down and agreed to pay the balance in 12 monthly payments. Find the amount of each payment.

33. **Test scores.** Elmer had scores of 89, 71, 93, and 87 on his four mathematics tests. What was his mean score?

34. Which of the numbers 5, 9, 13, 17, 22, 27, 31, and 45 are prime numbers? Which are composite numbers?

35. Use the divisibility test to determine which, if any, of the numbers 2, 3, and 5 are factors of 54,204.

36. Find the prime factorization for 264.

In Exercises 37 and 38, find the greatest common factor (GCF) for the given numbers.

37. 36 and 84

38. 16, 24, and 72

In Exercises 39 and 40, find the least common multiple (LCM) for the given numbers.

39. 12 and 30

40. 6, 15, and 45

25. $1008

26. 67 r43

27. 103 r176

28. 38

29. 56

30. 36

31. 8

32. $58

33. 85

34. Prime: 5, 13, 17, 31; composite: 9, 22, 27, 45

35. 2 and 3

36. $2^3 \times 3 \times 11$

37. 12

38. 8

39. 60

40. 90

MathWordPuzzle 1

ACROSS

1. 5, 17, and 53 are examples
6. Not out
7. The tenth month
8. The GCF of eight and six
9. They're usually married to pas
11. It could be olive or peanut
12. Spanish yes
13. It evenly divides a number

DOWN

1. Apple or pumpkin
2. He helps the Dr.
3. Cuts the grass
4. A kind of system
5. Related to an Ave.
8. A kind of powder
9. Missing in action
10. Jordan?
11. The associative property _____ addition
12. _____ what!

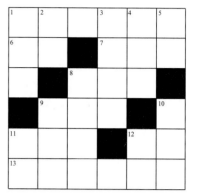

SOLUTION TO MATHWORD PUZZLE 1

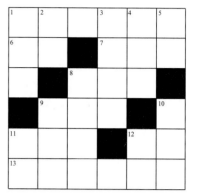

254

INTRODUCTION TO FRACTIONS

INTRODUCTION

Saturday is always chaotic at Home Base Building Supply. Not only is the store bustling with activity, all five phone lines are constantly in use. Most of the questions are from people in the midst of some do-it-yourself project.

Nadia started working at Home Base right after she received her A.S. degree. At first she was concerned that she did not have enough experience in all the trades that are represented at the store. Since she started, Nadia has learned to answer most questions about plumbing, carpentry, and electricity. She is surprised at the great number of questions that really require only a math background. While working full-time, Nadia has been taking night classes. Next year she will be getting her B.A. in business administration. She has enjoyed working the floor, but she is ready to move into administration. ▪

© Michael Newman/PhotoEdit

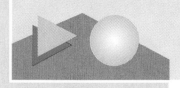

Pretest for Chapter 5

Name _____

Section _____ Date _____

ANSWERS

1. $\dfrac{5}{7}$ ← Numerator ← Denominator

2. See problem

3. $9\dfrac{1}{4}$

4. $\dfrac{46}{7}$

5. Yes

6. $\dfrac{3}{5}$

7. 14

8. $\dfrac{3}{5}$

9. $\dfrac{9}{12}, \dfrac{10}{12}$

10. $\dfrac{27}{72}, \dfrac{40}{72}$, and $\dfrac{42}{72}$

An Introduction to Fractions

This pretest will point out any difficulties you may be having with fractions. Do all the problems. Then check your answers with those in the back of the book.

1. What fraction names the shaded part of the following diagram? Identify the numerator and denominator.

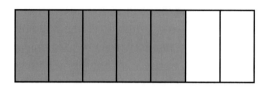

2. From the following group of numbers, list the proper fractions, improper fractions, and mixed numbers:

$$\dfrac{5}{6}, \dfrac{8}{7}, \dfrac{13}{9}, 2\dfrac{3}{5}, \dfrac{3}{8}, 7\dfrac{2}{9}, \dfrac{15}{8}, \dfrac{9}{9}, \dfrac{20}{21}, 3\dfrac{2}{7}, \dfrac{16}{5}, \dfrac{5}{11}$$

Proper: $\dfrac{5}{6}, \dfrac{3}{8}, \dfrac{20}{21}, \dfrac{5}{11}$

Improper: $\dfrac{8}{7}, \dfrac{13}{9}, \dfrac{15}{8}, \dfrac{9}{9}, \dfrac{16}{5}$

Mixed numbers: $2\dfrac{3}{5}, 7\dfrac{2}{9}, 3\dfrac{2}{7}$

3. Convert $\dfrac{37}{4}$ to a mixed number.

4. Convert $6\dfrac{4}{7}$ to a improper fraction.

5. Are $\dfrac{8}{12}$ and $\dfrac{20}{30}$ equivalent fractions?

6. Reduce $\dfrac{18}{30}$ to lowest terms.

7. Find the missing numerator: $\dfrac{2}{5} = \dfrac{?}{35}$

8. Which fraction is larger, $\dfrac{3}{5}$ or $\dfrac{4}{7}$?

Write equivalent fractions with a least common denominator.

9. $\dfrac{3}{4}$ and $\dfrac{5}{6}$

10. $\dfrac{3}{8}, \dfrac{5}{9}$, and $\dfrac{7}{12}$

© 1998 McGraw-Hill Companies

The Language of Fractions

5.1 OBJECTIVES

1. Identify the numerator and denominator of a fraction.
2. Use fractions to name parts of a whole.

Our word *fraction* comes from the Latin stem *fractio,* which means "breaking into pieces."

The previous chapters dealt with whole numbers and the operations that are performed on them. We are now ready to consider a new kind of number, a **fraction.**

> ### Definition of a Fraction
>
> Whenever a unit or a whole quantity is divided into parts, we call those parts **fractions** of the unit.

In Figure 1, the whole has been divided into five equal parts. We use the symbol $\frac{2}{5}$ to represent the shaded portion of the whole.

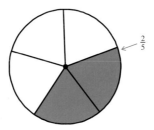

Figure 1

Common fraction is technically the correct term. We will normally just use *fraction* in these materials if there is no room for confusion.

The symbol $\frac{2}{5}$ is called a **common fraction,** or more simply a fraction. A fraction is written in the form a/b or $\frac{a}{b}$, where a and b represent whole numbers and b cannot be equal to 0.

We give the numbers a and b special names. The **denominator,** b, is the number on the bottom. This tells us into how many parts the unit or whole has been divided. The **numerator,** a, is the number on the top. This tells us how many parts of the unit are used.

In Figure 1, the *denominator* is 5; the unit or whole (the circle) has been divided into five parts. The *numerator* is 2. We have taken two parts of the unit.

$$\frac{2}{5} \xleftarrow{\text{\quad}} \text{Numerator}$$
$$\xleftarrow{\text{\quad}} \text{Denominator}$$

● Example 1

Labeling Fraction Components

The fraction $\frac{4}{7}$ names the shaded part of the rectangle in Figure 2.

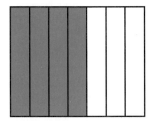

Figure 2

The unit or whole is divided into seven parts, so the denominator is 7. We have shaded four of those parts, and so we have a numerator of 4.

● ● ●   **CHECK YOURSELF 1**

What fraction names the shaded part of this diagram? Identify the numerator and denominator.

Fractions can also be used to name a part of a collection or a set of objects.

● Example 2

Naming a Fractional Part

The fraction $\dfrac{5}{6}$ names the shaded part of Figure 3. We have shaded five of the six objects.

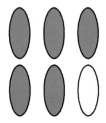

Figure 3

● ● ● **CHECK YOURSELF 2**

What fraction names the shaded part of this diagram?

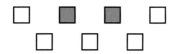

●Example 3

Naming a Fractional Part

Of course, the fraction $\dfrac{8}{23}$ names the part of the class that is not women.

In a class of 23 students, 15 are women. We can name the part of the class that is women as $\dfrac{15}{23}$.

● ● ● **CHECK YOURSELF 3**

Seven replacement parts out of a shipment of 50 were faulty. What fraction names the portion of the shipment that was faulty?

$\dfrac{a}{b}$ names the *quotient* when a is divided by b. Of course, b cannot be 0.

A fraction can also be thought of as indicating division. The symbol $\dfrac{a}{b}$ also means $a \div b$.

●Example 4

Interpreting Division as a Fraction

$\dfrac{2}{3}$ names the quotient when 2 is divided by 3. So $\dfrac{2}{3} = 2 \div 3$.

Note: $\dfrac{2}{3}$ can be read as "two-thirds" or as "2 divided by 3."

● ● ● **CHECK YOURSELF 4**

Using the numbers 5 and 9, write $\dfrac{5}{9}$ in another way.

Fractions are very important in applications. Example 5 illustrates how they might be used.

•Example 5

Naming a Fractional Part

Eight of the 11 starters on a football team are seniors. What fraction names the part of the team that is seniors?

8 of the total of 11 are seniors. We write this as

$$\frac{8}{11}$$

CHECK YOURSELF 5

Of 30 students in a class, 7 are absent. What fraction names the part of the class that is absent?

CHECK YOURSELF ANSWERS

1. $\dfrac{3 \longleftarrow \text{Numerator}}{8 \longleftarrow \text{Denominator}}$

2. $\dfrac{2}{7}$.

3. $\dfrac{7}{50}$.

4. $5 \div 9$.

5. $\dfrac{7}{30}$.

5.1 Exercises

Name _____

Section _____ Date _____

Identify the numerator and denominator of each fraction.

1. $\dfrac{6}{11}$

2. $\dfrac{5}{12}$

3. $\dfrac{3}{11}$

4. $\dfrac{9}{14}$

What fraction names the shaded part of each of the following figures?

5.

6.

7.

8.

9.

10.

11.

12.

13.

14.

15. $\dfrac{4}{5}$

16. $\dfrac{4}{7}$

17. $\dfrac{5}{8}$

18. $\dfrac{5}{9}$

19. Correct: $\dfrac{13}{20}$; wrong: $\dfrac{7}{20}$

20. $\dfrac{2}{5}$

21. Sold: $\dfrac{11}{17}$; not sold: $\dfrac{6}{17}$

22. Hamburgers: $\dfrac{5}{9}$; not: $\dfrac{4}{9}$

23. $2 \div 5$

24. $4 \div 5$

25.

15.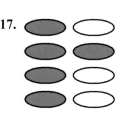

16.

17.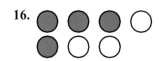

18.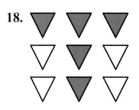

Solve the following applications.

19. **Test scores.** You missed 7 questions on a 20-question test. What fraction names the part you got correct? The part you got wrong?

20. **Basketball.** Of the 5 starters on a basketball team, 2 fouled out of a game. What fraction names the part of the starting team that fouled out?

21. **Car sales.** A used-car dealer sold 11 of the 17 cars in stock. What fraction names the portion sold? What fraction names the portion *not* sold?

22. **Purchases.** At lunch, 5 people out of a group of 9 had hamburgers. What fraction names the part of the group who had hamburgers? What fraction names the part who did *not* have hamburgers?

23. Using the numbers 2 and 5, show another way of writing $\dfrac{2}{5}$.

24. Using the numbers 4 and 5, show another way of writing $\dfrac{4}{5}$.

25. The U.S. Census information can be found in your library, or on the Web, at **www.census.gov.** Use the 1990 census to determine the following:

    (a) Fraction of the population of the United States contained in your state

    (b) Fraction of the population of the United States 65 or over

    (c) Fraction of the United States that is female

26. Using the information on the given label, determine the following:

(a) Fraction of calories that comes from fat

(b) Fraction of calories that comes from saturated fat

(c) Fraction of the total carbohydrates that comes from sugar

Nutrition Facts	
Serving Size 1 bar (22g)	
Serving Per Container 8	
Amount Per Serving	
Calories 90	Calories from Fat 20
	% Daily Value*
Total Fat 3g	4%
Saturated Fat 1.2	3%
Cholesterol 0mg	0%
Sodium 110mg	5%
Total Carbohydrate 17g	6%
Sugars 9g	
Protein 1g	

27. Based on a 2000-calorie diet, consult with your local health officials and determine what fraction of your daily diet each of the following should contribute.

(a) Fat **(b)** Sodium **(c)** Unsaturated fat **(d)** Calcium **(e)** Protein

Using this information, work with your classmates to plan a daily menu that will satisfy the requirements.

Getting Ready for Section 5.2
[Section 1.5]

Use the symbol < or > to complete each of the following statements.

a. 2   3

b. 7   4

c. 5   9

d. 10   7

e. 11   13

f. 18   17

Answers

1. 6 is the numerator; 11 is the denominator

3. 3 is the numerator; 11 is the denominator

5. $\frac{3}{4}$ **7.** $\frac{5}{6}$ **9.** $\frac{5}{5}$ **11.** $\frac{11}{12}$ **13.** $\frac{7}{12}$ **15.** $\frac{4}{5}$ **17.** $\frac{5}{8}$

19. $\frac{13}{20}, \frac{7}{20}$ **21.** $\frac{11}{17}, \frac{6}{17}$ **23.** $2 \div 5$ **25.**   **27.**

a. $2 < 3$ **b.** $7 > 4$ **c.** $5 < 9$ **d.** $10 > 7$ **e.** $11 < 13$ **f.** $18 > 17$

Name _____

Section _____ Date _____

A N S W E R S

1. Numerator: 3, denominator: 7

2. Numerator: 3, denominator: 20

3. $\dfrac{1}{4}$

4. $\dfrac{6}{7}$

5. $\dfrac{5}{8}$

6. $\dfrac{3}{3}$

7. $\dfrac{7}{10}$

8. $\dfrac{4}{5}$

9. $\dfrac{5}{7}$

10. Women: $\dfrac{7}{10}$; not women: $\dfrac{3}{10}$

11. Won: $\dfrac{8}{11}$; lost or tied: $\dfrac{3}{11}$

12. $6 \div 11$

Identify the numerator and denominator of each fraction.

1. $\dfrac{3}{7}$

2. $\dfrac{3}{20}$

What fraction names the shaded part of each of the figures below?

3.

4.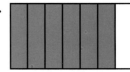

5.

6.

7.

8.

9.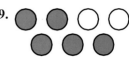

10. **Committees.** On a committee of 10 people, 7 of the members are women. What fraction names the part of the group that is women? What fraction names the part that is not women?

11. **Football.** A football team won 8 of its games in an 11-game season. What fraction names the portion of games won? What fraction names the portion lost or tied?

12. Using the numbers 6 and 11, show another way of writing $\dfrac{6}{11}$.

Proper Fractions, Improper Fractions, and Mixed Numbers

5.2 OBJECTIVES

1. Identify proper fractions and improper fractions.
2. Identify the whole-number part and fraction part of a mixed number.
3. Write improper fractions as mixed or whole numbers.
4. Write mixed numbers as improper fractions.

We can use the relative size of the numerator and denominator of a fraction to separate fractions into different categories.

Definition of a Proper Fraction

If the numerator is *less than* the denominator, the fraction names a number less than 1 and is called a **proper fraction.**

Definition of an Improper Fraction

If the numerator is *greater than or equal to* the denominator, the fraction names a number greater than or equal to 1 and is called an **improper fraction.**

• Example 1

Categorizing Fractions

(*a*) $\frac{2}{3}$ is a proper fraction because the numerator is less than the denominator (Figure 1).

(*b*) $\frac{4}{3}$ is an improper fraction because the numerator is larger than the denominator (Figure 2). Also, $\frac{8}{8}$ is an improper fraction becaused it names exactly 1 unit; the numerator is equal to the denominator.

$\frac{2}{3}$ names less than 1 unit and 2 < 3.

Numerator   Denominator

Figure 1

$\frac{4}{3}$ names more than 1 unit and 4 > 3.

Numerator   Denominator

Figure 2

●●● **CHECK YOURSELF 1**

List the proper fractions and the improper fractions in the following group:

$$\frac{5}{4}, \frac{10}{11}, \frac{3}{4}, \frac{8}{5}, \frac{6}{6}, \frac{13}{10}, \frac{7}{8}, \frac{15}{8}$$

Another way to write a fraction that is larger than 1 is called a **mixed number.**

> **Definition of a Mixed Number**
>
> A **mixed number** is the sum of a whole number and a proper fraction.

● Example 2

Identifying a Mixed Number

$2\frac{3}{4}$ means $2 + \frac{3}{4}$. In fact, we read the mixed number as "2 *and* three-fourths." The addition sign is usually not written.

$2\frac{3}{4}$ is a mixed number. It represents the sum of the whole number 2 and the fraction $\frac{3}{4}$. Look at the diagram below, which represents $2\frac{3}{4}$.

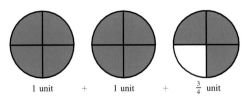

1 unit   +   1 unit   + $\frac{3}{4}$ unit

●●● **CHECK YOURSELF 2**

Give the mixed number that names the shaded portion of all of these diagrams.

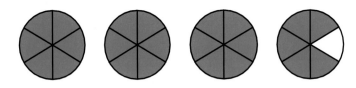

For our later work it will be important to be able to change back and forth between improper fractions and mixed numbers. Since an improper fraction represents a number that is greater than or equal to 1:

> An improper fraction can always be written as either a mixed number or as a whole number.

You can write the fraction $\frac{7}{5}$ as $7 \div 5$. We divide the numerator by the denominator.

In step 1, the quotient gives the whole-number portion of the mixed number.
Step 2 gives the fractional portion of the mixed number.

To do this, remember that you can think of a fraction as indicating division. The numerator is divided by the denominator. This leads us to the following rule:

To Change an Improper Fraction to a Mixed Number

STEP 1 Divide the numerator by the denominator.
STEP 2 If there is a remainder, write the remainder over the original denominator.

• Example 3

Converting a Fraction to a Mixed Number

Convert $\frac{17}{5}$ to a mixed number.

Divide 17 by 5.

$$\begin{array}{r} 3 \\ 5\overline{)17} \\ 15 \\ \hline 2 \end{array} \qquad \frac{17}{5} = 3\frac{2}{5}$$

Remainder ← 2
Original denominator ← 5
Quotient ↑ 3

In diagram form:

$$\frac{17}{5} = 3\frac{2}{5}$$

● ● ● **CHECK YOURSELF 3**

Convert $\frac{32}{5}$ to a mixed number.

● Example 4

Converting a Fraction to a Mixed Number

Convert $\dfrac{21}{7}$ to a mixed or a whole number.

Divide 21 by 7.

If there is *no* remainder, the improper fraction is equal to some whole number; in this case 3.

$$\begin{array}{r} 3 \\ 7\overline{)21} \\ \underline{21} \\ 0 \end{array} \qquad \frac{21}{7} = 3$$

● ● ● ● **CHECK YOURSELF 4**

Convert $\dfrac{48}{6}$ to a mixed or a whole number.

It is also easy to convert mixed numbers to improper fractions. Just use the following rule:

To Change a Mixed Number to an Improper Fraction

STEP 1 Multiply the denominator of the fraction by the whole-number portion of the mixed number.
STEP 2 Add the numerator of the fraction to that product.
STEP 3 Write that sum over the original denominator to form the improper fraction.

● Example 5

Converting Mixed Numbers to Improper Fractions

(*a*) Convert $3\dfrac{2}{5}$ to an improper fraction.

$$3\,\frac{2}{5} = \frac{(5 \times 3) + 2}{5}$$

Denominator × Whole number, Numerator, Denominator

Multiply the denominator by the whole number ($5 \times 3 = 15$). Add the numerator. We now have 17.

$$= \frac{17}{5}$$

Write 17 over the original denominator.

In diagram form:

Each of the three units has 5 fifths, so the whole-number portion is 5×3, or 15, fifths. Then add the 2 fifths from the fractional portion for 17 fifths.

(*b*) Convert $4\frac{5}{7}$ to an improper fraction.

Multiply the denominator, 7, by the whole number, 4, and add the numerator, 5.

$$4\frac{5}{7} = \frac{(7 \times 4) + 5}{7} = \frac{33}{7}$$

● ● ● ● **CHECK YOURSELF 5**

Convert $5\frac{3}{8}$ to an improper fraction.

One special kind of improper fraction should be mentioned at this point: a fraction with a denominator of 1.

> **Fractions with a Denominator of 1**
>
> Any fraction with a denominator of 1 is equal to the numerator alone.

This is the same as our earlier rule for dividing by 1.

●Example 6

Converting a Fraction to a Whole Number

Convert the improper fraction $\frac{5}{1}$ to a whole number.

Thinking in terms of division,
$\frac{5}{1} = 5 \div 1 = 5$

$$\frac{5}{1} = 5$$

● ● ● **CHECK YOURSELF 6**

Convert 9 to an improper fraction.

You probably do many conversions between mixed and whole numbers without even thinking about the process that you follow, as Example 7 illustrates.

• Example 7

Converting Quarter Dollars to Dollars

Maritza has 53 quarters in her bank. How many dollars does she have?

Because there are 4 quarters in each dollar, 53 quarters can be written as

$$\frac{53}{4}$$

Converting the amount to dollars is the same as rewriting it as a mixed number.

$$\frac{53}{4} = 13\frac{1}{4}$$

She has $13\frac{1}{4}$ dollars, which you would probably write as $13.25. (*Note:* We will discuss decimal point usage later in this text.)

● ● ● **CHECK YOURSELF 7**

Kevin is doing the inventory in the convenience store in which he works. He finds there are 11 half gallons (gal) of milk. Write the amount of milk as a mixed number of gallons.

● ● ● **CHECK YOURSELF ANSWERS**

1. Proper fractions: Improper fractions: **2.** $3\frac{5}{6}$.   **3.** $6\frac{2}{5}$.   **4.** 8.

$\dfrac{10}{11}, \dfrac{3}{4}, \dfrac{7}{8}$. $\dfrac{5}{4}, \dfrac{8}{5}, \dfrac{6}{6}, \dfrac{13}{10}, \dfrac{15}{8}$.

5. $\dfrac{43}{8}$.   **6.** $\dfrac{9}{1}$.   **7.** $\dfrac{11}{2} = 5\frac{1}{2}$ gal.

5.2 Exercises

Name

Section   Date

Identify each number as a proper fraction, an improper fraction, or a mixed number.

1. $\dfrac{3}{5}$

2. $\dfrac{9}{5}$

3. $2\dfrac{3}{5}$

4. $\dfrac{7}{9}$

5. $\dfrac{6}{6}$

6. $1\dfrac{1}{5}$

7. $\dfrac{11}{8}$

8. $\dfrac{8}{8}$

9. $4\dfrac{5}{7}$

10. $5\dfrac{3}{7}$

11. $\dfrac{13}{17}$

12. $\dfrac{16}{15}$

13. $\dfrac{31}{25}$

14. $\dfrac{28}{31}$

15. $6\dfrac{2}{5}$

16. $\dfrac{15}{14}$

Give the mixed number that names the shaded portion of each diagram.

17.

18.

19.

20.

Change to a mixed or a whole number.

21. $\dfrac{7}{2}$

22. $\dfrac{8}{3}$

23. $\dfrac{5}{4}$

24. $\dfrac{9}{7}$

A N S W E R S

1. Proper
2. Improper
3. Mixed number
4. Proper
5. Improper
6. Mixed number
7. Improper
8. Improper
9. Mixed number
10. Mixed number
11. Proper
12. Improper
13. Improper
14. Proper
15. Mixed number
16. Improper
17. $1\dfrac{3}{4}$
18. $1\dfrac{2}{3}$
19. $3\dfrac{5}{8}$
20. $2\dfrac{1}{4}$
21. $3\dfrac{1}{2}$
22. $2\dfrac{2}{3}$
23. $1\dfrac{1}{4}$
24. $1\dfrac{2}{7}$

25. $4\dfrac{2}{5}$

26. $3\dfrac{3}{8}$

27. $6\dfrac{4}{5}$

28. $4\dfrac{1}{6}$

29. $11\dfrac{4}{5}$

30. $8\dfrac{2}{7}$

31. $9\dfrac{1}{8}$

32. $12\dfrac{7}{12}$

33. 4

34. 20

35. 9

36. 8

37. $\dfrac{14}{3}$

38. $\dfrac{17}{6}$

39. $\dfrac{8}{1}$

40. $\dfrac{37}{8}$

41. $\dfrac{56}{9}$

42. $\dfrac{7}{1}$

43. $\dfrac{24}{7}$

44. $\dfrac{20}{9}$

45. $\dfrac{97}{13}$

46. $\dfrac{73}{10}$

47. $\dfrac{52}{5}$

48. $\dfrac{67}{5}$

49. $\dfrac{302}{3}$

50. $\dfrac{601}{4}$

51. $\dfrac{475}{4}$

52. $\dfrac{1003}{4}$

25. $\dfrac{22}{5}$ 26. $\dfrac{27}{8}$ 27. $\dfrac{34}{5}$ 28. $\dfrac{25}{6}$

29. $\dfrac{59}{5}$ 30. $\dfrac{58}{7}$ 31. $\dfrac{73}{8}$ 32. $\dfrac{151}{12}$

33. $\dfrac{24}{6}$ 34. $\dfrac{160}{8}$ 35. $\dfrac{9}{1}$ 36. $\dfrac{8}{1}$

Change to an improper fraction.

37. $4\dfrac{2}{3}$ 38. $2\dfrac{5}{6}$ 39. 8 40. $4\dfrac{5}{8}$

41. $6\dfrac{2}{9}$ 42. 7 43. $3\dfrac{3}{7}$ 44. $2\dfrac{2}{9}$

45. $7\dfrac{6}{13}$ 46. $7\dfrac{3}{10}$ 47. $10\dfrac{2}{5}$ 48. $13\dfrac{2}{5}$

49. $100\dfrac{2}{3}$ 50. $150\dfrac{1}{4}$ 51. $118\dfrac{3}{4}$ 52. $250\dfrac{3}{4}$

Solve the following applications.

53. Savings. Clayton has 64 quarters in his bank. How many dollars does he have?

54. Savings. Amy has 19 quarters in her purse. How many dollars does she have?

55. Inventory. Manuel counted 35 half gallons of orange juice in his store. Write the amount of orange juice as a mixed number of gallons.

56. Inventory. Sarah has 19 half gallons of turpentine in her paint store. Write the amount of turpentine as a mixed number of gallons.

57. Suppose the national debt had to be paid by individuals.
    (a) How would the amount each individual owed be determined?
    (b) Would this be a proper or improper fraction?.

58. In statistics, we describe the probability of an event by dividing the number of ways the event can happen by the total number of possibilities. Using this definition, determine the probability of each of the following:
    (a) Obtaining a head on a single toss of a coin
    (b) Obtaining a prime number when a die is tossed

● Getting Ready for Section 5.3 [Section 2.1]

Multiply and then complete each statement, using the symbol = (equal to) or ≠ (not equal to).

a. 2×3 1×6 **b.** 3×5 4×4

c. 4×5 3×7 **d.** 6×7 3×14

e. 5×10 25×2 **f.** 16×3 7×7

Answers

1. Proper   **3.** Mixed number   **5.** Improper   **7.** Improper
9. Mixed number   **11.** Proper   **13.** Improper   **15.** Mixed number
17. $1\frac{3}{4}$ **19.** $3\frac{5}{8}$ **21.** $3\frac{1}{2}$ **23.** $1\frac{1}{4}$ **25.** $4\frac{2}{5}$ **27.** $6\frac{4}{5}$ **29.** $11\frac{4}{5}$
31. $9\frac{1}{8}$ **33.** 4. There is no remainder in the division.   **35.** 9   **37.** $\frac{14}{3}$
39. $\frac{8}{1}$ **41.** $\frac{56}{9}$ **43.** $3\frac{3}{7} = \frac{(7 \times 3) + 3}{7} = \frac{24}{7}$ **45.** $\frac{97}{13}$
47. $10\frac{2}{5} = \frac{(5 \times 10) + 2}{5} = \frac{52}{5}$ **49.** $\frac{302}{3}$ **51.** $\frac{475}{4}$ **53.** $16   **55.** $17\frac{1}{2}$ gal
57.   **a.** =   **b.** ≠   **c.** ≠   **d.** =   **e.** =   **f.** ≠

273

ANSWERS

53. $16

54. $4.75

55. $17\frac{1}{2}$ gal

56. $9\frac{1}{2}$ gal

57.

58. (a) $\frac{1}{2}$
   (b) $\frac{3}{6} = \frac{1}{2}$

a. =

b. ≠

c. ≠

d. =

e. =

f. ≠

Name _____

Section _____ Date _____

A N S W E R S

1. Improper

2. Proper

3. Mixed number

4. Improper

5. Improper

6. Mixed number

7. Proper

8. Proper

9. $2\frac{2}{3}$

10. $3\frac{3}{4}$

11. $1\frac{2}{3}$

12. $1\frac{1}{6}$

13. $6\frac{2}{7}$

14. $6\frac{3}{8}$

15. $42\frac{6}{7}$

16. $44\frac{1}{3}$

17. 8

18. 11

19. $\frac{23}{7}$

20. $\frac{15}{1}$

21. $\frac{79}{12}$

22. $\frac{51}{7}$

23. $\frac{53}{5}$

24. $\frac{35}{3}$

25. $\frac{403}{4}$

26. $\frac{1001}{5}$

27. 7.50 or $\$7\frac{1}{2}$

Identify each number as a proper fraction, an improper fraction, or a mixed number.

1. $\frac{7}{4}$

2. $\frac{8}{9}$

3. $3\frac{3}{4}$

4. $\frac{11}{11}$

5. $\frac{9}{9}$

6. $3\frac{2}{5}$

7. $\frac{27}{32}$

8. $\frac{19}{20}$

Give the mixed number that names the shaded portions of each diagram.

9.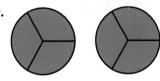

10.

Change to a mixed or a whole number.

11. $\frac{5}{3}$

12. $\frac{7}{6}$

13. $\frac{44}{7}$

14. $\frac{51}{8}$

15. $\frac{300}{7}$

16. $\frac{133}{3}$

17. $\frac{64}{8}$

18. $\frac{11}{1}$

Change to an improper fraction.

19. $3\frac{2}{7}$

20. 15

21. $6\frac{7}{12}$

22. $7\frac{2}{7}$

23. $10\frac{3}{5}$

24. $11\frac{2}{3}$

25. $100\frac{3}{4}$

26. $200\frac{1}{5}$

27. **Purchases.** Damien used his savings of 30 quarters to pay for a model car. What was the cost of the car in dollars?

Equivalent Fractions

5.3 OBJECTIVES

1. Determine whether two fractions are equivalent.
2. Write a fraction that is equivalent to a given fraction.

It is possible to represent the same portion of the whole by different fractions. Look at Figure 1, representing $\frac{3}{6}$ and $\frac{1}{2}$. The two fractions are simply different names for the same number. They are called **equivalent fractions** for this reason.

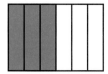

Figure 1

Any fraction has a large number of equivalent fractions. For instance, $\frac{2}{3}$, $\frac{4}{6}$, and $\frac{6}{9}$ are all equivalent fractions since they name the same part of a unit. This is illustrated in Figure 2.

Figure 2

Many more fractions are equivalent to $\frac{2}{3}$. All these fractions can be used interchangeably. An easy way to find out if two fractions are equivalent is to use cross products.

$$\frac{a}{b} = \frac{c}{d}$$

We call $a \times d$ and $b \times c$ the **cross products.**

Testing for Equality

If the cross products for two fractions are equal, the two fractions are equivalent.

• Example 1

Identifying Equivalent Fractions

Are $\frac{3}{24}$ and $\frac{4}{32}$ equivalent fractions? The cross products are 3×32, or 96, and 24×4, or 96. Since the cross products are equal, the fractions are equivalent.

● ● ●   **CHECK YOURSELF 1**

Are $\dfrac{3}{8}$ and $\dfrac{9}{24}$ equivalent fractions?

● Example 2

Identifying Equivalent Fractions Using Cross Products

Are $\dfrac{2}{5}$ and $\dfrac{3}{7}$ equivalent fractions?

The cross products are 2×7 and 5×3.

$2 \times 7 = 14$ and $5 \times 3 = 15$

Since $14 \neq 15$, the fractions are *not* equivalent.

● ● ●   **CHECK YOURSELF 2**

Are $\dfrac{7}{8}$ and $\dfrac{8}{9}$ equivalent fractions?

In writing equivalent fractions, we use the following important principle.

We are really multiplying by $\dfrac{c}{c}$, or 1, and multiplying by 1 does not change the value of a number.

Fundamental Principle of Fractions

For the fraction $\dfrac{a}{b}$ and any nonzero number c,

$$\frac{a}{b} = \frac{a \times c}{b \times c}$$

The fundamental principle of fractions tells us that we can multiply the numerator and denominator by the same nonzero number. The result is an equivalent fraction. For instance,

Multiply the numerator and denominator by 2, 3, and 4.

$\dfrac{1}{2} = \dfrac{1 \times 2}{2 \times 2} = \dfrac{2}{4}$ $\dfrac{1}{2} = \dfrac{1 \times 3}{2 \times 3} = \dfrac{3}{6}$ $\dfrac{1}{2} = \dfrac{1 \times 4}{2 \times 4} = \dfrac{4}{8}$

Multiply the numerator and denominator by 5, 6, and 7.

$\dfrac{1}{2} = \dfrac{1 \times 5}{2 \times 5} = \dfrac{5}{10}$ $\dfrac{1}{2} = \dfrac{1 \times 6}{2 \times 6} = \dfrac{6}{12}$ $\dfrac{1}{2} = \dfrac{1 \times 7}{2 \times 7} = \dfrac{7}{14}$

● ● ●   **CHECK YOURSELF ANSWERS**

1. Yes. The cross products are equal.   **2.** No. The cross products are *not* equal.

Are the pairs of fractions equivalent?

1. $\dfrac{1}{3}, \dfrac{3}{5}$

2. $\dfrac{3}{5}, \dfrac{9}{15}$

3. $\dfrac{1}{7}, \dfrac{4}{28}$

4. $\dfrac{2}{3}, \dfrac{3}{5}$

5. $\dfrac{5}{6}, \dfrac{15}{18}$

6. $\dfrac{3}{4}, \dfrac{16}{20}$

7. $\dfrac{2}{21}, \dfrac{4}{25}$

8. $\dfrac{20}{24}, \dfrac{5}{6}$

9. $\dfrac{2}{7}, \dfrac{3}{11}$

10. $\dfrac{12}{15}, \dfrac{36}{45}$

11. $\dfrac{16}{24}, \dfrac{40}{60}$

12. $\dfrac{15}{20}, \dfrac{20}{25}$

1. No

2. Yes

3. Yes

4. No

5. Yes

6. No

7. No

8. Yes

9. No

10. Yes

11. Yes

12. No

a. 2

b. 5

c. 1

d. 2

e. 5

f. 4

Getting Ready for Section 5.4 [Section 3.1]

Divide.

a. $8 \div 4$

b. $10 \div 2$

c. $9 \div 9$

d. $12 \div 6$

e. $15 \div 3$

f. $20 \div 5$

Answers

1. $1 \times 5 = 5$; $3 \times 3 = 9$. The fractions are not equivalent.

3. Yes   **5.** Yes   **7.** No   **9.** No

11. $16 \times 60 = 960$, and $24 \times 40 = 960$. The fractions are equivalent.

a. 2   **b.** 5   **c.** 1   **d.** 2   **e.** 5   **f.** 4

Name

Section   Date

1. No

2. Yes

3. Yes

4. Yes

5. No

6. No

Are the pairs of fractions equivalent?

1. $\dfrac{1}{3}, \dfrac{6}{11}$

2. $\dfrac{3}{4}, \dfrac{12}{16}$

3. $\dfrac{3}{4}, \dfrac{18}{24}$

4. $\dfrac{8}{11}, \dfrac{24}{33}$

5. $\dfrac{15}{21}, \dfrac{24}{27}$

6. $\dfrac{30}{42}, \dfrac{21}{27}$

Simplifying Fractions

5.4 OBJECTIVE

Use the fundamental principle to simplify fractions.

For our work in this section, we begin by reviewing the fundamental principle of fractions presented in Section 5.3:

> For the fraction $\dfrac{a}{b}$ and any nonzero number c, $\dfrac{a}{b} = \dfrac{a \times c}{b \times c}$.

However, we will use the principle in another way, one involving division. Using division, the principle can also be written as

$$\frac{a}{b} = \frac{a \div c}{b \div c} \qquad \text{where } c \text{ is nonzero}$$

In other words, the fundamental principle of fractions also tells us that we can *divide* the numerator and denominator by the same nonzero number. Let's see how this is applied.

Simplifying a fraction or *reducing a fraction to lower terms* means finding an equivalent fraction with a *smaller* numerator and denominator than those of the original fraction. Dividing the numerator and denominator by the same nonzero number will do exactly that.

Consider Example 1.

● Example 1

Simplifying Fractions

Simplify each fraction.

We apply the fundamental principle to divide the numerator and denominator by 5.

$(a)\ \dfrac{5}{15} = \dfrac{5 \div 5}{15 \div 5} = \dfrac{1}{3}$

$\dfrac{5}{15}$ and $\dfrac{1}{3}$ are equivalent fractions.

Check this by finding the cross products.

We divide the numerator and denominator by 2.

$(b)\ \dfrac{4}{8} = \dfrac{4 \div 2}{8 \div 2} = \dfrac{2}{4}$

$\dfrac{4}{8}$ and $\dfrac{2}{4}$ are equivalent fractions.

● ● ●   **CHECK YOURSELF 1**

Write two fractions that are equivalent to $\dfrac{30}{45}$.

a. Divide the numerator and denominator by 5.
b. Divide the numerator and denominator by 15.

We say that a fraction is in **simplest form,** or in **lowest terms,** if the numerator and denominator have no common factors other than 1. This means that the fraction has the smallest possible numerator and denominator.

In Example 1,

This means that the numerator and denominator can have no additional common factors other than 1. The fraction must be in lowest terms.

$\dfrac{1}{3}$ is in simplest form.

$\dfrac{2}{4}$ is *not* in simplest form.   Do you see that $\dfrac{2}{4}$ can also be written as $\dfrac{1}{2}$?

In this case, the numerator and denominator are *not* as small as possible. The numerator and denominator have a common factor of 2.

To write a fraction in simplest form or to *reduce a fraction to lowest terms,* divide the numerator and denominator by their greatest common factor (GCF).

● **Example 2**

Simplifying Fractions

Write $\dfrac{10}{15}$ in simplest form.

From our work in Chapter 4, we know that the greatest common factor of 10 and 15 is 5. To write $\dfrac{10}{15}$ in simplest form, divide the numerator and denominator by 5.

$$\frac{10}{15} = \frac{10 \div 5}{15 \div 5} = \frac{2}{3}$$

The resulting fraction, $\dfrac{2}{3}$, is in lowest terms.

● ● ●   **CHECK YOURSELF 2**

Write $\dfrac{12}{18}$ in simplest form by dividing the numerator and denominator by the GCF.

Many students prefer another method of reducing fractions, which uses the prime factorization of the numerator and denominator. Example 3 will be reduced by this method.

● Example 3

Factoring to Simplify a Fraction

To simplify $\dfrac{10}{15}$, we factor as shown.

$$\frac{10}{15} = \frac{2 \times \cancel{5}}{3 \times \cancel{5}} = \frac{2}{3}$$

We have a common factor of 5 in the original numerator and denominator, so we divide by 5. Note the use of the slash to indicate that division.

● ● ● CHECK YOURSELF 3

Write $\dfrac{12}{18}$ in simplest form.

● Example 4

Factoring to Simplify a Fraction

(*a*) Simplify $\dfrac{24}{42}$.

To simplify $\dfrac{24}{42}$, factor.

Divide by the common factors of 2 and 3.

$$\frac{24}{42} = \frac{\cancel{2} \times 2 \times 2 \times \cancel{3}}{\cancel{2} \times \cancel{3} \times 7} = \frac{4}{7}$$

Note: The numerator of the simplified fraction is the *product* of the prime factors remaining in the numerator after the division by 2 and 3.

(*b*) Simplify $\dfrac{120}{180}$.

To reduce $\dfrac{120}{180}$ to lowest terms, write the prime factorizations of the numerator and denominator. Then divide by any common factors.

$$\frac{120}{180} = \frac{\cancel{2} \times 2 \times 2 \times \cancel{3} \times \cancel{5}}{\cancel{2} \times 2 \times \cancel{3} \times 3 \times \cancel{5}} = \frac{2}{3}$$

● ● ● ● **CHECK YOURSELF 4**

Write each of the following fractions in simplest form.

a. $\dfrac{60}{75}$ **b.** $\dfrac{210}{252}$

There is another way to organize your work in simplifying fractions. It again uses the fundamental principle to divide the numerator and denominator by any common factors. Let's illustrate with the fractions considered in Example 4.

● Example 5

Using Common Factors to Simplify Fractions

(a) $\dfrac{24}{42} = \dfrac{\overset{12}{\cancel{24}}}{\underset{21}{\cancel{42}}} = \dfrac{\overset{4}{\cancel{12}}}{\underset{7}{\cancel{21}}} = \dfrac{4}{7}$

Divide by the common factor of 2.

Divide by the common factor of 3.

The original numerator and denominator are divisible by 2, and so we divide by that factor to arrive at $\dfrac{12}{21}$. Our divisibility tests tell us that a common factor of 3 still exists. (Do you remember why?) Divide again for the result $\dfrac{4}{7}$, which is in lowest terms.

Note: If we had seen the GCF of 6 at first, we could have divided by 6 and arrived at the same result in one step.

(b) $\dfrac{120}{180} = \dfrac{\overset{\overset{2}{20}}{\cancel{\cancel{120}}}}{\underset{\underset{3}{30}}{\cancel{\cancel{180}}}} = \dfrac{2}{3}$

Our first step is to divide by the common factor of 6. We then have $\dfrac{20}{30}$. There is still a common factor of 10, so we again divide.

Again, we could have removed the GCF of 60 (or 6×10) in one step if we had recognized it.

● ● ● ● **CHECK YOURSELF 5**

Using the method of Examples 4 and 5, write each of the fractions in simplest form.

a. $\dfrac{60}{75}$ **b.** $\dfrac{84}{196}$

To Review: In using division to reduce a fraction to lowest terms, we simply continue to divide by common factors until no common factors other than 1 exist in the numerator and denominator.

You have seen several approaches to reducing fractions. All are essentially the same; they are just different ways of writing down the steps. Experiment with the various methods, and use the one you are most comfortable with.

Will you be able to simplify or reduce all fractions? Division requires common factors in the numerator and denominator, and they may or may not exist. Let's look at an example.

● Example 6

Factoring to Simplify a Fraction

Reduce $\dfrac{14}{25}$ to lowest terms.

$$\frac{14}{25} = \frac{2 \times 7}{5 \times 5}$$

Looking at the factors of 14 and 25, we see that no common factors other than 1 exist. The fraction *cannot be reduced*. It is already in lowest terms.

● ● ●   CHECK YOURSELF 6

Write each of the following fractions in simplest form.

a. $\dfrac{16}{21}$

b. $\dfrac{105}{352}$

We mentioned earlier that it is very important to remember that the use of the fundamental principle means *dividing* the numerator and denominator by a common *factor*. Compare the two statements below.

$$\frac{12}{15} = \frac{4 \times \cancel{3}}{5 \times \cancel{3}} = \frac{4}{5}$$

This is correct. We have *divided* by the *common factor* of 3.

Now look at the following.

$$\frac{4}{5} = \frac{1 + \cancel{3}}{2 + \cancel{3}} \stackrel{?}{=} \frac{1}{2}$$

CAUTION

Note that cancellation *cannot* be performed when the numerator and/or denominator is separated by an addition or subtraction sign.

Since $\dfrac{4}{5}$ is not the same as $\dfrac{1}{2}$, we definitely have a problem now! Do you see what went wrong? In removing the "added 3" we have *subtracted* 3 in the numerator and denominator. This is *not* a legal step.

● ● ● **CHECK YOURSELF ANSWERS**

1. (a) $\dfrac{6}{9}$; (b) $\dfrac{2}{3}$.

2. 6 is the GCF of 12 and 18, so $\dfrac{12}{18} = \dfrac{12 \div 6}{18 \div 6} = \dfrac{2}{3}$.

3. $\dfrac{12}{18} = \dfrac{\cancel{2} \times 2 \times \cancel{3}}{\cancel{2} \times 3 \times \cancel{3}} = \dfrac{2}{3}$.

4. (a) $\dfrac{60}{75} = \dfrac{2 \times 2 \times \cancel{3} \times \cancel{5}}{\cancel{3} \times \cancel{5} \times 5} = \dfrac{4}{5}$; (b) $\dfrac{210}{252} = \dfrac{\cancel{2} \times \cancel{3} \times 5 \times \cancel{7}}{\cancel{2} \times 2 \times 3 \times \cancel{3} \times \cancel{7}} = \dfrac{5}{6}$.

5. (a) Divide by the common factors of 3 and 5. $\dfrac{60}{75} = \dfrac{4}{5}$.

  (b) Divide by the common factors of 4 and 7. $\dfrac{84}{196} = \dfrac{3}{7}$.

6. (a) Cannot be reduced; (b) cannot be reduced.

Write each fraction in simplest form.

1. $\dfrac{8}{12}$

2. $\dfrac{12}{15}$

3. $\dfrac{10}{14}$

4. $\dfrac{15}{50}$

5. $\dfrac{12}{18}$

6. $\dfrac{28}{35}$

7. $\dfrac{35}{40}$

8. $\dfrac{21}{24}$

9. $\dfrac{11}{44}$

10. $\dfrac{10}{25}$

11. $\dfrac{12}{36}$

12. $\dfrac{18}{48}$

13. $\dfrac{24}{27}$

14. $\dfrac{30}{50}$

15. $\dfrac{32}{40}$

16. $\dfrac{17}{51}$

17. $\dfrac{75}{105}$

18. $\dfrac{62}{93}$

19. $\dfrac{48}{60}$

20. $\dfrac{48}{66}$

21. $\dfrac{105}{135}$

22. $\dfrac{54}{126}$

ANSWERS

1. $\dfrac{2}{3}$

2. $\dfrac{4}{5}$

3. $\dfrac{5}{7}$

4. $\dfrac{3}{10}$

5. $\dfrac{2}{3}$

6. $\dfrac{4}{5}$

7. $\dfrac{7}{8}$

8. $\dfrac{7}{8}$

9. $\dfrac{1}{4}$

10. $\dfrac{2}{5}$

11. $\dfrac{1}{3}$

12. $\dfrac{3}{8}$

13. $\dfrac{8}{9}$

14. $\dfrac{3}{5}$

15. $\dfrac{4}{5}$

16. $\dfrac{1}{3}$

17. $\dfrac{5}{7}$

18. $\dfrac{2}{3}$

19. $\dfrac{4}{5}$

20. $\dfrac{8}{11}$

21. $\dfrac{7}{9}$

22. $\dfrac{3}{7}$

23. $\dfrac{66}{110}$

24. $\dfrac{280}{320}$

25. $\dfrac{16}{21}$

26. $\dfrac{21}{32}$

27. $\dfrac{31}{52}$

28. $\dfrac{42}{55}$

Solve the following applications.

29. **Coins.** A quarter is what fractional part of a dollar? Simplify your result.

30. **Coins.** A dime is what fractional part of a dollar? Simplify your result.

31. **Time.** What fractional part of an hour is 15 minutes (min)? Simplify your result.

32. **Time.** What fractional part of a day is 6 hours (h)? Simplify your result.

33. **Length.** A meter is equal to 100 centimeters (cm). What fractional part of a meter is 70 cm? Simplify your result.

34. **Length.** A kilometer is equal to 1000 meters (m). What fractional part of a kilometer is 300 m? Simplify your result.

35. **Auto repairs.** Susan did a tune-up on her automobile. She found that two of her eight spark plugs were fouled. What fraction represents the number of fouled plugs? Reduce to lowest terms.

36. **Testing.** Samantha answered 18 of 20 problems correctly on a test. What fractional part did she answer correctly? Reduce your answer to lowest terms.

37. **Baseball.** The local baseball team won 36 of the 58 games they played. What fractional part did they win? Reduce your answer to lowest terms.

38. **Salary.** Sharon earned $250 at her after-school job. A new bike costs $120. What fractional part of her money will remain after she purchases the bike? Reduce your answer to lowest terms.

39. A student is attempting to reduce the fraction $\dfrac{8}{12}$ to lowest terms. He produces the following argument:

$$\frac{8}{12} = \frac{4+4}{8+4} = \frac{4}{8} = \frac{1}{2}$$

What is the fallacy in this argument? What is the correct answer?

40. Can any of the following fractions be simplified?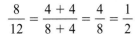

(a) $\dfrac{824}{73}$ (b) $\dfrac{59}{11}$ (c) $\dfrac{135}{17}$

What characteristic do you notice about the denominator of each fraction? What rule would you make up based on your observations?

 **Getting Ready for Section 5.5
[Section 4.4]**

Find the least common multiple (LCM) for the following numbers.

a. 5 and 6   **b.** 3 and 8

c. 6 and 8   **d.** 10 and 15

e. 5, 6, and 9   **f.** 12, 15, and 20

Answers

1. $\dfrac{2}{3}$ **3.** $\dfrac{5}{7}$ **5.** $\dfrac{2}{3}$ **7.** $\dfrac{7}{8}$ **9.** $\dfrac{1}{4}$ **11.** $\dfrac{1}{3}$ **13.** $\dfrac{8}{9}$ **15.** $\dfrac{4}{5}$

17. $\dfrac{5}{7}$ **19.** $\dfrac{4}{5}$ **21.** $\dfrac{7}{9}$ **23.** $\dfrac{3}{5}$ **25.** $\dfrac{16}{21}$ is already in simplest form.

27. $\dfrac{31}{52}$ is already in simplest form.   **29.** $\dfrac{1}{4}$ **31.** $\dfrac{1}{4}$ **33.** $\dfrac{7}{10}$ **35.** $\dfrac{1}{4}$

37. $\dfrac{18}{29}$ **39.**   **a.** 30   **b.** 24   **c.** 24   **d.** 30   **e.** 90

f. 60

ANSWERS

39.

40.

a. 30

b. 24

c. 24

d. 30

e. 90

Name

Section   Date

1. $\dfrac{2}{3}$

2. $\dfrac{3}{4}$

3. $\dfrac{7}{8}$

4. $\dfrac{2}{3}$

5. $\dfrac{8}{9}$

6. $\dfrac{5}{6}$

7. $\dfrac{6}{7}$

8. $\dfrac{7}{8}$

9. $\dfrac{3}{4}$

10. $\dfrac{3}{4}$

11. $\dfrac{10}{33}$

12. $\dfrac{2}{5}$

13. $\dfrac{29}{41}$

14. $\dfrac{24}{49}$

15. $\dfrac{4}{5}$

16. $\dfrac{7}{11}$

Write each fraction in simplest form.

1. $\dfrac{12}{18}$ 　　　　　　**2.** $\dfrac{9}{12}$

3. $\dfrac{14}{16}$ 　　　　　　**4.** $\dfrac{32}{48}$

5. $\dfrac{80}{90}$ 　　　　　　**6.** $\dfrac{30}{36}$

7. $\dfrac{48}{56}$ 　　　　　　**8.** $\dfrac{49}{56}$

9. $\dfrac{48}{64}$ 　　　　　　**10.** $\dfrac{36}{48}$

11. $\dfrac{50}{165}$ 　　　　　　**12.** $\dfrac{126}{315}$

13. $\dfrac{29}{41}$ 　　　　　　**14.** $\dfrac{24}{49}$

Solve the following applications.

15. Environment. A landfill can accept as many as 120 trucks each day. If the city garbage fleet has a total of 150 trucks, what fraction of the garbage fleet can use the landfill each day? Reduce to lowest terms.

16. Construction. If 8 of a total of 22 rolls of construction wire have been used, what fractional part of the wire is left? Reduce to lowest terms.

Using Your Calculator to Simplify Fractions

If you have a calculator that supports fraction arithmetic, you should learn to use it to check your work. We'll look at two different types of these calculators.

SCIENTIFIC CALCULATORS

Scientific calculators include the TI-34, the Casio fx-250 or fx115, and the Sharp 506g or 509g.

Before doing Example 1, find the button on your scientific calculator that is labeled $\boxed{\textbf{a b/c}}$. This is the button that will be used to enter fractions.

● Example 1

Using a Scientific Calculator to Simplify Fractions

Simplify the fraction $\frac{24}{68}$.

There are four steps in simplifying fractions using a scientific calculator.

1. Enter the numerator, 24.
2. Press the $\boxed{\textbf{a b/c}}$ key.
3. Enter the denominator, 68.
4. Press $\boxed{=}$.

The calculator will display the simplified fraction, 6/17.

● ● ● **CHECK YOURSELF 1**

Simplify the fraction $\frac{51}{81}$.

GRAPHING CALCULATORS

Let's simplify the same fraction, $\frac{24}{68}$, using a graphing calculator, such as the TI-82 or TI-83:

1. Enter the fraction as a division problem: 24 ÷ 68. The calculator will display 24/68.
2. Press the $\boxed{\textbf{MATH}}$ key.
3. Select $\boxed{\textbf{1:Frac.}}$.
4. Press $\boxed{\text{Enter}}$.

The calculator displays the simplified fraction.

Note: Some scientific calculators cannot handle denominators larger than 999.

The graphing calculator is particularly useful for simplifying fractions with large values in the numerator and denominator.

● Example 2

Using a Graphing Calculator to Simplify Fractions

Simplify $\dfrac{546}{637}$.

Using our calculator, we find that

$$\frac{546}{637} = \frac{6}{7}$$

● ● ● **CHECK YOURSELF 2**

Simplify $\dfrac{649}{885}$.

● ● ● **CHECK YOURSELF ANSWERS**

1. $\dfrac{17}{27}$ 2. $\dfrac{11}{15}$

Calculator Exercises

Name

Section   Date

ANSWERS

1. $\dfrac{7}{10}$

2. $\dfrac{11}{12}$

3. $\dfrac{2}{3}$

4. $\dfrac{7}{11}$

5. $\dfrac{5}{7}$

6. $\dfrac{3}{5}$

7. $\dfrac{13}{17}$

8. $\dfrac{8}{11}$

9. $\dfrac{17}{27}$

Use your calculator to simplify the following fractions.

1. $\dfrac{28}{40}$ 2. $\dfrac{121}{132}$ 3. $\dfrac{96}{144}$

4. $\dfrac{385}{605}$ 5. $\dfrac{445}{623}$ 6. $\dfrac{153}{255}$

7. $\dfrac{299}{391}$ 8. $\dfrac{152}{209}$ 9. $\dfrac{289}{459}$

Answers

1. $\dfrac{7}{10}$ 3. $\dfrac{2}{3}$ 5. $\dfrac{5}{7}$ 7. $\dfrac{13}{17}$ 9. $\dfrac{17}{27}$

Building Fractions

5.5 OBJECTIVES

1. Use the fundamental principle to build fractions.
2. Write a group of fractions as equivalent fractions with a common denominator.
3. Compare the size of a group of fractions.

Often we must **build a fraction** or **raise a fraction to higher terms.** This means finding an equivalent fraction with a numerator and denominator *larger* than those of the original fraction.

$\frac{1}{2}, \frac{2}{4}, \frac{3}{6}, \frac{4}{8}$, and $\frac{5}{10}$ are all equivalent fractions. We say that $\frac{2}{4}$ is in higher terms than $\frac{1}{2}$. We have *multiplied* the numerator and denominator by 2. In the same way, $\frac{3}{6}$ is in higher terms than $\frac{1}{2}$, as is $\frac{4}{8}$, and so on. Of course this uses the fundamental principle of fractions introduced in Section 5.3. It is restated here.

Do you see that this is just the opposite of simplifying a fraction?

Fundamental Principle of Fractions

$$\frac{a}{b} = \frac{a \times c}{b \times c} \qquad \text{for any nonzero number } c$$

In words, when the numerator and denominator of a fraction are multiplied by the same nonzero number, the result is an equivalent fraction.

• Example 1

Writing Equivalent Fractions

We can, of course, write as many equivalent fractions as we want by using different numbers as multipliers. For instance,

$$\frac{4}{5} = \frac{4 \times 4}{5 \times 4} = \frac{16}{20}$$

Write an equivalent fraction for $\frac{4}{5}$.

Let's use 3 as our multiplier.

$$\frac{4}{5} = \frac{4 \times 3}{5 \times 3} = \frac{12}{15}$$

$\frac{12}{15}$ is equivalent to $\frac{4}{5}$.

Note: Again, multiplying the numerator and denominator by 3 is the same as multiplying the fraction by $\frac{3}{3}$, which is just 1. Since multiplying by 1 does not change the value of the fraction, the result is equivalent to the original fraction.

● ● ●   **CHECK YOURSELF 1**

Write two fractions equivalent to $\dfrac{3}{7}$.

Many applications of our rule will involve building a fraction to an equivalent fraction with some specific denominator.

● Example 2

Building Equivalent Fractions

(*a*) Build $\dfrac{3}{4}$ to an equivalent fraction with denominator 12.

We can write

Think, "What must 4 be multiplied by to give 12?" The answer is 3, and so we multiply both the numerator and denominator by that number.

$$\dfrac{3}{4} = \dfrac{?}{12}$$
$$\underset{\times 3}{}$$

Applying our rule, we have

$$\overset{\times 3}{}$$
$$\dfrac{3}{4} = \dfrac{9}{12}$$
$$\underset{\times 3}{}$$

(*b*) Build $\dfrac{5}{8}$ to an equivalent fraction with a denominator of 48.

The original denominator must be multiplied by 6.

Think, "What must 8 be multiplied by to give 48?"

$$\overset{\times 6}{}$$
$$\dfrac{5}{8} = \dfrac{30}{48}$$
$$\underset{\times 6}{}$$

Remember: Multiply the numerator by the *same* number as you multiply the denominator by, in this case 6.

● ● ●   **CHECK YOURSELF 2**

a. Write $\dfrac{2}{3}$ as an equivalent fraction with a denominator of 6.

b. Write $\dfrac{3}{5}$ as an equivalent fraction with a denominator of 35.

Thus far, we have been working with individual fractions. Much of your work with fractions will require building each of a group of fractions to an equivalent fraction with a common (the same) denominator. When fractions have a common denominator, they are called **like fractions.**

$\frac{2}{3}$ and $\frac{3}{4}$ have different denominators.

$\frac{3}{5}$ and $\frac{4}{5}$ have the common denominator 5.

$\frac{3}{5}$ and $\frac{4}{5}$ are like fractions.

$\frac{2}{3}$ and $\frac{3}{4}$ are *not* like fractions.

Suppose you are asked to compare the sizes of the fractions $\frac{3}{7}$ and $\frac{4}{7}$. Since each unit in the diagram is divided into seven parts, it is easy to see that $\frac{4}{7}$ is larger than $\frac{3}{7}$.

$\frac{4}{7}$ $\frac{3}{7}$

Four parts of seven are a greater portion than three parts. Now compare the size of the fractions $\frac{2}{5}$ and $\frac{3}{7}$.

$\frac{2}{5}$ $\frac{3}{7}$

We *cannot* compare fifths with sevenths! $\frac{2}{5}$ and $\frac{3}{7}$ are *not* like fractions. Since they name different ways of dividing the whole, the task of deciding which fraction is larger is not nearly so easy to answer.

In order to compare the sizes of fractions, we change them to equivalent fractions having a *common denominator*. This common denominator must be a multiple of each of the original denominators.

•Example 3

Finding Common Denominators to Order Fractions

Compare the sizes of $\dfrac{2}{5}$ and $\dfrac{3}{7}$.

The original denominators are 5 and 7. Since 35 is a common multiple of 5 and 7, let's use 35 as our common denominator.

$$\frac{2}{5} = \frac{14}{35} \quad \overset{\times 7}{\underset{\times 7}{}}$$

Think, "What must we multiply 5 by to get 35?" The answer is 7. Multiply the numerator and denominator by that number.

$$\frac{3}{7} = \frac{15}{35} \quad \overset{\times 5}{\underset{\times 5}{}}$$

Multiply the numerator and denominator by 5.

15 of 35 parts represents a greater portion of the whole than 14 parts.

Since $\dfrac{2}{5} = \dfrac{14}{35}$ and $\dfrac{3}{7} = \dfrac{15}{35}$, we see that $\dfrac{3}{7}$ is larger than $\dfrac{2}{5}$.

●●● **CHECK YOURSELF 3**

Which is larger, $\dfrac{5}{9}$ or $\dfrac{4}{7}$?

Let's consider an example that uses the inequality notation.

•Example 4

Using an Inequality Symbol with Two Fractions

Use the symbol $<$ or $>$ to complete the statement below.

$$\frac{5}{8} \ \rule{2cm}{0.4pt} \ \frac{3}{5}$$

Once again we must compare the sizes of the two fractions, and this is done by converting the fractions to equivalent fractions with a common denominator. Here we will use 40 as that denominator.

$$\frac{5}{8} \quad \xrightarrow{\times 5} \quad \frac{25}{40} \qquad \frac{3}{5} \quad \xrightarrow{\times 8} \quad \frac{24}{40}$$

Since $\frac{5}{8}$ $\left(\text{or } \frac{25}{40}\right)$ is larger than $\frac{3}{5}$ $\left(\text{or } \frac{24}{40}\right)$, we write

$$\frac{5}{8} > \frac{3}{5}$$

●●● **CHECK YOURSELF 4**

Use the symbol < or > to complete the statement.

$$\frac{5}{9} \underline{\qquad} \frac{6}{11}$$

Comparing sizes is only one reason for changing fractions so that they have common denominators. Common denominators are also necessary in adding and subtracting fractions, as you will see in a later chapter.

Let's look at the process of finding a common denominator in greater detail. The best choice for the common denominator is the least common multiple (LCM) of the given denominators. Remember from Section 4.4 that this is the *smallest* number that is a multiple of each of the denominators. In our work with fractions, this will be called the **least common denominator (LCD).**

Using the smallest possible number for the common denominator simply lets us work with the fractions that make the calculation easiest.

To Review: How can you find the least common multiple of a group of numbers? Consider Example 5.

● Example 5

Finding the Least Common Multiple (LCM)

Find the least common multiple of 6, 8, and 12.
We write each number as a product of its prime factors.

$$
\begin{array}{l}
6 = 2 \times 3 \\
8 = 2 \times 2 \times 2 \\
12 = \underline{2 \times 2 \times 3} \\
 2 \times 2 \times 2 \times 3
\end{array}
$$

Note that 2 appears 3 times as a factor of 8, and so it must appear 3 times as a factor in forming the LCM.

The least common multiple is $2 \times 2 \times 2 \times 3$, or 24.

CHECK YOURSELF 5

Find the least common multiple of 9, 24, and 30.

We can use the LCM to write equivalent fractions. Example 6, in which a group of fractions with denominators 6, 8, and 12 is changed to equivalent fractions with denominator 24, illustrates the process.

●Example 6

Rewriting Fractions with the Least Common Denominator (LCD)

Write $\frac{5}{6}$, $\frac{3}{8}$, and $\frac{7}{12}$ as equivalent fractions with the LCD.

In Example 5, we found that the LCM for 6, 8, and 12 was 24.

$$\frac{5}{6} = \frac{20}{24}$$ (×4, ×4)

Think, "What must we multiply 6 by to get 24?" The answer is 4, so we multiply the numerator and denominator by 4.

$$\frac{3}{8} = \frac{9}{24}$$ (×3, ×3)

Repeat the process to multiply the numerator and denominator by 3.

$$\frac{7}{12} = \frac{14}{24}$$ (×2, ×2)

Repeat the process again to multiply the numerator and denominator by 2.

$\frac{20}{24}$, $\frac{9}{24}$, and $\frac{14}{24}$ all have the common denominator 24, and they are equivalent to the original fractions.

CHECK YOURSELF 6

Write $\frac{2}{3}$, $\frac{3}{4}$, and $\frac{5}{6}$ as equivalent fractions with the least common denominator.

Here is a similar example.

• Example 7

Rewriting Fractions with a Common Denominator

Convert $\dfrac{3}{8}$, $\dfrac{5}{12}$, and $\dfrac{7}{15}$ to equivalent fractions with the least common denominator.

Remember: Since 2 appears 3 times as a factor of 8, it must be used 3 times as a factor in forming the LCD.

Step 1 Find the LCD.

$$8 = 2 \times 2 \times 2$$
$$12 = 2 \times 2 \qquad \times 3$$
$$\underline{15 = \qquad\qquad\qquad 3 \times 5}$$
$$2 \times 2 \times 2 \times 3 \times 5 \text{ or } 120 \text{ is the LCD}$$

Step 2 Write equivalent fractions with that common denominator.

$120 \div 8 = 15$

$$\overset{\times 15}{\underset{\times 15}{\dfrac{3}{8} = \dfrac{45}{120}}}$$

We must multiply by 15 to change the original denominator, 8, to 120.

$120 \div 12 = 10$

$$\overset{\times 10}{\underset{\times 10}{\dfrac{5}{12} = \dfrac{50}{120}}}$$

Multiply the numerator and denominator by 10.

$120 \div 15 = 8$

$$\overset{\times 8}{\underset{\times 8}{\dfrac{7}{15} = \dfrac{56}{120}}}$$

Multiply the numerator and denominator by 8.

$\dfrac{45}{120}$, $\dfrac{50}{120}$, and $\dfrac{56}{120}$ are equivalent to the original fractions, and they now have the least common denominator.

● ● ● CHECK YOURSELF 7

Convert $\dfrac{4}{9}$, $\dfrac{5}{18}$, and $\dfrac{7}{12}$ to equivalent fractions with the least common denominator.

●●● **CHECK YOURSELF ANSWERS**

1. Two possible equivalent fractions are $\dfrac{6}{14}$ (multiply the numerator and denominator by 2) and $\dfrac{9}{21}$ (multiply by 3). Of course there are many more possibilities.

2. (a) $\dfrac{4}{6}$; **(b)** $\dfrac{21}{35}$.   **3.** $\dfrac{5}{9} = \dfrac{35}{63}$ and $\dfrac{4}{7} = \dfrac{36}{63}$, so $\dfrac{4}{7}$ is the larger fraction.

4. $\dfrac{5}{9} > \dfrac{6}{11}$.   **5.** 360.   **6.** $\dfrac{8}{12}, \dfrac{9}{12},$ and $\dfrac{10}{12}$.   **7.** $\dfrac{16}{36}, \dfrac{10}{36},$ and $\dfrac{21}{36}$.

Find the missing numerators.

1. $\dfrac{1}{2} = \dfrac{?}{8}$

2. $\dfrac{1}{3} = \dfrac{?}{21}$

3. $\dfrac{3}{7} = \dfrac{?}{21}$

4. $\dfrac{2}{9} = \dfrac{?}{36}$

5. $\dfrac{2}{5} = \dfrac{?}{60}$

6. $\dfrac{5}{6} = \dfrac{?}{48}$

7. $\dfrac{2}{7} = \dfrac{?}{35}$

8. $\dfrac{5}{8} = \dfrac{?}{96}$

9. $\dfrac{7}{11} = \dfrac{?}{99}$

10. $\dfrac{5}{9} = \dfrac{?}{81}$

11. $\dfrac{3}{8} = \dfrac{?}{32}$

12. $\dfrac{5}{11} = \dfrac{?}{33}$

13. $\dfrac{7}{9} = \dfrac{?}{108}$

14. $\dfrac{5}{16} = \dfrac{?}{352}$

15. $\dfrac{3}{10} = \dfrac{?}{200}$

16. $\dfrac{5}{16} = \dfrac{?}{144}$

Arrange the given fractions from smallest to largest.

17. $\dfrac{12}{17}, \dfrac{9}{10}$

18. $\dfrac{4}{9}, \dfrac{5}{11}$

19. $\dfrac{5}{8}, \dfrac{3}{5}$

20. $\dfrac{9}{10}, \dfrac{8}{9}$

21. $\dfrac{3}{8}, \dfrac{1}{3}, \dfrac{1}{4}$

22. $\dfrac{7}{12}, \dfrac{5}{18}, \dfrac{1}{3}$

ANSWERS

1. 4
2. 7
3. 9
4. 8
5. 24
6. 40
7. 10
8. 60
9. 63
10. 45
11. 12
12. 15
13. 84
14. 110
15. 60
16. 45
17. $\dfrac{12}{17}, \dfrac{9}{10}$
18. $\dfrac{4}{9}, \dfrac{5}{11}$
19. $\dfrac{3}{5}, \dfrac{5}{8}$
20. $\dfrac{8}{9}, \dfrac{9}{10}$
21. $\dfrac{1}{4}, \dfrac{1}{3}, \dfrac{3}{8}$
22. $\dfrac{5}{18}, \dfrac{1}{3}, \dfrac{7}{12}$

23. $\dfrac{11}{12}, \dfrac{4}{5}, \dfrac{5}{6}$

24. $\dfrac{5}{8}, \dfrac{9}{16}, \dfrac{13}{32}$

Complete the statements, using the symbol $<$ or $>$.

25. $\dfrac{5}{6}$ _____ $\dfrac{2}{5}$

26. $\dfrac{3}{4}$ _____ $\dfrac{10}{11}$

27. $\dfrac{4}{9}$ _____ $\dfrac{3}{7}$

28. $\dfrac{7}{10}$ _____ $\dfrac{11}{15}$

29. $\dfrac{7}{20}$ _____ $\dfrac{9}{25}$

30. $\dfrac{5}{12}$ _____ $\dfrac{7}{18}$

31. $\dfrac{5}{16}$ _____ $\dfrac{7}{20}$

32. $\dfrac{7}{12}$ _____ $\dfrac{9}{15}$

Write as equivalent fractions with the LCD as a common denominator.

33. $\dfrac{2}{5}, \dfrac{1}{4}$

34. $\dfrac{5}{6}, \dfrac{4}{5}$

35. $\dfrac{5}{8}, \dfrac{5}{12}$

36. $\dfrac{5}{14}, \dfrac{8}{21}$

37. $\dfrac{1}{2}, \dfrac{1}{3}, \dfrac{1}{4}$

38. $\dfrac{1}{5}, \dfrac{1}{3}, \dfrac{1}{6}$

39. $\dfrac{2}{15}, \dfrac{5}{7}, \dfrac{3}{5}$

40. $\dfrac{5}{8}, \dfrac{3}{10}, \dfrac{7}{12}$

Solve the following applications.

41. **Drill bits.** Three drill bits are marked $\dfrac{3}{8}, \dfrac{5}{16}$, and $\dfrac{11}{32}$. Which drill bit is largest?

42. **Bolt size.** Bolts can be purchased with diameters of $\dfrac{3}{8}, \dfrac{1}{4}$, or $\dfrac{3}{16}$ inches (in.). Which is smallest?

43. **Plywood size.** Plywood comes in thicknesses of $\frac{5}{8}$, $\frac{3}{4}$, $\frac{1}{2}$, and $\frac{3}{8}$ in. Which size is thickest?

44. **Doweling.** Doweling is sold with diameters of $\frac{1}{2}$, $\frac{9}{16}$, $\frac{5}{8}$, and $\frac{3}{8}$ in. Which size is smallest?

45.

46.

45. Josephine is asked to create a fraction equivalent to $\frac{1}{4}$. Her answer is $\frac{4}{7}$.

What did she do wrong? What would be a correct answer?

46. A sign on a busy highway says Exit 5A is $\frac{3}{4}$ mile away and Exit 5B is $\frac{5}{8}$ mile away. Which exit is first?

Answers

1. 4   **3.** 9   **5.** 24   **7.** 10   **9.** 63   **11.** 12   **13.** 84   **15.** 60

17. $\frac{12}{17}$, $\frac{9}{10}$ **19.** $\frac{3}{5}$, $\frac{5}{8}$ **21.** $\frac{1}{4}$, $\frac{1}{3}$, $\frac{3}{8}$ **23.** $\frac{4}{5}$, $\frac{5}{6}$, $\frac{11}{12}$ **25.** $\frac{5}{6} > \frac{2}{5}$

27. $\frac{4}{9} > \frac{3}{7}$ **29.** $\frac{7}{20} < \frac{9}{25}$ **31.** $\frac{5}{16} < \frac{7}{20}$ **33.** $\frac{8}{20}$, $\frac{5}{20}$ **35.** $\frac{15}{24}$, $\frac{10}{24}$

37. $\frac{6}{12}$, $\frac{4}{12}$, $\frac{3}{12}$ **39.** $\frac{14}{105}$, $\frac{75}{105}$, $\frac{63}{105}$ **41.** $\frac{3}{8}$ **43.** $\frac{3}{4}$ in.   **45.**

Find the missing numerators.

Name

Section   Date

ANSWERS

1. 32

2. 12

3. 24

4. 15

5. 63

6. 30

7. 105

8. 60

9. 28

10. $\dfrac{7}{12}, \dfrac{5}{8}$

11. $\dfrac{5}{8}, \dfrac{2}{3}, \dfrac{3}{4}$

12. $\dfrac{3}{8}, \dfrac{5}{12}, \dfrac{2}{3}$

13. <

14. <

15. >

16. >

17. $\dfrac{15}{35}, \dfrac{28}{35}$

18. $\dfrac{9}{48}, \dfrac{10}{48}$

19. $\dfrac{40}{48}, \dfrac{9}{48}, \dfrac{30}{48}$

20. $\dfrac{27}{36}, \dfrac{15}{36}, \dfrac{10}{36}$

---

1. $\dfrac{4}{7} = \dfrac{?}{56}$

2. $\dfrac{3}{4} = \dfrac{?}{16}$

3. $\dfrac{3}{4} = \dfrac{?}{32}$

4. $\dfrac{3}{11} = \dfrac{?}{55}$

5. $\dfrac{7}{8} = \dfrac{?}{72}$

6. $\dfrac{3}{8} = \dfrac{?}{80}$

7. $\dfrac{3}{5} = \dfrac{?}{175}$

8. $\dfrac{3}{20} = \dfrac{?}{400}$

Arrange the given fractions from smallest to largest.

9. $\dfrac{4}{7} = \dfrac{?}{49}$

10. $\dfrac{7}{12}, \dfrac{5}{8}$

11. $\dfrac{5}{8}, \dfrac{3}{4}, \dfrac{2}{3}$

12. $\dfrac{2}{3}, \dfrac{3}{8}, \dfrac{5}{12}$

Complete the statements using either < or > .

13. $\dfrac{2}{3}$ —— $\dfrac{3}{4}$

14. $\dfrac{9}{16}$ —— $\dfrac{7}{12}$

15. $\dfrac{9}{10}$ —— $\dfrac{13}{15}$

16. $\dfrac{9}{20}$ —— $\dfrac{13}{30}$

Write as equivalent fractions with the LCD as a common denominator.

17. $\dfrac{3}{7}, \dfrac{4}{5}$

18. $\dfrac{3}{16}, \dfrac{5}{24}$

19. $\dfrac{5}{6}, \dfrac{3}{16}, \dfrac{5}{8}$

20. $\dfrac{3}{4}, \dfrac{5}{12}, \dfrac{5}{18}$

Name _____

Section _____ Date _____

The purpose of the Self-Test is to help you check your progress and review for a chapter test in class. Allow yourself about 1 hour to take the test. When you are done, check your answers in the back of the book. If you missed any answers, be sure to go back and review the appropriate sections in the chapter and do the supplementary exercises provided.

[5.1] For Exercises 1 to 3, what fraction names the shaded part of each diagram? Identify the numerator and denominator.

1.

2.

3.

[5.2] **4.** Identify the proper fractions, improper fractions, and mixed numbers in the following group.

$$\frac{10}{11}, \frac{9}{5}, \frac{7}{7}, \frac{8}{1}, 2\frac{3}{5}, \frac{1}{8}$$

Proper $\frac{10}{11}, \frac{1}{8}$ Improper $\frac{9}{5}, \frac{7}{7}, \frac{8}{1}$ Mixed number $2\frac{3}{5}$

[5.2] **5.** Give the mixed number that names the shaded part of the following diagram.

[5.2] In Exercises 6 to 9, convert the fractions to mixed or whole numbers.

6. $\dfrac{17}{4} =$

7. $\dfrac{74}{8} =$

8. $\dfrac{18}{6} =$

9. $\dfrac{15}{1} =$

ANSWERS

1. $\dfrac{5}{6}$ 5 is the numerator; 6 is the denominator.

2. $\dfrac{5}{8}$ 5 is the numerator; 8 is the denominator.

3. $\dfrac{3}{5}$ 3 is the numerator; 5 is the denominator.

4. See exercise

5. $4\dfrac{1}{4}$

6. $4\dfrac{1}{4}$

7. $9\dfrac{1}{4}$

8. 3

9. 15

10. $\dfrac{37}{7}$

11. $\dfrac{35}{8}$

12. $\dfrac{74}{9}$

13. Yes

14. Yes

15. No

16. $\dfrac{7}{9}$

17. $\dfrac{3}{7}$

18. $\dfrac{8}{23}$

19. 28

20. 42

21. 105

22. $\dfrac{4}{7}, \dfrac{5}{8}$

23. $\dfrac{5}{14}, \dfrac{8}{21}$

24. $\dfrac{4}{15}, \dfrac{1}{3}, \dfrac{2}{5}$

25. $>$

26. $<$

27. $\dfrac{8}{20}, \dfrac{15}{20}$

28. $\dfrac{20}{72}, \dfrac{21}{72}$

29. $\dfrac{9}{24}, \dfrac{6}{24}, \dfrac{20}{24}$

30. $\dfrac{27}{60}, \dfrac{44}{60}, \dfrac{35}{60}$

[5.2]   In Exercises 10 to 12, convert the mixed numbers to improper fractions.

10. $5\dfrac{2}{7} =$ **11.** $4\dfrac{3}{8} =$ **12.** $8\dfrac{2}{9} =$

[5.3]   In Exercises 13 to 15, use the cross-product method to find out whether or not the pair of fractions is equivalent.

13. $\dfrac{2}{7}, \dfrac{8}{28}$ **14.** $\dfrac{8}{20}, \dfrac{12}{30}$ **15.** $\dfrac{3}{20}, \dfrac{2}{15}$

[5.4]   In Exercises 16 to 18, write the fractions in simplest form.

16. $\dfrac{21}{27} =$ **17.** $\dfrac{36}{84} =$ **18.** $\dfrac{8}{23} =$

[5.5]   In Exercises 19 to 21, find the missing numerators.

19. $\dfrac{4}{5} = \dfrac{?}{35}$ **20.** $\dfrac{3}{7} = \dfrac{?}{98}$ **21.** $\dfrac{7}{8} = \dfrac{?}{120}$

[5.5]   In Exercises 22 to 24, arrange the fractions from smallest to largest.

22. $\dfrac{4}{7}, \dfrac{5}{8}$ **23.** $\dfrac{8}{21}, \dfrac{5}{14}$ **24.** $\dfrac{2}{5}, \dfrac{1}{3}, \dfrac{4}{15}$

[5.5]   In Exercises 25 and 26, complete the statements using the symbol $<$ or $>$.

25. $\dfrac{3}{7}$ _____ $\dfrac{7}{18}$ **26.** $\dfrac{7}{12}$ _____ $\dfrac{11}{18}$

[5.5]   In Exercises 27 to 30, write the fractions as equivalent fractions with the LCD as a common denominator.

27. $\dfrac{2}{5}, \dfrac{3}{4}$ **28.** $\dfrac{5}{18}, \dfrac{7}{24}$

29. $\dfrac{3}{8}, \dfrac{1}{4}, \dfrac{5}{6}$ **30.** $\dfrac{9}{20}, \dfrac{11}{15}, \dfrac{7}{12}$

CHAPTER 6

THE MULTIPLICATION AND DIVISION OF FRACTIONS

INTRODUCTION

Colors, angles, shapes, light—all are topics of math and physics, but together they make up the primary elements of interior design.

Khalid struggled a little in his math and physics classes, but he has been well-rewarded for his diligence. When he set out to become an interior designer, Khalid had an image of taking only design classes. He felt that he had a great deal of talent and needed only to have that talent confirmed. Khalid was half right—he did have a great deal of talent. But now he understands why certain design elements look right and how to improve an existing design. As a result of his education, Khalid is a better designer and a better businessman.

© Steven Peters/Tony Stone Images

Pretest for Chapter 6

Name _____

Section _____ Date _____

ANSWERS

1. $\dfrac{15}{28}$

2. $4\dfrac{2}{3}$

3. $4\dfrac{1}{5}$

4. $\dfrac{2}{3}$

5. $12\dfrac{3}{5}$

6. $\dfrac{4}{5}$

7. $\dfrac{2}{21}$

8. $\dfrac{3}{4}$

9. $20\dfrac{5}{8}$ in.²

10. 14

Multiplication and Division of Fractions

This pretest will point out any difficulties you may be having with multiplying and dividing fractions. Do all the problems. Then check your answers with those in the back of the book. Perform each operation, and write each answer in simplest form.

1. $\dfrac{5}{8} \cdot \dfrac{6}{7}$ **2.** $6 \times \dfrac{7}{9}$

3. $2\dfrac{2}{5} \times 1\dfrac{3}{4}$ **4.** $\dfrac{18}{21} \times \dfrac{7}{9}$

5. $2\dfrac{2}{5} \times 5\dfrac{1}{4}$ **6.** $\dfrac{6}{25} \div \dfrac{3}{10}$

7. $\dfrac{2}{3} \div 7$ **8.** $1\dfrac{5}{6} \div 2\dfrac{4}{9}$

9. Area. A sheet of paper is $5\dfrac{1}{2}$ in. long by $3\dfrac{3}{4}$ in. wide. What is its area?

10. Size of cut. A piece of wood that is $15\dfrac{3}{4}$ in. long is to be cut into blocks $1\dfrac{1}{8}$ in. long. How many blocks can be cut?

6.1 Multiplying Fractions

6.1 OBJECTIVES

1. Multiply two fractions.
2. Multiply two fractions and give the product in lowest terms.

Multiplication is the easiest of the four operations with fractions. We can illustrate multiplication by picturing fractions as parts of a whole or unit. Using this idea, we show the fractions $\frac{4}{5}$ and $\frac{2}{3}$ in Figure 1.

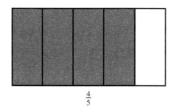 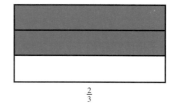

$\frac{4}{5}$ $\frac{2}{3}$

Figure 1

A fraction followed by the word "of" means that we want to multiply by that fraction.

Suppose now that we wish to find $\frac{2}{3}$ of $\frac{4}{5}$. We can combine the diagrams as shown in Figure 2. The part of the whole representing the product $\frac{2}{3} \times \frac{4}{5}$ is the purple region in Figure 2. The unit has been divided into 15 parts and 8 of those parts are used, so $\frac{2}{3} \times \frac{4}{5}$ must be $\frac{8}{15}$.

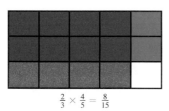

$\frac{2}{3} \times \frac{4}{5} = \frac{8}{15}$

Figure 2

The following rule is suggested by the diagrams.

This gives the numerator of the product.

This gives the denominator of the product.

To Multiply Fractions

STEP 1 Multiply the numerators.
STEP 2 Multiply the denominators.
STEP 3 Simplify the resulting fraction if possible.

Example 1 will require using steps 1 and 2.

• Example 1

Multiplying Two Fractions

Multiply.

We multiply fractions in this way *not* because it is easy, but because it works!

$$\frac{2}{3} \times \frac{4}{5} = \frac{2 \times 4}{3 \times 5} = \frac{8}{15}$$

$$\frac{5}{8} \times \frac{7}{9} = \frac{5 \times 7}{8 \times 9} = \frac{35}{72}$$

● ● ● CHECK YOURSELF 1

Multiply.

a. $\dfrac{7}{8} \times \dfrac{3}{10}$ **b.** $\dfrac{5}{7} \times \dfrac{3}{4}$

Step 3 indicates that the product of fractions should always be simplified to lowest terms. Consider the following.

• Example 2

Multiplying Two Fractions

Multiply and write the result in lowest terms.

$$\frac{3}{4} \times \frac{2}{9} = \frac{3 \times 2}{4 \times 9} = \frac{6}{36} = \frac{1}{6}$$

Noting that $\dfrac{6}{36}$ is not in simplest form, we divide numerator and denominator by 6 to write the product in lowest terms.

● ● ● CHECK YOURSELF 2

Multiply and write the result in lowest terms.

$$\frac{5}{7} \times \frac{3}{10}$$

● ● ● CHECK YOURSELF ANSWERS

1. **(a)** $\dfrac{7}{8} \times \dfrac{3}{10} = \dfrac{7 \times 3}{8 \times 10} = \dfrac{21}{80}$; **(b)** $\dfrac{5}{7} \times \dfrac{3}{4} = \dfrac{5 \times 3}{7 \times 4} = \dfrac{15}{28}$.

2. $\dfrac{5}{7} \times \dfrac{3}{10} = \dfrac{5 \times 3}{7 \times 10} = \dfrac{15}{70} = \dfrac{3}{14}$.

A N S W E R S

Multiply. Be sure to simplify each product.

1. $\dfrac{3}{4} \times \dfrac{5}{11}$ **2.** $\dfrac{2}{7} \times \dfrac{5}{9}$ **3.** $\dfrac{3}{4} \times \dfrac{7}{5}$

4. $\dfrac{2}{3} \times \dfrac{8}{5}$ **5.** $\dfrac{3}{5} \times \dfrac{5}{7}$ **6.** $\dfrac{6}{11} \times \dfrac{8}{6}$

7. $\dfrac{6}{13} \times \dfrac{4}{9}$ **8.** $\dfrac{5}{9} \times \dfrac{6}{11}$ **9.** $\dfrac{3}{11} \times \dfrac{7}{9}$

10. $\dfrac{7}{9} \times \dfrac{3}{5}$ **11.** $\dfrac{3}{10} \times \dfrac{5}{9}$ **12.** $\dfrac{5}{21} \times \dfrac{14}{25}$

13. $\dfrac{7}{9} \times \dfrac{6}{5}$ **14.** $\dfrac{8}{13} \times \dfrac{26}{5}$ **15.** Find $\dfrac{3}{4}$ of $\dfrac{6}{7}$.

16. Find $\dfrac{2}{3}$ of $\dfrac{7}{10}$.   **17.** What is $\dfrac{2}{7}$ of $\dfrac{7}{9}$?   **18.** What is $\dfrac{5}{8}$ of $\dfrac{12}{15}$?

 ## Getting Ready for Section 6.2 [Section 5.2]

Convert these mixed numbers to improper fractions.

a. $2\dfrac{1}{4}$ **b.** $2\dfrac{3}{4}$ **c.** $5\dfrac{1}{6}$

d. $7\dfrac{3}{8}$ **e.** $3\dfrac{2}{7}$ **f.** $6\dfrac{5}{7}$

1. $\dfrac{15}{44}$

2. $\dfrac{10}{63}$

3. $1\dfrac{1}{20}$

4. $1\dfrac{1}{15}$

5. $\dfrac{3}{7}$

6. $\dfrac{8}{11}$

7. $\dfrac{8}{39}$

8. $\dfrac{10}{33}$

9. $\dfrac{7}{33}$

10. $\dfrac{7}{15}$

11. $\dfrac{1}{6}$

12. $\dfrac{2}{15}$

13. $\dfrac{14}{15}$

14. $3\dfrac{1}{5}$

15. $\dfrac{9}{14}$

16. $\dfrac{7}{15}$

17. $\dfrac{2}{9}$

18. $\dfrac{1}{2}$

a. $\dfrac{9}{4}$

b. $\dfrac{11}{4}$

c. $\dfrac{31}{6}$

d. $\dfrac{59}{8}$

e. $\dfrac{23}{7}$

f. $\dfrac{47}{7}$

Answers

1. $\dfrac{15}{44}$ **3.** $\dfrac{3}{4} \times \dfrac{7}{5} = \dfrac{21}{20} = 1\dfrac{1}{20}$ **5.** $\dfrac{3}{7}$ **7.** $\dfrac{8}{39}$ **9.** $\dfrac{7}{33}$

11. $\dfrac{3}{10} \times \dfrac{5}{9} = \dfrac{15}{90} = \dfrac{1}{6}$ **13.** $\dfrac{14}{15}$

15. $\dfrac{3}{4}$ of $\dfrac{6}{7} = \dfrac{3}{4} \times \dfrac{6}{7} = \dfrac{3 \times 6}{4 \times 7} = \dfrac{18}{28} = \dfrac{9}{14}$ **17.** $\dfrac{2}{9}$ **a.** $\dfrac{9}{4}$ **b.** $\dfrac{11}{4}$

c. $\dfrac{31}{6}$ **d.** $\dfrac{59}{8}$ **e.** $\dfrac{23}{7}$ **f.** $\dfrac{47}{7}$

6.1 Supplementary Exercises

Name

Section   Date

A N S W E R S

1. $\dfrac{21}{44}$

2. $\dfrac{6}{35}$

3. $\dfrac{15}{44}$

4. $\dfrac{7}{27}$

5. $\dfrac{5}{14}$

6. $\dfrac{2}{5}$

7. $\dfrac{3}{8}$

8. $5\dfrac{5}{6}$

9. $3\dfrac{3}{10}$

10. $4\dfrac{4}{7}$

11. $\dfrac{7}{12}$

Multiply. Be sure to simplify each product.

1. $\dfrac{3}{4} \times \dfrac{7}{11}$ **2.** $\dfrac{2}{5} \times \dfrac{3}{7}$ **3.** $\dfrac{5}{8} \times \dfrac{6}{11}$

4. $\dfrac{4}{9} \times \dfrac{7}{12}$ **5.** $\dfrac{6}{7} \times \dfrac{5}{12}$ **6.** $\dfrac{5}{9} \times \dfrac{18}{25}$

7. $\dfrac{5}{6} \times \dfrac{9}{20}$ **8.** $7 \times \dfrac{5}{6}$ **9.** $\dfrac{3}{5} \times \dfrac{11}{2}$

10. Find $\dfrac{4}{7}$ of 8.   **11.** What is $\dfrac{2}{3}$ of $\dfrac{7}{8}$?

Multiplying Fractions and Mixed Numbers

6.2 OBJECTIVES

1. Multiply two mixed numbers.
2. Solve applications that involve mixed number multiplication.

To find the product of a fraction and a whole number, write the whole number as a fraction (the whole number divided by 1) and apply the multiplication rule as before. Example 1 illustrates this approach.

• Example 1

Multiplying a Whole Number and a Fraction

Do the indicated multiplication.

Remember that $5 = \dfrac{5}{1}$.

$(a)\ 5 \times \dfrac{3}{4} = \dfrac{5}{1} \times \dfrac{3}{4} = \dfrac{5 \times 3}{1 \times 4}$

$= \dfrac{15}{4} = 3\dfrac{3}{4}$

We have written the resulting improper fraction as a mixed number.

$(b)\ \dfrac{5}{12} \times 6 = \dfrac{5}{12} \times \dfrac{6}{1}$

$= \dfrac{5 \times 6}{12 \times 1}$

$= \dfrac{30}{12} = 2\dfrac{6}{12}$

$= 2\dfrac{1}{2}$

Write the product as a mixed number, then reduce the fractional portion to simplest form.

● ● ● **CHECK YOURSELF 1**

Multiply.

a. $\dfrac{3}{16} \times 8$

b. $4 \times \dfrac{5}{7}$

When mixed numbers are involved in multiplication, the problem requires an additional step. First, change any mixed numbers to improper fractions. Then apply our multiplication rule for fractions.

● Example 2

Multiplying a Mixed Number and a Fraction

$$1\frac{1}{2} \times \frac{3}{4} = \frac{3}{2} \times \frac{3}{4}$$

Change the mixed number to an improper fraction.

Here $1\frac{1}{2} = \frac{3}{2}$.

$$= \frac{3 \times 3}{2 \times 4}$$

Multiply as before.

$$= \frac{9}{8} = 1\frac{1}{8}$$

The product is usually written in mixed-number form.

● ● ● **CHECK YOURSELF 2**

Multiply.

$$\frac{5}{8} \times 3\frac{1}{2}$$

If two mixed numbers are involved, change both of the mixed numbers to improper fractions. Our third example illustrates.

● Example 3

Multiplying Two Mixed Numbers

Multiply.

$$3\frac{2}{3} \times 2\frac{1}{2} = \frac{11}{3} \times \frac{5}{2}$$

Change the mixed numbers to improper fractions.

$$= \frac{11 \times 5}{3 \times 2} = \frac{55}{6} = 9\frac{1}{6}$$

CAUTION

Be Careful! Students sometimes think of

$$3\frac{2}{3} \times 2\frac{1}{2} \qquad \text{as} \qquad (3 \times 2) + \left(\frac{2}{3} \times \frac{1}{2}\right)$$

This is *not* the correct multiplication pattern. You must first change the mixed numbers to improper fractions.

● ● ● **CHECK YOURSELF 3**

Multiply.

$$2\frac{1}{3} \times 3\frac{1}{2}$$

Again, be sure to reduce the product of the mixed numbers to simplest form. Consider Example 4.

●Example 4

Multiplying Two Mixed Numbers

Multiply.

$$5\frac{1}{3} \times 1\frac{1}{4} = \frac{16}{3} \times \frac{5}{4}$$

$$= \frac{16 \times 5}{3 \times 4} = \frac{80}{12} = 6\frac{8}{12} \qquad \text{Write the product as a mixed number.}$$

$$= 6\frac{2}{3} \qquad \text{Reduce the fractional portion to lowest terms.}$$

● ● ● **CHECK YOURSELF 4**

Multiply.

$$3\frac{3}{4} \times 1\frac{1}{5}$$

If a multiplication problem involves a whole number and a mixed number, write the mixed number as an improper fraction. Write the whole number as a fraction (the whole number over 1). Then multiply as before, as shown in Example 5.

●Example 5

Multiplying a Whole Number and a Mixed Number

Multiply.

$$8 \times 5\frac{1}{4} = \frac{8}{1} \times \frac{21}{4}$$

$$= \frac{8 \times 21}{1 \times 4} = \frac{168}{4} = 42$$

● ● ● **CHECK YOURSELF 5**

Multiply.

$$5\frac{1}{3} \times 9$$

Many applications can be solved by multiplying mixed numbers. The next example shows the process.

●Example 6

An Application Involving Multiplication of Mixed Numbers

Lisa worked $10\frac{1}{4}$ hours (h) each day for 5 days. How many hours did she work?

Multiply the number of hours per day by the number of days to find the total hours worked:

$$10\frac{1}{4} \times 5 = \frac{41}{4} \times \frac{5}{1} = \frac{205}{4} = 51\frac{1}{4}$$

Lisa worked $51\frac{1}{4}$ h.

● ● ● **CHECK YOURSELF 6**

Jesse bought three pieces of fabric. Each piece measured $4\frac{3}{5}$ yards (yd). How much fabric did Jesse buy? (*Hint:* Multiply 3 times $4\frac{3}{5}$.)

● ● ● **CHECK YOURSELF ANSWERS**

1. (a) $1\frac{1}{2}$; **(b)** $2\frac{6}{7}$.   **2.** $\frac{5}{8} \times 3\frac{1}{2} = \frac{5}{8} \times \frac{7}{2} = \frac{35}{16} = 2\frac{3}{16}$.   **3.** $8\frac{1}{6}$.

4. Write $3\frac{3}{4}$ as $\frac{15}{4}$ and $1\frac{1}{5}$ as $\frac{6}{5}$. Multiply as before. The product is $4\frac{1}{2}$.

5. 48.   **6.** $13\frac{4}{5}$ yd.

6.2 Exercises

Name

Section   Date

ANSWERS

Multiply. Be sure to simplify each product.

1. $1\frac{2}{5} \times \frac{3}{4}$

2. $1\frac{2}{3} \times \frac{1}{3}$

3. $\frac{5}{8} \times 1\frac{3}{4}$

4. $\frac{7}{8} \times 1\frac{3}{5}$

5. $3\frac{1}{3} \times \frac{9}{11}$

6. $\frac{2}{3} \times 2\frac{2}{5}$

7. $3\frac{1}{3} \times \frac{3}{7}$

8. $\frac{2}{5} \times 3\frac{1}{4}$

9. $2\frac{1}{3} \times 2\frac{1}{6}$

10. $2\frac{1}{3} \times 2\frac{1}{2}$

11. $1\frac{3}{4} \times 2\frac{3}{5}$

12. $3\frac{2}{3} \times 1\frac{1}{7}$

13. $3\frac{2}{5} \times 1\frac{2}{3}$

14. $3\frac{3}{4} \times 1\frac{1}{5}$

15. $5 \times \frac{4}{7}$

16. $\frac{7}{9} \times 5$

17. $\frac{3}{7} \times 14$

18. $9 \times \frac{5}{6}$

19. $15 \times \frac{5}{6}$

20. $\frac{7}{8} \times 16$

21. $1\frac{2}{5} \times 1\frac{1}{4}$

22. $3\frac{1}{8} \times 5\frac{1}{3}$

1. $1\frac{1}{20}$
2. $\frac{5}{9}$
3. $1\frac{3}{32}$
4. $1\frac{2}{5}$
5. $2\frac{8}{11}$
6. $1\frac{3}{5}$
7. $1\frac{3}{7}$
8. $1\frac{3}{10}$
9. $5\frac{1}{18}$
10. $5\frac{5}{6}$
11. $4\frac{11}{20}$
12. $4\frac{4}{21}$
13. $5\frac{2}{3}$
14. $4\frac{1}{2}$
15. $2\frac{6}{7}$
16. $3\frac{8}{9}$
17. 6
18. $7\frac{1}{2}$
19. $12\frac{1}{2}$
20. 14
21. $1\frac{3}{4}$
22. $16\frac{2}{3}$

© 1998 McGraw-Hill Companies

23. $7\dfrac{1}{2}$

24. $6\dfrac{2}{3}$

25. $23\dfrac{1}{3}$

26. 95

27. $4\dfrac{1}{3}$

28. $40\dfrac{1}{3}$

29. 36

30. 21

31. 4 cups

32. $3\dfrac{3}{4}$ cups

33. $2\dfrac{7}{8}$ ft²

34. $12\dfrac{1}{12}$ yd

35.

23. $2\dfrac{1}{2} \times 3$ **24.** $2 \times 3\dfrac{1}{3}$

25. $5 \times 4\dfrac{2}{3}$ **26.** $6\dfrac{1}{3} \times 15$

27. $1\dfrac{4}{9} \times 3$ **28.** $11 \times 3\dfrac{2}{3}$

29. $2\dfrac{2}{5} \times 15$ **30.** $6 \times 3\dfrac{1}{2}$

For each of the following applications, multiply to find the solution.

31. Recipes. A recipe calls for $\dfrac{2}{3}$ cup of sugar for each serving. How much sugar is needed for 6 servings?

32. Recipes. Mom-Mom's French toast requires $\dfrac{3}{4}$ cup of batter for each serving. If 5 people are expected for breakfast, how much batter is needed?

33. Gardening. A patch of dirt needs $3\dfrac{5}{6}$ square feet (ft²) of sod to cover it. If Nick decides to cover only $\dfrac{3}{4}$ of the dirt, how much sod does he need?

34. Construction. A driveway requires $4\dfrac{5}{6}$ yards (yd) of concrete to cover it. If Sheila wants to enlarge her driveway to $2\dfrac{1}{2}$ times its current size, how much concrete will she need?

35. A recipe calls for the following ingredients:

$\dfrac{7}{8}$ cup of flour, $\dfrac{3}{4}$ cup of sugar, $\dfrac{2}{3}$ cup of milk, and $\dfrac{5}{6}$ teaspoon of salt.

This recipe makes 8 servings. What amount of each quantity would you use if you wanted to serve 2 people?

36. Find the area of the following:

   (a) A 32¢ postage stamp   **(b)** A driver's license for your state

   (c) An $8\frac{1}{2}$ in. by 11 in. piece of paper   **(d)** A legal size envelope

Getting Ready for Section 6.3
[Section 5.4]

Write each fraction in simplest form.

a. $\dfrac{14}{16}$ **b.** $\dfrac{27}{30}$

c. $\dfrac{20}{35}$ **d.** $\dfrac{18}{24}$

e. $\dfrac{32}{40}$ **f.** $\dfrac{40}{60}$

Answers

1. $1\frac{1}{20}$ **3.** $1\frac{3}{32}$ **5.** $2\frac{8}{11}$ **7.** $3\frac{1}{3} \times \frac{3}{7} = \frac{10}{3} \times \frac{3}{7} = \frac{30}{21} = 1\frac{9}{21} = 1\frac{3}{7}$

9. $5\frac{1}{18}$ **11.** $4\frac{11}{20}$ **13.** $3\frac{2}{5} \times 1\frac{2}{3} = \frac{17}{5} \times \frac{5}{3} = \frac{85}{15} = 5\frac{10}{15} = 5\frac{2}{3}$

15. $2\frac{6}{7}$ **17.** 6 **19.** $12\frac{1}{2}$ **21.** $1\frac{3}{4}$ **23.** $7\frac{1}{2}$ **25.** $23\frac{1}{3}$ **27.** $4\frac{1}{3}$

29. 36 **31.** 4 cups   **33.** $2\frac{7}{8}$ ft^2 **35.**   **a.** $\frac{7}{8}$ **b.** $\frac{9}{10}$

c. $\frac{4}{7}$ **d.** $\frac{3}{4}$ **e.** $\frac{4}{5}$ **f.** $\frac{2}{3}$

36.

a. $\dfrac{7}{8}$

b. $\dfrac{9}{10}$

c. $\dfrac{4}{7}$

d. $\dfrac{3}{4}$

e. $\dfrac{4}{5}$

f. $\dfrac{2}{3}$

Name

Section   Date

ANSWERS

1. $\dfrac{1}{2}$

2. $\dfrac{9}{20}$

3. $1\dfrac{2}{9}$

4. $1\dfrac{2}{7}$

5. $4\dfrac{1}{3}$

6. $4\dfrac{1}{3}$

7. $16\dfrac{1}{3}$

8. $9\dfrac{3}{4}$

9. $9\dfrac{1}{3}$

10. $17\dfrac{1}{2}$

11. 27

12. 62

13. $6\dfrac{3}{5}$

14. $3\dfrac{3}{4}$

15. $1\dfrac{7}{9}$

16. $21\dfrac{1}{3}$

Multiply. Be sure each product is in simplest form.

1. $\dfrac{2}{5} \times 1\dfrac{1}{4}$

2. $2\dfrac{1}{4} \times \dfrac{1}{5}$

3. $1\dfrac{7}{15} \times \dfrac{5}{6}$

4. $\dfrac{3}{8} \times 3\dfrac{3}{7}$

5. $3\dfrac{1}{4} \times 1\dfrac{1}{3}$

6. $2\dfrac{3}{5} \times 1\dfrac{2}{3}$

7. $3\dfrac{1}{2} \times 4\dfrac{2}{3}$

8. $4\dfrac{1}{3} \times 2\dfrac{1}{4}$

9. $2\dfrac{1}{3} \times 4$

10. $5 \times 3\dfrac{1}{2}$

11. $6 \times 4\dfrac{1}{2}$

12. $7\dfrac{3}{4} \times 8$

13. $\dfrac{3}{5} \times 11$

14. $6 \times \dfrac{5}{8}$

15. $2 \times \dfrac{8}{9}$

16. $4 \times 5\dfrac{1}{3}$

Using Your Calculator to Multiply Fractions

 Scientific Calculator

To multiply fractions on a scientific calculator, you enter the first fraction, using the [**a b/c**] key, then press the multiplication sign, next enter the second fraction, then press the equals sign.

• Example 1

Multiplying Two Fractions

Find the product

$$\frac{7}{15} \times \frac{5}{21}$$

The keystroke sequence is

7 [**a b/c**] 15 [×] 5 [**a b/c**] 21 [=]

The result is $\frac{1}{9}$.

● ● ● **CHECK YOURSELF 1**

Find the product

$$\frac{24}{33} \times \frac{22}{39}$$

 GRAPHING CALCULATOR

When using a graphing calculator, you must choose the fraction option [**Frac.**] from the [**MATH**] menu before pressing [Enter].

For the fraction problem in Example 1, $\frac{7}{15} \times \frac{5}{21}$, the keystroke sequence is

7 [÷] 15 [×] 5 [÷] 21 [**Frac.**] [Enter]

Again, the result will be $\frac{1}{9}$.

● ● ● **CHECK YOURSELF ANSWER**

$$\frac{16}{39}$$

Name

Section   Date

A N S W E R S

1. $\dfrac{1}{2}$

2. $\dfrac{1}{6}$

3. $\dfrac{2}{5}$

4. $\dfrac{8}{35}$

5. $\dfrac{2}{7}$

6. $\dfrac{2}{7}$

7. $\dfrac{3}{2}$ or $1\dfrac{1}{2}$

8. $\dfrac{6}{7}$

9. $\dfrac{1}{3}$

10. $\dfrac{5}{24}$

11. $\dfrac{1}{10}$

12. $\dfrac{6}{77}$

13. $\dfrac{1}{5}$

14. $\dfrac{63}{170}$

Find the following products using your calculator.

1. $\dfrac{15}{20} \times \dfrac{8}{12}$

2. $\dfrac{7}{8} \times \dfrac{4}{21}$

3. $\dfrac{36}{55} \times \dfrac{33}{54}$

4. $\dfrac{28}{42} \times \dfrac{12}{35}$

5. $\dfrac{18}{84} \times \dfrac{36}{27}$

6. $\dfrac{6}{35} \times \dfrac{20}{12}$

7. $\dfrac{27}{26} \times \dfrac{13}{9}$

8. $\dfrac{32}{35} \times \dfrac{15}{16}$

9. $\dfrac{7}{12} \times \dfrac{36}{63}$

10. $\dfrac{8}{27} \times \dfrac{45}{64}$

11. $\dfrac{12}{45} \times \dfrac{27}{72}$

12. $\dfrac{18}{132} \times \dfrac{36}{63}$

13. $\dfrac{27}{72} \times \dfrac{24}{45}$

14. $\dfrac{81}{136} \times \dfrac{84}{135}$

Answers

1. $\dfrac{1}{2}$ **3.** $\dfrac{2}{5}$ **5.** $\dfrac{2}{7}$ **7.** $\dfrac{3}{2}$ or $1\dfrac{1}{2}$ **9.** $\dfrac{1}{3}$ **11.** $\dfrac{1}{10}$ **13.** $\dfrac{1}{5}$

Simplifying Before Multiplying Fractions and Mixed Numbers

6.3 OBJECTIVES

1. Use simplification in multiplying fractions.
2. Estimate products by rounding.
3. Use multiplication to solve applications.

In Sections 6.1 and 6.2, you saw that many of the products were not in lowest terms. When multiplying fractions, it is usually easier to simplify, that is, remove any common factors in the numerator and denominator, *before multiplying*. Remember that to simplify means to *divide* by the same common factor.

● Example 1

Simplifying Before Multiplying Two Fractions

Simplify and then multiply.

Once again we are applying the fundamental principle to divide the numerator and denominator by 3.

Since we divide by any common factors before we multiply, the resulting product *is in simplest form.*

$$\frac{3}{5} \times \frac{4}{9} = \frac{\overset{1}{\cancel{3}} \times 4}{5 \times \underset{3}{\cancel{9}}}$$

To simplify, we divide the *numerator* and *denominator* by the common factor 3. Remember that $\overset{1}{\cancel{3}}$ means $3 \div 3 = 1$. and $\underset{3}{\cancel{9}}$ means $9 \div 3 = 3$.

$$= \frac{1 \times 4}{5 \times 3}$$

$$= \frac{4}{15}$$

● ● ● CHECK YOURSELF 1

Simplify and then multiply.

$$\frac{7}{8} \times \frac{5}{21}$$

Our work in Example 1 leads to the following general rule about simplifying fractions in multiplication.

Simplifying Fractions Before Multiplying

In multiplying two or more fractions, we can divide any factor of the numerator and any factor of the denominator by the same nonzero number to simplify the product.

Example 2 further illustrates the use of this rule.

• Example 2

Simplifying Before Multiplying Two Fractions

Simplify and then multiply.

$$\frac{6}{25} \times \frac{20}{9} = \frac{\overset{2}{\cancel{6}} \times \overset{4}{\cancel{20}}}{\underset{5}{\cancel{25}} \times \underset{3}{\cancel{9}}}$$

We simplify by removing the common factor of 3 from 6 and 9 and then the common factor of 5 from 20 and 25.

$$= \frac{2 \times 4}{5 \times 3}$$

$$= \frac{8}{15}$$

As you may have observed, simplifying before you multiply makes the problem much easier. You'll get the same answer if you multiply and then reduce to lowest terms. It's just a lot more work that way.

CHECK YOURSELF 2

Simplify and then multiply.

$$\frac{5}{12} \times \frac{8}{15}$$

Simplifying is also useful when the multiplication involves whole or mixed numbers, as is shown in Example 3.

• Example 3

Simplifying Before Multiplying a Fraction and a Whole Number

Simplify and then multiply.

$$5 \times \frac{4}{25} = \frac{5}{1} \times \frac{4}{25}$$

Remember that the whole number 5 is written as $\frac{5}{1}$ for our first step.

$$= \frac{\overset{1}{\cancel{5}} \times 4}{1 \times \underset{5}{\cancel{25}}}$$

To simplify, divide by the common factor of 5.

$$= \frac{4}{5}$$

● ● ● **CHECK YOURSELF 3**

Simplify and then multiply.

$$\frac{9}{16} \times 12$$

When mixed numbers are involved, the process is similar. Consider Example 4.

● Example 4

Simplifying Before Multiplying Two Mixed Numbers

Multiply.

$$2\frac{2}{3} \times 2\frac{1}{4} = \frac{8}{3} \times \frac{9}{4}$$

First, convert the mixed numbers to improper fractions.

$$= \frac{\overset{2}{\cancel{8}} \times \overset{3}{\cancel{9}}}{\underset{1}{\cancel{3}} \times \underset{1}{\cancel{4}}}$$

To simplify, divide by the common factors of 3 and 4.

$$= \frac{2 \times 3}{1 \times 1}$$

Multiply as before.

$$= \frac{6}{1} = 6$$

● ● ● **CHECK YOURSELF 4**

Simplify and then multiply.

$$3\frac{1}{3} \times 2\frac{2}{5}$$

The ideas of our previous examples will also allow us to find the product of more than two fractions.

•Example 5

Simplifying Before Multiplying Three Numbers

Simplify and then multiply.

Remember our earlier rule: We can divide any factor of the numerator and any factor of the denominator by the same nonzero number.

$$\frac{2}{3} \times 1\frac{4}{5} \times \frac{5}{8} = \frac{2}{3} \times \frac{9}{5} \times \frac{5}{8}$$

Write any mixed or whole numbers as improper fractions.

$$= \frac{\overset{1}{2} \times \overset{3}{9} \times \overset{1}{5}}{\underset{1}{3} \times \underset{1}{5} \times \underset{4}{8}}$$

To simplify, divide by the common factors in the numerator and denominator.

$$= \frac{3}{4}$$

●●● **CHECK YOURSELF 5**

Simplify and then multiply.

$$\frac{5}{8} \times 4\frac{4}{5} \times \frac{1}{6}$$

We encountered estimation by rounding in our earlier work with whole numbers. Estimation can also be used to check the "reasonableness" of an answer when we are working with fractions or mixed numbers.

•Example 6

Estimating the Product of Two Mixed Numbers

Estimate the product of

$$3\frac{1}{8} \times 5\frac{5}{6}$$

Round each mixed number to the nearest whole number.

$$3\frac{1}{8} \rightarrow 3$$

$$5\frac{5}{6} \rightarrow 6$$

Our estimate of the product is then

$$3 \times 6 = 18$$

Note: The actual product in this case is $18\frac{11}{48}$, which certainly seems reasonable in view of our estimate.

● ● ● **CHECK YOURSELF 6**

Estimate the product.

$$2\frac{7}{8} \times 8\frac{1}{3}$$

Let's look at some applications of our work with the multiplication of fractions. In solving these word problems, we will use the same approach we used earlier with whole numbers. Let's review the four-step process introduced in Section 1.8.

Solving Applications Involving the Multiplication of Fractions

STEP 1 Read the problem carefully to determine the given information and what you are asked to find.
STEP 2 Decide upon the operation or operations to be used.
STEP 3 Write down the complete statement necessary to solve the problem and do the calculations.
STEP 4 Check to make sure that you have answered the question of the problem and that your answer seems reasonable.

Let's work through some examples, using these steps.

● **Example 7**

An Application Involving the Multiplication of Fractions

A grocery store survey shows that $\frac{2}{3}$ of the customers will buy meat. Of these, $\frac{3}{4}$ will buy at least one package of beef. What portion of the store's customers will buy beef?

Step 1 We know that $\frac{2}{3}$ of the customers will buy meat and that $\frac{3}{4}$ of these customers will buy beef.

Remember: In this problem, "of" means to multiply.

Step 2 We wish to know $\frac{3}{4}$ of $\frac{2}{3}$. The operation here is multiplication.

Step 3 Multiplying, we have

$$\frac{\overset{1}{\cancel{3}}}{\underset{2}{\cancel{4}}} \times \frac{\overset{1}{\cancel{2}}}{\underset{1}{\cancel{3}}} = \frac{1}{2}$$

Step 4 From step 3 we have the result: $\frac{1}{2}$ of the store's customers will buy beef.

● ● ● **CHECK YOURSELF 7**

A supermarket survey shows that $\dfrac{2}{5}$ of the customers will buy lunch meat. Of these, $\dfrac{3}{4}$ will buy boiled ham. What portion of the store's customers will buy boiled ham?

●**Example 8**

An Application Involving the Multiplication of Mixed Numbers

A sheet of notepaper is $6\dfrac{3}{4}$ inches (in.) wide by $8\dfrac{2}{3}$ in. long. Find the area of the paper.

Recall that the area of a rectangle is the product of its length and its width.

Solution Multiply the given length by the width. This will give the desired area.

$$8\dfrac{2}{3} \times 6\dfrac{3}{4} = \dfrac{\overset{13}{\cancel{26}}}{\underset{1}{\cancel{3}}} \times \dfrac{\overset{9}{\cancel{27}}}{\underset{2}{\cancel{4}}}$$

$$= \dfrac{117}{2} = 58\dfrac{1}{2} \text{ in.}^2$$

● ● ● **CHECK YOURSELF 8**

A window is $4\dfrac{1}{2}$ feet (ft) high by $2\dfrac{1}{3}$ ft wide. What is its area?

●**Example 9**

An Application Involving the Multiplication of a Mixed Number and a Fraction

A state park contains $38\dfrac{2}{3}$ acres. According to the plan for the park, $\dfrac{3}{4}$ of the park is to be left as a wildlife preserve. How many acres will this be?

The word "of" indicates multiplication.

Solution We want to find $\dfrac{3}{4}$ of $38\dfrac{2}{3}$ acres. We then multiply as shown:

$$\dfrac{3}{4} \times 38\dfrac{2}{3} = \dfrac{\overset{1}{\cancel{3}}}{\underset{1}{\cancel{4}}} \times \dfrac{\overset{29}{\cancel{116}}}{\underset{1}{\cancel{3}}} = 29 \text{ acres}$$

• • • CHECK YOURSELF 9

A backyard has $25\frac{3}{4}$ square yards (yd^2) of open space. If Patrick wants to build a vegetable garden covering $\frac{2}{3}$ of the open space, how many square yards will this be?

• Example 10

An Application Involving the Multiplication of a Whole Number and a Mixed Number

Shirley drives at an average speed of 52 miles per hour (mi/h) for $3\frac{1}{4}$ h. How far has she traveled at the end of $3\frac{1}{4}$ h?

Solution

Remember: Distance is the product of speed and time.

$$52 \times 3\frac{1}{4} = \frac{52}{1} \times \frac{13}{4}$$

Speed (mi/h)   Time (h)

$$= \frac{\overset{13}{\cancel{52}} \times 13}{1 \times \underset{1}{\cancel{4}}}$$

$$= 169 \text{ mi}$$

• • • CHECK YOURSELF 10

The scale on a map is 1 inch (in.) = 60 miles (mi). What is the distance in miles between two towns that are $3\frac{1}{2}$ in. apart on the map?

• Example 11

An Application Involving the Multiplication of Mixed Numbers

Lin is going to pour a new concrete patio. He wants the patio to be $\frac{1}{9}$ yard (yd) thick (that's 4 in.), $5\frac{1}{2}$ yd long, and $4\frac{1}{2}$ yd wide. He can order only whole numbers of cubic yards (yd^3) of concrete. How much concrete should he order?

Solution Lin needs to determine the volume of concrete he needs:

Volume = Length × Width × Height

$$= 5\frac{1}{2} \times 4\frac{1}{2} \times \frac{1}{9}$$

$$= \frac{11}{2} \times \frac{9}{2} \times \frac{1}{9}$$

$$= \frac{11 \times \overset{1}{9} \times 1}{2 \times 2 \times \underset{1}{9}}$$

$$= \frac{11}{4} \text{ or } 2\frac{3}{4} \text{ yd}^3$$

Since Lin must order whole-number amounts, he needs to order 3 yd^3.

● ● ● CHECK YOURSELF 11

Maria is ordering concrete for a new sidewalk that is to be $\frac{1}{9}$ yd thick, $22\frac{1}{2}$ yd long, and $1\frac{1}{3}$ yd wide. How much concrete should she order if she must order a whole number of cubic yards?

● ● ● CHECK YOURSELF ANSWERS

1. $\dfrac{7}{8} \times \dfrac{5}{21} = \dfrac{\overset{1}{7} \times 5}{8 \times \underset{3}{21}} = \dfrac{5}{24}$.   **2.** $\dfrac{5}{12} \times \dfrac{8}{15} = \dfrac{\overset{1}{5} \times \overset{2}{8}}{\underset{3}{12} \times \underset{3}{15}} = \dfrac{2}{9}$.   **3.** $6\dfrac{3}{4}$.

4. $3\dfrac{1}{3} \times 2\dfrac{2}{5} = \dfrac{10}{3} \times \dfrac{12}{5} = \dfrac{\overset{2}{10} \times \overset{4}{12}}{\underset{1}{3} \times \underset{1}{5}} = \dfrac{8}{1} = 8$.   **5.** $\dfrac{1}{2}$.   **6.** 24.   **7.** $\dfrac{3}{10}$.

8. $10\dfrac{1}{2}$ ft^2.   **9.** $17\dfrac{1}{6}$ yd^2.   **10.** 210 mi.

11. The answer, $3\dfrac{1}{3}$ yd^3, is rounded up to 4 yd^3.

6.3 Exercises

Multiply:

1. $\dfrac{3}{7} \times \dfrac{1}{9}$ **2.** $\dfrac{3}{7} \times \dfrac{5}{9}$ **3.** $\dfrac{6}{11} \times \dfrac{33}{12}$

4. $\dfrac{12}{25} \times \dfrac{11}{18}$ **5.** $\dfrac{10}{12} \times \dfrac{16}{25}$ **6.** $\dfrac{14}{15} \times \dfrac{10}{21}$

7. $\dfrac{21}{25} \times \dfrac{30}{7}$ **8.** $\dfrac{18}{28} \times \dfrac{35}{22}$ **9.** $3\dfrac{2}{3} \times \dfrac{9}{10}$

10. $\dfrac{4}{9} \times 3\dfrac{3}{5}$ **11.** $5\dfrac{1}{3} \times \dfrac{7}{8}$ **12.** $\dfrac{10}{27} \times 3\dfrac{3}{5}$

13. $1\dfrac{1}{3} \times 1\dfrac{1}{5}$ **14.** $2\dfrac{2}{5} \times 3\dfrac{3}{4}$ **15.** $2\dfrac{2}{7} \times 2\dfrac{1}{3}$

16. $7\dfrac{1}{5} \times 4\dfrac{1}{6}$ **17.** $3\dfrac{3}{7} \times 2\dfrac{5}{8}$ **18.** $4\dfrac{3}{8} \times 1\dfrac{5}{7}$

19. $6 \times 2\dfrac{1}{3}$ **20.** $1\dfrac{3}{10} \times 5$ **21.** $4\dfrac{2}{7} \times 8$

1.	$\dfrac{1}{21}$
2.	$\dfrac{5}{21}$
3.	$\dfrac{3}{2}$
4.	$\dfrac{22}{75}$
5.	$\dfrac{8}{15}$
6.	$\dfrac{4}{9}$
7.	$3\dfrac{3}{5}$
8.	$1\dfrac{1}{44}$
9.	$3\dfrac{3}{10}$
10.	$1\dfrac{3}{5}$
11.	$4\dfrac{2}{3}$
12.	$1\dfrac{1}{3}$
13.	$1\dfrac{3}{5}$
14.	9
15.	$5\dfrac{1}{3}$
16.	30
17.	9
18.	$7\dfrac{1}{2}$
19.	14
20.	$6\dfrac{1}{2}$
21.	$34\dfrac{2}{7}$

22. $8 \times 2\frac{3}{4}$

23. $\frac{7}{12} \times \frac{3}{4} \times \frac{8}{15}.$

24. $\frac{5}{18} \times \frac{2}{3} \times \frac{9}{10}$

25. $4\frac{1}{5} \times \frac{10}{21} \times \frac{9}{20}$

26. $\frac{7}{8} \times 5\frac{1}{3} \times \frac{5}{14}$

27. $3\frac{1}{3} \times \frac{4}{5} \times 1\frac{1}{8}$

28. $4\frac{1}{2} \times 5\frac{5}{6} \times \frac{8}{15}$

29. Find $\frac{2}{3}$ of $\frac{3}{7}.$

30. What is $\frac{5}{6}$ of $\frac{9}{10}$?

31. What is $\frac{3}{5}$ of 15?

32. Find $\frac{4}{7}$ of 28.

33. Find $\frac{3}{4}$ of $2\frac{2}{5}.$

34. What is $\frac{5}{8}$ of $1\frac{5}{7}$?

35. What is $\frac{6}{7}$ of $2\frac{4}{5}$?

36. Find $\frac{4}{9}$ of $3\frac{3}{5}.$

Estimate the following products.

37. $3\frac{1}{5} \times 4\frac{2}{3}$

38. $5\frac{1}{7} \times 2\frac{2}{13}$

39. $11\frac{3}{4} \times 5\frac{1}{4}$

40. $3\frac{4}{5} \times 5\frac{6}{7}$

41. $8\frac{2}{9} \times 7\frac{11}{12}$

42. $\frac{9}{10} \times 2\frac{2}{7}$

Solve the following applications.

43. **Map scales.** The scale on a map is 1 inch (in.) = 200 miles (mi). What actual distance, in miles, does $\frac{3}{8}$ in. represent?

44. **Salary.** You make $90 a day on a job. What will you receive for working $\frac{3}{4}$ of a day?

45. **Size.** A lumberyard has a stack of 80 sheets of plywood. If each sheet is $\frac{3}{4}$ in. thick, how high will the stack be?

46. **Family budget.** A family uses $\frac{2}{5}$ of its monthly income for housing and utilities on average. If the family's monthly income is $1750, what is spent for housing and utilities? What amount remains?

47. **Elections.** Of the eligible voters in an election, $\frac{3}{4}$ were registered. Of those registered, $\frac{5}{9}$ actually voted. What fraction of those people who were eligible voted?

48. **Surveys.** A survey has found that $\frac{7}{10}$ of the people in a city own pets. Of those who own pets, $\frac{2}{3}$ have dogs. What fraction of those surveyed own dogs?

49. **Area.** A kitchen has dimensions $3\frac{1}{3}$ by $3\frac{3}{4}$ yards (yd). How many square yards (yd^2) of linoleum must be bought to cover the floor?

50. **Distance.** If you drive at an average speed of 52 miles per hour (mi/h) for $1\frac{3}{4}$ h, how far will you travel in $1\frac{3}{4}$ h?

A N S W E R S

43. 75 mi

44. $67.50

45. 60 in.

46. $700, $1050

47. $\frac{5}{12}$

48. $\frac{7}{15}$

49. $12\frac{1}{2}$ yd^2

50. 91 mi

51. 2520 mi

52. $8\frac{1}{3}$ acres

53. 66 in.

54. $\frac{9}{13}$ yd^2

55. $42\frac{9}{64}$ in.^{3}

56. 27 ft^3

57. 4 in.^{2}

58. $2\frac{1}{4}$ in.^{2}

59.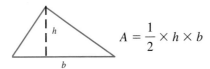

51. Distance. A jet flew at an average speed of 540 mi/h on a $4\frac{2}{3}$-h flight. What was the distance flown?

52. Area. A piece of land that has $11\frac{2}{3}$ acres is being subdivided for home lots. It is estimated that $\frac{2}{7}$ of the area will be used for roads. What amount remains to be used for lots?

53. Circumference. To find the approximate circumference or distance around a circle, we multiply its diameter by $\frac{22}{7}$. What is the circumference of a circle with a diameter of 21 in.?

54. Area. The length of a rectangle is $\frac{6}{7}$ yd, and its width is $\frac{21}{26}$ yd. What is its area in square yards?

55. Volume. Find the volume of a box that measures $2\frac{1}{4}$ in. by $3\frac{7}{8}$ in. by $4\frac{5}{6}$ in.

56. Topsoil. Nico wishes to purchase mulch to cover his garden. The garden measures $7\frac{7}{8}$ feet (ft) by $10\frac{1}{8}$ ft. He wants the mulch to be $\frac{1}{3}$ ft deep. How much mulch should Nico order if he must order a whole number of cubic feet?

The formula for the area of a triangle is

$$A = \frac{1}{2} \times h \times b$$

where h is the height of the triangle and b is the base.

57. Find the area of a triangle with a height of $2\frac{2}{5}$ in. and a base of $3\frac{1}{3}$ in.

58. Find the area of a triangle with a height of $1\frac{7}{8}$ in. and base of $2\frac{2}{5}$ in.

59. Obtain a map of your state and, using the legend provided, determine the distance from your state capital and any other city. Would this be the actual distance you would travel by car if you made the journey? Why or why not?

Getting Ready for Section 6.4
[Sections 6.1 and 6.2]

Multiply.

a. $\dfrac{2}{3} \times \dfrac{3}{2}$ **b.** $\dfrac{4}{5} \times \dfrac{5}{4}$ **c.** $\dfrac{4}{7} \times \dfrac{7}{4}$

d. $\dfrac{3}{10} \times \dfrac{10}{3}$ **e.** $5 \times \dfrac{1}{5}$ **f.** $9 \times \dfrac{1}{9}$

A N S W E R S

a. 1

b. 1

c. 1

d. 1

e. 1

f. 1

Answers

1. $\dfrac{1}{21}$ **3.** $\dfrac{\cancel{6}^{1}}{\cancel{11}_{1}} \times \dfrac{\cancel{33}^{3}}{\cancel{12}_{2}} = \dfrac{3}{2} = 1\dfrac{1}{2}$ **5.** $\dfrac{8}{15}$ **7.** $3\dfrac{3}{5}$ **9.** $3\dfrac{3}{10}$

11. $5\dfrac{1}{3} \times \dfrac{7}{8} = \dfrac{\cancel{16}^{2}}{3} \times \dfrac{7}{\cancel{8}_{1}} = \dfrac{14}{3} = 4\dfrac{2}{3}$ **13.** $1\dfrac{3}{5}$ **15.** $5\dfrac{1}{3}$ **17.** 9

19. $6 \times 2\dfrac{1}{3} = \dfrac{\cancel{6}^{2}}{1} \times \dfrac{7}{\cancel{3}_{1}} = 14$ **21.** $34\dfrac{2}{7}$ **23.** $\dfrac{7}{30}$ **25.** $\dfrac{9}{10}$ **27.** 3

29. $\dfrac{2}{7}$ **31.** $\dfrac{3}{5}$ of 15 is $\dfrac{3}{5} \times 15 = \dfrac{3}{\cancel{5}_{1}} \times \dfrac{\cancel{15}^{3}}{1} = 9$ **33.** $1\dfrac{4}{5}$ **35.** $2\dfrac{2}{5}$ **37.** 15

39. 60   **41.** 64   **43.** 75 mi   **45.** 60 in.   **47.** $\dfrac{5}{12}$ **49.** $12\dfrac{1}{2}$ yd^2

51. 2520 mi   **53.** 66 in.   **55.** $42\dfrac{9}{64}$ in.^{3} **57.** 4 in.^{2} **59.**

a. 1   **b.** 1   **c.** 1   **d.** 1   **e.** 1   **f.** 1

6.3 Supplementary Exercises

Name

Section   Date

Multiply:

1. $\dfrac{3}{7} \times \dfrac{7}{8}$ **2.** $\dfrac{4}{9} \times \dfrac{3}{7}$ **3.** $\dfrac{11}{18} \times \dfrac{10}{22}$

4. $\dfrac{7}{18} \times \dfrac{12}{25}$ **5.** $2\dfrac{2}{5} \times \dfrac{3}{4}$ **6.** $4\dfrac{2}{7} \times \dfrac{14}{15}$

A N S W E R S

1. $\dfrac{3}{8}$

2. $\dfrac{4}{21}$

3. $\dfrac{5}{18}$

4. $\dfrac{14}{75}$

5. $1\dfrac{4}{5}$

6. 4

7. $3\dfrac{3}{5}$

8. $3\dfrac{1}{9}$

9. 3

10. 13

11. 34

12. $\dfrac{1}{20}$

13. 3

14. 12

15. 22

16. $7\dfrac{1}{2}$

17. $2\dfrac{1}{7}$

18. $1\dfrac{1}{3}$

19. $\$45$

20. 156 mi

21. 408 mi

22. $\dfrac{1}{10}$

23. $\dfrac{5}{8}$

24. $25\dfrac{1}{2}$ yd²

7. $3\dfrac{1}{5} \times 1\dfrac{1}{8}$

8. $1\dfrac{2}{5} \times 2\dfrac{2}{9}$

9. $1\dfrac{1}{9} \times 2\dfrac{7}{10}$

10. $6 \times 2\dfrac{1}{6}$

11. $3\dfrac{2}{5} \times 10$

12. $\dfrac{2}{5} \times \dfrac{3}{4} \times \dfrac{1}{6}$

13. $2\dfrac{1}{2} \times \dfrac{9}{10} \times 1\dfrac{1}{3}$

14. $2\dfrac{2}{5} \times 1\dfrac{7}{8} \times 2\dfrac{2}{3}$

15. What is $\dfrac{2}{3}$ of 33?

16. Find $\dfrac{3}{10}$ of 25.

17. Find $\dfrac{3}{4}$ of $2\dfrac{6}{7}$.

18. What is $\dfrac{3}{4}$ of $1\dfrac{7}{9}$?

Solve the following applications.

19. **Earnings.** Maria earns \$72 per day. If she works $\dfrac{5}{8}$ of a day, how much will she earn?

20. **Miles traveled.** David drove at an average speed of 65 mi/h for $2\dfrac{2}{5}$ h. How many miles did he travel?

21. **Distance.** The scale on a map is 1 in. = 120 mi. What actual distance, in miles, does $3\dfrac{2}{5}$ in. on the map represent?

22. **Student numbers.** At a college, $\dfrac{2}{5}$ of the students take a science course. Of the students taking science, $\dfrac{1}{4}$ take biology. What fraction of the students take biology?

23. **Student workers.** A student survey found that $\dfrac{3}{4}$ of the students have jobs while going to school. Of those who have jobs, $\dfrac{5}{6}$ work more than 20 h per week. What fraction of those surveyed work more than 20 h per week?

24. **Area.** A living room has dimensions $5\dfrac{2}{3}$ by $4\dfrac{1}{2}$ yd. How much carpeting must be purchased to cover the room?

6.4 Dividing Fractions

6.4 OBJECTIVES

1. Divide fractions.
2. Divide mixed numbers.
3. Use division to solve applications.

We are now ready to look at the operation of division on fractions. First we will need a new concept, the **reciprocal** of a fraction.

> ### The Reciprocal of a Fraction
>
> We invert, or turn over, a fraction to write its **reciprocal.**

In general, the reciprocal of the fraction $\frac{a}{b}$ is $\frac{b}{a}$.
Neither a nor b can be 0.

● Example 1

Finding the Reciprocal of a Fraction

Find the reciprocal of (a) $\frac{3}{4}$ and (b) 5.

(a) The reciprocal of $\frac{3}{4}$ is $\frac{4}{3}$.   Just invert, or turn over, the fraction.

(b) The reciprocal of 5, or $\frac{5}{1}$, is $\frac{1}{5}$.   Write 5 as $\frac{5}{1}$ and then turn over the fraction.

●●● **CHECK YOURSELF 1**

Write the reciprocal of $\frac{5}{8}$.

To find the reciprocal of a mixed number, first write the mixed number as an improper fraction and then invert that fraction. Example 2 illustrates this approach.

● Example 2

Finding the Reciprocal of a Mixed Number

Find the reciprocal of $1\frac{2}{3}$.

Write $1\frac{2}{3}$ as $\frac{5}{3}$, *then* invert.

The reciprocal of $1\frac{2}{3}$, or $\frac{5}{3}$, is $\frac{3}{5}$.

● ● ● **CHECK YOURSELF 2**

What is the reciprocal of $3\frac{1}{4}$?

An important property relating a number and its reciprocal follows.

> **An Important Multiplication Fact**
>
> The product of any nonzero number and its reciprocal is 1.

Example 3 illustrates this property.

● Example 3

Finding the Product of a Fraction and Its Reciprocal

Show that the product of $\frac{3}{4}$ and its reciprocal is 1.

The reciprocal of $\frac{3}{4}$ is $\frac{4}{3}$.

$$\frac{3}{4} \times \frac{4}{3} = \frac{\overset{1}{\cancel{3}} \times \overset{1}{\cancel{4}}}{\underset{1}{\cancel{4}} \times \underset{1}{\cancel{3}}} = 1$$

● ● ● **CHECK YOURSELF 3**

Find the product of $\frac{5}{7}$ and its reciprocal.

We are now ready to use the reciprocal to find a rule for dividing fractions. Recall that we can represent the operation of division in several ways. We used the symbol ÷ earlier. Remember that a fraction also indicates division. For instance,

$3 \div 5$ and $\frac{3}{5}$ both mean "3 divided by 5."

$$3 \div 5 = \frac{3}{5}$$

In this statement, 5 is called the *divisor.* It follows the division sign ÷ and is written *below* the fraction bar.

Using this information, we can write a statement involving fractions and division as a *complex fraction,* which has a fraction as both its numerator and denominator, as Example 4 illustrates.

● Example 4

Writing a Quotient as a Complex Fraction

Write $\dfrac{2}{3} \div \dfrac{4}{5}$ as a complex fraction.

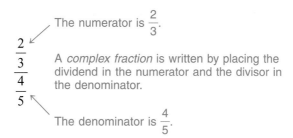

The numerator is $\dfrac{2}{3}$.

A *complex fraction* is written by placing the dividend in the numerator and the divisor in the denominator.

The denominator is $\dfrac{4}{5}$.

● ● ● CHECK YOURSELF 4

Write $\dfrac{2}{5} \div \dfrac{3}{4}$ as a complex fraction.

Let's continue with the same division problem.

● Example 5

Dividing Two Fractions

(1) $\dfrac{2}{5} \div \dfrac{3}{4} = \dfrac{\dfrac{2}{5}}{\dfrac{3}{4}}$

Write the original quotient as a complex fraction.

$= \dfrac{\dfrac{2}{5} \times \dfrac{4}{3}}{\dfrac{3}{4} \times \dfrac{4}{3}}$

Multiply the numerator and denominator by $\dfrac{4}{3}$, the reciprocal of the denominator. This does *not* change the value of the fraction.

$= \dfrac{\dfrac{2}{5} \times \dfrac{4}{3}}{1}$

The denominator becomes 1.

(2) $= \dfrac{2}{5} \times \dfrac{4}{3}$

Recall that a number divided by 1 is just that number.

We see from lines (1) and (2) that

Do you see a rule suggested?

$$\dfrac{2}{5} \div \dfrac{3}{4} = \dfrac{2}{5} \times \dfrac{4}{3}$$

We would certainly like to be able to divide fractions easily without all the work of the last example. Look carefully at the example. The following rule is suggested.

To Divide Fractions

To divide one fraction by another, invert the divisor (the fraction after the division sign) and multiply.

● ● ●   **CHECK YOURSELF 5**

Write $\dfrac{3}{5} \div \dfrac{7}{8}$ as a multiplication problem.

Example 6 applies the rule for dividing fractions.

● Example 6

Dividing Two Fractions

Divide.

Remember: The number inverted is the divisor. It *follows* the division sign.

$$\frac{1}{3} \div \frac{4}{7} = \frac{1}{3} \times \frac{7}{4}$$

We invert the divisor, $\dfrac{4}{7}$, then multiply.

$$= \frac{1 \times 7}{3 \times 4} = \frac{7}{12}$$

● ● ●   **CHECK YOURSELF 6**

Divide.

$$\frac{2}{5} \div \frac{3}{4}$$

Let's look at another similar example.

● Example 7

Dividing Two Fractions

Divide.

$$\frac{5}{8} \div \frac{3}{5} = \frac{5}{8} \times \frac{5}{3} = \frac{5 \times 5}{8 \times 3} = \frac{25}{24} = 1\frac{1}{24}$$

Write the quotient as a mixed number if necessary.

CHECK YOURSELF 7

Divide.

$$\frac{5}{6} \div \frac{3}{8}$$

Simplifying will also be useful in dividing fractions. Consider the next example.

• Example 8

Dividing Two Fractions

Divide.

Be careful! We must invert the divisor *before any simplification*.

$$\frac{3}{5} \div \frac{6}{7} = \frac{3}{5} \times \frac{7}{6}$$

Invert the divisor *first!* Then you can divide by the common factor of 3.

$$= \frac{\overset{1}{\cancel{3}} \times 7}{5 \times \underset{2}{\cancel{6}}} = \frac{7}{10}$$

CHECK YOURSELF 8

Divide.

$$\frac{4}{9} \div \frac{8}{15}$$

When mixed or whole numbers are involved, the process is similar. Simply change the mixed or whole numbers to improper fractions as the first step. Then proceed with the division rule. Example 9 illustrates this approach.

• Example 9

Dividing Two Mixed Numbers

Divide.

$$2\frac{3}{8} \div 1\frac{3}{4} = \frac{19}{8} \div \frac{7}{4}$$

Write the mixed numbers as improper fractions.

$$= \frac{19}{\underset{2}{\cancel{8}}} \times \frac{\overset{1}{\cancel{4}}}{7}$$

Invert the divisor and multiply as before.

$$= \frac{19}{14} = 1\frac{5}{14}$$

⦿⦿⦿ **CHECK YOURSELF 9**

Divide.

$$3\frac{1}{5} \div 2\frac{2}{5}$$

Example 10 illustrates the division process when a whole number is involved.

•Example 10

Dividing a Mixed Number and a Whole Number

Divide and simplify.

Write the whole number 6 as $\frac{6}{1}$.

$$1\frac{4}{5} \div 6 = \frac{9}{5} \div \frac{6}{1}$$

$$= \frac{\overset{3}{9}}{5} \times \frac{1}{\underset{2}{6}} \qquad \text{Invert the divisor, then divide by the common factor of 3.}$$

$$= \frac{3}{10}$$

⦿⦿⦿ **CHECK YOURSELF 10**

Divide.

$$8 \div 4\frac{4}{5}$$

As was the case with multiplication, our work with the division of fractions will be used in the solution of a variety of applications. The steps of the problem-solving process remain the same.

•Example 11

An Application Involving the Division of Mixed Numbers

A kilometer, abbreviated km, is a metric unit of distance. It is about $\frac{6}{10}$ mi.

Jack traveled 140 kilometers (km) in $2\frac{1}{3}$ hours (h). What was his average speed?

Solution

The important formula is
Speed = distance ÷ time.

$$\text{Speed} = 140 \text{ km} \div 2\frac{1}{3} \text{ h}$$

Distance → 140   Time → 2⅓

We know the distance traveled and the time for that travel. To find the *average* speed, we must use division. Do you remember why?

$$= \frac{140}{1} \text{ km} \div \frac{7}{3} \text{ h}$$

$$= \frac{\overset{20}{\cancel{140}}}{1} \times \frac{3}{\underset{1}{\cancel{7}}} \text{ km/h}$$

km/h is read "kilometers per hour." This is a unit of speed.

$$= 60 \text{ km/h}$$

● ● ● ● **CHECK YOURSELF 11**

A light plane flew 280 mi in $1\frac{3}{4}$ h. What was its average speed?

● **Example 12**

An Application Involving the Division of Mixed Numbers

An electrician needs pieces of wire $2\frac{3}{5}$ inches (in.) long. If she has a $20\frac{4}{5}$-in. piece of wire, how many of the shorter pieces can she cut?

Solution

We must divide the length of the longer piece by the desired length of the shorter piece.

$$20\frac{4}{5} \div 2\frac{3}{5} = \frac{104}{5} \div \frac{13}{5}$$

$$= \frac{\overset{8}{\cancel{104}}}{\underset{1}{\cancel{5}}} \times \frac{\overset{1}{\cancel{5}}}{\underset{1}{\cancel{13}}}$$

$$= 8 \text{ pieces}$$

● ● ● ● **CHECK YOURSELF 12**

A piece of plastic water pipe 63 in. long is to be cut into lengths of $3\frac{1}{2}$ in. How many of the shorter pieces can be cut?

Some applications require both multiplication and division. Example 13 is such an application.

• Example 13

An Application Involving the Division of Mixed Numbers

A parcel of land that is $2\frac{1}{2}$ miles (mi) long and $1\frac{1}{3}$ mi wide is to be divided into tracts that are each $\frac{1}{3}$ square mile (mi^2). How many of these tracts will the parcel make?

Solution   The area of the parcel is its length times its width:

$$\text{Area} = 2\frac{1}{2} \times 1\frac{1}{3}$$

$$= \frac{5}{2} \times \frac{4}{3}$$

$$= \frac{10}{3} \text{ mi}^2$$

We need to divide the total area of the parcel into $\frac{1}{3}$-mi^2 tracts.

$$\frac{10}{3} \div \frac{1}{3} = \frac{10}{3} \times \frac{3}{1} = 10$$

The land will provide 10 tracts, each with an area of $\frac{1}{3}$ mi^2.

● ● ● **CHECK YOURSELF 13**

A parcel of land that is $3\frac{1}{3}$ mi long and $2\frac{1}{2}$ mi wide is to be divided into $\frac{1}{3}$-mi^2 tracts. How many of these tracts will the parcel make?

● ● ● **CHECK YOURSELF ANSWERS**

1. $\frac{8}{5}$.   2. $3\frac{1}{4}$ is $\frac{13}{4}$, so the reciprocal is $\frac{4}{13}$.   3. $\frac{5}{7} \times \frac{7}{5} = 1$.   4. $\frac{\frac{2}{5}}{\frac{3}{4}}$.

5. $\frac{3}{5} \cdot \frac{8}{7}$.   6. $\frac{8}{15}$.   7. $2\frac{2}{9}$.   8. $\frac{4}{9} \div \frac{8}{15} = \frac{4}{9} \times \frac{15}{8} = \frac{\overset{1}{4} \times \overset{5}{15}}{\underset{3}{9} \times \underset{2}{8}} = \frac{5}{6}$.

9. $3\frac{1}{5} \div 2\frac{2}{5} = \frac{16}{5} \div \frac{12}{5} = \frac{\overset{4}{16}}{\underset{1}{5}} \times \frac{\overset{1}{5}}{\underset{3}{12}} = \frac{4}{3} = 1\frac{1}{3}$.   10. $1\frac{2}{3}$.   11. 160 mi/h.

12. 18 pieces.   13. 25 tracts.

Divide. Write each result in simplest form.

1. $\dfrac{1}{5} \div \dfrac{3}{4}$

2. $\dfrac{2}{5} \div \dfrac{1}{3}$

3. $\dfrac{2}{5} \div \dfrac{3}{4}$

4. $\dfrac{5}{8} \div \dfrac{3}{4}$

5. $\dfrac{8}{9} \div \dfrac{4}{3}$

6. $\dfrac{5}{9} \div \dfrac{8}{11}$

7. $\dfrac{7}{10} \div \dfrac{5}{9}$

8. $\dfrac{8}{9} \div \dfrac{11}{15}$

9. $\dfrac{8}{15} \div \dfrac{2}{5}$

10. $\dfrac{5}{27} \div \dfrac{15}{54}$

11. $\dfrac{5}{27} \div \dfrac{25}{36}$

12. $\dfrac{9}{28} \div \dfrac{27}{35}$

13. $\dfrac{4}{5} \div 4$

14. $27 \div \dfrac{3}{7}$

15. $12 \div \dfrac{2}{3}$

16. $\dfrac{5}{8} \div 5$

17. $\dfrac{12}{17} \div 6$

18. $\dfrac{3}{4} \div 9$

19. $3 \div \dfrac{5}{8}$

20. $6 \div \dfrac{9}{10}$

21. $4\dfrac{1}{2} \div 6$

22. $6 \div 2\dfrac{1}{2}$

23. $9 \div 2\dfrac{1}{4}$

24. $3\dfrac{1}{3} \div 5$

ANSWERS

1. $\dfrac{4}{15}$

2. $1\dfrac{1}{5}$

3. $\dfrac{8}{15}$

4. $\dfrac{5}{6}$

5. $\dfrac{2}{3}$

6. $\dfrac{55}{72}$

7. $1\dfrac{13}{50}$

8. $1\dfrac{7}{33}$

9. $1\dfrac{1}{3}$

10. $\dfrac{2}{3}$

11. $\dfrac{4}{15}$

12. $\dfrac{5}{12}$

13. $\dfrac{1}{5}$

14. 63

15. 18

16. $\dfrac{1}{8}$

17. $\dfrac{2}{17}$

18. $\dfrac{1}{12}$

19. $4\dfrac{4}{5}$

20. $6\dfrac{2}{3}$

21. $\dfrac{3}{4}$

22. $2\dfrac{2}{5}$

23. 4

24. $\dfrac{2}{3}$

25. $15 \div 3\frac{1}{3}$

26. $2\frac{4}{7} \div 12$

27. $1\frac{3}{5} \div \frac{4}{15}$

28. $\frac{9}{14} \div 2\frac{4}{7}$

29. $\frac{7}{12} \div 2\frac{1}{3}$

30. $1\frac{3}{8} \div \frac{5}{12}$

31. $5\frac{3}{5} \div \frac{7}{15}$

32. $\frac{7}{18} \div 5\frac{5}{6}$

33. $1\frac{1}{3} \div 1\frac{1}{7}$

34. $3\frac{1}{2} \div 2\frac{4}{5}$

35. $3\frac{3}{4} \div 1\frac{3}{8}$

36. $5\frac{1}{3} \div 2\frac{2}{5}$

37. $2\frac{1}{3} \div 1\frac{5}{9}$

38. $8\frac{3}{4} \div 3\frac{1}{8}$

Solve the following applications.

39. **Wire cutting.** A wire $5\frac{1}{4}$ feet (ft) long is to be cut into 7 pieces of the same length. How long will each piece be?

40. **Quantity.** A potter uses $\frac{2}{3}$ pound (lb) of clay in making a bowl. How many bowls can be made from 16 lb of clay?

41. **Speed.** Virginia made a trip of 95 miles (mi) in $1\frac{1}{4}$ hours (h). What was her average speed?

42. **Unit pricing.** A piece of land measures $3\frac{3}{4}$ acres and is for sale at $60,000. What is the price per acre?

43. **Number of servings.** A roast weighs $3\frac{1}{4}$ lb. How many $\frac{1}{4}$-lb servings will the roast provide?

44. **Number of books.** A bookshelf is 55 inches (in.) long. If the books have an average thickness of $1\frac{1}{4}$ in., how many books can be put on the shelf?

ANSWERS

45.	51 packages
46.	26 shirts
47.	64 sheets
48.	$\frac{1}{2}$ mi²
49.	(a) $\frac{8}{3}$ cup; (b) 72 oranges
50.	

45. **Quantity.** A butcher wants to wrap $\frac{3}{8}$-lb packages of ground beef from a cut of meat weighing $19\frac{1}{8}$ lb. How many packages can be prepared?

46. **Quantity.** A manufacturer has $45\frac{1}{2}$ yards (yd) of imported cotton fabric. A shirt pattern uses $1\frac{3}{4}$ yd. How many shirts can be made?

47. **Number of pieces.** A stack of $\frac{3}{4}$-in.-thick plywood is 48 in. high. How many sheets of plywood are in the stack?

48. **Area.** A landfill occupies land that measures $10\frac{2}{3}$ mi by $6\frac{3}{4}$ mi. If there are 144 cells in the landfill, what is the area of each cell?

49. In squeezing oranges for fresh juice, 3 oranges yield about $\frac{1}{3}$ of a cup.

    (a) How much juice could you expect to obtain from a bag containing 24 oranges?
    (b) If you needed 8 cups of orange juice, how many bags of oranges should you buy?

50. A farmer died and left 17 cows to be divided among 3 workers. The first worker was to receive $\frac{1}{2}$ of the cows, the second worker was to receive $\frac{1}{3}$ of the cows, and the third worker was to receive $\frac{1}{9}$ of the cows. The executor of the farmer's estate realized that 17 cows could not be divided by halves, thirds, or ninths and so added a neighbor's cow to the farmer's. With 18 cows, the executor gave 9 cows to the first worker, 6 cows to the second worker, and 2 cows to the third worker. This accounted for the 17 cows, so the executor returned the borrowed cow to the neighbor. Explain why this works.

Answers

1. $\frac{4}{15}$ 3. $\frac{8}{15}$ 5. $\frac{2}{3}$ 7. $1\frac{13}{50}$ 9. $1\frac{1}{3}$ 11. $\frac{4}{15}$ 13. $\frac{1}{5}$

15. 18   17. $\frac{2}{17}$ 19. $4\frac{4}{5}$ 21. $\frac{3}{4}$ 23. 4   25. $4\frac{1}{2}$ 27. 6

29. $\frac{1}{4}$ 31. 12   33. $1\frac{1}{6}$ 35. $2\frac{8}{11}$ 37. $1\frac{1}{2}$ 39. $\frac{3}{4}$ ft

41. 76 mi/h   43. 13 servings   45. 51 packages   47. 64 sheets

49. (a) $\frac{8}{3}$ cup; (b) 72 oranges   50.

Name _____

Section _____ Date _____

1. $\dfrac{28}{45}$

2. $\dfrac{21}{40}$

3. $1\dfrac{1}{6}$

4. $\dfrac{2}{3}$

5. $3\dfrac{3}{4}$

6. $1\dfrac{2}{3}$

7. 20

8. $\dfrac{2}{15}$

9. $\dfrac{1}{14}$

10. $33\dfrac{1}{3}$

11. $\dfrac{2}{3}$

12. $4\dfrac{1}{2}$

13. $1\dfrac{1}{3}$

14. 6

15. $\dfrac{1}{3}$

16. $2\dfrac{1}{4}$

17. $1\dfrac{1}{3}$

18. $1\dfrac{1}{2}$

19. $\dfrac{3}{4}$ ft

20. 16 blouses

21. 56 mi/h

Divide.

1. $\dfrac{4}{9} \div \dfrac{5}{7}$

2. $\dfrac{3}{8} \div \dfrac{5}{7}$

3. $\dfrac{7}{8} \div \dfrac{3}{4}$

4. $\dfrac{3}{10} \div \dfrac{9}{20}$

5. $\dfrac{9}{10} \div \dfrac{6}{25}$

6. $\dfrac{7}{8} \div \dfrac{21}{40}$

7. $18 \div \dfrac{9}{10}$

8. $\dfrac{14}{15} \div 7$

9. $\dfrac{6}{7} \div 12$

10. $30 \div \dfrac{9}{10}$

11. $5\dfrac{1}{3} \div 8$

12. $12 \div 2\dfrac{2}{3}$

13. $6\dfrac{2}{3} \div 5$

14. $2\dfrac{2}{3} \div \dfrac{4}{9}$

15. $\dfrac{7}{12} \div 1\dfrac{3}{4}$

16. $4\dfrac{1}{5} \div \dfrac{28}{15}$

17. $3\dfrac{3}{5} \div 2\dfrac{7}{10}$

18. $2\dfrac{5}{8} \div 1\dfrac{3}{4}$

Solve the following applications.

19. Length of wire. A piece of wire $3\dfrac{3}{4}$ ft long is to be cut into five pieces of the same length. How long will each piece be?

20. Quantity. A blouse pattern requires $1\dfrac{3}{4}$ yd of fabric. How many blouses can be made from a piece of silk that is 28 yd long?

21. Speed. If you drive 126 mi in $2\dfrac{1}{4}$ h, what is your average speed?

Using Your Calculator to Divide Fractions

Dividing fractions on your calculator is almost exactly the same as multiplying them. You simply press the ÷ key instead of the × key.

SCIENTIFIC CALCULATOR

Dividing fractions on a scientific calculator requires only that one enters the problem followed by the equal sign.

● Example 1

Dividing Two Fractions

Find the quotient

$$\frac{23}{24} \div \frac{13}{16}$$

The keystroke sequence is

23 **a b/c** 24 ÷ 13 **a b/c** 16 =

Some scientific calculators display the results of dividing fractions as improper fractions, and some display them as mixed numbers. Thus the answer to $\frac{23}{24} \div \frac{13}{16}$ might be displayed as $\frac{46}{39}$ or $1\frac{7}{39}$. Be sure you understand that the two answers are equivalent.

● ● ●   **CHECK YOURSELF 1**

Find the quotient

$$\frac{12}{17} \div \frac{3}{16}$$

GRAPHING CALCULATOR

When using a graphing calculator, you must choose the fraction option **Frac.** from the **MATH** menu before pressing Enter .

The keystroke sequence for the fraction problem in Example 1, $\frac{23}{24} \div \frac{13}{16}$, is

23 ÷ 24 ÷ (13 ÷ 16) **Frac.** Enter

● ● ● **CHECK YOURSELF ANSWER**

$\frac{64}{17}$ or $3\frac{13}{17}$

Calculator Exercises

Name

Section   Date

A N S W E R S

1. $\frac{3}{2}$

2. $\frac{2}{3}$

3. $\frac{4}{3}$ or $1\frac{1}{3}$

4. $\frac{4}{3}$ or $1\frac{1}{3}$

5. $\frac{10}{3}$ or $3\frac{1}{3}$

6. $\frac{1}{9}$

7. $\frac{16}{21}$

8. $\frac{3}{2}$ or $1\frac{1}{2}$

9. $\frac{1}{2}$

10. $\frac{1}{14}$

11. $\frac{3}{4}$

12. $\frac{4}{3}$ or $1\frac{1}{3}$

Find the following quotients using your calculator.

1. $\frac{1}{5} \div \frac{2}{15}$

2. $\frac{13}{17} \div \frac{39}{34}$

3. $\frac{5}{7} \div \frac{15}{28}$

4. $\frac{3}{7} \div \frac{9}{28}$

5. $\frac{20}{9} \div \frac{10}{15}$

6. $\frac{13}{15} \div \frac{39}{5}$

7. $\frac{20}{27} \div \frac{35}{36}$

8. $\frac{2}{3} \div \frac{4}{9}$

9. $\frac{15}{18} \div \frac{45}{27}$

10. $\frac{19}{63} \div \frac{38}{9}$

11. $\frac{25}{45} \div \frac{100}{135}$

12. $\frac{86}{24} \div \frac{258}{96}$

Name

Section   Date

A N S W E R S

1. $\dfrac{10}{21}$

2. $\dfrac{9}{16}$

3. $\dfrac{1}{6}$

4. $2\dfrac{1}{7}$

5. $3\dfrac{3}{7}$

6. $9\dfrac{1}{5}$

7. $\dfrac{2}{3}$

8. $\dfrac{8}{15}$

9. $\dfrac{4}{15}$

10. 4

11. $6\dfrac{1}{9}$

12. $7\dfrac{1}{5}$

13. $4\dfrac{10}{11}$

14. $\dfrac{1}{2}$

15. $\dfrac{7}{12}$ yd

The purpose of the Self-Test is to help you check your progress and review for a chapter test in class. Allow yourself about 1 hour to take the test. When you are done, check your answers in the back of the book. If you missed any answers, be sure to go back and review the appropriate sections in the chapter and do the supplementary exercises provided there.

[6.1]   In Exercises 1 to 14, multiply.

1. $\dfrac{2}{3} \times \dfrac{5}{7} =$

2. $\dfrac{9}{10} \times \dfrac{5}{8} =$

3. $\dfrac{7}{16} \times \dfrac{8}{21} =$

4. $5 \times \dfrac{3}{7} =$

5. $2\dfrac{2}{3} \times 1\dfrac{2}{7} =$

6. $3\dfrac{5}{6} \times 2\dfrac{2}{5} =$

7. $\dfrac{3}{4} \times \dfrac{8}{9} =$

8. $\dfrac{20}{21} \times \dfrac{14}{25} =$

9. $\dfrac{16}{35} \times \dfrac{14}{24} =$

10. $5\dfrac{1}{3} \times \dfrac{3}{4} =$

[6.2]   11. $4\dfrac{1}{6} \times 1\dfrac{7}{15} =$

12. $6\dfrac{2}{3} \times 1\dfrac{2}{25} =$

13. $9 \times \dfrac{6}{11} =$

14. $\dfrac{3}{5} \times \dfrac{7}{9} \times 1\dfrac{1}{14} =$

[6.3]   In Exercises 15 to 18, solve each application.

15. **Material use.** You had $\dfrac{7}{8}$ yards (yd) of upholstery material but used only $\dfrac{2}{3}$ of that for a project. How much material did you use?

16. Cost of apples. What is the cost of $2\dfrac{3}{4}$ pounds (lb) of apples if the price per pound is 48 cents?

17. Quantity of material. A room measures $5\dfrac{1}{3}$ by $3\dfrac{3}{4}$ yd. How many square yards (yd²) of linoleum must be purchased to cover the floor?

18. Distance. The scale on a map is 1 inch (in.) = 80 miles (mi). If two towns are $2\dfrac{3}{8}$ in. apart on the map, what is the actual distance in miles between the towns?

[6.4] In Exercises 19 to 26, divide.

19. $\dfrac{6}{7} \div \dfrac{3}{4}$

20. $\dfrac{7}{12} \div \dfrac{14}{15}$

21. $\dfrac{2}{9} \div 3$

22. $12 \div \dfrac{3}{4}$

23. $2\dfrac{2}{5} \div \dfrac{8}{11}$

24. $\dfrac{7}{8} \div 1\dfrac{5}{16}$

25. $1\dfrac{3}{4} \div 1\dfrac{3}{8}$

26. $5\dfrac{3}{5} \div 2\dfrac{1}{10}$

In Exercises 27 to 30, solve each application.

[6.4] **27. Number of homes.** A $31\dfrac{1}{3}$-acre piece of land is subdivided into home lots. Each home lot is to be $\dfrac{2}{3}$ acre. How many homes can be built?

28. Mean speed. If you drive 88 mi in $1\dfrac{5}{6}$ hours (h), what is your mean speed?

29. Amount of material. A stack of $\dfrac{5}{8}$-in.-thick plywood is 40 in. high. How many sheets of plywood are in the stack?

30. Number of books. A bookshelf is 66 in. long. If the thickness of each book on the shelf is $1\dfrac{3}{8}$ in., how many books can be placed on the shelf?

CHAPTER 7

ADDITION AND SUBTRACTION OF FRACTIONS

INTRODUCTION

Creating something useful and beautiful with your hands is one of the most satisfying ways of spending free time. Some people who start out working with wood as a hobby end up doing carpentry work as a career. Other people are introduced to carpentry first as a job. Roberto learned his carpentry trade in this manner.

Roberto went to the career counselor at the local community college to find a career. Aptitude tests showed that he liked both working with his hands and solving problems. Following the counselor's advice, Roberto entered the carpentry apprentice program that fall. He is now quite happy with that recommendation; carpentry has become both his job and his hobby. ━━━━━━━━━━━━━━ ■

© 1993, COMSTOCK, Inc.

Pretest for Chapter 7

ANSWERS

1. $\dfrac{5}{7}$

2. 48

3. $\dfrac{17}{30}$

4. $2\dfrac{1}{8}$

5. $\dfrac{4}{15}$

6. $\dfrac{5}{72}$

7. $6\dfrac{17}{24}$

8. $1\dfrac{1}{36}$

9. $18\dfrac{1}{4}$ yd²

10. $4\dfrac{3}{8}$ points

Addition and Subtraction of Fractions

This pretest will point out any difficulties you may be having in adding and subtracting fractions. Do all the problems. Then check your answers with those in the back of the book.

1. $\dfrac{3}{7} + \dfrac{2}{7} =$

2. Find the least common denominator for fractions with the denominators 16 and 24.

3. $\dfrac{3}{10} + \dfrac{4}{15} =$

4. $\dfrac{4}{5} + \dfrac{7}{8} + \dfrac{9}{20} =$

5. $\dfrac{8}{15} - \dfrac{4}{15} =$

6. $\dfrac{7}{24} - \dfrac{2}{9} =$

7. $3\dfrac{5}{6} + 2\dfrac{7}{8} =$

8. $2\dfrac{1}{9} - 1\dfrac{1}{12} =$

9. Carpet Purchase. A house plan calls for $12\dfrac{3}{4}$ yd² of carpeting in the living room and $5\dfrac{1}{2}$ yd² in the hallway. How much carpeting will be needed?

10. Stock Prices. A stock is listed at $50\dfrac{1}{4}$ points at the start of a week. By the end of the week it is at $54\dfrac{5}{8}$ points. How much did it gain during the week?

Adding Fractions with a Common Denominator

7.1 OBJECTIVES

1. Add two like fractions.
2. Add a group of like fractions.

For instance, we can add two nickels and three nickels to get five nickels. We *cannot* directly add two nickels and three dimes!

Recall from our work in Chapter 1 that adding can be thought of as combining groups of the *same kinds* of objects. This is also true when you think about adding fractions.

Fractions can be added only if they name a number of the *same parts* of a whole. This means that we can add fractions only when they are **like fractions,** that is, when they have the *same (a common)* denominator.

As long as we are dealing with like fractions, addition is an easy matter. Just use the following rule.

To Add Like Fractions

Step 1 Add the numerators.
Step 2 Place the sum over the common denominator.
Step 3 Simplify the resulting fraction when necessary.

Our first example illustrates the use of this rule.

● Example 1

Adding Like Fractions

Add.

$$\frac{1}{5} + \frac{3}{5}$$

Step 1 Add the numerators.

$$1 + 3 = 4$$

Step 2 Write that sum over the common denominator, 5. We are done at this point because the answer, $\frac{4}{5}$, is in the simplest possible form.

Step 1　　　　　Step 2

$$\frac{1}{5} + \frac{3}{5} = \frac{1 + 3}{5} = \frac{4}{5}$$

Let's illustrate with a diagram.

Combining 1 of the 5 parts with 3 of the 5 parts gives a total of 4 of the 5 equal parts.

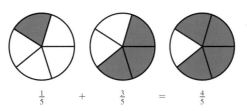

$$\frac{1}{5} \quad + \quad \frac{3}{5} \quad = \quad \frac{4}{5}$$

● ● ●   **CHECK YOURSELF 1**

Add.

$$\frac{2}{9} + \frac{5}{9}$$

Be Careful! In adding fractions, *do not* follow the rule for multiplying fractions. To multiply $\frac{1}{5} \times \frac{3}{5}$, you would multiply both the numerator and the denominator:

$$\frac{1}{5} \times \frac{3}{5} = \frac{3}{25}$$

Also, since to add fractions, you must have a common denominator, you *do not* add the denominators. Thus

$$\frac{1}{5} + \frac{3}{5} \neq \frac{4}{10}$$

Step 3 of the addition rule for like fractions tells us to *simplify* the sum. The sum of fractions should always be written in lowest terms. Consider Example 2.

•Example 2

Adding Like Fractions That Require Simplifying

Add and simplify.

Step 3

$$\frac{3}{12} + \frac{5}{12} = \frac{8}{12} = \frac{2}{3}$$

The sum $\frac{8}{12}$ is *not* in lowest terms.
Divide the numerator and denominator
by 4 to simplify the result.

● ● ●   **CHECK YOURSELF 2**

Add.

$$\frac{4}{15} + \frac{6}{15}$$

If the sum of two fractions is an improper fraction, we will usually write that sum as a mixed number.

• Example 3

Adding Like Fractions That Result in Mixed Numbers

Add as before. Then convert the sum to a mixed number.

Add $\dfrac{5}{9} + \dfrac{8}{9} = \dfrac{13}{9} = 1\dfrac{4}{9}$ Write the sum $\dfrac{13}{9}$ as a mixed number.

● ● ● CHECK YOURSELF 3

Add.

$$\dfrac{7}{12} + \dfrac{10}{12}$$

We can also easily extend our addition rule to find the sum of more than two fractions as long as they all have the same denominator. This is shown in Example 4.

• Example 4

Adding a Group of Like Fractions

Add.

$$\dfrac{2}{7} + \dfrac{3}{7} + \dfrac{6}{7} = \dfrac{11}{7}$$ Add the numerators: $2 + 3 + 6 = 11$.

$$= 1\dfrac{4}{7}$$

● ● ● CHECK YOURSELF 4

Add.

$$\dfrac{1}{8} + \dfrac{3}{8} + \dfrac{5}{8}$$

Many applications can be solved by adding fractions.

•Example 5

An Application Involving the Adding of Like Fractions

Noel walked $\frac{9}{10}$ miles (mi) to Jensen's house and then walked $\frac{7}{10}$ mi to school. How far did Noel walk?

Solution To find the total distance Noel walked, add the two distances.

$$\frac{9}{10} + \frac{7}{10} = \frac{16}{10} = 1\frac{6}{10} = 1\frac{3}{5}$$

Noel walked $1\frac{3}{5}$ mi.

● ● ● **CHECK YOURSELF 5**

Emir bought $\frac{7}{16}$ pounds (lb) of candy at one store and $\frac{11}{16}$ lb at another store. How much candy did Emir buy?

● ● ● **CHECK YOURSELF ANSWERS**

1. $\frac{2}{9} + \frac{5}{9} = \frac{2+5}{9} = \frac{7}{9}$. **2.** $\frac{2}{3}$. **3.** $1\frac{5}{12}$. **4.** $1\frac{1}{8}$. **5.** $1\frac{1}{8}$ lb.

Add. Simplify when possible.

1. $\dfrac{3}{5} + \dfrac{1}{5}$

2. $\dfrac{4}{7} + \dfrac{1}{7}$

3. $\dfrac{4}{11} + \dfrac{6}{11}$

4. $\dfrac{5}{16} + \dfrac{4}{16}$

5. $\dfrac{2}{10} + \dfrac{3}{10}$

6. $\dfrac{5}{12} + \dfrac{1}{12}$

7. $\dfrac{3}{7} + \dfrac{4}{7}$

8. $\dfrac{3}{20} + \dfrac{7}{20}$

9. $\dfrac{9}{30} + \dfrac{11}{30}$

10. $\dfrac{4}{9} + \dfrac{5}{9}$

11. $\dfrac{13}{48} + \dfrac{23}{48}$

12. $\dfrac{17}{60} + \dfrac{31}{60}$

13. $\dfrac{3}{7} + \dfrac{6}{7}$

14. $\dfrac{3}{5} + \dfrac{4}{5}$

15. $\dfrac{7}{10} + \dfrac{9}{10}$

16. $\dfrac{5}{8} + \dfrac{7}{8}$

17. $\dfrac{11}{12} + \dfrac{10}{12}$

18. $\dfrac{13}{18} + \dfrac{11}{18}$

19. $\dfrac{1}{8} + \dfrac{1}{8} + \dfrac{3}{8}$

20. $\dfrac{1}{10} + \dfrac{3}{10} + \dfrac{3}{10}$

21. $\dfrac{1}{9} + \dfrac{4}{9} + \dfrac{5}{9}$

22. $\dfrac{7}{12} + \dfrac{11}{12} + \dfrac{1}{12}$

A N S W E R S

1. $\dfrac{4}{5}$

2. $\dfrac{5}{7}$

3. $\dfrac{10}{11}$

4. $\dfrac{9}{16}$

5. $\dfrac{1}{2}$

6. $\dfrac{1}{2}$

7. 1

8. $\dfrac{1}{2}$

9. $\dfrac{2}{3}$

10. 1

11. $\dfrac{3}{4}$

12. $\dfrac{4}{5}$

13. $1\dfrac{2}{7}$

14. $1\dfrac{2}{5}$

15. $1\dfrac{3}{5}$

16. $1\dfrac{1}{2}$

17. $1\dfrac{3}{4}$

18. $1\dfrac{1}{3}$

19. $\dfrac{5}{8}$

20. $\dfrac{7}{10}$

21. $1\dfrac{1}{9}$

22. $1\dfrac{7}{12}$

23. $\dfrac{3}{10} + \dfrac{7}{10} + \dfrac{5}{10}$

24. $\dfrac{9}{20} + \dfrac{7}{20} + \dfrac{11}{20}$

Solve the following applications.

25. **Money.** You collect 3 dimes, 2 dimes, and then 4 dimes. How much money do you have as a fraction of a dollar?

26. **Money.** You collect 7 nickels, 4 nickels, and then 5 nickels. How much money do you have as a fraction of a dollar?

27. **Work.** You work 7 hours (h) one day, 5 h the second day, and 6 h the third day. How long did you work, as a fraction of a 24-h day?

28. **Time.** One task took 7 minutes (min), a second task took 12 min, and a third task took 21 min. How long did the three tasks take, as a fraction of an hour?

29. **Perimeter.** What is the perimeter of a rectangle if each of the lengths is $\dfrac{7}{10}$ inches (in.) and each of the widths is $\dfrac{2}{10}$ in.?

30. **Perimeter.** Find the perimeter of a rectangular picture if each width is $\dfrac{7}{9}$ in. and each length is $\dfrac{5}{9}$ in.

Getting Ready for Section 7.2
[Section 4.2]

Find the prime factorization for each number.

a. 12   **b.** 20

c. 36   **d.** 48

e. 60   **f.** 98

Answers

1. $\dfrac{4}{5}$ **3.** $\dfrac{10}{11}$ **5.** $\dfrac{2}{10} + \dfrac{3}{10} = \dfrac{5}{10} = \dfrac{1}{2}$ **7.** $\dfrac{3}{7} + \dfrac{4}{7} = \dfrac{7}{7} = 1$

9. $\dfrac{2}{3}$ **11.** $\dfrac{3}{4}$ **13.** $1\dfrac{2}{7}$ **15.** $\dfrac{7}{10} + \dfrac{9}{10} = \dfrac{16}{10} = 1\dfrac{6}{10} = 1\dfrac{3}{5}$ **17.** $1\dfrac{3}{4}$

19. $\dfrac{5}{8}$ **21.** $1\dfrac{1}{9}$ **23.** $1\dfrac{1}{2}$ **25.** $\dfrac{9}{10}$ of a dollar   **27.** $\dfrac{3}{4}$ day   **29.** $\dfrac{9}{5}$ in.

a. $2 \times 2 \times 3$ **b.** $2 \times 2 \times 5$ **c.** $2 \times 2 \times 3 \times 3$ **d.** $2 \times 2 \times 2 \times 2 \times 3$

e. $2 \times 2 \times 3 \times 5$ **f.** $2 \times 7 \times 7$

Name

Section   Date

ANSWERS

1. $\dfrac{4}{7}$

2. $\dfrac{5}{9}$

3. $\dfrac{3}{4}$

4. $\dfrac{2}{5}$

5. $\dfrac{2}{3}$

6. 1

7. $1\dfrac{2}{13}$

8. $1\dfrac{2}{9}$

9. $1\dfrac{1}{3}$

10. $\dfrac{7}{9}$

11. $1\dfrac{2}{9}$

12. $1\dfrac{1}{5}$

13. $\dfrac{9}{10}$ of a dollar

14. $\dfrac{10}{11}$ in.

Add. Simplify when possible.

1. $\dfrac{3}{7} + \dfrac{1}{7}$

2. $\dfrac{2}{9} + \dfrac{3}{9}$

3. $\dfrac{5}{8} + \dfrac{1}{8}$

4. $\dfrac{1}{10} + \dfrac{3}{10}$

5. $\dfrac{8}{15} + \dfrac{2}{15}$

6. $\dfrac{4}{7} + \dfrac{3}{7}$

7. $\dfrac{8}{13} + \dfrac{7}{13}$

8. $\dfrac{17}{18} + \dfrac{5}{18}$

9. $\dfrac{19}{24} + \dfrac{13}{24}$

10. $\dfrac{1}{9} + \dfrac{2}{9} + \dfrac{4}{9}$

11. $\dfrac{2}{9} + \dfrac{5}{9} + \dfrac{4}{9}$

12. $\dfrac{4}{15} + \dfrac{7}{15} + \dfrac{7}{15}$

Solve the following applications.

13. Money. You collect 11 nickels on Monday, 4 nickels on Tuesday, and 3 nickels on Wednesday. How much money do you have as a fraction of a dollar?

14. Perimeter. What is the perimeter of the given figure?

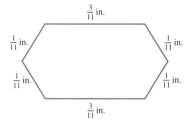

Finding the Least Common Denominator for Unlike Fractions

7.2 OBJECTIVES

1. Find the LCD of two fractions.
2. Find the LCD of a group of fractions.

Only *like* fractions can be added.

We can now add because we have like fractions.

In Section 7.1, you dealt with like fractions (fractions with a common denominator). What about a sum that deals with **unlike fractions**, such as $\frac{1}{3} + \frac{1}{4}$?

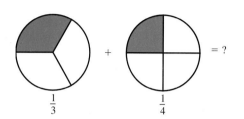

$\frac{1}{3}$ + $\frac{1}{4}$ = ?

We cannot add unlike fractions because they have different denominators.

In order to add unlike fractions, write them as equivalent fractions with a common denominator. In this case, let's use 12 as the denominator.

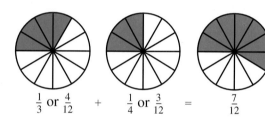

$\frac{1}{3}$ or $\frac{4}{12}$ + $\frac{1}{4}$ or $\frac{3}{12}$ = $\frac{7}{12}$

We have chosen 12 because it is a *multiple* of 3 and 4.

$\frac{1}{3}$ is equivalent to $\frac{4}{12}$.

$\frac{1}{4}$ is equivalent to $\frac{3}{12}$.

Any common multiple of the denominators will work in forming equivalent fractions. For instance, we can write $\frac{1}{3}$ as $\frac{8}{24}$ and $\frac{1}{4}$ as $\frac{6}{24}$. Our work is simplest, however, if we use the smallest possible number for the common denominator. This is called the **least common denominator (LCD)**.

The LCD is the least common multiple of the denominators of the fractions. This is the *smallest* number that is a multiple of all the denominators. For example, the LCD for $\frac{1}{3}$ and $\frac{1}{4}$ is 12, *not* 24 (see Section 4.5). In fact, finding the least common denominator is so important for our work in this chapter that it is worth a review.

To Find the Least Common Denominator

STEP 1   Write the prime factorization for each of the denominators.
STEP 2   Find all the prime factors that appear in any one of the prime factorizations.
STEP 3   Form the product of those prime factors, using each factor the greatest number of times it occurs in any one factorization.

• Example 1

Finding the Least Common Denominator (LCD)

Find the LCD of fractions with denominators 6 and 8.

Solution Our first step in adding fractions with denominators 6 and 8 will be to determine the least common denominator. Factor 6 and 8.

$6 = 2 \times 3$
$8 = 2 \times 2 \times 2$

Note that since 2 appears 3 times as a factor of 8, it is used 3 times in writing the LCD.

The LCD is $2 \times 2 \times 2 \times 3$, or 24.

● ● ● CHECK YOURSELF 1

Find the LCD of fractions with denominators 9 and 12.

The process is similar if more than two denominators are involved.

• Example 2

Finding the Least Common Denominator (LCD)

Find the LCD of fractions with denominators 6, 9, and 15.

Solution To add fractions with denominators 6, 9, and 15, we need to find the LCD. Factor the three numbers.

$6 = 2 \times 3$
$9 = 3 \times 3$
$15 = 3 \times 5$

2 and 5 appear only once in any one factorization.
3 appears twice as a factor of 9.

The LCD is $2 \times 3 \times 3 \times 5$, or 90.

● ● ● CHECK YOURSELF 2

Find the LCD of fractions with denominators 5, 8, and 20.

● ● ● CHECK YOURSELF ANSWERS

1. $9 = 3 \times 3$
$12 = 2 \times 2 \times 3$
The LCD is $2 \times 2 \times 3 \times 3 = 36$.

2. 40.

A N S W E R S

Find the least common denominator (LCD) for fractions with the given denominators.

1. 3 and 4

2. 3 and 5

3. 4 and 8

4. 6 and 12

5. 9 and 27

6. 10 and 30

7. 8 and 12

8. 15 and 40

9. 14 and 21

10. 15 and 20

11. 20 and 30

12. 24 and 36

13. 30 and 50

14. 36 and 48

15. 48 and 80

16. 60 and 84

17. 3, 4, and 5

18. 3, 4, and 6

19. 8, 10, and 15

20. 6, 22, and 33

21. 5, 10, and 25

22. 8, 24, and 48

23. 14, 24, and 28

24. 8, 18, and 30

1.	12
2.	15
3.	8
4.	12
5.	27
6.	30
7.	24
8.	120
9.	42
10.	60
11.	60
12.	72
13.	150
14.	144
15.	240
16.	420
17.	60
18.	12
19.	120
20.	66
21.	50
22.	48
23.	168
24.	360

Getting Ready for Section 7.3
[Section 5.5]

Find the missing numerators.

a. $\dfrac{1}{3} = \dfrac{?}{9}$ **b.** $\dfrac{1}{5} = \dfrac{?}{20}$ **c.** $\dfrac{1}{6} = \dfrac{?}{30}$

d. $\dfrac{2}{3} = \dfrac{?}{24}$ **e.** $\dfrac{3}{8} = \dfrac{?}{32}$ **f.** $\dfrac{5}{9} = \dfrac{?}{45}$

Answers

1. 12 **3.** 8 **5.** 27
7. $8 = 2 \times 2 \times 2$; $12 = 2 \times 2 \times 3$; The LCD is $2 \times 2 \times 2 \times 3 = 24$ **9.** 42
11. $20 = 2 \times 2 \times 5$; $30 = 2 \times 3 \times 5$; The LCD is $2 \times 2 \times 3 \times 5 = 60$
13. 150 **15.** 240 **17.** 60 **19.** 120 **21.** 50 **23.** 168 **a.** 3
b. 4 **c.** 5 **d.** 16 **e.** 12 **f.** 25

7.2 Supplementary Exercises

Name _____

Section _____ Date _____

Find the least common denominator (LCD) for fractions with the given denominators.

1. 5 and 7 **2.** 5 and 15

3. 6 and 24 **4.** 12 and 18

5. 20 and 24 **6.** 25 and 40

7. 4, 5, and 9 **8.** 3, 4, and 11

9. 2, 5, and 8 **10.** 3, 6, and 8

11. 12, 18, and 24 **12.** 4, 15, and 18

 **7.3**

Adding Unlike Fractions

See Section 7.2 if you wish to review how we arrived at 24.

7.3 OBJECTIVES

1. Add any two fractions.
2. Add any group of fractions.
3. Solve an application involving the addition of fractions.

We are now ready to add unlike fractions. In this case, the fractions must be renamed as equivalent fractions that have the same denominator. We will use the following rule.

To Add Unlike Fractions

STEP 1 Find the LCD of the fractions.
STEP 2 Change each unlike fraction to an equivalent fraction with the LCD as a common denominator.
STEP 3 Add the resulting like fractions as before.

Our first example shows the use of this rule.

• Example 1

Adding Unlike Fractions

Add the fractions $\dfrac{1}{6}$ and $\dfrac{3}{8}$.

Step 1 We find that the LCD of 6 and 8 is 24.

Step 2 Convert the fractions so that they have the denominator 24.

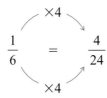

$$\frac{1}{6} = \frac{4}{24}$$

How many sixes are in 24? There are 4. So multiply the numerator and denominator by 4.

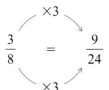

$$\frac{3}{8} = \frac{9}{24}$$

How many eights are in 24? There are 3. So multiply the numerator and denominator by 3.

Step 3 We can now add the equivalent like fractions.

$$\frac{1}{6} + \frac{3}{8} = \frac{4}{24} + \frac{9}{24} = \frac{13}{24}$$

Add the numerators and place that sum over the common denominator.

●●● **CHECK YOURSELF 1**

Add.

$$\frac{3}{5} + \frac{1}{3}$$

Here is a similar example. Remember that the sum should always be written in simplest form.

●Example 2

Adding Unlike Fractions that Require Simplifying

Add the fractions $\frac{7}{10}$ and $\frac{2}{15}$.

Step 1 The LCD of 10 and 15 is 30.

Step 2 $\frac{7}{10} = \frac{21}{30}$

$\frac{2}{15} = \frac{4}{30}$ Do you see how the equivalent fractions are formed?

Step 3 $\frac{7}{10} + \frac{2}{15} = \frac{21}{30} + \frac{4}{30}$

$= \frac{25}{30} = \frac{5}{6}$ Add the resulting like fractions. Be sure the sum is in simplest form.

●●● **CHECK YOURSELF 2**

Add.

$$\frac{1}{6} + \frac{7}{12}$$

We can easily add more than two fractions by using the same procedure. Example 3 illustrates this approach.

● Example 3

Adding a Group of Unlike Fractions

Add $\dfrac{5}{6} + \dfrac{2}{9} + \dfrac{4}{15}$.

Go back and review if you need to.

Step 1 We found the LCD of 90 for these same denominators in Section 7.2.

Step 2 $\dfrac{5}{6} = \dfrac{75}{90}$ Multiply the numerator and denominator by 15.

$\dfrac{2}{9} = \dfrac{20}{90}$ Multiply the numerator and denominator by 10.

$\dfrac{4}{15} = \dfrac{24}{90}$ Multiply the numerator and denominator by 6.

Step 3 $\dfrac{75}{90} + \dfrac{20}{90} + \dfrac{24}{90} = \dfrac{119}{90}$ Now add.

$= 1\dfrac{29}{90}$ *Remember,* if the sum is an improper fraction, it should be changed to a mixed number.

● ● ●   **CHECK YOURSELF 3**

(*Hint:* You found the LCD in Section 7.2.)

Add.

$$\dfrac{2}{5} + \dfrac{3}{8} + \dfrac{7}{20}$$

Many of the measurements you deal with in everyday life involve fractions. Let's look at some typical situations.

● Example 4

An Application Involving the Addition of Unlike Fractions

Jack ran $\dfrac{1}{2}$ mi on Monday, $\dfrac{2}{3}$ mi on Wednesday, and $\dfrac{3}{4}$ mi on Friday. How far did he run during the week?

Solution The three distances that Jack ran are the given information in the problem. We want to find a total distance, so we must add for the solution.

$\dfrac{1}{2} + \dfrac{2}{3} + \dfrac{3}{4} = \dfrac{6}{12} + \dfrac{8}{12} + \dfrac{9}{12}$ Since we have no common denominator, we must convert to equivalent fractions before we can add.

$= \dfrac{23}{12} = 1\dfrac{11}{12}$ mi

Jack ran $1\dfrac{11}{12}$ mi during the week.

● ● ● **CHECK YOURSELF 4**

Susan is designing an office complex. She needs $\frac{2}{5}$ acre for buildings, $\frac{1}{3}$ acre for driveways and parking, and $\frac{1}{6}$ acre for walks and landscaping. How much land does she need?

● **Example 5**

An Application Involving the Addition of Unlike Fractions

Sam bought three packages of spices weighing $\frac{1}{4}$, $\frac{5}{8}$, and $\frac{1}{2}$ pounds (lb). What was the total weight?

Solution   We need to find the total weight, so we must add.

The abbreviation for pounds is "lb" from the Latin *libra,* meaning "balance" or "scales."

$$\frac{1}{4} + \frac{5}{8} + \frac{1}{2} = \frac{2}{8} + \frac{5}{8} + \frac{4}{8}$$

Write each fraction with the denominator 8.

$$= \frac{11}{8} = 1\frac{3}{8} \text{ lb}$$

The total weight was $1\frac{3}{8}$ lb.

● ● ● **CHECK YOURSELF 5**

For three different recipes, Max needs $\frac{3}{8}$, $\frac{1}{2}$, and $\frac{5}{8}$ gallons (gal) tomato sauce. How many gallons should he buy altogether?

● ● ● **CHECK YOURSELF ANSWERS**

1. $\frac{14}{15}$.   2. $\frac{1}{6} + \frac{7}{12} = \frac{2}{12} + \frac{7}{12} = \frac{9}{12} = \frac{3}{4}$.   3. $1\frac{1}{8}$.   4. $\frac{9}{10}$ acre.
5. $1\frac{1}{2}$ gal.

Name

Section   Date

Add:

1. $\dfrac{2}{3} + \dfrac{1}{4}$

2. $\dfrac{3}{5} + \dfrac{1}{3}$

3. $\dfrac{1}{5} + \dfrac{3}{10}$

4. $\dfrac{1}{3} + \dfrac{1}{18}$

5. $\dfrac{3}{4} + \dfrac{1}{8}$

6. $\dfrac{4}{5} + \dfrac{1}{10}$

7. $\dfrac{1}{7} + \dfrac{3}{5}$

8. $\dfrac{1}{6} + \dfrac{2}{15}$

9. $\dfrac{3}{7} + \dfrac{3}{14}$

10. $\dfrac{7}{20} + \dfrac{9}{40}$

11. $\dfrac{7}{15} + \dfrac{2}{35}$

12. $\dfrac{3}{10} + \dfrac{3}{8}$

13. $\dfrac{5}{8} + \dfrac{1}{12}$

14. $\dfrac{5}{12} + \dfrac{3}{10}$

15. $\dfrac{5}{6} + \dfrac{7}{9}$

16. $\dfrac{4}{15} + \dfrac{3}{20}$

17. $\dfrac{5}{12} + \dfrac{7}{18}$

18. $\dfrac{4}{15} + \dfrac{9}{25}$

A N S W E R S

1. $\dfrac{11}{12}$

2. $\dfrac{14}{15}$

3. $\dfrac{1}{2}$

4. $\dfrac{7}{18}$

5. $\dfrac{7}{8}$

6. $\dfrac{9}{10}$

7. $\dfrac{26}{35}$

8. $\dfrac{3}{10}$

9. $\dfrac{9}{14}$

10. $\dfrac{23}{40}$

11. $\dfrac{11}{21}$

12. $\dfrac{27}{40}$

13. $\dfrac{17}{24}$

14. $\dfrac{43}{60}$

15. $1\dfrac{11}{18}$

16. $\dfrac{5}{12}$

17. $\dfrac{29}{36}$

18. $\dfrac{47}{75}$

19. $\dfrac{13}{16} + \dfrac{17}{24}$

20. $\dfrac{17}{30} + \dfrac{14}{25}$

21. $\dfrac{1}{5} + \dfrac{1}{3} + \dfrac{1}{4}$

22. $\dfrac{1}{3} + \dfrac{1}{4} + \dfrac{1}{6}$

23. $\dfrac{1}{5} + \dfrac{7}{10} + \dfrac{4}{15}$

24. $\dfrac{2}{3} + \dfrac{1}{4} + \dfrac{3}{8}$

25. $\dfrac{1}{9} + \dfrac{7}{12} + \dfrac{5}{8}$

26. $\dfrac{1}{3} + \dfrac{5}{12} + \dfrac{4}{5}$

27. $\dfrac{5}{12} + \dfrac{2}{21} + \dfrac{11}{28}$

Solve the following applications.

28. **Consumer buying.** Paul bought $\dfrac{1}{2}$ pounds (lb) of peanuts and $\dfrac{3}{8}$ lb of cashews. How many pounds of nuts did he buy?

29. **Countertop thickness.** A countertop consists of a board $\dfrac{3}{4}$ inches (in.) thick and tile $\dfrac{3}{8}$ in. thick. What is the overall thickness?

30. **Budgets.** Amy budgets $\dfrac{2}{5}$ of her income for housing and $\dfrac{1}{6}$ of her income for food. What fraction of her income is budgeted for these two purposes? What fraction of her income remains?

31. **Daily schedule.** A person spends $\dfrac{3}{8}$ day at work and $\dfrac{1}{3}$ day sleeping. What fraction of a day do these two activities use? What fraction of the day remains?

32. **Distance.** Jose walked $\dfrac{3}{4}$ miles (mi) to the store, $\dfrac{1}{2}$ mi to a friend's house, and then $\dfrac{2}{3}$ mi home. How far did he walk?

A N S W E R S

33. $1\frac{7}{8}$ in.

34. $\frac{5}{8}$

35. $\frac{1}{4}$

a. 5

b. 14

c. 18

d. 19

e. 15

f. 6

33. Perimeter. Find the perimeter of, or the distance around, the accompanying figure.

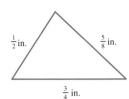

$\frac{1}{2}$ in. $\frac{5}{8}$ in.

$\frac{3}{4}$ in.

34. Budgeting. A budget guide states that you should spend $\frac{1}{4}$ of your salary for housing, $\frac{3}{16}$ for food, $\frac{1}{16}$ for clothing, and $\frac{1}{8}$ for transportation. What total portion of your salary will these four expenses account for?

35. Salary. Deductions from your paycheck are made roughly as follows: $\frac{1}{8}$ for federal tax, $\frac{1}{20}$ for state tax, $\frac{1}{20}$ for social security, and $\frac{1}{40}$ for a savings withholding plan. What portion of your pay is deducted?

Getting Ready for Section 7.4
[Section 1.7]

Find each of the following differences.

a. $17 - 12$

b. $29 - 15$

c. $43 - 25$

d. $58 - 39$

e. $92 - 77$

f. $82 - 76$

Answers

1. $\frac{11}{12}$ **3.** $\frac{1}{2}$ **5.** $\frac{7}{8}$ **7.** $\frac{26}{35}$ **9.** $\frac{9}{14}$ **11.** $\frac{11}{21}$ **13.** $\frac{17}{24}$

15. $\frac{5}{6} + \frac{7}{9} = \frac{15}{18} + \frac{14}{18} = \frac{29}{18} = 1\frac{11}{18}$ **17.** $\frac{29}{36}$ **19.** $1\frac{25}{48}$ **21.** $\frac{47}{60}$

23. $\frac{1}{5} + \frac{7}{10} + \frac{4}{15} = \frac{6}{30} + \frac{21}{30} + \frac{8}{30} = \frac{35}{30} = 1\frac{5}{30} = 1\frac{1}{6}$ **25.** $1\frac{23}{72}$

27. $\frac{19}{21}$ **29.** $1\frac{1}{8}$ in.   **31.** $\frac{17}{24}, \frac{7}{24}$ **33.** $1\frac{7}{8}$ in.   **35.** $\frac{1}{4}$ **a.** 5

b. 14   **c.** 18   **d.** 19   **e.** 15   **f.** 6

Name

Section   Date

A N S W E R S

1. $\dfrac{13}{20}$

2. $\dfrac{29}{30}$

3. $\dfrac{13}{15}$

4. $\dfrac{53}{60}$

5. $\dfrac{19}{24}$

6. $\dfrac{31}{72}$

7. $\dfrac{7}{12}$

8. $1\dfrac{5}{42}$

9. $1\dfrac{17}{90}$

10. $1\dfrac{17}{150}$

11. $\dfrac{7}{8}$

12. $\dfrac{19}{30}$

13. $1\dfrac{13}{72}$

14. $1\dfrac{49}{60}$

15. $\dfrac{9}{20}, \dfrac{11}{20}$

16. $1\dfrac{7}{12}$ gal

17. $\dfrac{3}{4}$ in.

18. $1\dfrac{3}{4}$ yd

Add.

1. $\dfrac{2}{5} + \dfrac{1}{4}$

2. $\dfrac{2}{3} + \dfrac{3}{10}$

3. $\dfrac{2}{5} + \dfrac{7}{15}$

4. $\dfrac{3}{10} + \dfrac{7}{12}$

5. $\dfrac{3}{8} + \dfrac{5}{12}$

6. $\dfrac{5}{36} + \dfrac{7}{24}$

7. $\dfrac{2}{15} + \dfrac{9}{20}$

8. $\dfrac{9}{14} + \dfrac{10}{21}$

9. $\dfrac{7}{15} + \dfrac{13}{18}$

10. $\dfrac{12}{25} + \dfrac{19}{30}$

11. $\dfrac{1}{2} + \dfrac{1}{4} + \dfrac{1}{8}$

12. $\dfrac{1}{3} + \dfrac{1}{5} + \dfrac{1}{10}$

13. $\dfrac{3}{8} + \dfrac{5}{12} + \dfrac{7}{18}$

14. $\dfrac{5}{6} + \dfrac{8}{15} + \dfrac{9}{20}$

Solve the following applications.

15. **Portion of income.** Marjory spends $\dfrac{1}{4}$ of her income for housing and $\dfrac{1}{5}$ of her income for food. What portion of her income goes for these two expenses? What amount remains?

16. **Paint usage.** Roberto used $\dfrac{3}{4}$ gallons (gal) of paint in his living room, $\dfrac{1}{3}$ gal in the dining room, and $\dfrac{1}{2}$ gal in a hallway. How much paint did he use?

17. **Plywood thickness.** A sheet of plywood consists of two outer sections that are $\dfrac{3}{16}$ inches (in.) thick and a center section that is $\dfrac{3}{8}$ in. thick. How thick is the plywood overall?

18. **Quantity.** A pattern calls for four pieces of fabric with lengths $\dfrac{1}{2}, \dfrac{3}{8}, \dfrac{1}{4}$, and $\dfrac{5}{8}$ yard (yd). How much fabric must be purchased to use the pattern?

Using Your Calculator to Add Fractions

Adding fractions on the calculator is very much like the multiplication and division you did in the previous chapter. The only thing that changes is the operation.

SCIENTIFIC CALCULATOR

Here's where the fraction calculator is a great tool for checking your work. No muss, no fuss, no searching for a common denominator. Just enter the fractions and get the right answer!

• Example 1

Adding Fractions

Find the sum.

$$\frac{3}{14} + \frac{7}{12}$$

The keystroke sequence is

3 $\boxed{\textbf{a b/c}}$ 14 $\boxed{+}$ 7 $\boxed{\textbf{a b/c}}$ 12 $\boxed{=}$

The result is $\frac{67}{84}$.

● ● ● CHECK YOURSELF 1

Find the sum.

$$\frac{5}{24} + \frac{11}{18}$$

GRAPHING CALCULATOR

Use your graphing calculator to find the sum.

$$\frac{7}{24} + \frac{17}{42}$$

The keystroke sequence is

7 $\boxed{\div}$ 24 $\boxed{+}$ 17 $\boxed{\div}$ 42 $\boxed{\blacktriangleright\textbf{Frac.}}$ $\boxed{\text{Enter}}$

The solution is $\frac{39}{56}$.

● ● ●   **CHECK YOURSELF ANSWER**

$\dfrac{59}{72}$.

Calculator Exercises

Name

Section   Date

A N S W E R S

1. $\dfrac{7}{6}$ or $1\dfrac{1}{6}$

2. $\dfrac{7}{4}$ or $1\dfrac{3}{4}$

3. $\dfrac{55}{36}$ or $1\dfrac{19}{36}$

4. $\dfrac{97}{66}$ or $1\dfrac{31}{66}$

5. $\dfrac{7}{12}$

6. $\dfrac{37}{56}$

7. $\dfrac{67}{60}$ or $1\dfrac{7}{60}$

8. $\dfrac{13}{16}$

9. $\dfrac{41}{60}$

10. $\dfrac{47}{40}$ or $1\dfrac{7}{40}$

11. $\dfrac{110}{63}$ or $1\dfrac{47}{60}$

12. $\dfrac{13}{15}$

13. $\dfrac{37}{36}$ or $1\dfrac{1}{36}$

14. $\dfrac{77}{72}$ or $1\dfrac{5}{72}$

Find the following sums using your calculator.

1. $\dfrac{2}{3} + \dfrac{1}{2}$

2. $\dfrac{11}{12} + \dfrac{5}{6}$

3. $\dfrac{3}{4} + \dfrac{7}{9}$

4. $\dfrac{7}{11} + \dfrac{5}{6}$

5. $\dfrac{5}{12} + \dfrac{1}{6}$

6. $\dfrac{2}{7} + \dfrac{3}{8}$

7. $\dfrac{8}{15} + \dfrac{7}{12}$

8. $\dfrac{7}{16} + \dfrac{9}{24}$

9. $\dfrac{1}{10} + \dfrac{7}{12}$

10. $\dfrac{7}{15} + \dfrac{17}{24}$

11. $\dfrac{8}{9} + \dfrac{6}{7}$

12. $\dfrac{7}{15} + \dfrac{2}{5}$

13. $\dfrac{11}{18} + \dfrac{5}{12}$

14. $\dfrac{5}{8} + \dfrac{4}{9}$

Answers

1. $\dfrac{7}{6}$ or $1\dfrac{1}{6}$ **3.** $\dfrac{55}{36}$ or $1\dfrac{19}{36}$ **5.** $\dfrac{7}{12}$ **7.** $\dfrac{67}{60}$ or $1\dfrac{7}{60}$ **9.** $\dfrac{41}{60}$

11. $\dfrac{110}{63}$ or $1\dfrac{47}{60}$ **13.** $\dfrac{37}{36}$ or $1\dfrac{1}{36}$

Subtracting Like and Unlike Fractions

7.4 OBJECTIVES

1. Subtract two like fractions.
2. Subtract any two fractions.
3. Solve an application of fraction subtraction.

If a problem involves like fractions, then subtraction, like addition, is not difficult.

To Subtract Like Fractions

STEP 1 Subtract the numerators.
STEP 2 Place the difference over the common denominator.
STEP 3 Simplify the resulting fraction when necessary.

Like fractions have the same denominator.

Note the similarity to our rule for adding like fractions.

•Example 1

Subtracting Like Fractions

Subtract.

Step 1   Step 2

$(a)\ \dfrac{4}{5} - \dfrac{2}{5} = \dfrac{4-2}{5} = \dfrac{2}{5}$

Subtract the numerators: $4 - 2 = 2$. Write the difference over the common denominator, 5. Step 3 is not necessary because the difference is in simplest form.

Illustrating with a diagram:

Subtracting 2 of the 5 parts from 4 of the 5 parts leaves 2 of the 5 parts.

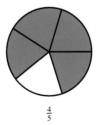

$\dfrac{4}{5}$ $-$ $\dfrac{2}{5}$ $=$ $\dfrac{2}{5}$

Always write the result in lowest terms.

$(b)\ \dfrac{5}{8} - \dfrac{3}{8} = \dfrac{5-3}{8} = \dfrac{2}{8} = \dfrac{1}{4}$

● ● ● **CHECK YOURSELF 1**

Subtract.

$\dfrac{11}{12} - \dfrac{5}{12}$

To subtract unlike fractions, which are fractions that do not have the same denominator, we have the following rule:

To Subtract Unlike Fractions

STEP 1   Find the LCD of the fractions.
STEP 2   Change each unlike fraction to an equivalent fraction with the LCD as a common denominator.
STEP 3   Subtract the resulting like fractions as before.

Of course, this is the same as our rule for adding fractions. We just subtract instead of add!

• Example 2

Subtracting Unlike Fractions

Subtract $\dfrac{5}{8} - \dfrac{1}{6}$.

Step 1   The LCD is 24.

Step 2   Convert the fractions so that they have the common denominator 24.

$$\frac{5}{8} = \frac{15}{24}$$
$$\frac{1}{6} = \frac{4}{24}$$

The first two steps are exactly the same as if we were adding.

Just as you did with the sums of fractions, you can use your calculator to check your result.

Step 3   Subtract the equivalent like fractions.

$$\frac{5}{8} - \frac{1}{6} = \frac{15}{24} - \frac{4}{24} = \frac{11}{24}$$

CAUTION

Be Careful! You *cannot* subtract the numerators and subtract the denominators.

$$\frac{5}{8} - \frac{1}{6} \qquad \text{is } not \qquad \frac{4}{2}$$

● ● ●   **CHECK YOURSELF 2**

Subtract.

$$\frac{7}{10} - \frac{1}{4}$$

The difference of two fractions should always be written in simplest form.

• Example 3

Subtracting Unlike Fractions

Subtract $\dfrac{7}{15} - \dfrac{3}{10}$.

$$\frac{7}{15} = \frac{14}{30} \quad \text{and} \quad \frac{3}{10} = \frac{9}{30}$$

Convert to like fractions with the LCD 30.

$$\frac{14}{30} - \frac{9}{30} = \frac{5}{30} = \frac{1}{6}$$

Subtract the like fractions and then reduce the difference to simplest form.

● ● ● **CHECK YOURSELF 3**

Subtract.

$$\frac{11}{15} - \frac{7}{12}$$

Let's look at an example that applies our work in subtracting unlike fractions.

● Example 4

An Application Involving the Subtraction of Unlike Fractions

You have $\frac{7}{8}$ yards (yd) of a handwoven linen. A pattern for a placemat calls for $\frac{1}{2}$ yd. Will you have enough left for two napkins that will use $\frac{1}{3}$ yd?

Solution   First, find out how much fabric is left over after the placemat is made.

$$\frac{7}{8} - \frac{1}{2} = \frac{7}{8} - \frac{4}{8} = \frac{3}{8} \text{ yd}$$

Remember that $\frac{3}{8}$ yd is left over and that $\frac{1}{3}$ yd is needed.

Now compare the size of $\frac{1}{3}$ and $\frac{3}{8}$.

$$\frac{3}{8} = \frac{9}{24} \text{ yd} \quad \text{and} \quad \frac{1}{3} = \frac{8}{24} \text{ yd}$$

Since $\frac{3}{8}$ yd is *more than* the $\frac{1}{3}$ yd that is needed, there is enough material for the placemat *and* two napkins.

● ● ● **CHECK YOURSELF 4**

A concrete walk will require $\frac{3}{4}$ cubic yard (yd^3) of concrete. If you have mixed $\frac{8}{9}$ yd^3, will enough concrete remain to do a project that will use $\frac{1}{6}$ yd^3?

Our next application involves measurement in inches. Note that on a ruler or yardstick, the marks divide each inch into $\frac{1}{2}$-in., $\frac{1}{4}$-in., and $\frac{1}{8}$-in. sections, and on some rulers, $\frac{1}{16}$-in. sections. We will use denominators of 2, 4, 8, and 16 in our measurement applications.

● Example 5

An Application Involving the Subtraction of Unlike Fractions

Jeannine is cutting two slats that are each to be $\frac{3}{16}$ in. in width from a piece of wood that is $\frac{3}{4}$ in. across. How much will be left?

Solution   The two $\frac{3}{16}$ in. pieces will total

$$2 \times \frac{3}{16} = \frac{6}{16} = \frac{3}{8} \text{ in.}$$

$$\frac{3}{4} = \frac{6}{8}$$

$$\frac{6}{8} - \frac{3}{8} = \frac{5}{8}$$

The remaining strip will be $\frac{5}{8}$ in. wide.

● ● ●   **CHECK YOURSELF 5**

Ricardo is cutting three strips from a piece of metal with a width of 1 in. Each strip has a width of $\frac{3}{16}$ in. How much metal will remain after the cuts?

___

● ● ●   **CHECK YOURSELF ANSWERS**

1. $\frac{11}{12} - \frac{5}{12} = \frac{6}{12} = \frac{1}{2}$.   2. $\frac{9}{20}$.

3. The LCD is 60. $\frac{11}{15} = \frac{44}{60}$ and $\frac{7}{12} = \frac{35}{60}$ $\frac{44}{60} - \frac{35}{60} = \frac{9}{60} = \frac{3}{20}$.

4. $\frac{5}{36}$ yd^3 will remain. You do *not* have enough concrete for both projects.

5. $\frac{7}{16}$ in.

Subtract.

1. $\dfrac{3}{5} - \dfrac{1}{5}$

2. $\dfrac{5}{7} - \dfrac{2}{7}$

3. $\dfrac{7}{9} - \dfrac{4}{9}$

4. $\dfrac{7}{10} - \dfrac{3}{10}$

5. $\dfrac{13}{20} - \dfrac{3}{20}$

6. $\dfrac{19}{30} - \dfrac{17}{30}$

7. $\dfrac{19}{24} - \dfrac{5}{24}$

8. $\dfrac{25}{36} - \dfrac{13}{36}$

9. $\dfrac{4}{5} - \dfrac{1}{3}$

10. $\dfrac{7}{9} - \dfrac{1}{6}$

11. $\dfrac{11}{15} - \dfrac{3}{5}$

12. $\dfrac{5}{6} - \dfrac{2}{7}$

13. $\dfrac{3}{8} - \dfrac{1}{4}$

14. $\dfrac{9}{10} - \dfrac{4}{5}$

15. $\dfrac{5}{12} - \dfrac{3}{8}$

16. $\dfrac{13}{15} - \dfrac{11}{20}$

17. $\dfrac{13}{25} - \dfrac{2}{15}$

18. $\dfrac{11}{12} - \dfrac{7}{10}$

ANSWERS

1. $\dfrac{2}{5}$

2. $\dfrac{3}{7}$

3. $\dfrac{1}{3}$

4. $\dfrac{2}{5}$

5. $\dfrac{1}{2}$

6. $\dfrac{1}{15}$

7. $\dfrac{7}{12}$

8. $\dfrac{1}{3}$

9. $\dfrac{7}{15}$

10. $\dfrac{11}{18}$

11. $\dfrac{2}{15}$

12. $\dfrac{23}{42}$

13. $\dfrac{1}{8}$

14. $\dfrac{1}{10}$

15. $\dfrac{1}{24}$

16. $\dfrac{19}{60}$

17. $\dfrac{29}{75}$

18. $\dfrac{13}{60}$

19. $\dfrac{15}{27} - \dfrac{7}{18}$

20. $\dfrac{7}{20} - \dfrac{4}{15}$

21. $\dfrac{13}{18} - \dfrac{7}{12}$

22. $\dfrac{17}{30} - \dfrac{5}{9}$

23. $\dfrac{33}{40} - \dfrac{7}{24}$

24. $\dfrac{13}{24} - \dfrac{5}{16}$

25. $\dfrac{15}{16} - \dfrac{5}{8} - \dfrac{1}{4}$

26. $\dfrac{9}{10} - \dfrac{1}{5} - \dfrac{1}{2}$

27. $\dfrac{5}{9} + \dfrac{7}{12} - \dfrac{5}{8}$

For Exercises 28 and 29, find the missing dimension (?) in the given figure.

28.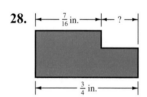

29.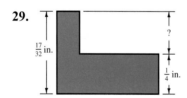

Solve the following applications.

30. **Cooking.** A hamburger that weighed $\dfrac{1}{4}$ pound (lb) before cooking was $\dfrac{3}{16}$ lb after cooking. How much weight was lost in cooking?

31. **Property.** Martin owned a $\dfrac{7}{8}$-acre piece of land. He sold $\dfrac{1}{3}$ acre. What amount of land remains?

32. $\frac{7}{16}$

33. No—only $\frac{1}{8}$ cup remains.

34. Yes— $\frac{1}{3}$ gal remains.

35.

36. $\frac{2}{3}$

37.

38. $2\frac{3}{4}$" or $2\frac{1}{16}$ ft

32. **Painting.** On Monday, $\frac{7}{8}$ of a house painting project remained to be done. John painted $\frac{1}{4}$ of the house on Tuesday and $\frac{3}{16}$ of the house on Wednesday. What portion of the project remained to be done?

33. **Baking.** Geraldo has $\frac{3}{4}$ cup of flour. Biscuits use $\frac{5}{8}$ cup. Will he have enough left over for a small pie crust that requires $\frac{1}{4}$ cup?

34. **Painting.** You have $\frac{5}{6}$ gallons (gal) of paint. You estimate that one wall will use $\frac{1}{2}$ gal. Can you also finish a smaller wall that will need $\frac{1}{4}$ gal?

35. A teenager makes a log of his activities during the course of a year.

Activity	Time	Fraction of the Year
Sleeping	10 hours per day	
Eating	2 hours per day	
Weekend activities	2 days per week	
Summer vacation	3 months per year	

He uses this data to claim that he has no time left over for school.

(a) Complete the above chart.

(b) What is the flaw in the teenager's reasoning?

36. Manny, Moe, and Jack each have equal shares in a automotive store. Moe decides to retire and sell his shares. He sells $\frac{1}{4}$ of his shares to Manny and the remainder to Jack. What is Jack's share of the store now?

37. Measure the length of an unstretched rubber band to the nearest eighth of an inch. Then stretch the band as far as you can (without breaking it), and again measure the length to the nearest eighth of an inch.

(a) What is the difference between the length of the stretched and unstretched rubber bands?

(b) Repeat this process for other rubber bands of different thicknesses. What is the relationship between the thickness of the rubber band and the distance it can be stretched?

38. A door is $4\frac{1}{4}$ ft wide. Two hooks are to be attached to the door so that they are $1\frac{1}{2}$ in. apart and the same distance from each edge. How far from the edge of the door should each hook be located?

a. $1\frac{2}{9}$

b. $1\frac{3}{7}$

c. $1\frac{1}{15}$

d. 1

e. $\frac{4}{3}$

f. $\frac{13}{5}$

g. $\frac{19}{6}$

h. $\frac{45}{8}$

 **Getting Ready for Section 7.5
[Section 5.3]**

Write each improper fraction as a mixed or whole number.

a. $\frac{11}{9}$ **b.** $\frac{10}{7}$ **c.** $\frac{16}{15}$ **d.** $\frac{10}{10}$

Write each mixed number as an improper fraction.

e. $1\frac{1}{3}$ **f.** $2\frac{3}{5}$ **g.** $3\frac{1}{6}$ **h.** $5\frac{5}{8}$

Answers

1. $\frac{2}{5}$ **3.** $\frac{5}{9} - \frac{2}{9} = \frac{3}{9} = \frac{1}{3}$ **5.** $\frac{1}{2}$ **7.** $\frac{7}{12}$

9. $\frac{4}{5} - \frac{1}{3} = \frac{12}{15} - \frac{5}{15} = \frac{7}{15}$ **11.** $\frac{2}{15}$ **13.** $\frac{1}{8}$ **15.** $\frac{1}{24}$ **17.** $\frac{29}{75}$

19. $\frac{1}{6}$ **21.** $\frac{13}{18} - \frac{7}{12} = \frac{26}{36} - \frac{21}{36} = \frac{5}{36}$

23. $\frac{33}{40} - \frac{7}{24} = \frac{99}{120} - \frac{35}{120} = \frac{64}{120} = \frac{8}{15}$ **25.** $\frac{1}{16}$ **27.** $\frac{37}{72}$ **29.** $\frac{9}{32}$ in.

31. $\frac{13}{24}$ acre   **33.** No. Only $\frac{1}{8}$ cup will be left over.   **35.**

37.   **a.** $1\frac{2}{9}$ **b.** $1\frac{3}{7}$ **c.** $1\frac{1}{15}$ **d.** 1 **e.** $\frac{4}{3}$ **f.** $\frac{13}{5}$

g. $\frac{19}{6}$ **h.** $\frac{45}{8}$

Name

Section   Date

Subtract.

1. $\dfrac{8}{9} - \dfrac{3}{9}$

2. $\dfrac{9}{10} - \dfrac{6}{10}$

3. $\dfrac{5}{8} - \dfrac{1}{8}$

4. $\dfrac{11}{12} - \dfrac{7}{12}$

5. $\dfrac{7}{8} - \dfrac{2}{3}$

6. $\dfrac{5}{6} - \dfrac{3}{5}$

7. $\dfrac{11}{18} - \dfrac{2}{9}$

8. $\dfrac{5}{6} - \dfrac{1}{4}$

9. $\dfrac{5}{8} - \dfrac{1}{6}$

10. $\dfrac{13}{18} - \dfrac{5}{12}$

11. $\dfrac{8}{21} - \dfrac{1}{14}$

12. $\dfrac{13}{18} - \dfrac{7}{15}$

13. $\dfrac{11}{12} - \dfrac{1}{4} - \dfrac{1}{3}$

14. $\dfrac{13}{15} + \dfrac{2}{3} - \dfrac{3}{5}$

Solve the following applications.

15. Baking. A recipe calls for $\dfrac{1}{3}$ cup of milk. You have $\dfrac{3}{4}$ cup. How much milk will be left over?

ANSWERS

1. $\dfrac{5}{9}$

2. $\dfrac{3}{10}$

3. $\dfrac{1}{2}$

4. $\dfrac{1}{3}$

5. $\dfrac{5}{24}$

6. $\dfrac{7}{30}$

7. $\dfrac{7}{18}$

8. $\dfrac{7}{12}$

9. $\dfrac{11}{24}$

10. $\dfrac{11}{36}$

11. $\dfrac{13}{42}$

12. $\dfrac{23}{90}$

13. $\dfrac{1}{3}$

14. $\dfrac{14}{15}$

15. $\dfrac{5}{12}$ cup

16. **Construction.** The outside diameter of a pipe is $\dfrac{7}{8}$ inch (in.). The inside diameter is $\dfrac{9}{16}$ in. What is the wall thickness of the pipe? (*Hint:* Draw a picture of the pipe.)

17. **Homework.** You have $\dfrac{7}{8}$ of a homework assignment left to do over a weekend. On Saturday you do $\dfrac{1}{4}$ of the assignment. What portion remains to be done on Sunday?

18. **Building.** A concrete divider requires $\dfrac{1}{2}$ yd^3 of concrete. If you have mixed $\dfrac{7}{8}$ yd^3, will enough be left over, after you have poured the concrete for the divider, for a small project that requires $\dfrac{1}{4}$ yd^3?

Adding and Subtracting Mixed Numbers

7.5 OBJECTIVES

1. Add two mixed numbers.
2. Subtract two mixed numbers.
3. Solve a problem that involves addition and subtraction of mixed numbers.
4. Solve an application that involves mixed number addition or subtraction.

Once you know how to add fractions, adding mixed numbers should be no problem if you keep in mind that addition involves combining groups of the *same kind* of objects. Since mixed numbers consist of two parts—a whole number and a fraction—we can work with the whole numbers and the fractions separately. Consider the following:

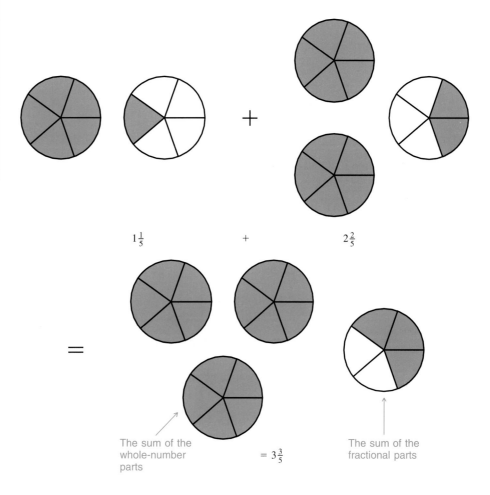

$$1\frac{1}{5} \qquad + \qquad 2\frac{2}{5}$$

The sum of the whole-number parts

$$= 3\frac{3}{5}$$

The sum of the fractional parts

This suggests the following general rule.

To Add Mixed Numbers

STEP 1 Add the whole-number parts.
STEP 2 Add the fractional parts.
STEP 3 Combine the results as a mixed number.

Note: Step 2 requires that the fractional parts have the same denominator.

Our first example illustrates the use of this rule.

• Example 1

Adding Mixed Numbers

Add.

$$3\frac{1}{5} + 4\frac{2}{5}$$

$$3\frac{1}{5} + 4\frac{2}{5} \qquad 3 + 4 = 7 \qquad \text{Add the whole numbers.}$$

$$3\frac{1}{5} + 4\frac{2}{5} \qquad \frac{1}{5} + \frac{2}{5} = \frac{3}{5} \qquad \text{Add the fractional parts.}$$

$$7 + \frac{3}{5} = 7\frac{3}{5} \qquad \text{Now combine the results.}$$

● ● ●   CHECK YOURSELF 1

Add $2\frac{3}{10} + 3\frac{4}{10}$.

• Example 2

Adding Mixed Numbers

Add.

We add as before, but there is a problem. The fractional portion is an improper fraction.

$$2\frac{5}{7} + 3\frac{6}{7} = 5\frac{11}{7}$$

$$= 5 + 1\frac{4}{7} \qquad \text{Write the improper fraction, } \frac{11}{7}, \text{ as a mixed number, } 1\frac{4}{7}.$$

$$= 6\frac{4}{7} \qquad \text{Add the remaining whole numbers.} \\ 5 + 1 = 6.$$

● ● ●   CHECK YOURSELF 2

Add $4\frac{3}{8} + 2\frac{7}{8}$.

When the fractional portions of the mixed numbers have different denominators, we must rename these fractions as equivalent fractions with the least common denominator in order to perform the addition in step 2. Consider Example 3.

• Example 3

Adding Mixed Numbers with Different Denominators

Add.

$$3\frac{1}{6} + 2\frac{3}{8} = 3\frac{4}{24} + 2\frac{9}{24}$$ The LCD of the fractions is 24. Rename
them with that denominator.

$$= 5\frac{13}{24}$$ Then add as before.

● ● ● **CHECK YOURSELF 3**

Add $5\frac{7}{10} + 3\frac{5}{6}$.

You follow the same procedure if more than two mixed numbers are involved in the problem.

• Example 4

Adding Mixed Numbers with Different Denominators

Add.

The LCD of the three
fractions is 40. Convert to
equivalent fractions.

$$2\frac{1}{5} + 3\frac{3}{4} + 4\frac{1}{8} = 2\frac{8}{40} + 3\frac{30}{40} + 4\frac{5}{40}$$

$$= 9\frac{43}{40}$$ The fractional portion is an improper
fraction. Write it as a mixed number.

$$= 9 + 1\frac{3}{40} = 10\frac{3}{40}$$ Add the whole numbers:
9 + 1 = 10

● ● ● **CHECK YOURSELF 4**

Add $5\frac{1}{2} + 4\frac{2}{3} + 3\frac{3}{4}$.

We can use a similar technique for *subtracting* mixed numbers if the fractional part being subtracted is the *smaller* of the two fractions. The rule is similar to that stated earlier for adding mixed numbers.

To Subtract Mixed Numbers

STEP 1 Subtract the whole-number parts.
STEP 2 Subtract the fractional parts.
STEP 3 Combine the results as a mixed number.

Example 5 illustrates the use of this rule.

•Example 5

Subtracting Mixed Numbers

Subtract.

$$5\frac{7}{12} - 3\frac{5}{12} = 2\frac{2}{12} = 2\frac{1}{6}$$

$$\left\{5 - 3\right\} \quad \left\{\frac{7}{12} - \frac{5}{12}\right\}$$

Subtract the whole numbers and then the fractional portions. Simplify your answer.

• • • CHECK YOURSELF 5

Subtract $8\frac{7}{8} - 5\frac{3}{8}$.

Again, we must rename the fractions if different denominators are involved. This approach is shown in Example 6.

•Example 6

Subtracting Mixed Numbers with Different Denominators

Subtract.

$$8\frac{7}{10} - 3\frac{3}{8} = 8\frac{28}{40} - 3\frac{15}{40}$$ Write the fractions with denominator 40.

$$= 5\frac{13}{40}$$ Subtract as before.

• • • CHECK YOURSELF 6

Subtract $7\frac{11}{12} - 3\frac{5}{8}$.

One method for handling the subtraction of mixed numbers requires writing the mixed numbers as improper fractions as the first step. This method is shown in Example 7.

• Example 7

Subtracting Mixed Numbers by Using Improper Fractions

Subtract.

$$4\frac{3}{8} - 2\frac{5}{8} = \frac{35}{8} - \frac{21}{8}$$ Write the mixed numbers as improper fractions for the first step.

$$= \frac{14}{8}$$ Now you can subtract.

$$= 1\frac{6}{8} = 1\frac{3}{4}$$ Simplify your answer.

● ● ● **CHECK YOURSELF 7**

Subtract by rewriting as improper fractions.

$$6\frac{3}{5} - 2\frac{4}{5}$$

Another method of subtracting mixed numbers uses borrowing.

• Example 8

Subtracting Mixed Numbers by Using Borrowing

$$4\frac{3}{8} - 2\frac{5}{8} = ?$$

Solution We cannot subtract the fractional portions because $\frac{5}{8}$ is larger than $\frac{3}{8}$.

To do the subtraction when the fractional part being subtracted is the *larger* of the fractions, we must *rename* the first mixed number.

$$4\frac{3}{8} = 4 + \frac{3}{8} = 3 + 1 + \frac{3}{8}$$ Borrow 1 from the 4.

$$= 3 + \frac{8}{8} + \frac{3}{8}$$ Think of that 1 as $\frac{8}{8}$.

$$= 3 + \frac{11}{8}$$ Add $\frac{8}{8}$ to the original $\frac{3}{8}$.

$$= 3\frac{11}{8}$$ We have renamed $4\frac{3}{8}$ as $3\frac{11}{8}$.

So

$$4\frac{3}{8} - 2\frac{5}{8} = 3\frac{11}{8} - 2\frac{5}{8}$$ By writing $4\frac{3}{8}$ as $3\frac{11}{8}$, you can subtract as before.

$$= 1\frac{6}{8} = 1\frac{3}{4}$$

● ● ●   **CHECK YOURSELF 8**

Subtract by using the method of Example 8.

$$6\frac{3}{5} - 2\frac{4}{5}$$

Note: You will probably find that the second method is best when the whole numbers in the problem are large.

If the fractions have different denominators, rewrite the fractions so that they have the least common denominator. Either method that we have considered can then be used.

● Example 9

Subtracting Mixed Numbers

Subtract $3\frac{1}{9} - 1\frac{5}{6}$.

Method 1: Rewriting as Improper Fractions

$$3\frac{1}{9} - 1\frac{5}{6} = \frac{28}{9} - \frac{11}{6}$$ Write the mixed numbers as improper fractions.

$$= \frac{56}{18} - \frac{33}{18}$$ Convert the fractions so that they have the denominator 18.

$$= \frac{23}{18} = 1\frac{5}{18}$$

Method 2: Using Borrowing

$$3\frac{1}{9} - 1\frac{5}{6} = 3\frac{2}{18} - 1\frac{15}{18}$$ Write the fractions with the LCD of 18. Do you see that we cannot subtract the way things stand?

$$= 2\frac{20}{18} - 1\frac{15}{18}$$ Borrow 1 from the 3 of the minuend. Think of that 1 as $\frac{18}{18}$ and combine it

$$= 1\frac{5}{18}$$ with the original $\frac{2}{18}$.

● ● ● **CHECK YOURSELF 9**

Subtract using both methods.

$$5\frac{1}{6} - 3\frac{3}{4}$$

To subtract a mixed number from a whole number, we again have two possible approaches.

● Example 10

Subtracting Mixed Numbers

Subtract.

$$6 - 2\frac{3}{4}$$

Method 1: Rewriting as Improper Fractions

Note:

$$6 = \frac{6}{1} = \frac{24}{4}$$

Multiply the numerator and denominator by 4 to form a common denominator.

$$6 - 2\frac{3}{4} = \frac{24}{4} - \frac{11}{4}$$

Write both the whole number and the mixed number as improper fractions with a common denominator.

$$= \frac{13}{4}$$

$$= 3\frac{1}{4}$$

Method 2: Using Borrowing

$$6 = 5 + 1 \qquad \text{Borrow 1 from 6.}$$

$$6 = 5 + 1 = 5 + \frac{4}{4} = 5\frac{4}{4} \qquad \text{Rewrite 1 as } \frac{4}{4}.$$

So,

$$6 - 2\frac{3}{4} = 5\frac{4}{4} - 2\frac{3}{4} = 3\frac{1}{4}$$

● ● ● **CHECK YOURSELF 10**

Subtract $7 - 3\frac{2}{5}$.

Some students prefer a vertical arrangement when adding or subtracting mixed numbers. Look at the following.

Subtract $6\frac{3}{8} - 2\frac{5}{6}$.

This vertical arrangement is not new. It is just another way of writing a problem when you are adding or subtracting mixed numbers.

Experiment with all the arrangements that have been shown, and decide which you like best.

Write the fractional parts as like fractions. We cannot subtract in this form. Do you see why?

Rename $6\frac{9}{24}$ as $5\frac{33}{24}$. The subtraction can now be done.

$$6\frac{3}{8} \longrightarrow 6\frac{9}{24} \longrightarrow 5\frac{33}{24}$$
$$-2\frac{5}{6} \longrightarrow -2\frac{20}{24} \longrightarrow -2\frac{20}{24}$$
$$3\frac{13}{24}$$

• Example 11

An Application of the Subtraction of Mixed Numbers

Linda was $48\frac{1}{4}$ inches (in.) tall on her sixth birthday. By her seventh year she was $51\frac{5}{8}$ in. tall. How much had she grown during the year?

Solution Since we want the difference in height, we must subtract.

$$51\frac{5}{8} - 48\frac{1}{4} = 51\frac{5}{8} - 48\frac{2}{8} = 3\frac{3}{8} \text{ in.}$$

Linda grew $3\frac{3}{8}$ in. during the year.

● ● ● CHECK YOURSELF 11

You use $4\frac{3}{4}$ yards (yd) of fabric from a 50-yd bolt. How much fabric remains on the bolt?

Often we will have to use more than one operation to find the solution to a problem. Consider Example 12.

● Example 12

An Application Involving the Subtraction of Mixed Numbers

A rectangular poster is to have a total length of $12\frac{1}{4}$ in. We want a $1\frac{3}{8}$-in. border on the top and a 2-in. border on the bottom. What is the length of the printed part of the poster?

Solution First, we will draw a sketch of the posts:

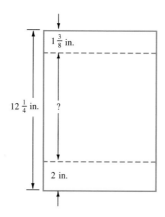

Now, we will use that sketch to find the total width of the top and bottom borders.

$$1\frac{3}{8} + 2 = 3\frac{3}{8} \text{ in.}$$

Now *subtract* that sum (the top and bottom borders) from the total length of the poster.

$$12\frac{1}{4} - 3\frac{3}{8} = 8\frac{7}{8} \text{ in.}$$

The length of the printed part is $8\frac{7}{8}$ in.

● ● ● **CHECK YOURSELF 12**

You cut one shelf $3\frac{3}{4}$ feet (ft) long and one $4\frac{1}{2}$ ft long from a 12-ft piece of lumber. Can you cut another shelf 4 ft long?

● ● ● **CHECK YOURSELF ANSWERS**

1. $5\frac{7}{10}$.   **2.** $4\frac{3}{8} + 2\frac{7}{8} = 6\frac{10}{8} = 6 + 1\frac{2}{8} = 7\frac{2}{8} = 7\frac{1}{4}$.

3. $5\frac{7}{10} + 3\frac{5}{6} = 5\frac{21}{30} + 3\frac{25}{30} = 8\frac{46}{30} = 9\frac{16}{30} = 9\frac{8}{15}$.

4. $13\frac{11}{12}$.   **5.** $3\frac{1}{2}$.

6. $7\frac{11}{12} - 3\frac{5}{8} = 7\frac{22}{24} - 3\frac{15}{24} = 4\frac{7}{24}$.

7. $6\frac{3}{5} - 2\frac{4}{5} = \frac{33}{5} - \frac{14}{5} = \frac{19}{5} = 3\frac{4}{5}$.

8. $6\frac{3}{5} - 2\frac{4}{5} = 5\frac{8}{5} - 2\frac{4}{5} = 3\frac{4}{5}$.

9. $1\frac{5}{12}$.   **10.** $3\frac{3}{5}$.   **11.** $45\frac{1}{4}$ yd.

12. No, only $3\frac{3}{4}$ ft is "left over."

Do the indicated operations.

1. $2\dfrac{2}{9} + 3\dfrac{5}{9}$

2. $5\dfrac{2}{9} + 6\dfrac{4}{9}$

3. $2\dfrac{1}{9} + 5\dfrac{5}{9}$

4. $1\dfrac{1}{6} + 5\dfrac{5}{6}$

5. $6\dfrac{5}{9} + 4\dfrac{7}{9}$

6. $5\dfrac{8}{9} + 4\dfrac{4}{9}$

7. $1\dfrac{1}{3} + 2\dfrac{1}{5}$

8. $2\dfrac{1}{4} + 1\dfrac{1}{6}$

9. $5\dfrac{3}{8} + 3\dfrac{5}{12}$

10. $4\dfrac{5}{9} + 6\dfrac{3}{4}$

11. $3\dfrac{5}{6} + 2\dfrac{3}{4}$

12. $3\dfrac{2}{5} + 3\dfrac{5}{8}$

13. $2\dfrac{1}{4} + 3\dfrac{5}{8} + 1\dfrac{1}{6}$

14. $3\dfrac{1}{5} + 2\dfrac{1}{2} + 5\dfrac{1}{4}$

15. $3\dfrac{3}{5} + 4\dfrac{1}{4} + 5\dfrac{3}{10}$

16. $4\dfrac{5}{6} + 3\dfrac{2}{3} + 7\dfrac{5}{9}$

17. $5\dfrac{5}{6} + 3\dfrac{4}{5} + 7\dfrac{2}{3}$

18. $5\dfrac{7}{12} + 2\dfrac{3}{8} + 4\dfrac{1}{2}$

1. $5\dfrac{7}{9}$

2. $11\dfrac{2}{3}$

3. $7\dfrac{2}{3}$

4. 7

5. $11\dfrac{1}{3}$

6. $10\dfrac{1}{3}$

7. $3\dfrac{8}{15}$

8. $3\dfrac{5}{12}$

9. $8\dfrac{19}{24}$

10. $11\dfrac{11}{36}$

11. $6\dfrac{7}{12}$

12. $7\dfrac{1}{40}$

13. $7\dfrac{1}{24}$

14. $10\dfrac{19}{20}$

15. $13\dfrac{3}{20}$

16. $16\dfrac{1}{18}$

17. $17\dfrac{3}{10}$

18. $12\dfrac{11}{24}$

19. $4\frac{1}{2}$

20. $2\frac{2}{3}$

21. $1\frac{3}{5}$

22. $3\frac{2}{7}$

23. $1\frac{5}{12}$

24. $4\frac{19}{30}$

25. $2\frac{5}{6}$

26. $3\frac{41}{60}$

27. $3\frac{29}{36}$

28. $6\frac{17}{21}$

29. $2\frac{3}{4}$

30. $2\frac{1}{3}$

31. $1\frac{5}{9}$

32. $3\frac{5}{9}$

33. $6\frac{7}{8}$

34. 3

35. $2\frac{19}{24}$

36. $1\frac{17}{30}$

37. $41\frac{3}{8}$ in.

38. $6\frac{5}{8}$ lb

19. $7\frac{7}{8} - 3\frac{3}{8}$

20. $3\frac{5}{6} - 1\frac{1}{6}$

21. $3\frac{2}{5} - 1\frac{4}{5}$

22. $5\frac{3}{7} - 2\frac{1}{7}$

23. $3\frac{2}{3} - 2\frac{1}{4}$

24. $5\frac{4}{5} - 1\frac{1}{6}$

25. $6\frac{3}{10} - 3\frac{7}{15}$

26. $8\frac{5}{12} - 4\frac{11}{15}$

27. $7\frac{5}{12} - 3\frac{11}{18}$

28. $9\frac{3}{7} - 2\frac{13}{21}$

29. $5 - 2\frac{1}{4}$

30. $4 - 1\frac{2}{3}$

31. $7 - 5\frac{4}{9}$

32. $9 - 5\frac{4}{9}$

33. $3\frac{3}{4} + 5\frac{1}{2} - 2\frac{3}{8}$

34. $1\frac{5}{6} + 3\frac{5}{12} - 2\frac{1}{4}$

35. $2\frac{3}{8} + 2\frac{1}{4} - 1\frac{5}{6}$

36. $1\frac{1}{15} + 3\frac{3}{10} - 2\frac{4}{5}$

Solve the following applications.

37. **Plumbing.** A plumber needs pieces of pipe $15\frac{5}{8}$ and $25\frac{3}{4}$ inches (in.) long. What is the total length of the pipe that is needed?

38. **Postage.** Marcus has to figure the postage for sending two packages. One weighs $3\frac{7}{8}$ pounds (lb), and the other weighs $2\frac{3}{4}$ lb. What is the total weight?

39. Working hours. Franklin worked $2\frac{1}{4}$ hours (h) on Monday, $5\frac{3}{4}$ h on Wednesday and $4\frac{1}{2}$ h on Friday. What was the total number of hours that he worked?

40. Distance. Robin ran $5\frac{1}{3}$ mi on Sunday, $2\frac{1}{4}$ mi on Tuesday, and $3\frac{1}{2}$ mi on Friday. How far did she run during the week?

41. Perimeter. Find the perimeter of the figure below.

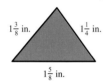

$1\frac{3}{8}$ in. $1\frac{1}{4}$ in.

$1\frac{5}{8}$ in.

42. Perimeter. Find the perimeter of the figure below.

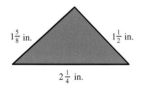

$1\frac{5}{8}$ in. $1\frac{1}{2}$ in.

$2\frac{1}{4}$ in.

43. Consumer purchases. Senta is working on a project that uses three pieces of fabric with lengths $\frac{3}{4}$, $1\frac{1}{4}$, and $\frac{5}{8}$ yd. She needs to allow for $\frac{1}{8}$ yd of waste. How much fabric should she buy?

44. Construction. The framework of a wall is $3\frac{1}{2}$ in. thick. We apply $\frac{5}{8}$-in. wallboard and $\frac{1}{4}$-in. paneling to the inside. Siding that is $\frac{3}{4}$ in. thick is applied to the outside. What is the finished thickness of the wall?

45. Stocks. A stock was listed at $34\frac{3}{8}$ points on Monday. By closing time Friday, it was at $28\frac{3}{4}$. How much did it drop during the week?

46. Cooking. A roast weighed $4\frac{1}{4}$ lb before cooking and $3\frac{3}{8}$ lb after cooking. How many pounds were lost during the cooking?

ANSWERS

39. $12\frac{1}{2}$ h

40. $11\frac{1}{12}$ mi

41. $4\frac{1}{4}$ in.

42. $5\frac{3}{8}$ in.

43. $2\frac{3}{4}$ yd

44. $5\frac{1}{8}$ in.

45. $5\frac{5}{8}$ points

46. $\frac{7}{8}$ lb

47. Interest. The interest rate on an auto loan in May was $12\frac{3}{8}$%. By September the rate was up to $14\frac{1}{4}$%. How much did the interest rate increase over the period?

48. Quantity of material. A roll of paper contains $30\frac{1}{4}$ yd. If $16\frac{7}{8}$ yd is cut from the roll, how much paper remains?

49. Geometry. Find the missing dimension in the figure below.

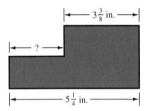

50. Carpentry. A $4\frac{1}{4}$-in. bolt is placed through a board that is $3\frac{1}{2}$ in. thick. How far does the bolt extend beyond the board?

51. Working hours. Ben can work 20 h per week on a part-time job. He works $5\frac{1}{2}$ h on Monday and $3\frac{3}{4}$ h on Tuesday. How many more hours can he work during the week?

52. Geometry. Find the missing dimension in the figure below.

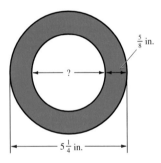

53. Carpeting. The Whites used $20\frac{3}{4}$ square yards (yd²) of carpet for their living room, $15\frac{1}{2}$ yd² for the dining room, and $6\frac{1}{4}$ yd² for a hallway. How much will remain if a 50-yd² roll of carpeting is used?

A N S W E R S

54. $2\frac{5}{12}$ mi

55. $2\frac{3}{4}$ h

56. $3\frac{3}{8}$ mi

57. $\frac{2}{5}$

58. $\frac{3}{8}$

54. Construction. A construction company has bids for paving roads of $1\frac{1}{2}$, $\frac{3}{4}$, and $3\frac{1}{3}$ miles (mi) for the month of July. With their present equipment, they can pave 8 mi in 1 month. How much more work can they take on in July?

55. Travel. On an 8-h trip, Jack drives $2\frac{3}{4}$ h and Pat drives $2\frac{1}{2}$ h. How many hours are left to drive?

56. Distance. A runner has told herself that she will run 20 mi each week. She runs $5\frac{1}{2}$ mi on Sunday, $4\frac{1}{4}$ mi on Tuesday, $4\frac{3}{4}$ mi on Wednesday, and $2\frac{1}{8}$ mi on Friday. How far must she run on Saturday to meet her goal?

57. Environment. If paper takes up $\frac{1}{2}$ of the space in a landfill and plastic takes up $\frac{1}{10}$ of the space, how much of the landfill is used for other materials?

58. Environment. If paper takes up $\frac{1}{2}$ of the space in a landfill and organic waste takes up $\frac{1}{8}$ of the space, how much of the landfill is used for other materials?

Answers

1. $5\frac{7}{9}$ **3.** $7\frac{2}{3}$ **5.** $11\frac{1}{3}$ **7.** $3\frac{8}{15}$ **9.** $5\frac{3}{8} + 3\frac{5}{12} = 5\frac{9}{24} + 3\frac{10}{24} = 8\frac{19}{24}$

11. $6\frac{7}{12}$ **13.** $7\frac{1}{24}$ **15.** $13\frac{3}{20}$ **17.** $17\frac{3}{10}$ **19.** $4\frac{1}{2}$

21. $3\frac{2}{5} - 1\frac{4}{5} = 2\frac{7}{5} - 1\frac{4}{5} = 1\frac{3}{5}$ **23.** $1\frac{5}{12}$

25. $6\frac{3}{10} - 3\frac{7}{15} = 6\frac{9}{30} - 3\frac{14}{30} = 5\frac{39}{30} - 3\frac{14}{30} = 2\frac{25}{30} = 2\frac{5}{6}$ **27.** $3\frac{29}{36}$

29. $2\frac{3}{4}$ **31.** $7 - 5\frac{4}{9} = 6\frac{9}{9} - 5\frac{4}{9} = 1\frac{5}{9}$ **33.** $6\frac{7}{8}$ **35.** $2\frac{19}{24}$

37. $41\frac{3}{8}$ in. **39.** $12\frac{1}{2}$ h **41.** $4\frac{1}{4}$ in. **43.** $2\frac{3}{4}$ yd **45.** $5\frac{5}{8}$ points

47. $1\frac{7}{8}\%$ **49.** $1\frac{7}{8}$ in. **51.** $10\frac{3}{4}$ h **53.** $7\frac{1}{2}$ yd^2 **55.** $2\frac{3}{4}$ h

57. $\frac{2}{5}$ of the landfill

A N S W E R S

© 1998 McGraw-Hill Companies

Do the indicated operations.

1. $6\dfrac{8}{11}$

2. $8\dfrac{1}{2}$

3. 10

4. $4\dfrac{7}{12}$

5. $6\dfrac{5}{24}$

6. $9\dfrac{11}{30}$

7. $3\dfrac{7}{8}$

8. $5\dfrac{4}{5}$

9. $7\dfrac{1}{24}$

10. $3\dfrac{1}{2}$

11. $7\dfrac{2}{7}$

12. $5\dfrac{1}{16}$

13. $3\dfrac{13}{18}$

14. $5\dfrac{35}{36}$

15. $3\dfrac{1}{4}$

16. $3\dfrac{9}{10}$

17. $5\dfrac{3}{10}$

18. $5\dfrac{11}{20}$

19. $5\dfrac{11}{16}$ lb

---

1. $4\dfrac{5}{11} + 2\dfrac{3}{11}$

2. $4\dfrac{5}{8} + 3\dfrac{7}{8}$

3. $6\dfrac{7}{10} + 3\dfrac{3}{10}$

4. $2\dfrac{1}{3} + 2\dfrac{1}{4}$

5. $4\dfrac{5}{6} + 1\dfrac{3}{8}$

6. $5\dfrac{9}{10} + 3\dfrac{7}{15}$

7. $1\dfrac{1}{4} + \dfrac{1}{8} + 2\dfrac{1}{2}$

8. $\dfrac{1}{5} + 3\dfrac{1}{2} + 2\dfrac{1}{10}$

9. $2\dfrac{1}{4} + 3\dfrac{5}{8} + 1\dfrac{1}{6}$

10. $5\dfrac{3}{4} - 2\dfrac{1}{4}$

11. $11\dfrac{4}{7} - 4\dfrac{2}{7}$

12. $8\dfrac{3}{8} - 3\dfrac{5}{16}$

13. $5\dfrac{5}{6} - 2\dfrac{1}{9}$

14. $9\dfrac{7}{18} - 3\dfrac{5}{12}$

15. $9 - 5\dfrac{3}{4}$

16. $6 - 2\dfrac{1}{10}$

17. $3\dfrac{4}{5} + 2\dfrac{5}{6} - 1\dfrac{1}{3}$

18. $5\dfrac{1}{4} + 3\dfrac{1}{5} - 2\dfrac{9}{10}$

Solve the following applications.

19. Weight. Eddie bought two packages of meat weighing $2\dfrac{7}{16}$ and $3\dfrac{1}{4}$ pounds (lb). What was the total weight of his purchase?

20. **Cooking.** A recipe calls for $2\frac{1}{3}$ cups of milk and $3\frac{3}{4}$ cups of water. What is the total amount of liquid used?

21. **Travel.** On a trip the Wilsons stopped for gas three times, using $10\frac{1}{2}$, $9\frac{3}{10}$, and $9\frac{9}{10}$ gallons (gal) of gas. How much gasoline did they use?

22. **Working hours.** Michele worked $6\frac{3}{4}$ hours (h) on Monday, $5\frac{1}{2}$ h on Wednesday, $7\frac{1}{4}$ h on Thursday, and 6 h on Friday. How many hours did she work during the week?

23. **Construction.** A post $6\frac{1}{4}$ feet (ft) long is to be set $1\frac{1}{2}$ ft into the ground. How much of the post will be above ground?

24. **Construction.** An $18\frac{3}{4}$ inch (in.) length of wire is cut from a piece 60 in. long. How much of the wire remains?

25. **Quantity of material.** If you cut $5\frac{3}{4}$ yard (yd) and $6\frac{7}{8}$-yd lengths from a 30-yd roll of wallpaper, how much paper remains on the roll?

26. **Printing.** A poster has an overall length of 15 in. If there is a $1\frac{3}{4}$-in. border on the top and a $2\frac{1}{2}$-in. border on the bottom, what is the length of the printed portion of the poster?

27. **Quantity of material.** Gene picked $22\frac{3}{4}$, $23\frac{7}{8}$, and $25\frac{1}{2}$ pounds (lb) of beans in the hours that he worked before noon. He must pick 100 lb during the day to earn a bonus. How many more pounds must he pick to earn the bonus?

28. **Construction.** A house plan calls for $25\frac{1}{2}$ yd^2 of carpeting in the living room, $5\frac{2}{3}$ yd^2 in a bathroom, and $17\frac{2}{9}$ yd^2 in the family room. A floor-covering shop has on hand a 50-yd^2 roll of the desired carpet. Is it enough for the house? If so, how much will be left over?

	ANSWERS
20.	$6\frac{1}{12}$ cups
21.	$29\frac{7}{10}$ gal
22.	$25\frac{1}{2}$ h
23.	$4\frac{3}{4}$ ft
24.	$41\frac{1}{4}$ in.
25.	$17\frac{3}{8}$ yd
26.	$10\frac{3}{4}$ in.
27.	$27\frac{7}{8}$ lb
28.	Yes; $1\frac{11}{18}$ yd^2

Special Group Activity. Tom and Susan like eating in ethnic restaurants, so they were thrilled when Marco's Cafe, an Indian restaurant, opened in their neighborhood. The first time they ate there, Susan had a bowl of Mulligatawny soup. She loved it. She decided that it would be a great soup to serve her friends, so she asked Marco for the recipe. Marco said that was no problem. He had already had so many requests for the recipe that he made up a handout. A copy of it is reproduced here (try it if you are adventurous):

Mulligatawny Soup

This recipe makes 10 gallons; recommended serving size is a 12-ounce bowl.

Saute the following in a steam kettle until the onions are translucent:

10 lb	diced onion
10 lb	diced celery
1/2 cup	garlic puree
1 cup	madras curry
2 cups	mild curry

Add the following and bring to a boil:

4 cups	white wine
1/3 cup	sugar
1 #10 can	diced tomato
1 gallon	fresh apple juice
1/3 cup	lemon juice
2 gallons	water
1 #10 can	diced carrots
16 oz	chicken stock

Finish with:

Roux (1 lb butter and 1 lb flour) and 8 quarts cream (temper into hot liquid). Season to taste with hash spice.

How many servings does this recipe make?

Visit local grocery stores to find out how much each item costs. Calculate the total cost for the 10 gallons of soup. What is the cost for each 12-ounce serving? (This is called the marginal cost—it does not include the overhead for running the restaurant.)

What is roux? Call a restaurant to find whether they would use the same definition.

Susan does want to make this soup for a dinner party she is having. Rewrite the recipe so that it will serve six 12-ounce bowls. Use reasonable measures, like teaspoons, and cups. Completing the following chart will help. For some items you may have to experiment.

How many ounces in a #10 can?
How many cups in a gallon?
How many ounces in a pound?
How many teaspoons in a cup?
How many cups in a pound of diced onion?

Using Your Calculator to Add and Subtract Mixed Numbers

We have already seen how to add, multiply, and divide fractions using our calculators. Now we will use our calculators to add and subtract mixed numbers.

SCIENTIFIC CALCULATOR

To enter a mixed number on a scientific calculator, press the fraction key between both the whole number and the numerator and denominator. For example, to enter $3\frac{7}{12}$, press

3 $\boxed{\textbf{a b/c}}$ 7 $\boxed{\textbf{a b/c}}$ 12

• Example 1

Adding Mixed Numbers

Add.

$$3\frac{7}{12} + 2\frac{11}{16}$$

The keystroke sequence is

3 $\boxed{\textbf{a b/c}}$ 7 $\boxed{\textbf{a b/c}}$ 12 $\boxed{+}$ 2 $\boxed{\textbf{a b/c}}$ 11 $\boxed{\textbf{a b/c}}$ 16 $\boxed{=}$

The result is $6\frac{13}{48}$.

GRAPHING CALCULATOR

As with multiplying and dividing fractions, when using a graphing calculator, you must choose the fraction option from the math menu before pressing $\boxed{\text{Enter}}$.

For the problem in Example 1, $3\frac{7}{12} + 2\frac{11}{16}$, the keystroke sequence is

3 $\boxed{+}$ 7 $\boxed{÷}$ 12 $\boxed{+}$ 2 $\boxed{+}$ 11 $\boxed{÷}$ 16 $\boxed{\blacktriangleright\textbf{Frac.}}$ $\boxed{\text{Enter}}$

● ● ● ● CHECK YOURSELF 1

Find the sum.

$$4\frac{3}{7} + 4\frac{5}{6}$$

● ● ● **CHECK YOURSELF ANSWER**

$9\dfrac{11}{42}.$

Calculator Exercises

Name

Section   Date

A N S W E R S

1. $5\dfrac{11}{12}$

2. $14\dfrac{5}{6}$

3. $8\dfrac{1}{9}$

4. $7\dfrac{1}{14}$

5. $16\dfrac{11}{12}$

6. $20\dfrac{19}{24}$

7. $37\dfrac{31}{54}$

8. $180\dfrac{113}{135}$

9. $2\dfrac{1}{6}$

10. $3\dfrac{3}{22}$

11. $3\dfrac{13}{48}$

12. $7\dfrac{2}{15}$

13. $4\dfrac{5}{6}$

14. $32\dfrac{91}{180}$

15. 22

16. $12\dfrac{1}{15}$

Add or Subtract the following.

1. $3\dfrac{2}{3} + 2\dfrac{1}{4}$

2. $6\dfrac{1}{6} + 8\dfrac{2}{3}$

3. $5\dfrac{4}{9} + 2\dfrac{2}{3}$

4. $2\dfrac{3}{7} + 4\dfrac{9}{14}$

5. $11\dfrac{2}{3} + 5\dfrac{1}{4}$

6. $6\dfrac{3}{8} + 14\dfrac{5}{12}$

7. $14\dfrac{13}{18} + 22\dfrac{23}{27}$

8. $82\dfrac{41}{45} + 97\dfrac{25}{27}$

9. $4\dfrac{7}{9} - 2\dfrac{11}{18}$

10. $7\dfrac{8}{11} - 4\dfrac{13}{22}$

11. $5\dfrac{11}{16} - 2\dfrac{5}{12}$

12. $18\dfrac{5}{24} - 11\dfrac{3}{40}$

13. $6\dfrac{2}{3} - 1\dfrac{5}{6}$

14. $131\dfrac{43}{45} - 99\dfrac{27}{60}$

15. $10\dfrac{2}{3} + 4\dfrac{1}{5} + 7\dfrac{2}{15}$

16. $7\dfrac{1}{5} + 3\dfrac{2}{3} + 1\dfrac{1}{5}$

Answers

1. $5\dfrac{11}{12}$ **3.** $8\dfrac{1}{9}$ **5.** $16\dfrac{11}{12}$ **7.** $37\dfrac{31}{54}$ **9.** $2\dfrac{1}{6}$ **11.** $3\dfrac{13}{48}$

13. $4\dfrac{5}{6}$ **15.** 22

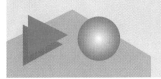

A N S W E R S

The purpose of the Self-Test is to help you check your progress and review for a chapter test in class. Allow yourself about 1 hour to take the test. When you are done, check your answers in the back of the book. If you missed any answers, be sure to go back and review the appropriate sections in the chapter and do the supplementary exercises provided there.

[7.1]   In Exercises 1 to 3, add.

1. $\dfrac{3}{10} + \dfrac{6}{10}$ **2.** $\dfrac{5}{12} + \dfrac{3}{12}$ **3.** $\dfrac{13}{18} + \dfrac{11}{18}$

[7.2]   In Exercises 4 and 5, find the least common denominator for fractions with the given denominators.

4. 12 and 15   **5.** 3, 4, and 18

[7.3]   In Exercises 6 to 11, add.

6. $\dfrac{2}{5} + \dfrac{4}{10}$ **7.** $\dfrac{1}{6} + \dfrac{3}{7}$ **8.** $\dfrac{3}{8} + \dfrac{5}{12}$

9. $\dfrac{11}{15} + \dfrac{9}{20}$ **10.** $\dfrac{1}{4} + \dfrac{5}{8} + \dfrac{7}{10}$ **11.** $\dfrac{5}{24} + \dfrac{3}{8}$

Solve the following application.

12. Total ingredients. A recipe calls for $\dfrac{1}{2}$ cup of raisins, $\dfrac{1}{4}$ cup of walnuts, and $\dfrac{2}{3}$ cup of rolled oats. What is the total amount of these ingredients?

[7.4]   In Exercises 13 to 15, subtract.

13. $\dfrac{7}{9} - \dfrac{4}{9}$ **14.** $\dfrac{7}{18} - \dfrac{5}{18}$ **15.** $\dfrac{11}{12} - \dfrac{3}{20}$

Solve the following application.

16. Amount of time. You have $\dfrac{5}{6}$ hours (h) to take a three-part test. You use $\dfrac{1}{3}$ h for the first section and $\dfrac{1}{4}$ h for the second. How much time do you have left to finish the last section of the test?

1. $\dfrac{9}{10}$

2. $\dfrac{2}{3}$

3. $1\dfrac{1}{3}$

4. 60

5. 36

6. $\dfrac{4}{5}$

7. $\dfrac{25}{42}$

8. $\dfrac{19}{24}$

9. $1\dfrac{11}{60}$

10. $1\dfrac{23}{40}$

11. $\dfrac{7}{12}$

12. $1\dfrac{5}{12}$ cups

13. $\dfrac{1}{3}$

14. $\dfrac{1}{9}$

15. $\dfrac{23}{30}$

16. $\dfrac{1}{4}$ h

405

17. $7\frac{7}{10}$

18. $10\frac{1}{4}$

19. $7\frac{11}{12}$

20. $12\frac{3}{40}$

21. $1\frac{3}{4}$

22. $1\frac{11}{18}$

23. $3\frac{23}{24}$

24. $1\frac{8}{15}$

25. $9\frac{1}{7}$

26. $13\frac{11}{20}$

27. $5\frac{3}{4}$ h

28. $24\frac{3}{4}$ in.

29. $4\frac{3}{4}$ in.

30. $30\frac{5}{6}$ yd

[7.5] In Exercises 17 to 26, perform the indicated operations.

17. $5\frac{3}{10} + 2\frac{4}{10}$

18. $7\frac{3}{8} + 2\frac{7}{8}$

19. $4\frac{1}{6} + 3\frac{3}{4}$

20. $6\frac{3}{8} + 5\frac{7}{10}$

21. $7\frac{3}{8} - 5\frac{5}{8}$

22. $3\frac{5}{6} - 2\frac{2}{9}$

23. $7\frac{1}{8} - 3\frac{1}{6}$

24. $7 - 5\frac{7}{15}$

25. $4\frac{2}{7} + 3\frac{3}{7} + 1\frac{3}{7}$

26. $3\frac{1}{2} + 4\frac{3}{4} + 5\frac{3}{10}$

In Exercises 27 to 30, solve each application.

27. Total overtime. A worker has $2\frac{1}{6}$ h of overtime on Tuesday, $1\frac{3}{4}$ h on Wednesday, and $1\frac{5}{6}$ h on Friday. What is the total overtime for the week?

28. Construction. A fence post is $86\frac{1}{4}$ in. long. To have $61\frac{1}{2}$ in. of the post aboveground, how deep should the post be set into the ground?

29. Geometry. Find the missing dimension in the accompanying figure.

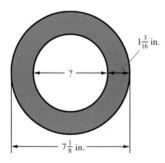

30. Fabric remaining. You cut two pieces of fabric $2\frac{3}{4}$ yd long and one piece $3\frac{2}{3}$ yd long from a bolt of fabric containing 40 yd. How much fabric remains on the bolt?

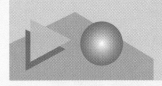
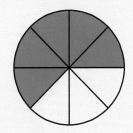

$\frac{5}{8}$ ← Numerator
← Denominator

$\frac{2}{3}$ and $\frac{11}{15}$ are proper fractions.

$\frac{7}{5}, \frac{21}{20}$, and $\frac{8}{8}$ are improper fractions.

$2\frac{1}{3}$ and $5\frac{7}{8}$ are mixed numbers. Note that $2\frac{1}{3}$ means $2 + \frac{1}{3}$.

The Language of Fractions

Fraction Fractions name a number of equal parts of a unit or whole. A fraction is written in the form $\frac{a}{b}$, where a and b are whole numbers and b cannot be zero.

Denominator The number of equal parts the whole is divided into.

Numerator The number of parts of the whole that are used.

Proper Fraction A fraction whose numerator is less than its denominator. It names a number less than 1.

Improper Fractions A fraction whose numerator is greater than or equal to its denominator. It names a number greater than or equal to 1.

Mixed Number The sum of a whole number and a proper fraction.

Converting Mixed Numbers and Improper Fractions

To change $\frac{22}{5}$ to a mixed number:

$$\begin{array}{r} 4 \\ 5\overline{)22} \\ \underline{20} \\ 2 \end{array}$$ ← Quotient
← Remainder

$$\frac{22}{5} = 4\frac{2}{5}$$

To Change an Improper Fraction into a Mixed Number

1. Divide the numerator by the denominator. The quotient is the whole-number portion of the mixed number.
2. If there is a remainder, write the remainder over the original denominator. This gives the fractional portion of the mixed number.

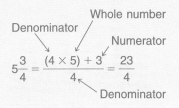

$$5\frac{3}{4} = \frac{(4 \times 5) + 3}{4} = \frac{23}{4}$$

To Change a Mixed Number to an Improper Fraction

1. Multiply the denominator of the fraction by the whole-number portion of the mixed number.
2. Add the numerator of the fraction to that product.
3. Write that sum over the original denominator to form the improper fraction.

407

Equivalent Fractions

Equivalent Fractions Two fractions that are equivalent (have equal value) are different names for the same number.

Cross Products

$$\frac{a}{b} \diagdown \diagup = \diagup \diagdown \frac{c}{d} \qquad a \times d \text{ and } b \times c \text{ are called the}$$
cross products.

$\dfrac{2}{3} = \dfrac{4}{6}$ because

$2 \times 6 = 3 \times 4.$

If the cross products for two fractions are equal, the two fractions are equivalent.

The Fundamental Principle

For the fraction $\dfrac{a}{b}$, and any nonzero number c,

$$\frac{a}{b} = \frac{a \times c}{b \times c}$$

$\dfrac{1}{2} = \dfrac{1 \times 5}{2 \times 5} = \dfrac{5}{10}$

$\dfrac{1}{2}$ and $\dfrac{5}{10}$ are
equivalent fractions.

$\dfrac{2}{3}$ is in simplest form.

$\dfrac{12}{18}$ is *not* in simplest form.
The numerator and
denominator have the
common factor 6.

In words: We can multiply the numerator and denominator of a fraction by the same nonzero number. The result will be an equivalent fraction.

Simplest Form A fraction is in simplest form, or in lowest terms, if the numerator and denominator have no common factors other than 1. This means that the fraction has the smallest possible numerator and denominator.

To Write a Fraction in Simplest Form

$\dfrac{10}{15} = \dfrac{10 \div 5}{15 \div 5} = \dfrac{2}{3}$

Divide the numerator and denominator by any common factor greater than 1 to reduce a fraction to an equivalent fraction in lower terms.

To Build a Fraction

$\dfrac{3}{4} = \dfrac{3 \times 2}{4 \times 2} = \dfrac{6}{8}$

Multiply the numerator and denominator by any whole number greater than 1 to raise a fraction to an equivalent fraction in higher terms.

Multiplying Fractions

$\dfrac{5}{8} \times \dfrac{3}{7} = \dfrac{5 \times 3}{8 \times 7} = \dfrac{15}{56}$

1. Multiply numerator by numerator. This gives the numerator of the product.
2. Multiply denominator by denominator. This gives the denominator of the product.
3. Simplify the resulting fraction if possible.

$\dfrac{5}{9} \times \dfrac{3}{10} = \dfrac{\overset{1}{\cancel{5}} \times \overset{1}{\cancel{3}}}{\underset{3}{\cancel{9}} \times \underset{2}{\cancel{10}}} = \dfrac{1}{6}$

In multiplying fractions it is usually easiest to divide by any common factors in the numerator and denominator *before* multiplying.

Dividing Fractions

$$\frac{3}{7} \div \frac{4}{5} = \frac{3}{7} \times \frac{5}{4} = \frac{15}{28}$$

Invert the divisor and multiply.

Multiplying or Dividing Mixed Numbers

$$6\frac{2}{3} \times 3\frac{1}{5} = \frac{\overset{4}{\cancel{20}}}{3} \times \frac{16}{\underset{1}{\cancel{5}}}$$

$$= \frac{64}{3} = 21\frac{1}{3}$$

Convert any mixed or whole numbers to improper fractions. Then multiply or divide the fractions as before.

Finding the Least Common Denominator

To Find the LCD of a Group of Fractions

To find the LCD of fractions with denominators 4, 6, and 15:

$$
\begin{aligned}
4 &= 2 \times 2 \\
6 &= 2 \quad\ \times 3 \\
15 &= \qquad\qquad 3 \times 5 \\
\hline
 & 2 \times 2 \times 3 \times 5
\end{aligned}
$$

The LCD = $2 \times 2 \times 3 \times 5$, or 60.

1. Write the prime factorization for each of the denominators.
2. Find all the prime factors that appear in any one of the prime factorizations.
3. Form the product of those prime factors, using each factor the greatest number of times it occurs in any one factorization.

Adding Fractions

To Add Like Fractions

$$\frac{5}{18} + \frac{7}{18} = \frac{12}{18} = \frac{2}{3}$$

1. Add the numerators.
2. Place the sum over the common denominator.
3. Simplify the resulting fraction if necessary.

To Add Unlike Fractions

$$\frac{3}{4} + \frac{7}{10} = \frac{15}{20} + \frac{14}{20}$$

$$= \frac{29}{20} = 1\frac{9}{20}$$

1. Find the LCD of the fractions.
2. Change each fraction to an equivalent fraction with the LCD as a common denominator.
3. Add the resulting like fractions as before.

Subtracting Fractions

To Subtract Like Fractions

$$\frac{17}{20} - \frac{7}{20} = \frac{10}{20} = \frac{1}{2}$$

1. Subtract the numerators.
2. Place the difference over the common denominator.
3. Simplify the resulting fraction if necessary.

To Subtract Unlike Fractions

$$\frac{8}{9} - \frac{5}{6} = \frac{16}{18} - \frac{15}{18} = \frac{1}{18}$$

1. Find the LCD of the fractions.
2. Change each fraction to an equivalent fraction with the LCD as a common denominator.
3. Subtract the resulting like fractions as before.

Adding or Subtracting Mixed Numbers

$$5\frac{1}{2} - 3\frac{3}{4} = 5\frac{2}{4} - 3\frac{3}{4}$$

Rename

$$= 4\frac{6}{4} - 3\frac{3}{4} = 1\frac{3}{4}$$

$$\left\{4 - 3\right\} \quad \left\{\frac{6}{4} - \frac{3}{4}\right\}$$

1. Add or subtract the whole-number parts.

2. Add or subtract the fractional parts. *Note:* Subtracting may require renaming the first mixed number.

3. Combine the results as a mixed number.

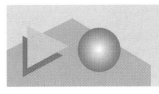
A N S W E R S

You should now be reviewing the material in Chapters 5–7. The following exercises will help in that process. Work all the exercises carefully. Then check your answers in the back of the book. References to the chapter and section for each exercise are provided there. If you made an error, go back and review the related material and do the supplementary exercises for that section.

[5.1] Give the fractions that name the shaded portions of the following diagrams. Identify the numerator and the denominator.

1.

Fraction: $\frac{3}{8}$

Numerator: 3

Denominator: 8

2.

Fraction: $\frac{5}{6}$

Numerator: 5

Denominator: 6

[5.2] **3.** From the following group of numbers:

$$\frac{2}{3}, \frac{5}{4}, 2\frac{3}{7}, \frac{45}{8}, \frac{7}{7}, 3\frac{4}{5}, \frac{9}{1}, \frac{7}{10}, \frac{12}{5}, 5\frac{2}{9}$$

List the proper fractions. _____ $\frac{2}{3}, \frac{7}{10}$

List the improper fractions. _____ $\frac{5}{4}, \frac{45}{8}, \frac{7}{7}, \frac{9}{1}, \frac{12}{5}$

List the mixed numbers. _____ $2\frac{3}{7}, 3\frac{4}{5}, 5\frac{2}{9}$

[5.3] Convert to mixed or whole numbers.

4. $\frac{41}{6}$ **5.** $\frac{32}{8}$

Convert to improper fractions.

6. $7\frac{5}{8}$ **7.** $4\frac{3}{10}$

8. No

9. Yes

10. $\dfrac{2}{3}$

11. $\dfrac{3}{5}$

12. $\dfrac{7}{9}$

13. $\dfrac{16}{21}$

14. 15

15. 32

16. $\dfrac{7}{12}, \dfrac{5}{8}$

17. $\dfrac{7}{10}, \dfrac{4}{5}, \dfrac{5}{6}$

18. >

19. =

20. <

21. $\dfrac{4}{24}, \dfrac{21}{24}$

22. $\dfrac{36}{120}, \dfrac{75}{120}, \dfrac{70}{120}$

23. $\dfrac{1}{9}$

24. $\dfrac{1}{6}$

25. $1\dfrac{1}{2}$

26. $2\dfrac{1}{8}$

27. $9\dfrac{3}{5}$

28. $11\dfrac{1}{3}$

29. 8

30. $\dfrac{2}{3}$

31. $\dfrac{5}{6}$

32. $\dfrac{3}{16}$

[5.4] Determine whether each of the following pairs of fractions are equivalent.

8. $\dfrac{5}{8}, \dfrac{7}{12}$ **9.** $\dfrac{8}{15}, \dfrac{32}{60}$

[5.5] Write each fraction in simplest form.

10. $\dfrac{24}{36}$ **11.** $\dfrac{45}{75}$ **12.** $\dfrac{140}{180}$ **13.** $\dfrac{16}{21}$

[5.6] Find the missing numerators.

14. $\dfrac{3}{5} = \dfrac{?}{25}$ **15.** $\dfrac{4}{5} = \dfrac{?}{40}$

Arrange the fractions in order from smallest to largest.

16. $\dfrac{5}{8}, \dfrac{7}{12}$ **17.** $\dfrac{5}{6}, \dfrac{4}{5}, \dfrac{7}{10}$

Complete the following statements, using the symbol <, =, or >.

18. $\dfrac{5}{12}$ —— $\dfrac{3}{8}$ **19.** $\dfrac{3}{7}$ —— $\dfrac{9}{21}$ **20.** $\dfrac{9}{16}$ —— $\dfrac{7}{12}$

Write as equivalent fractions with the LCD as a common denominator.

21. $\dfrac{1}{6}, \dfrac{7}{8}$ **22.** $\dfrac{3}{10}, \dfrac{5}{8}, \dfrac{7}{12}$

[6.1]–[6.3] Multiply.

23. $\dfrac{7}{15} \times \dfrac{5}{21}$ **24.** $\dfrac{10}{27} \times \dfrac{9}{20}$ **25.** $4 \times \dfrac{3}{8}$

26. $3\dfrac{2}{5} \times \dfrac{5}{8}$ **27.** $5\dfrac{1}{3} \times 1\dfrac{4}{5}$ **28.** $1\dfrac{5}{12} \times 8$

29. $3\dfrac{1}{5} \times \dfrac{7}{8} \times 2\dfrac{6}{7}$

[6.4] Divide.

30. $\dfrac{5}{12} \div \dfrac{5}{8}$ **31.** $\dfrac{7}{15} \div \dfrac{14}{25}$ **32.** $\dfrac{9}{20} \div 2\dfrac{2}{5}$

33. $3\frac{3}{8} \div 2\frac{1}{4}$ **34.** $3\frac{3}{7} \div 8$

[6.3]–[6.4]   Solve the following applications.

35. Distance. The scale on a map is 1 inch (in.) = 80 miles (mi). If two cities are $2\frac{3}{4}$ in. apart on the map, what is the actual distance between the cities?

36. Cost of linoleum. A kitchen measures $5\frac{1}{3}$ by $4\frac{1}{4}$ yards (yd). If you purchase linoleum costing $9 per square yard (yd)^2, what will it cost to cover the floor?

37. Mean speed. If you drive 117 mi in $2\frac{1}{4}$ h, what is your mean speed?

38. Number of lots. An 18-acre piece of land is to be subdivided into home lots that are each $\frac{3}{8}$ acre. How many lots can be formed?

[7.1]–[7.3]   Add.

39. $\frac{2}{9} + \frac{5}{9}$ **40.** $\frac{7}{10} + \frac{9}{10}$ **41.** $\frac{5}{6} + \frac{11}{18}$

42. $\frac{5}{18} + \frac{7}{12}$ **43.** $\frac{3}{5} + \frac{1}{4} + \frac{5}{6}$

[7.4]–[7.5]   Perform the indicated operations.

44. $\frac{5}{8} - \frac{3}{8}$ **45.** $\frac{11}{18} - \frac{2}{9}$ **46.** $\frac{7}{10} - \frac{7}{12}$

47. $\frac{11}{27} - \frac{5}{18}$ **48.** $\frac{4}{9} + \frac{5}{12} - \frac{3}{8}$ **49.** $6\frac{5}{7} + 3\frac{4}{7}$

50. $5\frac{7}{10} + 3\frac{11}{12}$ **51.** $2\frac{1}{2} + 3\frac{5}{6} + 3\frac{3}{8}$ **52.** $7\frac{7}{9} - 3\frac{4}{9}$

53. $9\frac{1}{6} - 3\frac{1}{8}$ **54.** $6\frac{5}{12} - 3\frac{5}{8}$ **55.** $2\frac{1}{3} + 5\frac{1}{6} - 2\frac{4}{5}$

413

33. $1\frac{1}{2}$

34. $\frac{3}{7}$

35. 220 mi

36. $204

37. 52 mi/h

38. 48 lots

39. $\frac{7}{9}$

40. $1\frac{3}{5}$

41. $1\frac{4}{9}$

42. $\frac{31}{36}$

43. $1\frac{41}{60}$

44. $\frac{1}{4}$

45. $\frac{7}{18}$

46. $\frac{7}{60}$

47. $\frac{7}{54}$

48. $\frac{35}{72}$

49. $10\frac{2}{7}$

50. $9\frac{37}{60}$

51. $9\frac{17}{24}$

52. $4\frac{1}{3}$

53. $6\frac{1}{24}$

54. $2\frac{19}{24}$

55. $4\frac{7}{10}$

56. $69\frac{1}{16}$ in.

57. $19\frac{9}{16}$ in.

58. $1\frac{3}{8}$ in.

59. $53\frac{9}{16}$ in.

60. Yes, $\frac{5}{12}$ yd

[7.5]   Solve each of the following applications.

56. Length. Bradley needs two shelves, one $32\frac{3}{8}$ in. long and the other $36\frac{11}{16}$ in. long. What is the total length of shelving that is needed?

57. Perimeter. Find the perimeter of the figure below.

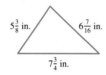

58. Height. At the beginning of a year Miguel was $51\frac{3}{4}$ in. tall. In June, he measured $53\frac{1}{8}$ in. How much did he grow during that period?

59. Length. A bookshelf that is $42\frac{5}{16}$ in. long is cut from a board with a length of 8 ft. If $\frac{1}{8}$ in. is wasted in the cut, what length board remains?

60. Measuring wallpaper. Amelia buys an 8-yd roll of wallpaper on sale. After measuring, she finds that she needs the following amounts of the paper: $2\frac{1}{3}$, $1\frac{1}{2}$, and $3\frac{3}{4}$ yd. Does she have enough for the job? If so, how much will be left over?

This test is provided to help you in the process of reviewing Chapters 5 through 7. Answers are provided in the back of the book. If you missed any answers, be sure to go back and review the appropriate chapter sections.

In Exercise 1, give the fraction that names the shaded portion of the diagram. Identify the numerator and denominator.

1.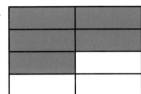

Fraction: _____ $\frac{5}{8}$ _____

Numerator: _____ 5 _____

Denominator: _____ 8 _____

In Exercise 2, identify the proper fractions, improper fractions, and mixed numbers from the following group.

$$\frac{7}{12}, \frac{10}{8}, 3\frac{1}{5}, \frac{9}{9}, \frac{7}{1}, \frac{3}{7}, 2\frac{2}{3}$$

2. Proper: _____ $\frac{7}{12}, \frac{3}{7}$ _____   Improper: _____ $\frac{10}{8}, \frac{9}{9}, \frac{7}{1}$ _____

   Mixed numbers: _____ $3\frac{1}{5}, 2\frac{2}{3}$ _____

In Exercises 3 and 4, convert to mixed or whole numbers.

3. $\frac{14}{5}$ 4. $\frac{28}{7}$

In Exercises 5 and 6, convert to improper fractions.

5. $4\frac{1}{3}$ 6. $7\frac{7}{8}$

In Exercises 7 and 8, find out whether the pair of fractions is equivalent.

7. $\frac{7}{21}, \frac{8}{24}$ 8. $\frac{7}{12}, \frac{8}{15}$

A N S W E R S

1. See exercise

2. See exercise

3. $2\frac{4}{5}$

4. 4

5. $\frac{13}{3}$

6. $\frac{63}{8}$

7. Yes

8. No

In Exercises 9 and 10, write the fraction in simplest form.

9. $\dfrac{28}{42}$ **10.** $\dfrac{36}{96}$

In Exercises 11 and 12, arrange the fractions in order from smallest to largest.

11. $\dfrac{5}{9}, \dfrac{6}{11}$ **12.** $\dfrac{7}{10}, \dfrac{3}{5}, \dfrac{8}{15}$

In Exercises 13 and 14, write the fractions as equivalent fractions with the LCD as a common denominator.

13. $\dfrac{5}{8}, \dfrac{7}{12}$ **14.** $\dfrac{2}{3}, \dfrac{5}{9}, \dfrac{3}{4}$

In Exercises 15 to 19, multiply.

15. $\dfrac{5}{9} \times \dfrac{8}{15}$ **16.** $\dfrac{20}{21} \times \dfrac{7}{25}$ **17.** $1\dfrac{1}{8} \times 4\dfrac{4}{5}$

18. $8 \times 2\dfrac{5}{6}$ **19.** $\dfrac{2}{3} \times 1\dfrac{4}{5} \times \dfrac{5}{8}$

In Exercises 20 to 23, divide.

20. $\dfrac{5}{8} \div \dfrac{15}{32}$ **21.** $2\dfrac{5}{8} \div \dfrac{7}{12}$

22. $4\dfrac{1}{6} \div 5$ **23.** $2\dfrac{2}{7} \div 1\dfrac{11}{21}$

In Exercises 24 and 25, solve each application.

24. Cost of carpet. Your living room measures $6\dfrac{2}{3}$ by $4\dfrac{1}{2}$ yards (yd). If you purchase carpeting at $18 per square yard (yd²), what will it cost to carpet the room?

25. Sheets of plywood. If a stack of $\dfrac{5}{8}$-inch (in.) plywood measures 55 in. in height, how many sheets of plywood are in the stack?

In Exercises 26 to 28, add.

26. $\dfrac{4}{15} + \dfrac{8}{15}$ **27.** $\dfrac{7}{25} + \dfrac{8}{15}$ **28.** $\dfrac{2}{5} + \dfrac{3}{4} + \dfrac{5}{8}$

In Exercises 29 to 31, perform the indicated operations.

29. $\dfrac{17}{20} - \dfrac{7}{20}$ **30.** $\dfrac{5}{9} - \dfrac{5}{12}$ **31.** $\dfrac{5}{18} + \dfrac{4}{9} - \dfrac{1}{6}$

In Exercises 32 to 37, perform the indicated operations.

32. $3\dfrac{5}{7} + 2\dfrac{4}{7}$ **33.** $4\dfrac{7}{8} + 3\dfrac{1}{6}$ **34.** $8\dfrac{1}{9} - 3\dfrac{5}{9}$

35. $7\dfrac{7}{8} - 3\dfrac{5}{6}$ **36.** $9 - 5\dfrac{3}{8}$ **37.** $3\dfrac{1}{6} + 3\dfrac{1}{4} - 2\dfrac{7}{8}$

In Exercises 38 to 40, solve each application.

38. Hours worked. In his part-time job, Manuel worked $3\dfrac{5}{6}$ hours (h) on Monday, $4\dfrac{3}{10}$ h on Wednesday, and $6\dfrac{1}{2}$ h on Friday. Find the number of hours that he worked during the week.

39. Bolt extension. A $6\dfrac{1}{2}$-in. bolt is placed through a wall that is $5\dfrac{7}{8}$ in. thick. How far does the bolt extend beyond the wall?

40. Time and distance. On a 6-hour (h) trip, Carlos drove $1\dfrac{3}{4}$ h and then Maria drove for another $2\dfrac{1}{3}$ h. How many hours remained on the trip?

41. Sapphire and Danny are building a 16 ft × 18 ft patio outside their back door. They are going to pour 24 concrete footings. Each footing will be 24 in. × 24 in. × 9 in. Across each set of four footings will be a 16 ft 4 × 4 (which in reality is 3 1/2 in. by 3 1/2 in.).

Lying across the 4 × 4s will be 2 × 6 cedar boards (again, they are actually 1 1/2 in. × 5 1/2 in.). Each board is 18 ft long. There will be a 1/2 in. gap between each two boards.

26. $\dfrac{4}{5}$

27. $\dfrac{61}{75}$

28. $1\dfrac{31}{40}$

29. $\dfrac{1}{2}$

30. $\dfrac{5}{36}$

31. $\dfrac{5}{9}$

32. $6\dfrac{2}{7}$

33. $8\dfrac{1}{24}$

34. $4\dfrac{5}{9}$

35. $4\dfrac{1}{24}$

36. $3\dfrac{5}{8}$

37. $3\dfrac{13}{24}$

38. $14\dfrac{19}{30}$ h

39. $\dfrac{5}{8}$ in.

40. $1\dfrac{11}{12}$ h

41.

(a) What is the total volume of concrete in cubic inches?

(b) Every cubic foot is 12 in. $\times$ 12 in. $\times$ 12 in., so 1 cubic foot = 12 cubic inches (in.^{3}) = 1728 in.^{3} What is the total volume in cubic feet?

(c) What is the total volume of the concrete in cubic yards?

(d) How many of these 18-ft boards will be needed to finish the entire 16-ft deck?

(e) What is the total area of the deck?

(f) The lineal feet of lumber represents the total length of all the boards used. What is the lineal feet for this project?

(g) A board foot of lumber is 144 in.^{3} (1 in. $\times$ 12 in. $\times$ 12 in.) of lumber (the actual size of the board is ignored; assume that a 1 $\times$ 6 is 1 in. $\times$ 6 in.). Find the total board feet of lumber used for the deck.

(h) Call a lumber yard and estimate the cost of the deck. Assume that you will use three 9 penny nails everywhere a 1 $\times$ 6 crosses a 4 $\times$ 4. Write out a list of all the items needed, and show the cost for each item.

(i) Assuming all the materials are delivered and you and a friend are going to assemble the deck, how long will it take you? How did you arrive at that figure?

CHAPTER 8

ADDITION, SUBTRACTION, AND MULTIPLICATION OF DECIMALS

INTRODUCTION

When you look into the cockpit of a plane, you have to be impressed with the number of gauges that face the pilot. It is remarkable that the pilot can keep all that information straight. It is even more remarkable to realize that the pilot has to know how to calculate with pencil and paper much of the information available.

When Gwen decided that she was going to become a pilot, she went to the local airport to sign up for classes. Upon registration, she received a packet that described the tests she was going to have to pass to get the permit. She was amazed to see that the test was almost entirely made up of math questions. It took Gwen 1 month of reviewing to prepare for the text. Now that she has her pilot's license, she understands why she needed to be able to do all that math! ⬛

© Tony Stone Images/David Frazier

Pretest
for Chapter 8

Name _____

Section _____ Date _____

ANSWERS

1. Thousandths

2. 2.371; two and three hundred seventy-one thousandths

3. (a) 14.28; (b) 63.29

4. 2.375

5. $2.36

6. (a) 0.86037; (b) 536.2

7. 2.36

8. $11.95

9. 25.12 yd

10. (a) 153.86 in²; (b) 7.5 in²; (c) 6 ft²

Addition, Subtraction, and Multiplication of Decimals

This pretest will point out any difficulties you may be having in adding, subtracting, or multiplying decimals. Do all the exercises. Then check your answers with those in the back of the book.

1. Give the place value of 5 in the decimal 13.4658.

2. Write $2\dfrac{371}{1000}$ in decimal form and in words.

3. (a) $4.26 + 3.18 + 6.84$ **(b)** $56 + 5.16 + 1.8 + 0.33$

4. $4.6 - 2.225 =$

5. Consumer spending. You have $20 in cash and make purchases of $6.89 and $10.75. How much cash do you have left?

6. (a) 0.357×2.41 **(b)** 0.5362×1000

7. Round 2.35878 to the nearest hundredth.

8. Fuel costs. You fill up your car with 9.2 gal of fuel at $1.299 per gallon. What is the cost of the fill up (to the nearest cent)?

9. Find the circumference of the given circle. Use 3.14 for π.

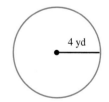
4 yd

10. Find the area of the given figures.

(a) 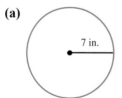 7 in.   **(b)** 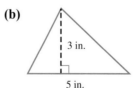 3 in. 5 in.   **(c)** 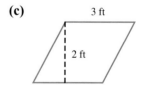 3 ft 2 ft

Place Value in Decimal Fractions

8.1

8.1 OBJECTIVES

1. Identify place value in a decimal fraction.
2. Write a decimal fraction in words.
3. Write a decimal fraction as a mixed number.
4. Compare the size of several decimal fractions.

Remember that the powers of 10 are 1, 10, 100, 1000, and so on. You might want to review Section 2.6 before going on.

In Chapters 5 to 7, we looked at common fractions. Let's turn now to a special kind of fraction, a **decimal fraction.** A decimal fraction is a fraction whose denominator is a *power of 10.* $\frac{3}{10}$, $\frac{45}{100}$, and $\frac{123}{1000}$ are examples of decimal fractions.

Earlier we talked about the idea of place value. Recall that in our decimal place-value system, each place has *one-tenth* the value of the place to its left.

●Example 1

Identifying Place Values

Label the place values for the number 538.

5   3   8
↑   ↑   ↑
Hundreds   Tens   Ones

The ones place value is one-tenth of the tens place value; the tens place value is one-tenth of the hundreds place value; and so on.

● ● ●   CHECK YOURSELF 1

Label the place values for the number 2793.

We now want to extend this idea *to the right* of the ones place. Write a period to the *right* of the ones place. This is called the **decimal point.** Each digit to the right of that decimal point will represent a fraction whose denominator is a power of 10. The first place to the right of the decimal point is the tenths place:

The decimal point separates the whole-number part and the fractional part of a decimal fraction.

$$0.1 = \frac{1}{10}$$

●Example 2

Writing a Number in Decimal Form

Write the mixed number $3\frac{2}{10}$ in decimal form.

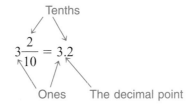

●●● **CHECK YOURSELF 2**

Write $5\frac{3}{10}$ in decimal form.

As you move farther to the *right,* each place value must be one-tenth of the value before it. The second place value is hundredths $\left(0.01 = \frac{1}{100}\right)$. The next place is thousandths, the fourth position is the ten thousandths place, and so on. Figure 1 illustrates the value of each position as we move to the right of the decimal point.

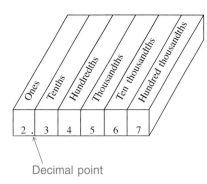

Figure 1

For convenience we will shorten the term "decimal fraction" to "decimal" from this point on.

Identifying Place Values

What is the place value for 4 and 6 in the decimal 2.34567? The place value of 4 is hundredths, and the place value of 6 is ten thousandths.

●●● **CHECK YOURSELF 3**

What is the place value of 5 in the decimal of Example 3?

Understanding place values will allow you to read and write decimals by using the following steps.

Reading or Writing Decimals in Words

If there are *no* nonzero digits to the left of the decimal point, start directly with step 3.

STEP 1   Read the digits *to the left* of the decimal point as a whole number.
STEP 2   Read the decimal point as the word "and."
STEP 3   Read the digits *to the right* of the decimal point as a whole number followed by the place value of the rightmost digit.

• Example 4

Writing a Decimal Number in Words

Write each decimal number in words.

5.03 is read "five and three hundredths."

Hundredths — The rightmost digit, 3, is in the hundredths position.

12.057 is read "twelve and fifty-seven thousandths."

Thousandths — The rightmost digit, 7, is in the thousandths position.

0.5321 is read "five thousand three hundred twenty-one ten thousandths."

An informal way of reading decimals is to simply read the digits in order and use the word "point" to indicate the decimal point. 2.58 can be read "two point five eight." 0.689 can be read "zero point six eight nine."

When the decimal has no whole-number part, we have chosen to write a 0 to the left of the decimal point. This simply makes sure that you don't miss the decimal point. However, both 0.5321 and .5321 are correct.

● ● ● CHECK YOURSELF 4

Write 2.58 in words.

The number of digits to the right of the decimal point is called the number of **decimal places** in a decimal number. So, 0.35 has two decimal places.

One quick way to write a decimal as a common fraction is to remember that the number of decimal places must be the same as the number of zeros in the denominator of the common fraction.

• Example 5

Writing a Decimal Number as a Mixed Number

Write each decimal as a common fraction or mixed number.

$$0.35 = \frac{35}{100}$$

Two places   Two zeros

The same method can be used with decimals that are greater than 1. Here the result will be a mixed number.

The 0 to the right of the decimal point is a "placeholder" that is not needed in the common-fraction form.

$$2.058 = 2\frac{58}{1000}$$

Three places | Three zeros

CHECK YOURSELF 5

Write as common fractions or mixed numbers.

a. 0.528 **b.** 5.08

It is often useful to compare the sizes of two decimal fractions. One approach to comparing decimals uses the following fact.

Adding zeros to the right *does not change* the value of a decimal. 0.53 is the same as 0.530. Look at the fractional form:

Remember: By the fundamental principle of fractions, multiplying the numerator and denominator of a fraction by the same nonzero number does not change the value of the fraction.

$$\frac{53}{100} = \frac{530}{1000}$$

The fractions are equivalent. We have multiplied the numerator and denominator by 10.

Let's see how this is used to compare decimals in our final example.

• Example 6

Comparing the Sizes of Two Decimal Numbers

Which is larger?

0.84   or   0.842

Write 0.84 as 0.840. Then we see that 0.842 (or 842 thousandths) is greater than 0.840 (or 840 thousandths), and we can write

0.842 > 0.84

CHECK YOURSELF 6

Complete the statement below, using the symbol < or >.

0.588 _____ 0.59

CHECK YOURSELF ANSWERS

1.   2 7 9 3.

Thousands / \ Ones
   Hundreds   Tens

2. $5\frac{3}{10} = 5.3$.   **3.** Thousandths.

4. Two and fifty-eight hundredths.   **5. (a)** $\frac{528}{1000}$; **(b)** $5\frac{8}{100}$.   **6.** 0.588 < 0.59.

8.1 Exercises

Name

Section   Date

ANSWERS

For the decimal 8.57932:

1. What is the place value of 7?

2. What is the place value of 5?

3. What is the place value of 3?

4. What is the place value of 2?

Write in decimal form.

5. $\dfrac{23}{100}$

6. $\dfrac{371}{1000}$

7. $\dfrac{209}{10,000}$

8. $3\dfrac{5}{10}$

9. $23\dfrac{56}{1000}$

10. $7\dfrac{431}{10,000}$

Write in words.

11. 0.23

12. 0.371

13. 0.071

14. 0.0251

15. 12.07

16. 23.056

Write in decimal form.

17. Fifty-one thousandths

18. Two hundred fifty-three ten thousandths

19. Seven and three tenths

20. Twelve and two hundred forty-five thousandths

1.	Hundredths
2.	Tenths
3.	Ten thousandths
4.	Hundred thousandths
5.	0.23
6.	0.371
7.	0.0209
8.	3.5
9.	23.056
10.	7.0431
11.	Twenty-three hundredths
12.	Three hundred seventy-one thousandths
13.	Seventy-one thousandths
14.	Two hundred fifty-one ten thousandths
15.	Twelve and seven hundredths
16.	Twenty-three and fifty-six thousandths
17.	0.051
18.	0.0253
19.	7.3
20.	12.245

Write each of the following as a common fraction or mixed number.

21. 0.65   **22.** 0.00765

23. 5.231   **24.** 4.0171

Complete each of the following statements, using the symbol $<$, $=$, or $>$.

25. 0.69 _____ 0.689   **26.** 0.75 _____ 0.752

27. 1.23 _____ 1.230   **28.** 2.451 _____ 2.45

29. 10 _____ 9.9   **30.** 4.98 _____ 5

31. 1.459 _____ 1.46   **32.** 0.235 _____ 0.2350

33. **(a)** Explain in your own words why placing zeros to the right of the decimal point does not change the value of a number.

(b) What is the difference in value of the following: 0.120, 0.1200, 0.12000?

34. Arrange the following decimals in order from largest to smallest.

0.0600   0.609   0.690   0.0609   0.6191

Getting Ready for Section 8.2
[Section 1.5]

Round each number to the indicated place value.

a. 5378   nearest thousand

b. 25,189   nearest hundred

c. 219,473   nearest ten thousand

d. 3,438,000   nearest hundred thousand

e. 351,098   nearest ten thousand

f. 5,298,500   nearest hundred thousand

Answers

1. Hundredths   **3.** Ten thousandths   **5.** 0.23   **7.** 0.0209   **9.** 23.056
11. Twenty-three hundredths   **13.** Seventy-one thousandths

15. Twelve and seven hundredths   **17.** 0.051   **19.** 7.3   **21.** $\dfrac{65}{100}\left(\text{or } \dfrac{13}{20}\right)$

23. $5\dfrac{231}{1000}$ **25.** 0.69 > 0.689   **27.** 1.23 = 1.230   **29.** 10 > 9.9

31. 1.459 < 1.46   **33.**   **a.** 5000   **b.** 25,200   **c.** 220,000

d. 3,400,000   **e.** 350,000   **f.** 5,300,000

8.1 Supplementary Exercises

Name

Section   Date

For the decimal 9.63584:

1. What is the place value of 3?   **2.** What is the place value of 8?

Write in decimal form.

3. $\dfrac{47}{100}$ **4.** $\dfrac{47}{1000}$ **5.** $12\dfrac{251}{10,000}$

Write in words.

6. 0.47   **7.** 0.419   **8.** 2.0043

Write in decimal form.

9. Two hundred forty-five ten thousandths

10. Seven and thirty-two thousandths

Write as common fractions or mixed numbers.

11. 0.0067   **12.** 21.857

Complete each of the following statements, using the symbol <, =, or >.

13. 0.78 _____ 0.778   **14.** 0.53 _____ 0.532

15. 0.27 _____ 0.270   **16.** 2.31 _____ 2.308

Rounding Decimals

8.2 OBJECTIVES

1. Round a decimal to the nearest tenth.
2. Round a decimal to any specific decimal place.

Whenever a decimal represents a measurement made by some instrument (a rule or a scale), the decimals are not exact. They are accurate only to a certain number of places and are called **approximate numbers.** Usually, we want to make all decimals in a particular problem accurate to a specified decimal place or tolerance. This will require **rounding** the decimals. We can picture the process on a number line.

• Example 1

Rounding to the Nearest Tenth

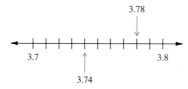

3.74 is closer to 3.7 than it is to 3.8.
3.78 is closer to 3.8.

3.74 is rounded down to the nearest tenth, 3.7. 3.78 is rounded up to 3.8.

● ● ● **CHECK YOURSELF 1**

Use the number line in Example 1 to round 3.77 to the nearest tenth.

Rather than using the number line, the following rule can be applied.

To Round a Decimal

STEP 1 Find the place to which the decimal is to be rounded.
STEP 2 If the next digit to the right is 5 or more, increase the digit in the place you are rounding by 1. Discard remaining digits to the right.
STEP 3 If the next digit to the right is less than 5, just discard that digit and any remaining digits to the right.

• Example 2

Rounding to the Nearest Tenth

Round 34.58 to the nearest tenth.

Many students find it easiest to mark this digit with an arrow.

34.58   Locate the digit you are rounding to. The 5 is in the tenths place.

Since the next digit to the right, (8), is 5 or more, increase the tenths digit by 1. Then discard the remaining digits.

34.58 is rounded to 34.6.

● ● ● **CHECK YOURSELF 2**

Round 48.82 to the nearest tenth.

● Example 3

Rounding to the Nearest Hundredth

Round 5.673 to the nearest hundredth.

5.673   The 7 is in the hundredths place.

The next digit to the right, (3), is less than 5. Leave the hundredths digit as it is, and discard the remaining digits to the right.

5.673 is rounded to 5.67.

● ● ● **CHECK YOURSELF 3**

Round 29.247 to the nearest hundredth.

● Example 4

Rounding to a Specified Decimal Place

Round 3.14159 to four decimal places.

The fourth place to the *right* of the decimal point is the ten thousandths place.

3.14159   The 5 is in the ten-thousandths place.

The next digit to the right, (9), is 5 or more, so increase the digit you are rounding to by 1. Discard the remaining digits to the right.

3.14159 is rounded to 3.1416.

● ● ● **CHECK YOURSELF 4**

Round 0.8235 to three decimal places.

● ● ● **CHECK YOURSELF ANSWERS**

1. 3.8.   **2.** 48.8.   **3.** 29.25.   **4.** 0.824.

Round to the indicated place.

1. 53.48   tenth

2. 6.785   hundredth

3. 21.534   hundredth

4. 5.842   tenth

5. 0.342   hundredth

6. 2.3576   thousandth

7. 2.71828   thousandth

8. 1.543   tenth

9. 0.0475   tenth

10. 0.85356   ten thousandth

11. 4.85344   ten thousandth

12. 52.8728   thousandth

13. 6.734   two decimal places

14. 12.5467   three decimal places

15. 6.58739   four decimal places

16. 503.824   two decimal places

Round 56.35829 to the nearest:

17. Tenth

18. Ten thousandth

19. Thousandth

20. Hundredth

21. Eva wants to round 76.24491 to the nearest hundredths. She first rounds 76.24491 to 76.245 and then rounds 76.245 to 76.25 and claims that this is the final answer. What is wrong with this approach?

ANSWERS

1. 53.5

2. 6.79

3. 21.53

4. 5.8

5. 0.34

6. 2.358

7. 2.718

8. 1.5

9. 0.0

10. 0.8536

11. 4.8534

12. 52.873

13. 6.73

14. 12.547

15. 6.5874

16. 503.82

17. 56.4

18. 56.3583

19. 56.358

20. 56.36

21.

a. 101

b. 116

c. 1333

d. 1660

e. 130

f. 1461

Getting Ready for Section 8.3
[Section 1.5]

Add.

a. 43
   +58

b. 79
   +37

c. 584
   +749

d. 675
   +985

e. 29
    58
   +43

f. 129
    538
   +794

Answers

1. 53.5   **3.** 21.53   **5.** 0.34   **7.** 2.718   **9.** 0.0   **11.** 4.8534
13. 6.73   **15.** 6.5874   **17.** 56.4   **19.** 56.358   **a.** 101   **b.** 116
c. 1333   **d.** 1660   **e.** 130   **f.** 1461

8.2 Supplementary Exercises

Name

Section   Date

ANSWERS

1. 0.7

2. 23.45

3. 5.880

4. 5.9

5. 5.8719

6. 27.3218

7. 4.83

8. 0.826

9. 43.63

10. 43.626

11. 43.6258

12. 43.6

Round to the indicated place.

1. 0.738   tenth

2. 23.454   hundredth

3. 5.8796   thousandth

4. 5.853   tenth

5. 5.87194   ten thousandth

6. 27.32178   ten thousandth

7. 4.8281   two decimal places

8. 0.8257   three decimal places

Round 43.62583 to the nearest:

9. Hundredth

10. Thousandth

11. Ten thousandth

12. Tenth

432

Adding Decimals

8.3 OBJECTIVES

1. Add decimals.
2. Use addition of decimals to solve an application.

Working with decimals rather than common fractions makes the basic operations much easier. Let's start by looking at addition. One method for adding decimals is to write the decimals as common fractions, add, and then change the sum back to a decimal.

$$0.34 + 0.52 = \frac{34}{100} + \frac{52}{100} = \frac{86}{100} = 0.86$$

It is much more efficient to leave the numbers in decimal form and perform the addition in the same way as we did with whole numbers. You can use the following rule.

To Add Decimals

STEP 1 Write the numbers being added in column form *with their decimal points aligned (in line) vertically.*
STEP 2 Add just as you would with whole numbers.
STEP 3 Place the decimal point of the sum in line with the decimal points of the addends.

Example 1 illustrates the use of this rule.

● Example 1

Adding Decimals

Add 0.13, 0.42, and 0.31.

Placing the decimal points in a vertical line ensures that we are adding digits of the same place value.

$$
\begin{array}{r}
0.13 \\
0.42 \\
+0.31 \\
\hline
0.86
\end{array}
$$

● ● ● **CHECK YOURSELF 1**

Add 0.23, 0.15, and 0.41.

In adding decimals, you can use the *carrying process* just as you did in adding whole numbers. Consider the following.

•Example 2

Adding Decimals Involving Carrying

Add 0.35, 1.58, and 0.67.

```
  1 2 ←—— Carries
  0.35
  1.58
 +0.67
  2.60
```

In the hundredths column:
$5 + 8 + 7 = 20$
Write 0 and carry 2 to the tenths column.

In the tenths column:
$2 + 3 + 5 + 6 = 16$
Write 6 and carry 1 to the ones column.

Note: The carrying process works with decimals, just as it did with whole numbers, because each place value is again *one-tenth* the value of the place to its left.

● ● ● **CHECK YOURSELF 2**

Add 23.546, 0.489, 2.312, and 6.135.

In adding decimals, the numbers may not have the same number of decimal places. Just fill in as many zeros as needed so that all of the numbers added have the same number of decimal places.

Recall that adding zeros to the right *does not change* the value of a decimal. 0.53 is the same as 0.530.

Let's see how this is used in Example 3.

•Example 3

Adding Decimals

Add 0.53, 4, 2.7, and 3.234.

Be sure that the decimal points are in a vertical line.

```
  0.53
  4.
  2.7
 +3.234
```

Note that for a whole number, the decimal is understood to be to its right. So 4 = 4.

Now fill in the missing zeros, and add as before.

```
   0.530
   4.000
   2.700
 + 3.234
  10.464
```

Now all the numbers being added have *three* decimal places.

● ● ● **CHECK YOURSELF 3**

Add 6, 2.583, 4.7, and 2.54.

Many applied problems require working with decimals. For instance, filling up at a gas station means reading decimal amounts.

•Example 4

An Application of the Addition of Decimals

On a trip the Chang family kept track of their gas purchases. If they bought 12.3, 14.2, 10.7, and 13.8 gallons (gal), how much gas did they use on the trip?

Solution

Since we want a total amount, addition is used for the solution.

```
  12.3
  14.2
  10.7
+ 13.8
  51.0 gal
```

● ● ● CHECK YOURSELF 4

The Higueras kept track of the gasoline they purchased on a recent trip. If they bought 12.4, 13.6, 9.7, 11.8, and 8.3 gal, how much gas did they buy on the trip?

Every day you deal with amounts of money. Since our system of money is a decimal system, most problems involving money also involve operations with decimals.

•Example 5

An Application of the Addition of Decimals

Andre makes deposits of $3.24, $15.73, $50, $28.79, and $124.38 during May. What is his total deposit for the month?

Solution

```
$   3.24      Simply add the amounts of money
   15.73      deposited as decimals. Note that
   50.00      we write $50 as $50.00.
   28.79
+ 124.38
$222.14  ←── The total deposit for May
```

● ● ● CHECK YOURSELF 5

Your textbooks for the fall term cost $33.50, $38.95, $23.15, $42, and $45.85. What was the total cost of textbooks for the term?

In Chapter 1, we defined *perimeter* as the distance around the outside of a straight-edged shape. Finding the perimeter often requires that we add decimal numbers.

• Example 6

An Application Involving the Addition of Decimals

Rachel is going to put a fence around the perimeter of her farm. Figure 1 is a picture of the land, measured in kilometers (km). How much fence does she need to buy?

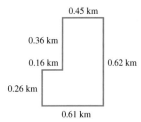

Figure 1

The perimeter is the sum of the lengths of the sides, so we add those lengths to find the total fencing needed.

$$0.16 + 0.36 + 0.45 + 0.62 + 0.61 + 0.26 = 2.46$$

Rachel needs 2.46 km of fence for the perimeter of her farm.

● ● ● **CHECK YOURSELF 6**

Manuel intends to build a walkway around the perimeter of his garden (Figure 2). What will the total length of the walkway be?

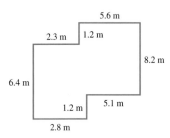

Figure 2

● ● ● **CHECK YOURSELF ANSWERS**

1. 0.79.　**2.** 32.482.　**3.**
```
   6.000
   2.583
   4.700
 + 2.540
  ------
  15.823.
```
4. 55.8 gal.　**5.** $183.45.

6. 32.8 m.

A N S W E R S

Add.

1.   0.28
    +0.79

2.   2.59
    +0.63

3.   1.045
    +0.23

4.   2.485
    +1.25

5.   0.62
    4.23
    +12.5

6.   0.50
    2.99
    +24.8

7.   5.28
    +19.455

8.   23.845
    + 7.29

9.   13.58
    7.239
    + 1.5

10.   8.625
    2.45
    +12.6

11.   25.3582
    6.5
    1.898
    + 0.69

12.   1.336
    15.6857
    7.9
    + 0.85

13. $0.43 + 0.8 + 0.561$

14. $1.25 + 0.7 + 0.259$

15. $5 + 23.7 + 8.7 + 9.85$

16. $28.3 + 6 + 8.76 + 3.8$

17. $25.83 + 1.7 + 3.92$

18. $4.8 + 32.59 + 4.76$

19. $42.731 + 1.058 + 103.24$

20. $27.4 + 213.321 + 39.38$

21. Add twenty-three hundredths, five tenths, and two hundred sixty-eight thousandths.

22. Add seven tenths, four hundred fifty-eight thousandths, and fifty-six hundredths.

1.	1.07
2.	3.22
3.	1.275
4.	3.735
5.	17.35
6.	28.29
7.	24.735
8.	31.135
9.	22.319
10.	23.675
11.	34.4462
12.	25.7717
13.	1.791
14.	2.209
15.	47.25
16.	46.86
17.	31.45
18.	42.15
19.	147.029
20.	280.101
21.	0.998
22.	1.718

23. Add five and three tenths, seventy-five hundredths, twenty and thirteen hundredths, and twelve and seven tenths.

24. Add thirty-eight and nine tenths, five and fifty-eight hundredths, seven, and fifteen and eight tenths.

Solve the following applications.

25. Gas purchase. On a 3-day trip, Dien bought 12.7, 15.9, and 13.8 gallons (gal) of gas. How many gallons of gas did he buy?

26. Distance. Felix ran 2.7 miles (mi) on Monday, 1.9 mi on Wednesday, and 3.6 mi on Friday. How far did he run during the week?

27. Rainfall. Rainfall was recorded in centimeters (cm) during the winter months as follows: December, 5.38 cm; January, 3.2 cm; and February, 4.79 cm. How much rain fell during those months?

28. Total length. A metal fitting has three sections, with lengths 2.5, 1.775, and 1.45 inches (in.). What is the total length of the fitting?

29. Total expenses. Nicole had the following expenses on a business trip: gas, $45.69; food, $123; lodging, $95.60; and parking and tolls, $8.65. What were her total expenses during the trip?

30. Textbook costs. Hok Sum's textbooks for one quarter cost $29.95, $47, $52.85, $33.35, and $10. What was his total cost for textbooks?

31. Checking. Bruce wrote checks of $50, $11.38, $112.57, and $9.73 during a single week. What was the total amount of the checks he wrote?

32. Checking. Frederika made deposits of $75.35, $58, $7.89, and $100 to her checking account in a single month. What was the total amount of her deposits?

33. Perimeter. Lupe is putting a fence around her yard. Her yard measures 8.16 yards (yd) long and 12.68 yd wide. How much fence should Lupe purchase?

34. Fencing. Find the amount of fencing needed to enclose Moira's yard if it is 14.56 yd long by 23.86 yd wide.

ANSWERS

35. 33.2 ft

36. 11.535 mi

37. $43

38. $50

39. $1209

40. $1008

35. Perimeter. Find the perimeter of the figure given below.

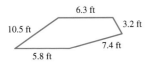

36. Fencing. The figure below gives the distance in miles (mi) of the boundary sections around a ranch. How much fencing is needed for the property?

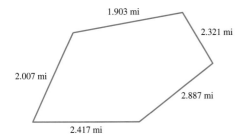

Estimation can be a useful tool when working with decimal fractions. To estimate a sum, one approach is to round the addends to the nearest whole number and add for your estimate. For instance, to estimate the sum below:

```
          Round
 19.8   ⟶        20
  3.5             4
 24.2            24
+10.4          +10      Add for
                58      the estimate.
```

Use estimation to solve the following applications.

37. Restaurant bill. Your restaurant bill includes $18.25 for dinners, $6.80 for salads, $8.75 for wine, $7.40 for dessert, and $1.70 for coffee. Estimate your bill by rounding each amount to the nearest dollar.

38. Car maintenance. Your bill for a car tune-up includes $7.80 for oil, $5.90 for a filter, $3.40 for spark plugs, $4.10 for points, and $28.70 for labor. Estimate your total cost.

39. Payroll. The payroll at a car repair shop for one week was $456.73, utilities were $123.89, advertising was $212.05, and payments to distributors were $415.78. Estimate the amount spent in 1 week.

40. Expenses. On a recent business trip your expenses were $343.78 for airfare, $412.78 for lodging, $148.89 for food, and $102.15 for other items. Estimate your total expenses.

41.

42.

a. 6

b. 29

c. 89

d. 666

e. 3224

f. 2756

41. Following are charges on a credit card account:

$8.97, $32.75, $15.95, $67.32, $215.78, $74.95, $83.90, and $257.28

(a) Estimate the total bill for the charges by rounding each number to the nearest dollar and adding the results.

(b) Estimate the total bill by adding the charges and then rounding to the nearest dollar.

(c) What are the advantages and disadvantages of the methods in (a) and (b)?

42. Find the next number in the following sequence: 3.125, 3.375, 3.625, . . .

Getting Ready for Section 8.4 [Section 1.7]

Subtract.

a.
$$\begin{array}{r} 43 \\ -37 \\ \hline \end{array}$$

b.
$$\begin{array}{r} 58 \\ -29 \\ \hline \end{array}$$

c.
$$\begin{array}{r} 247 \\ -158 \\ \hline \end{array}$$

d.
$$\begin{array}{r} 923 \\ -257 \\ \hline \end{array}$$

e.
$$\begin{array}{r} 4283 \\ -1059 \\ \hline \end{array}$$

f.
$$\begin{array}{r} 5324 \\ -2568 \\ \hline \end{array}$$

Answers

1. 1.07   **3.** 1.275   **5.** 17.35   **7.** 24.735   **9.** 22.319   **11.** 34.4462

13. 1.791   **15.**
$$\begin{array}{r} ^{22} \\ 5.00 \\ 23.70 \\ 8.70 \\ + 9.85 \\ \hline 47.25 \end{array}$$
17. 31.45   **19.** 147.029   **21.** 0.998

23.
$$\begin{array}{r} ^{1} \\ 5.30 \\ 0.75 \\ 20.13 \\ +12.70 \\ \hline 38.88 \end{array}$$
25. 42.4 gal   **27.** 13.37 cm   **29.** $272.94   **31.** $183.68

33. 41.68 yd   **35.** 33.2 ft   **37.** $43   **39.** $1209   **41.**
a. 6   **b.** 29   **c.** 89   **d.** 666   **e.** 3224   **f.** 2756

A N S W E R S

1.	1.50
2.	3.825
3.	18.25
4.	102.214
5.	36.298
6.	48.1644
7.	27.72
8.	549.021
9.	259.345
10.	35.3634
11.	1.475
12.	40.907

Add.

1. $\begin{array}{r} 0.57 \\ +0.93 \\ \hline \end{array}$

2. $\begin{array}{r} 2.875 \\ +0.95 \\ \hline \end{array}$

3. $\begin{array}{r} 13.86 \\ 3.8 \\ +\ 0.59 \\ \hline \end{array}$

4. $\begin{array}{r} 4.23 \\ +97.984 \\ \hline \end{array}$

5. $\begin{array}{r} 29.358 \\ 0.24 \\ +\ 6.7 \\ \hline \end{array}$

6. $\begin{array}{r} 36.8954 \\ 8.7 \\ 1.789 \\ +\ 0.78 \\ \hline \end{array}$

7. $23.5 + 3.26 + 0.96$

8. $528.271 + 1.85 + 18.9$

9. $13.675 + 9 + 0.87 + 235.8$

10. $27.2 + 0.815 + 0.3484 + 7$

11. Add five tenths, seventy-three hundredths, and two hundred forty-five thousandths.

12. Add thirty-five and six tenths, twenty-seven hundredths, four and five tenths, and five hundred thirty-seven thousandths.

13. 19.8 h

14. $52.85

15. 50.8 gal

16. 6.52 cm

17. 54.46 yd

18. 23.1 yd

Solve the following applications.

13. Employment. Roland worked 6.7 hours (h) on Monday, 5.9 h on Wednesday, and 7.2 h on Friday. How many hours did he work during the week?

14. Billing. Your bill from a service station includes $7.85 for oil, $19.25 for parts, and $25.75 for labor. What is the total charge?

15. Gas usage. On a vacation trip, the Villas bought the following gallon amounts of gasoline: 12.3, 15, 9.8, and 13.7. How much gasoline did they purchase during the trip?

16. Rainfall. Rainfall amounts during a week were 2.38, 0.45, 1.5, and 2.19 centimeters (cm). What was the total amount of rain during the week?

17. Fencing. How much fencing is needed for a property that is 11.78 yards (yd) long by 15.45 yd wide?

18. Perimeter. Find the perimeter of the figure given below.

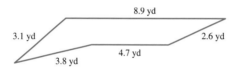

8.4 Subtracting Decimals

8.4 OBJECTIVES

1. Subtract one decimal number from another.
2. Use subtraction to solve an application.

Much of what we said in the last section about adding decimals is also true of subtraction. To subtract decimals, we use the following rule:

To Subtract Decimals

STEP 1 Write the numbers being subtracted in column form *with their decimal points aligned vertically*.

STEP 2 Subtract just as you would with whole numbers.

STEP 3 Place the decimal point of the difference in line with the decimal points of the numbers being subtracted.

Our first example illustrates the use of this rule.

● Example 1

Subtracting a Decimal

Subtract 1.23 from 3.58.

$$
\begin{array}{r}
3.58 \\
-1.23 \\
\hline
2.35
\end{array}
$$

Subtract in the hundredths, the tenths, and then the ones columns.

● ● ● CHECK YOURSELF 1

Subtract $9.87 - 5.45$.

Since each place value is one-tenth the value of the place to its left, borrowing, when you are subtracting decimals, works just as it did in subtracting whole numbers.

● Example 2

Subtraction of a Decimal That Involves Borrowing

Subtract 1.86 from 6.54.

$$
\begin{array}{r}
{}^{5}{}^{1}4{}^{1} \\
6.54 \\
-1.86 \\
\hline
4.68
\end{array}
$$

Here, borrow in the tenths and ones places to do the subtraction.

● ● ● **CHECK YOURSELF 2**

Subtract $35.35 - 13.89$.

In subtracting decimals, as in adding, we can add zeros to the right of the decimal point so that both decimals have the same number of decimal places.

• Example 3

Subtracting a Decimal

(*a*) Subtract 2.36 from 7.5.

When you are subtracting, align the decimal points, then add zeros to the right to align the digits.

$$
\begin{array}{r}
\overset{4\,1}{7.\cancel{5}0} \\
-\,2.36 \\
\hline
5.14
\end{array}
$$

We have added a 0 to 7.5. Next, borrow 1 tenth from the 5 tenths in the minuend.

(*b*) Subtract 3.657 from 9.

9 has been rewritten as 9.000.

$$
\begin{array}{r}
\overset{8\ \overset{9}{\cancel{1}}\,\overset{9}{\cancel{1}}\,\overset{9}{\cancel{1}}}{9.000} \\
-\,3.657 \\
\hline
5.343
\end{array}
$$

In this case, move left to the ones place to begin the borrowing process.

● ● ● **CHECK YOURSELF 3**

Subtract $5 - 2.345$.

We can apply the subtraction methods of Examples 1 to 3 in solving applications involving decimals.

• Example 4

An Application of the Subtraction of a Decimal Number

Jonathan was 98.3 centimeters (cm) tall on his sixth birthday. On his seventh birthday he was 104.2 cm. How much did he grow during the year?

Solution

We want to find the difference between the two measurements, so we subtract.

$$
\begin{array}{r}
104.2 \text{ cm} \\
-\ \ 98.3 \text{ cm} \\
\hline
5.9 \text{ cm}
\end{array}
$$

Jonathan grew 5.9 cm during the year.

● ● ● CHECK YOURSELF 4

A car's highway mileage before a tune-up was 28.8 miles per gallon (mi/gal). After the tune-up it measured 30.1 mi/gal. What was the increase in mileage?

The same method can be used in working with money.

● Example 5

An Application of the Subtraction of a Decimal Number

At the grocery store, Sally buys a roast that is marked $12.37. She pays for her purchase with a $20 bill. How much change does she get?

Solution

Sally's change will be the *difference* between the price of the roast and the $20 paid. We must use subtraction for the solution.

$$\begin{array}{r} \$20.00 \\ -\underline{12.37} \\ \$\ 7.63 \end{array}$$

Add zeros to write $20 as $20.00. Then subtract as before.

Sally will receive $7.63 in change after her purchase.

● ● ● CHECK YOURSELF 5

A stereo system that normally sells for $549.50 is discounted (or marked down) to $499.95 for a sale. What is the savings?

Keeping one's checkbook requires addition and subtraction of decimal numbers.

● Example 6

An Application Involving the Addition and Subtraction of Decimals

For the following check register, find the running balance.

Beginning balance	$234.15
Check # 301	23.88
Balance	_____
Check # 302	38.98
Balance	_____
Check # 303	114.66
Balance	_____
Deposit	175.75
Balance	_____
Check # 304	212.55
Ending balance	_____

Solution In order to keep a running balance, we add the deposits and subtract the checks.

Beginning balance	$234.15
Check # 301	23.88
Balance	210.27
Check # 302	38.98
Balance	171.29
Check # 303	114.66
Balance	56.63
Deposit	175.75
Balance	232.38
Check # 304	212.55
Ending balance	19.83

● ● ● **CHECK YOURSELF 6**

For the following check register, add the deposits and subtract the checks to find the balance.

	Beginning balance	$398.00
	Check # 401	19.75
a.	Balance	_____
	Check # 402	56.88
b.	Balance	_____
	Check # 403	117.59
c.	Balance	_____
	Deposit	224.67
d.	Balance	_____
	Check # 404	411.48
e.	Ending balance	_____

● ● ● **CHECK YOURSELF ANSWERS**

1. 4.42.   **2.** 21.46.   **3.** 2.655.   **4.** 1.3 mi/gal.   **5.** $49.55.
6. (a) $378.25; **(b)** $321.37; **(c)** $203.78; **(d)** $428.45; **(e)** $16.97

Name

Section   Date

Subtract.

1. 0.85
  −0.59

2. 5.68
  −2.65

3. 23.81
  − 6.57

4. 48.73
  −19.95

5. 17.134
  − 3.502

6. 40.092
  −21.595

7. 35.8
  − 7.45

8. 7.83
  −5.2

9. 3.82
  −1.565

10. 8.59
  −5.6

11. 7.32
  −4.7

12. 45.6
  − 8.75

13. 12
  − 5.35

14. 15
  − 8.85

15. 12.02
  − 2.545

16. 36.05
  − 3.675

17. $28 - 24.725$

18. $40 - 13.875$

A N S W E R S

1. 0.26

2. 3.03

3. 17.24

4. 28.78

5. 13.632

6. 18.497

7. 28.35

8. 2.63

9. 2.255

10. 2.99

11. 2.62

12. 36.85

13. 6.65

14. 6.15

15. 12.475

16. 32.375

17. 3.275

18. 26.125

19. Subtract 2.87 from 6.84.   **20.** Subtract 3.69 from 10.57.

21. Subtract 7.75 from 9.4.   **22.** Subtract 5.82 from 12.

23. Subtract 0.24 from 5.   **24.** Subtract 8.7 from 16.32.

Solve the following applications.

25. Discounts. A television set selling for $399.50 is discounted (or marked down) to $365.75. What is the savings?

26. Tubing radii. The outer radius of a piece of tubing is 2.8325 inches (in.). The inner radius is 2.775 in. What is the thickness of the wall of the tubing?

27. Temperature. If normal body temperature is 98.6°F and a person is running a temperature of 101.3°F, how much is that temperature above normal?

28. Amount of change. You pay your hotel bill of $84.58 with two $50 traveler's checks. What change will you receive?

29. Perimeter. Given the following figure, find dimension a.

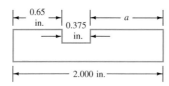

30. Credit cards. You make charges of $37.25, $8.78, and $53.45 on a credit card. If you make a payment of $73.50, how much do you still owe?

31. Distance. At the start of a trip, Laura notes that her odometer (mileage indicator) reads 15,785.3 miles (mi). At the end of the trip it reads 16,479.8 mi. How far did she drive?

32. Rainfall. Rainfall for the first 3 months of 1992 was recorded at 2.73, 1.41, and 1.48 inches (in.). If the normal rainfall for that period is 6.51 in., by how much was the 1992 amount above or below normal?

33. Checkbook balance. For the following check register, find the running balance.

Beginning balance	$896.74
Check # 501	$425.69
Balance	$471.05
Check # 502	$ 56.34
Balance	$414.71
Check # 503	$ 41.89
Balance	$372.82
Deposit	$123.91
Balance	$496.73
Check # 504	$356.98
Ending balance	$139.75

34. Checkbook balance. For the following check register, find the running balance.

Beginning balance	$456.00
Check # 601	$199.29
Balance	$256.71
Service charge	$ 18.00
Balance	$238.71
Check # 602	$ 85.78
Balance	$152.93
Deposit	$250.45
Balance	$403.38
Check # 603	$201.24
Ending balance	$202.14

35. Checkbook balance. For the following check register, find the running balance.

Beginning balance	$589.21
Check # 678	$175.63
Balance	$413.58
Check # 679	$ 56.92
Balance	$356.66
Deposit	$121.12
Balance	$477.78
Check # 680	$345.99
Ending balance	$131.79

36. Checkbook balance. For the following check register, find the running balance.

Beginning balance	$1,345.23
Check # 821	$ 234.99
Balance	$1,110.24
Check # 822	$ 555.77
Balance	$ 554.47
Deposit	$ 126.77
Balance	$ 681.24
Check # 823	$ 53.89
Ending balance	$ 627.35

Recall that a magic square is one in which the sum of every row, column, and diagonal is the same. Complete the magic squares below.

37.

1.6	0.2	1.2
0.6	1	1.4
0.8	1.8	0.4

38.

2.4	8.4	7.2
10.8	6	1.2
4.8	3.6	9.6

39. Find the next two numbers in each of the following sequences:
(a) **0.**75   0.62   0.5   0.39

(b) 1.0   1.5   0.9   3.5   0.8

40. (a) Determine the average amount of rainfall (to the nearest hundredth of an inch) in your town or city for each of the past 24 months.

(b) Determine the difference in rainfall amounts per month for each month from 1 year to the next.

Getting Ready for Section 8.5
[Section 6.1]

Multiply.

a. $\dfrac{3}{10} \times \dfrac{7}{10}$ **b.** $\dfrac{7}{10} \times \dfrac{9}{10}$ **c.** $\dfrac{3}{10} \times \dfrac{23}{100}$

d. $\dfrac{7}{10} \times \dfrac{33}{100}$ **e.** $\dfrac{53}{100} \times \dfrac{7}{100}$ **f.** $\dfrac{21}{100} \times \dfrac{17}{100}$

Answers

1. 0.26 **3.** $\overset{11\ 7\ 1}{23.81}$ **5.** 13.632 **7.** 28.35 **9.** 2.255 **11.** 2.62
$\underline{-\ \ 6.57}$
17.24

13. $\begin{array}{r} 12.00 \\ -\ 5.35 \\ \hline 6.65 \end{array}$ **15.** 12.475 **17.** 3.275 **19.** 3.97 **21.** 1.65 **23.** 4.76

25. $33.75 **27.** 2.7°F **29.** 0.975 in. **31.** 694.5 mi

33. End balance, $139.75 **35.** End balance, $131.79

37.

1.6	0.2	1.2
0.6	1	1.4
0.8	1.8	0.4

39. (a) 0.29, 0.20 **(b)** 5.5, 0.7

a. $\dfrac{21}{100}$ **b.** $\dfrac{63}{100}$ **c.** $\dfrac{69}{1000}$ **d.** $\dfrac{231}{1000}$ **e.** $\dfrac{371}{10,000}$ **f.** $\dfrac{357}{10,000}$

ANSWERS

a. $\dfrac{21}{100}$

b. $\dfrac{63}{100}$

c. $\dfrac{69}{1000}$

d. $\dfrac{231}{1000}$

e. $\dfrac{371}{10,000}$

f. $\dfrac{357}{10,000}$

Name

Section   Date

A N S W E R S

1. 5.52

2. 12.55

3. 7.378

4. 18.75

5. 2.79

6. 7.365

7. 3.905

8. 6.65

9. 12.125

10. 6.25

11. 2.18

12. 6.292

13. $68.75

14. 1687.5 mi

15. $7.18

16. 1.6 in.

Subtract.

1. 5.97
   -0.45

2. 27.53
   -14.98

3. 20.235
   -12.857

4. 27.3
   $-\ 8.55$

5. 4.59
   -1.8

6. 8.235
   -0.87

7. 5.25
   -1.345

8. 15
   $-\ 8.35$

9. 20
   $-\ 7.875$

10. Subtract 2.75 from 9.

11. Subtract 0.82 from 3.

12. Subtract 2.358 from 8.65.

Solve the following applications.

13. Savings. A stereo that lists for $498.50 is discounted (or marked down) to $429.75. What is the savings?

14. Distance. At the beginning of a trip, your odometer read 12,583.7 miles (mi). At the end of the trip, the reading was 14,271.2 mi. How many miles did you drive?

15. Purchases. You pay for purchases of $2.45, $9.78, and $0.59 with a $20 bill. How much money do you have left?

16. Geometry. Find dimension a in the figure below.

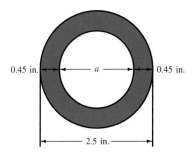

0.45 in. a 0.45 in.

2.5 in.

Using Your Calculator to Add and Subtract Decimals

Entering decimals in your calculator is similar to entering whole numbers. There is just one difference: The decimal point key ⚫ is used to place the decimal point as you enter the number.

⚫Example 1

Entering a Decimal Number into a Calculator

To enter 12.345, press

1 2 ⚫ 3 4 5

Display 12.345

⚫ ⚫ ⚫ CHECK YOURSELF 1

Enter 14.367 on your calculator.

⚫Example 2

Entering a Decimal Number into a Calculator

To enter 0.678, press

⚫ 6 7 8

Display 0.678

You don't have to press the 0 key for the digit 0 to the left of the decimal point.

⚫ ⚫ ⚫ CHECK YOURSELF 2

Enter 0.398 on your calculator.

The process of adding and subtracting decimals on your calculator is the same as we saw earlier in the sections about adding and subtracting whole numbers.

•Example 3

Adding Decimals

You don't need to worry about the fact that the decimals don't have the same number of places. If a whole number is involved, just enter that whole number. The decimal point key is not necessary.

To add $2.567 + 0.89 + 5$, enter

2.567 $\boxed{+}$ 0.89 $\boxed{+}$ $\boxed{5}$ $\boxed{=}$

Display $\boxed{8.457}$

● ● ● **CHECK YOURSELF 3**

Add on your calculator

$5.39 + 0.68 + 9.7$

Subtraction of decimals on the calculator is similar.

•Example 4

Subtracting a Decimal

To subtract $4.2 - 2.875$, enter

4.2 $\boxed{-}$ 2.875 $\boxed{=}$

Display $\boxed{1.325}$

● ● ● **CHECK YOURSELF 4**

Subtract on your calculator

$16.3 - 7.895$

Often both addition and subtraction are involved in a calculation. In this case, just enter the decimals and the operation signs, $+$ or $-$, as they appear in the problem.

• Example 5

Adding and Subtracting Decimals

Again there are differences in the operation of various calculators. Try this problem on yours to check that its operation sequence is correct.

To find $23.7 - 5.2 + 3.87 - 2.341$, enter

$23.7 \boxed{-} 5.2 \boxed{+} 3.87 \boxed{-} 2.341 \boxed{=}$

Display $\boxed{20.029}$

● ● ● **CHECK YOURSELF 5**

Use your calculator to find

$52.8 - 36.9 + 15.87 - 9.36$

● ● ● **CHECK YOURSELF ANSWERS**

1. 14.367.   **2.** 0.398.   **3.** 15.77.   **4.** 8.405.   **5.** 22.41.

Calculator Exercises

Name _____

Section _____ Date _____

Solve the following exercises using your calculator.

1. $5.87 + 3.6 + 9.25$

2. $3.456 + 10 + 2.8 + 5.62$

3. $28.21 + 387.6 + 3935.21$

4. $10,345.2 + 2308.35 + 153.58$

5. $4.59 - 2.389$

6. $19.375 - 14.2$

7. $27.85 - 3.45 - 2.8$

8. $8.8 - 4.59 - 2.325 + 8.5$

9. $14 + 3.2 - 9.35 - 3.375$

10. $8.7675 + 2.8 - 3.375 - 6$

ANSWERS

1. 18.72

2. 21.876

3. 4351.02

4. 12,807.13

5. 2.201

6. 5.175

7. 21.6

8. 10.385

9. 4.475

10. 2.1925

ANSWERS

11. $370.60

12. $77.64 overdrawn

13. $8465.25

14. $974.46

Solve the following applications using your calculator.

11. Checking balance. Your checking account has a balance of $532.89. You write checks of $50, $27.54, and $134.75 and make a deposit of $50. What is your ending balance?

12. Checking balance. Your checking account has a balance of $278.45. You make deposits of $200 and $135.46. You write checks for $389.34, $249, and $53.21. What is your ending balance? Be careful with this problem. A negative balance means that your account is overdrawn.

13. Car costs. You buy a car for $9548. If you buy additional options for $85.75, $236, and $95.50 and make a down payment of $1500, how much do you owe on the car?

14. Profit. A small store makes a profit of $934.20 in the first week of a given month, $1238.34 in the second week, and $853 in the third week. If the goal is a profit of $4000 for the month, what profit must the store make during the remainder of the month?

Answers

1. 18.72   **3.** 4351.02   **5.** 2.201   **7.** 21.6   **9.** 4.475   **11.** $370.60
13. $8465.25

Multiplying Decimals

8.5 OBJECTIVES

1. Multiply two or more decimals.
2. Use multiplication to solve an application.

To start our discussion of the multiplication of decimals, let's write the decimals in common-fraction form and then multiply.

•Example 1

Multiplying Two Decimals

$$0.32 \times 0.2 = \frac{32}{100} \times \frac{2}{10} = \frac{64}{1000} = 0.064$$

Here 0.32 has *two* decimal places, and 0.2 has *one* decimal place. The product 0.064 has *three* decimal places.

Note:

$2 + 1 = 3$

Places in 0.32 Place in 0.2 Places in the product 0.064

• • • CHECK YOURSELF 1

Find the product and the number of decimal places.

0.14×0.054

You do not need to write decimals as common fractions to multiply. Our work suggests the following rule.

To Multiply Decimals

STEP 1 Multiply the decimals as though they were whole numbers.
STEP 2 Add the number of decimal places in the numbers being multiplied.
STEP 3 Place the decimal point in the product so that the number of decimal places in the product is the sum of the number of decimal places in the factors.

Example 2 illustrates this rule.

•Example 2

Multiplying Two Decimals

Multiply 0.23 by 0.7.

$$
\begin{array}{r}
0.23 \\
\times\ 0.7 \\
\hline
0.161
\end{array}
$$
⟵ Two places
⟵ One place
⟵ Three places

● ● ● CHECK YOURSELF 2

Multiply 0.36×1.52.

You may have to affix zeros to the left in the product in order to place the decimal point. Consider our next example.

●Example 3

Multiplying Two Decimals

Multiply.

$$
\begin{array}{r}
0.136 \\
\times\ \ 0.28 \\
\hline
1088 \\
272\ \ \\
\hline
0.03808
\end{array}
$$
⟵ Three places
⟵ Two places

$3 + 2 = 5$

⟵ Five place

↑
Insert 0

Insert a 0 to mark off five decimal places.

● ● ● CHECK YOURSELF 3

Multiply 0.234×0.24.

Estimation is also helpful in multiplying decimals.

●Example 4

Estimating the Product of Two Decimals

Estimate the product 24.3×5.8.

Round

$$
\begin{array}{r}
24.3 \\
\times\ 5.8 \\
\end{array}
\qquad \rightarrow \qquad
\begin{array}{r}
24 \\
\times\ 6 \\
\hline
144
\end{array}
$$

Multiply for the estimate.

● ● ● **CHECK YOURSELF 4**

Estimate the product.

17.95×8.17

Let's look at some applications of our work in multiplying decimals.

● Example 5

An Application Involving the Multiplication of Two Decimals

A sheet of paper has dimensions 27.5 by 21.5 centimeters (cm). What is its area?

Solution   We multiply to find the required area.

Recall that area is length times width, so multiplication is the necessary operation.

$$
\begin{array}{r}
27.5 \text{ cm} \\
\times\, 21.5 \text{ cm} \\
\hline
137\,5 \\
275 \\
550 \\
\hline
591.25 \text{ cm}^2
\end{array}
$$

The area of the paper is 591.25 cm^2.

● ● ● **CHECK YOURSELF 5**

If 1 kilogram (kg) is 2.2 pounds (lb), how many pounds equal 5.3 kg?

● Example 6

An Application Involving the Multiplication of Two Decimals

Usually in problems dealing with money we round the result to the nearest cent (hundredth of a dollar).

Jack buys 8.7 gallons (gal) of gas at 98.9¢ per gallon. Find the cost of the gas.

Solution   We multiply the cost per gallon by the number of gallons. Then we round the result to the nearest cent.

$$
\begin{array}{r}
98.9 \\
\times\, 8.7 \\
\hline
69\,23 \\
791\,2 \\
\hline
860.43
\end{array}
$$
   The product 860.43 (cents) is rounded to 860 (cents), or $8.60.

The cost of Jack's gas will be $8.60.

● ● ● **CHECK YOURSELF 6**

One liter (L) is approximately 0.265 gal. On a trip to Europe, the Bernards purchased 88.4 L of gas for their rental car. How many gallons of gas did they purchase, to the nearest tenth of a gallon?

Sometimes we will have to use more than one operation for a solution, as Example 7 shows.

● **Example 7**

An Application Involving Two Operations

Steve purchased a television set for $299.50. He agreed to pay for the set by making payments of $27.70 for 12 months. How much extra did he pay on the installment plan?

Solution First we multiply to find the amount actually paid.

$$
\begin{array}{r}
\$\ 27.70 \\
\times\ \underline{\hspace{4pt}12} \\
55\ 40 \\
\underline{277\ 0\hspace{8pt}} \\
\$332.40 \longleftarrow \text{Amount paid}
\end{array}
$$

Now subtract the listed price. The difference will give the extra amount Steve paid.

$$
\begin{array}{r}
\$332.40 \\
\underline{-299.50} \\
\$\ \ 32.90 \longleftarrow \text{Extra amount}
\end{array}
$$

Steve will pay an additional $32.90 on the installment plan.

● ● ● **CHECK YOURSELF 7**

Sandy's new car had a list price of $7385. She paid $1500 down and will pay $205.35 per month for 36 months on the balance. How much extra will she pay with this loan arrangement?

● ● ● **CHECK YOURSELF ANSWERS**

1. 0.00756, 5 decimal places.   **2.** 0.5472.   **3.** 0.05616.   **4.** 144.
5. 11.66 lb.   **6.** 23.4 gal.   **7.** $1507.60.

8.5 Exercises

Multiply.

1. 2.3
    ×3.4

2. 6.5
    ×4.3

3. 8.4
    ×5.2

4. 9.2
    ×4.6

5. 2.56
    × 72

6. 56.7
    × 35

7. 0.78
    × 2.3

8. 9.5
    ×0.45

9. 15.7
    ×2.35

10. 28.3
    ×0.59

11. 0.354
    × 0.8

12. 0.624
    × 0.85

13. 3.28
    ×5.07

14. 0.582
    × 6.3

15. 5.238
    × 0.48

16. 0.372
    ×   58

17. 1.053
    ×0.552

18. 2.375
    × 0.28

19. 0.0056
    × 0.082

20. 1.008
    ×0.046

21. 0.8×2.376

22. 3.52×58

23. 0.3085×4.5

24. 0.028×0.685

#	Answer
1.	7.82
2.	27.95
3.	43.68
4.	42.32
5.	184.32
6.	1984.5
7.	1.794
8.	4.275
9.	36.895
10.	16.697
11.	0.2832
12.	0.5304
13.	16.6296
14.	3.6666
15.	2.51424
16.	21.576
17.	0.581256
18.	0.665
19.	0.0004592
20.	0.046368
21.	1.9008
22.	204.16
23.	1.38825
24.	0.01918

25. $39.92

26. $1483.80

27. 20.85 lb

28. $252.45

29. $142.50

30. $18.85

31. $1,266.30

32. $337.60

33. 604.8 cm²

34. $261.30

35. 13.46 cm

36. 18.6 gal

37. $120.82

Solve the following applications.

25. Total cost. Kurt bought four shirts on sale for $9.98 each. What was the total cost of the purchase?

26. Total payments. Juan makes monthly payments of $123.65 on his car. What will he pay in 1 year?

27. Weight. If 1 gallon (gal) of water weighs 8.34 pounds (lb), how much will 2.5 gal weigh?

28. Salary. Tony worked 37.4 hours (h) in 1 week. If his hourly rate of pay is $6.75, what was his pay for the week?

29. Interest. To find the amount of interest on a loan at $9\frac{1}{2}$ percent, we have to multiply the amount of the loan by 0.095. Find the interest on a $1500 loan for 1 year.

30. Cost. A beef roast weighing 5.8 lb costs $3.25/lb. What is the cost of the roast?

31. State tax. Tom's state income tax is found by multiplying his income by 0.054. If Tom's income is $23,450, find the amount of his tax.

32. Salary. Claudia earns $6.40 per hour (h). For overtime (each hour over 40 h) she earns $9.60. If she works 48.5 h in a week, what pay should she receive?

33. Area. A sheet of typing paper has dimensions of 21.6 by 28 centimeters (cm). What is its area?

34. Car rental. A rental car costs $24 per day plus 18 cents per mile (mi). If you rent a car for 5 days and drive 785 mi, what will the total car rental bill be?

35. Metrics. One inch (in.)is approximately 2.54 centimeters (cm). How many centimeters does 5.3 in. equal? Give your answer to the nearest hundredth of a centimeter.

36. Fuel consumption. A light plane uses 5.8 gal/h of fuel. How much fuel is used on a flight of 3.2 h? Give your answer to the nearest tenth of a gallon.

37. Cost. The Hallstons select a carpet costing $15.49 per square yard. (yd²). If they need 7.8 yd² of carpet, what is the cost to the nearest cent?

38. Car payment. Maureen's car payment is $242.38 per month for 4 years. How much will she pay altogether?

39. Area. A classroom is 7.9 meters (m) wide and 11.2 m long. Estimate its area.

40. Cost. You buy a roast that weighs 6.2 lb and costs $3.89 per pound. Estimate the cost of the roast.

 **Getting Ready for Section 8.6
[Section 2.1]**

Round each number to the indicated place value.

a. 5378   thousand

b. 25,189   hundred

c. 219,473   ten thousand

d. 3,438,000   hundred thousand

e. 351,098   ten thousand

f. 5,298,500   hundred thousand

Answers

1. 7.82   **3.** 43.68   **5.** 184.32   **7.** 1.794   **9.** 36.895   **11.** 0.2832
13. 16.6296   **15.** 2.51424   **17.** 0.581256   **19.** 0.0004592   **21.** 1.9008
23. 1.38825   **25.** $39.92   **27.** 20.85 lb   **29.** $142.50   **31.** $1266.30
33. 604.8 cm^2 **35.** 13.46 cm   **37.** $120.82   **39.** 88 m^2 **a.** 5000
b. 25,200   **c.** 220,000   **d.** 3,400,000   **e.** 350,000   **f.** 5,300,000

© 1998 McGraw-Hill Companies

8.5 Supplementary Exercises

Name

Section   Date

Multiply.

1.   5.8
      ×3.4

2.   7.3
      × 8

3.   0.42
      × 4.8

4. 6.293

5. 37.066

6. 2.1945

7. 1.1385

8. 0.4338

9. 1.08888

10. 0.001344

11. 0.0618

12. 0.0002592

13. $17.88

14. $93.50

15. $22.35

16. $13.26

17. $66.75

18. $79.70

19. 145.4 ft^2

20. $38.53

4. 2.17
$\times$ 2.9

5. 4.31
$\times$ 8.6

6. 3.85
$\times$0.57

7. 2.53
$\times$0.45

8. 0.482
$\times$ 0.9

9. 4.537
$\times$ 0.24

10. 0.0048
$\times$ 0.28

11. 12.36×0.005

12. 0.054×0.0048

Solve the following applications.

13. Cost of pens. Alan bought 12 pens on sale at $1.49 each. How much did he pay for the 12 pens?

14. Mileage expenses. A salesperson is allowed 22 cents per mile (mi) by her company as a business expense. If she drove 425 mi during a week, how much should she claim for mileage expense for that week?

15. Cost of fabric. How much will 7.5 yards (yd) of fabric cost if the cost per yard is $2.98?

16. Cost of gasoline. What amount will you pay for 13.6 gallons (gal) of gasoline if the cost per gallon is 97.5 cents?

17. Rental car. The cost of a rental car is $28.95 per day plus 27 cents per mile. What will it cost you to rent a car for a single day if you drive 140 mi?

18. Interest. John purchases a television set and agrees to make monthly payments of $39.60 for 12 months. If the list price of the set was $395.50, how much extra is John paying on this installment plan?

19. Area. A room is 15.3 feet (ft) long by 9.5 ft wide. What is its area to the nearest tenth of a square foot?

20. Taxes. To find the amount deducted from your paycheck for Social Security, multiply your salary by 0.076. If your weekly pay is $507, what amount to the nearest cent will be deducted for Social Securityt?

Using Your Calculator to Multiply Decimals

The steps for finding the product of decimals on a calculator are similar to the ones we used for multiplying whole numbers.

• Example 1

Multiplying Two Decimals

To multiply 34.2×1.387, enter

34.2 $\boxed{\times}$ 1.387 $\boxed{=}$

Display $\boxed{47.4354}$

● ● ● CHECK YOURSELF 1

Multiply.

92.7×2.36

To find the product of a group of decimals, just extend the process.

• Example 2

Multiplying a Group of Decimals

To multiply $2.8 \times 3.45 \times 3.725$, enter

2.8 $\boxed{\times}$ 3.45 $\boxed{\times}$ 3.725 $\boxed{=}$

Display $\boxed{35.9835}$

● ● ● CHECK YOURSELF 2

Multiply $3.1 \times 5.72 \times 6.475$.

You can also easily find powers of decimals with your calculator by using a procedure similar to that in Example 2.

• Example 3

Finding the Power of a Decimal Number

Remember: $(2.35)^3 =$ 2.35 × 2.35 × 2.35

Find $(2.35)^3$.

Solution Enter

2.35 ☒ 2.35 ☒ 2.35 ☐

Display ☐ 12.977875

Some calculators have keys that will find powers more quickly. Look for keys marked $\boxed{x^2}$ or $\boxed{y^x}$. These were discussed in Chapter 2. Other calculators have a power key marked $\boxed{\wedge}$.

CHECK YOURSELF 3

Multiply.

$6.2 \times 6.2 \times 6.2 \times 6.2$

• Example 4

Finding the Power of a Decimal Number Using Power Keys

Find $(2.35)^3$.

Solution Enter

2.35 $\boxed{\wedge}$ 3   or   2.3 $\boxed{y^x}$ 3

CHECK YOURSELF 4

Find $(6.2)^4$.

CHECK YOURSELF ANSWERS

1. 218.772.   **2.** 114.8147.   **3.** 1477.6336.   **4.** 1477.6336.

Calculator Exercises

A N S W E R S

1.	0.59
2.	0.1067
3.	293.975
4.	755.811
5.	111.38292
6.	196.66995
7.	7.0225
8.	0.000512
9.	61.629875
10.	0.271441
11.	8.8125 in^2
12.	$224.64
13.	$131.60
14.	$182.25

Compute.

1. 0.08×7.375

2. 21.34×0.005

3. 21.38×13.75

4. 58.05×13.02

5. $127.85 \times 0.055 \times 15.84$

6. $18.28 \times 143.45 \times 0.075$

7. $(2.65)^2$

8. $(0.08)^3$

9. $(3.95)^3$

10. $(0.521)^2$

11. Find the area of a rectangle with length 3.75 in and width 2.35 in.

12. Mark works 38.4 h in a given week. If his hourly rate of pay is $5.85, what will he be paid for the week?

13. If fuel oil costs 87.5¢ per gallon, what will 150.4 gal cost?

14. To find the interest on a loan for 1 year at 12.5 percent, multiply the amount of the loan by 0.125. What interest will you pay on a loan of $1458 at 12.5 percent for 1 year?

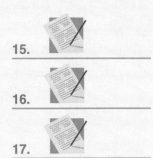
15. You are the office manager for Dr. Rogers. The increasing costs of making photocopies is a concern to Dr. Rogers. She wants to examine alternatives to the current financing plan. The office currently leases a copy machine for $110 per month and $0.025 per copy. A 3-year payment plan is available that costs $125 per month and $0.015 per copy.

   (a) If the office expects to run 100,000 copies per year, which is the better plan?

   (b) How much monay will the better plan save over the other plan?

16. In a bottling company, a machine can fill a 2-liter (L) bottle in 0.5 seconds (s) and move the next bottle into place in 0.1 s. How many 2-L bottles can be filled by the machine in 2 hours?

17. The owner of a bakery sells a finished cake for $8.99. The cost of baking 16 cakes is $75.63. Write a plan to find out how much profit the baker can make on each cake sold.

ANSWERS

1. 0.59   **3.** 293.975   **5.** 111.38292   **7.** 7.0225   **9.** 61.629875

11. 8.8125 in^2 **13.** $131.60   **15.**   **17.**

8.6 Finding the Circumference and Area of a Circle

8.6 OBJECTIVES

1. Use π to find the circumference of a circle.
2. Use π to find the area of a circle.

In Section 8.3, we again looked at the perimeter of a straight-edged figure. The distance around the outside of a circle is closely related to this concept of perimeter. We call the perimeter of a circle the **circumference.**

Circumference of a Circle

The *circumference* of a circle is the distance around that circle.

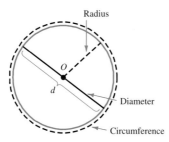

Figure 1

Let's begin by defining some terms. In the circle of Figure 1, *d* represents the **diameter.** This is the distance across the circle through its center (labeled with the letter *O*, for **origin**). The **radius** *r* is the distance from the center to a point on the circle. The diameter is always twice the radius.

It was discovered long ago that the ratio of the circumference of a circle to its diameter always stays the same. The ratio has a special name. It is named by the Greek letter π (pi). Pi is approximately $\frac{22}{7}$, or 3.14 rounded to two decimal places. We can write the following formula.

The formula comes from the ratio

$$\frac{C}{d} = \pi$$

Formula for the Circumference of a Circle

$$C = \pi d \tag{1}$$

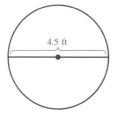

Figure 2

●Example 1

Finding the Circumference of a Circle

A circle has a diameter of 4.5 ft, as shown in Figure 2. Find its circumference, using 3.14 for π. If your calculator has a π key, use that key instead of a decimal approximation for π.

Since 3.14 is an approximation for pi, we can only say that the circumference is approximately 14.1 ft. The symbol ≈ means approximately.

Solution By Formula (1),

$C = \pi d$

$\approx 3.14 \times 4.5$ ft

≈ 14.1 ft   (rounded to one decimal place)

● ● ● **CHECK YOURSELF 1**

A circle has a diameter of $3\frac{1}{2}$ inches (in.). Find its circumference.

If you want to approximate π, you needn't worry about running out of decimal places. The value for pi has been calculated to over 100,000,000 decimal places on a computer (the printout was some 20,000 pages long).

Note: In finding the circumference of a circle, you can use whichever approximation for pi you choose. If you are using a calculator and want more accuracy, use the $\boxed{\pi}$ key.

There is another useful formula for the circumference of a circle.

Since $d = 2r$ (the diameter is twice the radius) and $C = \pi d$, we have $C = \pi(2r)$, or $C = 2\pi r$.

Formula for the Circumference of a Circle

$C = 2\pi r$ (2)

● **Example 2**

Finding the Circumference of a Circle

A circle has a radius of 8 in., as shown in Figure 3. Find its circumference using 3.14 for π.

8 in.

Figure 3

Solution From Formula (2),

$C = 2\pi r$

$\approx 2 \times 3.14 \times 8$ in.

≈ 50.2 in.   (rounded to one decimal place)

● ● ● **CHECK YOURSELF 2**

Find the circumference of a circle with a radius of 2.5 in.

Sometimes we will want to combine the ideas of perimeter and circumference to solve a problem.

• Example 3

Finding Perimeter

We wish to build a wrought-iron frame gate according to the diagram in Figure 4. How many feet (ft) of material will be needed?

The distance around the semicircle is $\frac{1}{2}\pi d$.

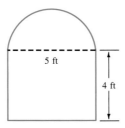

Figure 4

Solution The problem can be broken into two parts. The upper part of the frame is a semicircle (half a circle). The remaining part of the frame is just three sides of a rectangle.

Using a calculator with a $\boxed{\pi}$ key,

$1 \boxed{\div} 2 \boxed{\times} \boxed{\pi} \boxed{\times} 5$

Circumference (upper part) $\approx \dfrac{1}{2} \times 3.14 \times 5$ ft ≈ 7.9 ft

Perimeter (lower part) $= 4 + 5 + 4 = 13$ ft

Adding, we have

$7.9 + 13 = 20.9$ ft

We will need approximately 20.9 ft of material.

● ● ● **CHECK YOURSELF 3**

Find the perimeter of the following figure.

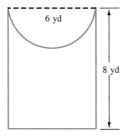

The number pi (π), which we used to find circumference, is also used in finding the area of a circle. If r is the radius of a circle, we have the following formula.

Formula (3) for the Area of a Circle

$A = \pi r^2$ (3)

This is read, "Area equals pi r squared." You can multiply the radius by itself and then by pi.

• Example 4

Find the Area of a Circle

A circle has a radius of 7 inches (in.) (see Figure 5). What is its area?

Figure 5

Solution   Use Formula (3), using 3.14 for π and $r = 7$ in.

$A \approx 3.14 \times (7 \text{ in.})^2$ Again the area is an approximation because we use 3.14, an approximation for π.

$\approx 153.86 \text{ in.}^2$

●●● **CHECK YOURSELF 4**

Find the area of a circle whose diameter is 4.8 centimeters (cm). Remember that the formula refers to the radius. Use 3.14 for π, and round your result to the nearest tenth of a square inch.

●●● **CHECK YOURSELF ANSWERS**

1. $C \approx 11$ in.   **2.** $C \approx 15.7$ in.   **3.** 31.42 yd.   **4.** ≈ 18.1 cm^2.

Find the perimeter or circumference of each figure. Use 3.14 for π, and round your answer to one decimal place.

1.

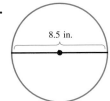

9 ft

2.

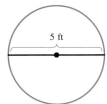

5 ft

3.

8.5 in.

4.

3.75 ft

In Exercises 5 and 6, use $\dfrac{22}{7}$ for π.

5.

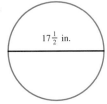

$17\frac{1}{2}$ in.

6.

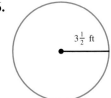

$3\frac{1}{2}$ ft

Find the perimeter of each figure. (The curves are semicircles.) Round answers to one decimal place.

7.

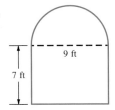

9 ft

7 ft

8.

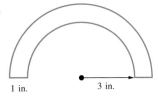

1 in.   3 in.

9.

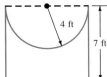

4 ft

7 ft

10.

10 in.

1.	56.5 ft
2.	15.7 ft
3.	26.7 in.
4.	23.6 ft
5.	55 in.
6.	22 ft
7.	37.1 ft
8.	24 in.
9.	34.6 ft
10.	35.7 in.

11. 153.9 in.^{2}

12. 113 ft^2

13. 38.5 yd^2

14. 201 ft^2

15. $9\dfrac{5}{8}$ yd^2

16. $7\dfrac{1}{14}$ in.^{2}

17. 9420 yd

18. $9.42

19. Yes; area = 2461.8 ft^2

20. $58.88

21. $150.72

Find the area of each figure. Use 3.14 for π, and round your answer to one decimal place.

11.

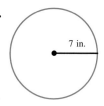
7 in.

12.

12 ft

13.

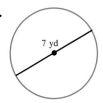
7 yd

14.

8 ft

In Exercises 19 and 20, use $\dfrac{22}{7}$ for π.

15.

$3\frac{1}{2}$ yd

16.

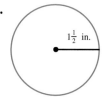
$1\frac{1}{2}$ in.

Solve the following applications.

17. Jogging. A path runs around a circular lake with a diameter of 1000 yards (yd). Robert jogs around the lake three times for his morning run. How far has he run?

18. Binding. A circular rug is 6 feet (ft) in diameter. Binding for the edge costs $1.50 per yard. What will it cost to bind around the rug?

19. Lawn care. A circular piece of lawn has a radius of 28 ft. You have a bag of fertilizer that will cover 2500 ft^2 of lawn. Do you have enough?

20. Cost. A circular coffee table has a diameter of 5 ft. What will it cost to have the top refinished if the company charges $3 per square foot for the refinishing?

21. Cost. A circular terrace has a radius of 6 ft. If it costs $12 per square yard to pave the terrace with brick, what will the total cost be?

22. Area. A house addition is in the shape of a semicircle (a half circle) with a radius of 9 ft. What is its area?

Find the area of the shaded part in each figure.

23.

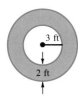

3 ft

2 ft

24.

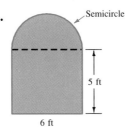

Semicircle

5 ft

6 ft

25.

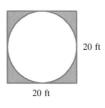

20 ft

20 ft

26.

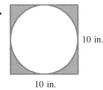

10 in.

10 in.

Getting Ready for Section 8.7
[Section 2.6]

Evaluate.

a. 5^2

b. 7^2

c. 3×4^2

d. 3×6^2

e. $(2.4)^2$

f. $3 \times (5.2)^2$

g. $\left(\dfrac{3}{4}\right)^2$

h. $3 \times \left(\dfrac{5}{2}\right)^2$

Answers

1. 56.5 ft   **3.** 26.7 in.   **5.** 55 in.   **7.** 37.1 ft   **9.** 34.6 ft

11. 153.9 in.^{2} **13.** 38.5 yd^2 **15.** $9\dfrac{5}{8}$ yd   **17.** 9420 yd

19. yes; area = 2461.8 ft^2 **21.** $150.72   **23.** 50.2 ft^2 **25.** 86 ft^2

a. 25   **b.** 49   **c.** 48   **d.** 108   **e.** 5.76   **f.** 81.12   **g.** $\dfrac{9}{16}$

h. $\dfrac{75}{4}$

ANSWERS

22. 127.2 ft^2

23. 50.2 ft^2

24. 44.1 ft^2

25. 86 ft^2

26. 21.5 in^2

a. 25

b. 49

c. 48

d. 108

e. 5.76

f. 81.12

g. $\dfrac{9}{16}$

h. $\dfrac{75}{4}$

Name

Section   Date

Find the perimeter or circumference of each figure. Use 3.14 for π, and round your answer to one decimal place.

1.

2.5 ft

2.

4 yd

In Exercises 3–5, use the π key on your calculator, or use 3.14 for π.

3.
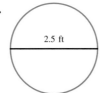
28 in.

4.

4.2 in.

5.
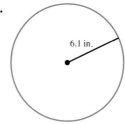
6.1 in.

Solve the following applications.

6. **Distance.** If a wheel has a diameter of 22 inches (in.), how far will it travel in four complete revolutions?

7. **Amount of fencing.** A wire fence is to be placed around each of 12 new shrubs for protection. If each fence forms a circle with a diameter of 3 feet (ft), how much fencing will be needed?

8. **Distance.** How far is one lap around the track shown below?

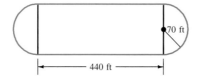
70 ft
440 ft

9. **Area.** Which of these two cake pans is larger: a round pan with a 9-in. diameter or a square pan with 8-in. sides? What is the difference in their areas?

Area and Unit Conversion

8.7 OBJECTIVES

1. Find the area of a parallelogram.
2. Find the area of a triangle.
3. Convert square units.

Two other figures that are frequently encountered are parallalograms and triangles.

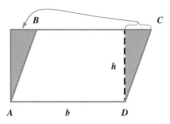

Figure 1

In Figure 1, *ABCD* is called a **parallelogram.** Its opposite sides are parallel and equal. Let's draw a line from *D* that forms a right angle with side *BC*. This cuts off one corner of the parallelogram. Now imagine that we move that corner over to the left side of the figure, as shown. This gives us a rectangle instead of a parallelogram. Since we haven't changed the area of the figure by moving the corner, the parallelogram has the same area as the rectangle, the product of the base and the height.

Formula for the Area of a Parallelogram

$$A = b \cdot h \tag{1}$$

• Example 1

Finding the Area of a Parallelogram

A parallelogram has the dimensions shown in Figure 2. What is its area?

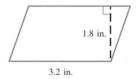

Figure 2

Solution   Use Formula (1), with $b = 3.2$ in. and $h = 1.8$ in.

$$A = b \cdot h$$
$$= 3.2 \text{ in.} \times 1.8 \text{ in.} = 5.76 \text{ in.}^2$$

● ● ● ● **CHECK YOURSELF 1**

If the base of a parallelogram is $3\frac{1}{2}$ in. and its height is $1\frac{1}{2}$ in., what is its area?

Another common geometric figure is the **triangle.** It has three sides. An example is triangle *ABC* in Figure 3.

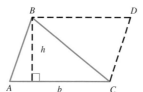

b is the base of the triangle.
h is the height, or the
altitude, of the triangle.

Figure 3

Once we have a formula for the area of a parallelogram, it is not hard to find the area of a triangle. If we draw the dotted lines from *B* to *D* and from *C* to *D* parallel to the sides of the triangle, we form a parallelogram. The area of the triangle is then one-half the area of the parallelogram [which is $b \cdot h$ by Formula (1)].

Formula for the Area of a Triangle

$$A = \frac{1}{2} \cdot b \cdot h \tag{2}$$

●Example 2

Finding the Area of a Triangle

A triangle has an altitude of 2.3 in., and its base is 3.4 in. (see Figure 4). What is its area?

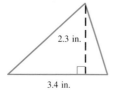

2.3 in.

3.4 in.

Figure 4

Solution   Use Formula (2), with $b = 3.4$ in. and $h = 2.3$ in.

$$A = \frac{1}{2} \cdot b \cdot h$$

$$= \frac{1}{2} \times 3.4 \text{ in.} \times 2.3 \text{ in.} = 3.91 \text{ in.}^2$$

● ● ● CHECK YOURSELF 2

A triangle has a base of 10 feet (ft) and an altitude of 6 ft. Find its area.

___

Sometimes we will want to convert from one square unit to another. For instance, look at 1 yd^2 in Figure 5.

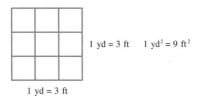

1 yd = 3 ft   1 yd^2 = 9 ft^2

1 yd = 3 ft

Figure 5

The table below gives some useful relationships.

Originally the acre was the area that could be plowed by a team of oxen in a day!

Square Units and Equivalents		
1 square foot (ft^2)	= 144 square inches (in.^{2})	
1 square yard (yd^2)	= 9 ft^2	
1 acre	= 4840 yd^2 = 43,560 ft^2	

● Example 3

Converting Between Feet and Yards in Finding Area

A room has the dimensions 12 by 15 ft. How many square yards of linoleum will be needed to cover the floor?

Solution

We first find the area in square feet, then convert to square yards.

$$A = 12 \text{ ft} \times 15 \text{ ft} = 180 \text{ ft}^2$$
$$= 180(1 \text{ ft}^2)$$
$$= \overset{20}{\cancel{180}} \frac{1}{\underset{1}{\cancel{9}}} \text{ yd}^2 \qquad \text{Replace 1 ft}^2 \text{ with } \frac{1}{9} \text{ yd}^2.$$
$$= 20 \text{ yd}^2$$

You can also make the conversion by multiplying by the unit ratio $\dfrac{1 \text{ yd}^2}{9 \text{ ft}^2}$, which is just 1.

● ● ● CHECK YOURSELF 3

A hallway is 27 ft long and 4 ft wide. How many square yards of carpeting will be needed to carpet the hallway?

___

Example 4 illustrates the use of a common unit of area, the acre.

•Example 4

Converting Between Yards and Acres in Finding Area

A rectangular field is 220 yd long and 110 yd wide. Find its area in acres.

Solution

$A = 220 \text{ yd} \times 110 \text{ yd} = 24{,}200 \text{ yd}^2$

$= 24{,}200(1 \text{ yd}^2)$ Replace 1 yd² with $\dfrac{1}{4840}$ acre.

$= \overset{5}{24{,}200} \dfrac{1}{\underset{1}{4840}} \text{ acre}$

$= 5 \text{ acres}$

● ● ●   **CHECK YOURSELF 4**

A proposed site for an elementary school is 220 yd long and 198 yd wide. Find its area in acres.

● ● ●   **CHECK YOURSELF ANSWERS**

1. $A = 3\dfrac{1}{2} \text{ in.} \times 1\dfrac{1}{2} \text{ in.}$ **2.** $A = \dfrac{1}{2} \times 10 \text{ ft} \times 6 \text{ ft}$ **3.** 12 yd².   **4.** 9 acres.

$= \dfrac{7}{2} \text{ in.} \times \dfrac{3}{2} \text{ in.}$ $= 30 \text{ ft}^2.$

$= 5\dfrac{1}{4} \text{ in.}^2$

Find the area of each figure.

1.

4 ft

7 ft

2.
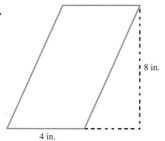
8 in.

4 in.

3.
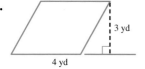
3 yd

4 yd

4.
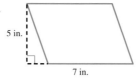
5 in.

7 in.

5.
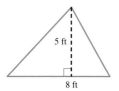
5 ft

8 ft

6.

6 ft

11 ft

7.
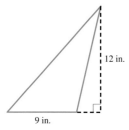
12 in.

9 in.

8.
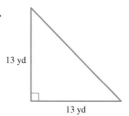
13 yd

13 yd

9.

2 ft

4 ft

5 ft

10.
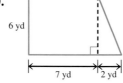
6 yd

7 yd 2 yd

ANSWERS

1. 28 ft²

2. 32 in.²

3. 12 yd²

4. 35 in.²

5. 20 ft²

6. 33 ft²

7. 54 in.²

8. 84.5 yd²

9. 24 ft²

10. 48 yd²

Solve the following applications.

11. Amount of material. A Tetra-Kite uses 12 pieces of plastic for its surface. Those pieces are triangular, with base 12 inches (in.) and height 12 in. How much material is needed for the kite?

12. Acreage. You buy a lot which is 110 yd². What is its size in acres?

13. Area. You are making posters 12 by 15 in. How many square feet of material will you need for four posters?

14. Cost. Andy is carpeting a recreation room 18 feet (ft) long and 12 ft wide. If the carpeting costs $15/yd², what will be the total cost of the carpet?

15. Acreage. A shopping center is rectangular, with dimensions of 550 by 440 yd. What is its size in acres?

16. Cost. An A-frame cabin has a triangular front with a base of 30 ft and a height of 20 ft. If the front is to be glass that costs $3 a square foot, what will the glass cost?

17. Papa Doc's delivers pizza. The 8-inch (in.)-diameter pizza is $8.99, and the price of a 16-in.-diameter pizza is $17.98. Write a plan to determine which is the better buy.

18. The distance from Philadelphia to Sea Isle City is 100 miles (mi). A car was driven this distance using tires with a radius of 14 in. How many revolutions of each tire occurred on the trip?

19. Find the area of the circumference (or perimeter) of each of the following:
(a) a penny **(b)** a nickel **(c)** a dime **(d)** a quarter **(e)** a half-dollar
(f) a silver dollar **(g)** a Susan B. Anthony dollar **(h)** a dollar bill
(i) one face of the pyramid on the back of a $1 bill.

20. An indoor track layout is shown below.

How much would it cost to lay down hardwood floor if the hardwood floor costs $10.50 per square meter?

21. Another formula for determining the area of a triangle is given by

$$A = \sqrt{s(s - a)(s - b)(s - c)}$$

where $s = \dfrac{1}{2}(a + b + c)$

a, b, c = lengths of sides of triangle

Use this formula to determine the area of the triangle

Check your results using the formula $A = \dfrac{1}{2}ab$.

10 cm

6 cm

8 cm

22. What is the effect on the area of a triangle if the base is doubled and the altitude is cut in half?

 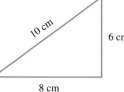
Getting Ready for Section 8.8
[Section 2.6]

Multiply.

a. 27
  ×10

b. 59
  ×10

c. 43
  ×100

d. 971
  ×100

e. 523
  ×1000

f. 498
  ×1000

Answers

1. 28 ft^2 **3.** 12 yd^2 **5.** 20 ft^2 **7.** 54 in.^{2} **9.** 24 ft^2 **11.** 864 in.^{2}

13. 5 ft^2 **15.** 50 acres   **17.**   **19.**   **21.**

a. 270   **b.** 590   **c.** 4300   **d.** 97,100   **e.** 523,000   **f.** 498,000

Name _____

Section _____ Date _____

Find the area of each figure. Round your answer to one decimal place.

1.
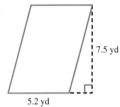
9 in.
6 in.

2.

6 in.
7 in.

3.
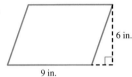
7.5 yd
5.2 yd

4.
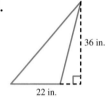
36 in.
22 in.

5.
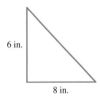
6 in.
8 in.

6.
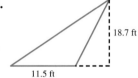
18.7 ft
11.5 ft

Solve the following applications.

7. **Cost of carpet.** You are carpeting a room that measures 13 feet (ft) by 11 ft. If the carpeting costs $11 per square yard, how much will it cost to carpet the room?

8. **Material cost.** A shopping center designs a triangular sign, 10 ft across its base and 8 ft high. If the material for the sign costs $3.50 a square foot, what will the material cost?

9. **Area.** How many square yards of carpet are needed to carpet a rectangular floor that measures 21 ft by 30 ft?

10. **Area.** How many square tiles, each 6 inches (in.) on a side, are needed to cover a floor that is 13 ft wide and 16 ft long?

Multiplying Decimals by Powers of Ten

8.8 OBJECTIVES

1. Multiply a decimal by a power of ten.
2. Use multiplication by a power of ten to solve an application.

The rule will be used to multiply by 10, 100, 1000, and so on.

The digits remain the same. Only the *position* of the decimal point is changed.

Multiplying by 10, 100, or any other larger power of 10 makes the number *larger*. Move the decimal point *to the right*.

There are enough applications involving multiplication by the powers of 10 to make it worthwhile to develop a special rule so you can do such operations quickly and easily. Look at the patterns in some of these special multiplications.

$$
\begin{array}{r}
0.679 \\
\times \quad 10 \\
\hline
6.790, \text{ or } 6.79
\end{array}
\qquad
\begin{array}{r}
23.58 \\
\times \quad 10 \\
\hline
235.80, \text{ or } 235.8
\end{array}
$$

Do you see that multiplying by 10 has moved the decimal point *one place to the right?* Now let's look at what happens when we multiply by 100.

$$
\begin{array}{r}
0.892 \\
\times \quad 100 \\
\hline
89,200, \text{ or } 89.2
\end{array}
\qquad
\begin{array}{r}
5.74 \\
\times \quad 100 \\
\hline
574.00, \text{ or } 574
\end{array}
$$

Multiplying by 100 shifts the decimal point *two places to the right*. The pattern of these examples gives us the following rule:

To Multiply by a Power of 10

Move the decimal point to the right the same number of places as there are zeros in the power of 10.

• Example 1

Multiplying by Powers of Ten

$2.356 \times 10 = 23.56$

      One   Move the decimal one
      zero   place to the right.

$34.788 \times 100 = 3478.8$

      Two   Move the decimal two
      zeros places to the right.

$3.67 \times 1000 = 3670.$

      Three Move the decimal three places to
      zeros the right. Note that we added a 0
              to place the decimal point correctly.

Remember that 10^5 is just a 1 followed by 5 zeros.

$$0.005672 \times 10^5 = 567.2$$

Five zeros

Move the decimal five places to the right

● ● ● **CHECK YOURSELF 1**

Multiply.

a. 43.875×100 **b.** 0.0083×10^3

Example 2 is just one of many applications that require multiplying by a power of 10.

●Example 2

An Application Involving Multiplication by a Power of 10

There are 1000 meters in a kilometer.

To convert from kilometers to meters, multiply by 1000. Find the number of meters (m) in 2.45 kilometers (km).

Solution

If the result is a whole number, there is no need to write the decimal point.

2.45 km = 2450. m   Just move the decimal point three places right to make the conversion. Note that we added a zero to place the decimal point correctly.

● ● ● **CHECK YOURSELF 2**

To convert from kilograms to grams, multiply by 1000. Find the number of grams (g) in 5.23 kilograms (kg).

● ● ● **CHECK YOURSELF ANSWERS**

1. (a) 4387.5; **(b)** 8.3.   **2.** 5230 g.

Multiply.

1. 5.89×10

2. 0.895×100

3. 23.79×100

4. 2.41×10

5. 0.045×10

6. 5.8×100

7. 0.431×100

8. 0.025×10

9. 0.471×100

10. $0.95 \times 10,000$

11. 0.7125×1000

12. 23.42×1000

13. 4.25×10^2

14. 0.36×10^3

15. 3.45×10^4

16. 0.058×10^5

Solve the following applications.

17. Cost. A store purchases 100 items at a cost of $1.38 each. Find the total cost of the order.

18. Conversion. To convert from meters (m) to centimeters (cm), multiply by 100. How many centimeters are there in 5.3 meters?

19. Conversion. How many grams (g) are there in 2.2 kilograms (kg)? Multiply by 1000 to make the conversion.

20. Cost. An office purchases 1000 pens at a cost of 17.8 cents each. What is the cost of the purchase in dollars?

1. 58.9

2. 89.5

3. 2379

4. 24.1

5. 0.45

6. 580

7. 43.1

8. 0.25

9. 47.1

10. 9500

11. 712.5

12. 23,420

13. 425

14. 360

15. 34,500

16. 5800

17. $138

18. 530 cm

19. 2200 g

20. $178.00

Answers

1. 58.9   **3.** 2379   **5.** 0.45   **7.** 43.1   **9.** 47.1   **11.** 712.5

13. 425   **15.** 34,500   **17.** $138   **19.** 2200 g

A N S W E R S

1. 48.2

2. 357

3. 78.5

4. 1485

5. 45,710

6. 57.3

7. 24,700

8. 15,300

9. $2439

10. 5800 m

Multiply.

1. 4.82×10

2. 3.57×100

3. 0.785×100

4. 14.85×100

5. 45.71×1000

6. 0.0573×1000

7. 0.247×10^5

8. 1.53×10^4

Solve the following applications.

9. Cost. A builder bought 100 lighting fixtures from a supplier at $24.39 each. What was the total cost of the order?

10. Conversion. How many meters (m) are in 5.8 kilometers (km)? Multiply by 1000 to make the conversion.

Using Your Calculator to Multiply by Powers of Ten

How many places can your calculator display? Most calculators can display either 8, 9, or 10 digits. To find the display capability of your calculator, just enter digits until the calculator can accept no more numbers. For example, try entering

1 $\boxed{-}$ 0.226592266

Does your calculator display 10 digits? Now turn the calculator upside down. What does it say?

What happens when your calculator wants to display an answer that is too big to fit in the display? Let's try an experiment to see. Enter

10 $\boxed{\times}$ 10 $\boxed{=}$.

Now continue to multiply this answer by 10. Many calculators will let you do this by simply pressing $\boxed{=}$. Others require you to "$\boxed{\times}$ 10" for each calculation. Multiply by 10 until the display is no longer a 1 followed by a series of zeros. The new display represents the power of 10 of the answer. It will be displayed as either

Display $\boxed{1^{\,10}}$

(which looks like 1 to the tenth power, but means 1 times 10^{10}) or

Display $\boxed{1 \text{ E } 10}$

(which also means 1 times 10^{10}).

Answers that are displayed in this way are said to be in **scientific notation.** This is a topic that you will study in your next math course. In this text we will avoid exercises with answers that are too large to display in the decimal notation that you already know. If you do get such an answer, you should go back and check your work. Do not be afraid to try experimenting with your calculator. It is amazing how much math you can (accidently) learn while playing!

•Example 1

Multiplying by a Power of Ten Using the Power Key on a Calculator

Find the product 3.485×10^4.

Use your calculator to enter

3.485 $\boxed{\times}$ 10 $\boxed{y^x}$ $\boxed{=}$ or   3.485 $\boxed{\times}$ 10 $\boxed{\wedge}$ 4 $\boxed{=}$

The result will be 34850. Note that the decimal point has moved four places (the power of 10) to the right.

● ● ● CHECK YOURSELF 1

Find the product 8.755×10^6.

● ● ● ● **CHECK YOURSELF ANSWER**

1. 8,755,000

Calculator Exercises

Name

Section   Date

A N S W E R S

1. 3365

2. 4128

3. 431,600

4. 8,163,000

5. 723,600,000

6. 52,340,000

7. 3,213,600

8. 412,340

9. 317,890

10. 61,356

Find the following products using your calculator.

1. 3.365×10^3 **2.** 4.128×10^3

3. 4.316×10^5 **4.** 8.163×10^6

5. 7.236×10^8 **6.** 5.234×10^7

7. 32.136×10^5 **8.** 41.234×10^4

9. 31.789×10^4 **10.** 61.356×10^3

Answers

1. 3365   **3.** 431,600   **5.** 723,600,000   **7.** 3,213,600   **9.** 317,890

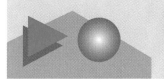

Self-Test
for Chapter 8

Name

Section   Date

The purpose of the Self-Test is to help you check your progress and review for a chapter test in class. Allow yourself about 1 hour to take the test. When you are done, check your answers in the back of the book. If you missed any answers, be sure to go back and review the appropriate sections in the chapter and do the supplementary exercises provided there.

[8.1]   **1.** Write $\dfrac{431}{1000}$ in decimal form.   **2.** Write $5\dfrac{13}{100}$ in decimal form.

3. Write 0.431 in words.   **4.** Write 5.13 in words.

In Exercises 5 and 6, complete the statements, using the symbol < or >.

5. 5.93 _____ 5.928   **6.** 2.149 _____ 2.15

[8.2]   In Exercises 7 and 8, round to the indicated place.

7. 2.571   tenth   **8.** 23.3448   two decimal places

[8.3]   In Exercises 9 to 12, add.

9.   1.238   **10.**   2.581
        +0.97   0.24
                                  +0.7

11. 31.7, 6, 2.81, and 0.254

12. Three and four tenths, four hundred five thousandths, and seven

In Exercises 13 and 14, solve the applications.

13. Gas mileage. The Wongs purchased 7, 12.7, 10, 11.3, and 9.8 gallons (gal) of gas on a vacation trip. How much gas did they buy on the trip?

	ANSWERS
1.	0.431
2.	5.13
3.	Four hundred thirty-one thousandths
4.	Five and thirteen hundredths
5.	>
6.	<
7.	2.6
8.	23.34
9.	2.208
10.	3.521
11.	40.764
12.	10.805
13.	50.8 gal

14. Perimeter. Find the perimeter (the distance around) of the figure below.

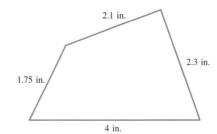

[8.4]   In Exercises 15 to 17, subtract.

15.   5.47   **16.**   8.75   **17.** Subtract 5.485
        −2.89   −3.875   from 12

In Exercises 18 to 20, solve the applications.

18. Distance traveled. Wally's car odometer read 8534.8 miles (mi) at the beginning of a trip. If it read 9472.1 mi upon his return, how far did he drive?

19. Earnings. Marion earns $360.40 per week. If $27.38 is deducted for Social Security taxes and $53.45 for federal tax withholding, what is her take-home pay?

20. Find dimension a in the following drawing.

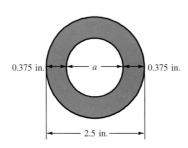

[8.5]   In Exercises 21 to 23, multiply.

21.   5.8   **22.**   2.71   **23.** 0.235 by 0.04
        ×3.7   ×0.58

In Exercises 24 and 25, solve the applications.

24. Area. Find the area of a rectangle that has dimensions 8.7 by 4.3 cm.

25. Amount paid. A washer-dryer combination has an advertised price of $535.90. You buy the set and agree to make payments of $20.50 for 36 months. How much extra are you paying for this installment plan?

26. Cost of items. A store buys 1000 items costing $0.47 per item. What is the total cost of the order?

[8.6] Find the perimeter of each figure. (The curves are semicircles.) Round answers to one decimal place.

27.

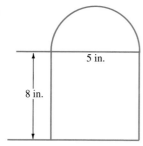

5 in.

8 in.

28.

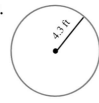

4.3 ft

[8.7] Find the area of the following figures.

29.

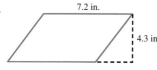

7.2 in.

4.3 in.

30.

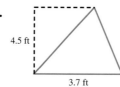

4.5 ft

3.7 ft

In Exercises 31 and 32, multiply.

31. 0.054×100

32. 8.432×10^4

© 1998 McGraw-Hill Companies

A N S W E R S

25.	$202.10
26.	$470
27.	41.7 in.
28.	27.0 ft
29.	139.32 in.^{2}
30.	8.325 ft^2
31.	5,4
32.	84,320

Special Group Activity. 1998. Using the digits 1, 9, 9, and 8 along with any mathematical operation (addition, subtraction, multiplication, division, exponentiation), how many different numbers between 0 and 25 can you create? Use the the chart below to do your calculations. Award yourself 2 points for each one you create using 1998 in order and 1 point for each number you create by scrambling the order of those digits. Two of the numbers have been created to get you started. Feel free to claim those points for yourself!

	1998 in Order	1998 in a Different Order	Total Points
0	$1^9 - (9 - 8) = 0$	$9 \times 9 - 81 = 0$	3
1			
2			
3			
4			
5			
6			
7			
8			
9	$(1 + 9) - 9 + 8 = 9$	$(9 \times 8) \div (9 \times 1) = 9$	3
10			
11			
12			
13			
14			
15			
16			
17			
18			
19			
20			
21			
22			
23			
24			
25			

Total Points _____

CHAPTER 9

DECIMALS, FRACTIONS, AND SQUARE ROOTS

INTRODUCTION

Bookkeeping and accounting used to be done exclusively with pencils and ledgers. Now, because of the popularity of computers and accounting software, many people count on their computers to do their accounting.

When Jean became an accountant, she expected that she would work with the books of a single, large company. Instead, she has almost 50 clients for whom she does their accounting. Many of these clients enter the information into their computers and bring only the disks to Jean. She says that many of the errors she finds could be avoided if people had better estimating skills. Jean learned early in her math classes that no answer (especially one obtained by use of a calculator or computer) should be accepted without checking to see if it is reasonable.

Copyright © '90 by Blair Seitz

Pretest for Chapter 9

Name _____

Section _____ Date _____

Division of Decimals

ANSWERS

1. 4.25

2. 1.67

3. $5.68

4. **(a)** 3.42;
   (b) 2.435

5. $7.60

6. 0.0534

7. **(a)** 0.375;
   (b) 0.29

8. $\dfrac{39}{50}$

9. $8^2 + 15^2 = 289$
   and $17^2 = 289$

10. 10

This pretest will point out any difficulties you may be having in divising decimals. Do all the problems. Then check your answers with those in the back of the book.

1. Divide $57\overline{)242.25}$.

2. Divide $108.59 \div 65$. Give the quotient to the nearest hundredth.

3. Cost. A shipment of 86 items cost $488.48. What is the cost per item?

4. Divide each of the following.

(a) $0.9576 \div 0.28$
(b) $1.6\overline{)3.896}$ (to the nearest thousandth)

5. Salary. Manny worked 27.5 h in a week and earned $209. What was his hourly rate of pay?

6. Divide $53.4 \div 1000$.

7. Find the decimal equivalent of each of the following.

(a) $\dfrac{3}{8}$
(b) $\dfrac{7}{24}$ (to the nearest hundredth)

8. Write 0.78 as a common fraction.

9. Show that the numbers 8, 15, and 17 form a perfect triple.

10. Use the Pythagorean theorem to find the length of the hypotenuse for $\triangle ABC$.

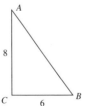

© 1998 McGraw-Hill Companies

496

9.1 Dividing Decimals by Whole Numbers

9.1 OBJECTIVES

1. Divide a decimal by a whole number.
2. Apply division to the solution of an application.

The division of decimals is very similar to our earlier work with dividing whole numbers. The only difference is in learning to place the decimal point in the quotient. Let's start with the case of dividing a decimal by a whole number. Here, placing the decimal point is easy. You can apply the following rule.

To Divide a Decimal by a Whole Number

STEP 1   Place the decimal point in the quotient *directly above* the decimal point of the dividend.

STEP 2   Divide as you would with whole numbers.

• Example 1

Dividing a Decimal by a Whole Number

Divide 29.21 by 23.

Do the division just as if you were dealing with whole numbers. Just remember to place the decimal point in the quotient *directly above* the one in the dividend.

$$
\begin{array}{r}
1.27 \\
23\overline{)29.21} \\
\underline{23} \\
6\,2 \\
\underline{4\,6} \\
1\,61 \\
\underline{1\,61} \\
0
\end{array}
$$

The quotient is 1.27.

● ● ●   **CHECK YOURSELF 1**

Divide 80.24 by 34.

Let's look at another example of dividing a decimal by a whole number.

• Example 2

Dividing a Decimal by a Whole Number

Divide 122.2 by 52.

Again place the decimal point of the quotient above that of the dividend.

```
        2.3
  52)122.2
     104
      18 2
      15 6
       2 6
```

We normally do not use a remainder when dealing with decimals. Add a 0 to the dividend and continue.

Remember that adding a 0 does not change the value of the dividend. It simply allows us to complete the division process in this case.

```
        2.35
  52)122.20  ←— Add a 0.
     104
      18 2
      15 6
       2 60
       2 60
          0
```

So $122.2 \div 52 = 2.35$. The quotient is 2.35.

●●● **CHECK YOURSELF 2**

Divide 234.6 by 68.

Often you will be asked to give a quotient to a certain place value. In this case, continue the division process to *one digit past* the indicated place value. Then round the result back to the desired accuracy.

When working with money, for instance, we normally give the quotient to the nearest hundredth of a dollar (the nearest cent). This means carrying the division out to the thousandths place and then rounding back.

• Example 3

Dividing a Decimal by a Whole Number and Rounding the Result

Find the quotient to *one place past* the desired place, and then round the result.

Find the quotient of $25.75 \div 15$ to the nearest hundredth.

```
        1.716
  15)25.750
     15
     10 7           Add a 0 to carry the division to the
     10 5           thousandths place.
        25
        15
       100
        90
        10
```

So $25.75 \div 15 = 1.72$ (to the nearest hundredth).

● ● ● **CHECK YOURSELF 3**

Find 99.26 ÷ 35 to the nearest hundredth.

As we mentioned, problems similar to the one in Example 3 often occur when working with money. Example 4 is one of the many applications of this type of division.

● Example 4

An Application Involving the Division of a Decimal by a Whole Number

A carton of 144 items costs $56.10. What is the price per item to the nearest cent?

Solution   To find the price per item, divide the total price by 144.

You might want to review the rules for rounding decimals in Section 8.2.

$$
\begin{array}{r}
0.389 \\
144\overline{)56.100} \\
\underline{43\ 2} \\
12\ 90 \\
\underline{11\ 52} \\
1\ 380 \\
\underline{1\ 296} \\
84
\end{array}
$$

Carry the division to the thousandths place and then round back.

The cost per item is rounded to $0.39, or 39¢.

● ● ● **CHECK YOURSELF 4**

An office paid $26.55 for 72 pens. What was the cost per pen to the nearest cent?

Other applications will also require that you round the result.

● Example 5

An Application Involving the Division of a Decimal by a Whole Number

Lee's grade points were 2.37 in fall term, 2.58 in winter, and 2.87 in spring. Find the mean of these grade-point scores to approximate his grade-point average for the year.

Solution   Do you remember the method for finding the mean? We add the numbers and then divide by the number of items.

$$2.37 + 2.58 + 2.87 = 7.82$$

Divide the total by 3, the number of items.

Now divide by 3:

```
    2.606
3)7.820
  6
  1 8
  1 8
    020
     18
      2
```

Now round 2.606 to 2.61.

We choose to round the average to the decimal place (hundredths) of the given grade points.

Lee's grade-point average is 2.61 (rounded to the nearest hundredth). Note that we have carried the division to the thousandths place and then rounded back.

● ● ● **CHECK YOURSELF 5**

Whitney's new car had mileages of 38.9, 41.2, and 40.7 miles per gallon (mi/gal) in three readings. What was the average mileage for the period? Give your answer to the nearest tenth of a mile per gallon.

● ● ● **CHECK YOURSELF ANSWERS**

1. 2.36.　**2.** 3.45.　**3.** 2.84.　**4.** $0.37, or 37¢.　**5.** 40.3 mi/gal.

Divide.

1. $16.68 \div 6$ **2.** $43.92 \div 8$

3. $1.92 \div 4$ **4.** $5.52 \div 6$

5. $5.48 \div 8$ **6.** $2.76 \div 8$

7. $13.89 \div 6$ **8.** $21.92 \div 5$

9. $185.6 \div 32$ **10.** $165.6 \div 36$

11. $79.9 \div 34$ **12.** $179.3 \div 55$

13. $52\overline{)13.78}$ **14.** $76\overline{)26.22}$

15. $45\overline{)144.63}$ **16.** $65\overline{)183.04}$

Divide and round the quotient to the indicated decimal place.

17. $23.8 \div 9$ tenth   **18.** $5.27 \div 8$ hundredth

19. $38.48 \div 46$ hundredth   **20.** $3.36 \div 36$ thousandth

21. $125.4 \div 52$ tenth   **22.** $2.563 \div 54$ thousandth

23. $0.927 \div 28$ thousandth   **24.** $5.8 \div 65$ hundredth

A N S W E R S

1. 2.78

2. 5.49

3. 0.48

4. 0.92

5. 0.685

6. 0.345

7. 2.315

8. 4.384

9. 5.8

10. 4.6

11. 2.35

12. 3.26

13. 0.265

14. 0.345

15. 3.214

16. 2.816

17. 2.6

18. 0.66

19. 0.84

20. 0.093

21. 2.4

22. 0.047

23. 0.033

24. 0.09

Solve the following applications.

25. Cost of CDs. Marv paid $40.41 for three CDs on sale. What was the cost per CD?

26. Contributions. Seven employees of an office donated $172.06 during a charity drive. What was the mean donation?

27. Book purchases. A shipment of 72 paperback books costs a store $190.25. What was the mean cost per book to the nearest cent?

28. Mean cost. A restaurant bought 50 glasses at a cost of $39.90. What was the cost per glass to the nearest cent?

29. Mean cost. The cost of a case of 48 items is $28.20. What is the cost of an individual item to the nearest cent?

30. Office supplies. An office bought 18 hand-held calculators for $284. What was the cost per calculator to the nearest cent?

31. Monthly payments. Al purchased a new refrigerator that cost $736.12 with interest included. He paid $100 as a down payment and agreed to pay the remainder in 18 monthly payments. What amount will he be paying per month?

32. Monthly payments. The cost of a television set with interest is $490.64. If you make a down payment of $50 and agree to pay the balance in 12 monthly payments, what will be the amount of each monthly payment?

33. Mileage. In five readings, Lucia's gas mileage was 32.3, 31.6, 29.5, 27.3, and 33.4 miles per gallon (mi/gal). What was her mean gas mileage to the nearest tenth of a mile per gallon?

34. Pollution. Pollution index readings were 53.3, 47.8, 41.9, 55.8, 43.7, 41.7, and 52.3 for a 7-day period. What was the mean reading (to the nearest tenth) for the 7 days?

35. Grade average. Jeremy's grade points for eight semesters were 2.81, 3.05, 3.62, 2.95, 3.15, 2.79, 3.45, and 3.53. Find the mean of these grade-point scores to the nearest hundredth of a point.

36. Ratings. The ratings for a weekly television program over a 6-week period were 19.7, 15.2, 18.5, 17.8, 16.3, and 18.6. What was the mean rating per show to the nearest tenth of a point?

Getting Ready for Section 9.2
[Section 8.2]

Round each decimal to the indicated place.

a. 5.43   tenth

b. 6.87   tenth

c. 27.428   hundredth

d. 30.583   hundredth

e. 0.0587   thousandth

f. 0.12545   thousandth

a. 5.4

b. 6.9

c. 27.43

d. 30.58

e. 0.059

f. 0.125

Answers

1.
```
     2.78
  6)16.68
    12
     4 6
     4 2
       48
       48
        0
```
3. 0.48

5.
```
    .685
 8)5.480
   4 8
     68
     64
     40
     40
      0
```
7. 2.315   **9.** 5.8   **11.** 2.35

13.
```
      .265
 52)13.780
    10 4
     3 38
     3 12
       260
       260
         0
```
15. 3.214   **17.** 2.6   **19.** 0.84

21.
```
      2.41
 52)125.40
    104
     21 4
     20 8
        60
        52
         8
```
125.4 ÷ 52 = 2.4 (to the nearest tenth)   **23.** 0.033

25. $13.47   **27.** $2.64   **29.** $0.59, or 59¢   **31.** $35.34

33. 30.8 mi/gal   **35.** 3.17   **a.** 5.4   **b.** 6.9   **c.** 27.43   **d.** 30.58

e. 0.059   **f.** 0.125

Supplementary Exercises

Name

Section   Date

ANSWERS	
1.	0.78
2.	3.85
3.	0.535
4.	0.825
5.	0.465
6.	0.235
7.	0.058
8.	0.0765
9.	5.5
10.	0.23
11.	1.53
12.	0.535
13.	$6.49
14.	$8.36
15.	$35.24
16.	$85.35
17.	2.97
18.	33.5 mi/gal

Divide.

1. $4.68 \div 6$

2. $34.65 \div 9$

3. $3.745 \div 7$

4. $4.95 \div 6$

5. $12.555 \div 27$

6. $15.98 \div 68$

7. $64)\overline{3.712}$

8. $58)\overline{4.437}$

Divide and round the quotient to the indicated decimal place.

9. $43.76 \div 8$ tenth

10. $2.106 \div 9$ hundredth

11. $98.23 \div 64$ hundredth

12. $27.81 \div 52$ thousandth

Solve the following applications.

13. **Unit pricing.** Marc bought three T-shirts on sale for $19.47. What was the price per shirt?

14. **Party expenses.** A group of eight people shared expenses of $66.88 for a party. What amount should each person pay?

15. **Monthly payments.** Phil buys a stereo system costing $945.76 with interest. He makes a $100 down payment and agrees to pay the balance in 24 monthly payments. What will he pay per month?

16. **Monthly payments.** Ann Marie bought a used car that cost, with the interest included, $1936.30. She paid $400 down and agreed to pay off the balance in equal monthly payments for 18 months. What will she pay per month?

17. **Grade average.** Yusef's grade points for six semesters were 2.74, 3.15, 2.86, 3.21, 2.84, and 3.04. What was his grade-point average to the nearest hundredth of a point?

18. **Mileage.** In four fill-ups, Lven's gas mileage was 31.5, 32.8, 37.3, and 32.2 mi/gal. What was his mileage to the nearest tenth of a mile per gallon?

Dividing by Decimals

9.2 OBJECTIVES

1. Divide a decimal by a decimal.
2. Use division of decimals to solve an application.

We want now to look at division *by* decimals. Here is an example using a fractional form.

● Example 1

Rewriting a Problem That Requires Dividing by a Decimal

Divide.

$$2.57 \div 3.4 = \frac{2.57}{3.4}$$ Write the division as a fraction.

$$= \frac{2.57 \times 10}{3.4 \times 10}$$ We multiply the numerator and denominator by 10 so the divisor is a whole number. This *does not change* the value of the fraction.

$$= \frac{25.7}{34}$$ Multiplying by 10, shift the decimal point in the numerator and denominator *one place to the right*.

$$= 25.7 \div 34$$ Our division problem is rewritten so that the divisor is a whole number.

It's always easier to rewrite a division problem so that you're dividing by a whole number. Dividing by a whole number makes it easy to place the decimal point in the quotient.

So

$$2.57 \div 3.4 = 25.7 \div 34$$ After we multiply the numerator and denominator by 10, we see that $2.57 \div 3.4$ is the same as $25.7 \div 34$.

● ● ●   CHECK YOURSELF 1

Rewrite the division problem so that the divisor is a whole number.

$$3.42 \div 2.5$$

● Example 2

Rewriting a Problem That Requires Dividing by a Decimal

Divide.

$$14.835 \div 2.14 = \frac{14.835}{2.14}$$

$$= \frac{14.835 \times 100}{2.14 \times 100}$$ To shift the decimal *two places right*, multiply by 100.

$$= \frac{1483.5}{214}$$ Now we are dividing by a whole number.

$$= 1483.5 \div 214$$

505

So

$$14.835 \div 2.14 = 1483.5 \div 214$$

● ● ● CHECK YOURSELF 2

Rewrite the division problem so that the divisor is a whole number.

$$24.536 \div 4.68$$

Of course, multiplying by any whole-number power of 10 greater than 1 is just a matter of shifting the decimal point to the right.

Do you see the rule suggested by these examples? In Example 1, we multiplied the numerator and the denominator (the dividend and the divisor) by 10. In Example 2, we multiplied by 100. Both operations made the divisor a whole number without altering the actual digits involved. All we did was shift the decimal point in the divisor and dividend the same number of places. This leads us to the following rule.

To Divide by a Decimal

STEP 1 Move the decimal point in the divisor *to the right,* making the divisor a whole number.
STEP 2 Move the decimal point in the dividend to the right *the same number of places.* Add zeros if necessary.
STEP 3 Place the decimal point in the quotient directly above the decimal point of the dividend.
STEP 4 Divide as you would with whole numbers.

Example 3 illustrates how to shift decimal points the appropriate number of places when dividing by a decimal.

● Example 3

Rewriting Problems That Require Dividing by a Decimal

(*a*) $2.3\overline{)15.85} \longrightarrow 2.3\,\overline{)15.8\,5}$

Shift the decimal points one place to the right. The divisor is now the whole number 23.

(*b*) $4.53\overline{)12.40} \longrightarrow 4.53\,\overline{)12.40}$

Shift the decimal points two places to the right.

(*c*) $3.245\overline{)34.5} \longrightarrow 3.245\,\overline{)34.500}$

Shift the decimal points three places to the right. As you can see, you may have to add zeros in the dividend.

(*d*) $0.34\overline{)58} \longrightarrow 0.34\,\overline{)58.00}$

The decimal point is assumed to be to the right of 58!

CHECK YOURSELF 3

Rewrite the division problems so that the divisors are whole numbers.

a. $3.7\overline{)5.93}$ **b.** $2.58\overline{)125.7}$

Let's look at another example of the use of our division rule.

● Example 4

Rounding the Result of Dividing by a Decimal

Divide 1.573 by 0.48 and give the quotient to the nearest tenth.

Write

$0.48\overline{)1.57\,3}$ Shift the decimal points two places
to the right to make the divisor a
whole number.

Now divide:

Once the division statement
is rewritten, place the
decimal point in the quotient
above that in the dividend.

$$
\begin{array}{r}
3.27 \\
48\overline{)157.30} \\
\underline{144} \\
13\,3 \\
\underline{9\,6} \\
3\,70 \\
\underline{3\,36} \\
34
\end{array}
$$

Note that we add a 0 to carry the division to the
hundredths place. In this case, we want to
find the quotient to the nearest tenth.

Round 3.27 to 3.3. So

$1.573 \div 0.48 = 3.3$ (to the nearest tenth)

CHECK YOURSELF 4

Divide, rounding the quotient to the nearest tenth.

$3.4 \div 1.24$

Let's look at some applications of our work in dividing by decimals.

• Example 5

Solving an Application Involving the Division of Decimals

Andrea worked 41.5 hours in a week and earned $239.87. What was her hourly rate of pay?

Solution To find the hourly rate of pay we must use division. We divide the number of hours worked into the total pay.

Note that we must add a zero to the dividend to complete the division process.

$$
\begin{array}{r}
5.78 \\
41.5\overline{)239.8\,70} \\
207\,5\,70 \\
32\,3\;7 \\
29\,0\;5 \\
3\,3\;20 \\
3\,3\;20 \\
0
\end{array}
$$

Andrea's hourly rate of pay is $5.78.

● ● ● CHECK YOURSELF 5

A developer wants to subdivide a 12.6-acre piece of land into 0.45-acre lots. How many lots are possible?

We may also have to round the quotient in some problems. Example 6 illustrates this approach.

• Example 6

Solving an Application Involving the Division of Decimals

Jesse drove 185 miles (mi) in 3.5 hours (h). What was his average speed (to the nearest mile per hour)?

Solution To find the average speed (miles per hour) we divide the distance by the time.

Remember: We assume that the decimal point is to the right of 185.

$185 = 185.0$

Then shift the decimal points.

$$
\begin{array}{r}
52.8 \\
3.5\overline{)185.0\,0} \\
175 \\
10\,0 \\
7\,0 \\
3\,0\,0 \\
2\,8\,0 \\
2\,0
\end{array}
$$

Once again we use the formula

Speed = distance ÷ time

Carry the division to the tenths place and then round 52.8 to 53. So Jesse's average speed was 53 mi/h.

● ● ● **CHECK YOURSELF 6**

To convert from centimeters to inches, you can divide by 2.54. Find the number of inches in 25 cm (to the nearest hundredth of an inch).

● Example 7

Solving an Application Involving the Division of Decimals

At the start of a trip the odometer read 34,563. At the end of the trip, it reads 36,235. If 86.7 gallons (gal) of gas were used, find the number of miles per gallon (to the nearest tenth).

Solution   First, find the number of miles traveled by subtracting the initial reading from the final reading.

36,235	Final reading
−34,563	Initial reading
1,672	Miles covered

Next, divide the miles traveled by the number of gallons used. This will give us the miles per gallon.

$$
\begin{array}{r}
19.28 \\
86.7\,)\overline{1672.0\ 00} \\
\underline{867} \\
805\ 0 \\
\underline{780\ 3} \\
24\ 7\ 0 \\
\underline{17\ 3\ 4} \\
7\ 3\ 60 \\
\underline{6\ 9\ 36} \\
4\ 24
\end{array}
$$

Round 19.28 to 19.3 mi/gal.

● ● ● **CHECK YOURSELF 7**

John starts his trip with an odometer reading 15,436 and ends with a reading of 16,238. If he used 45.9 gallons (gal) of gas, find the number of miles per gallon (mi/gal) (to the nearest tenth).

Recall from Section 3.6 that we find the *mean* of a set of values by adding them and then dividing by the number of values we have.

• Example 8

Solving a Financial Application

Charlene made the following five deposits to her checking account.

$123.45, $131.57, $144.00, $227.56, $2396.50

Find the mean amount of her deposits.

Solution To find the mean, we add the five values.

$123.45 + $131.57 + $144.00 + $227.56 + $2396.50 = $3023.08

Dividing by 5, we get

$3023.08 ÷ 5 = $604.616

However, since we're talking about money, we'll round to the nearest penny, so her mean deposit was $604.62.

● ● ● CHECK YOURSELF 8

Charlene has written seven checks this month in the amounts

$23.45, $25.77, $33.69, $41.67, $110.25, $166.95, $1896.45

Find the mean amount of the checks she wrote.

● ● ● CHECK YOURSELF ANSWERS

1. $34.2 \div 25$.   **2.** $2453.6 \div 468$.   **3.** **(a)** $3.7 \,\overline{)5.9\,3}$; **(b)** $2.58 \,\overline{)125.70}$.

4. 2.7.   **5.** 28 lots.   **6.** 9.84 in.   **7.** 17.5 mi/gal.   **8.** $328.32.

Name

Section   Date

A N S W E R S

Divide.

1. 0.6)‾11.07‾

2. 0.8)‾10.84‾

3. 3.8)‾7.22‾

4. 2.9)‾13.34‾

5. 5.2)‾11.622‾

6. 6.4)‾3.616‾

7. 0.27)‾1.8495‾

8. 0.038)‾0.8132‾

9. 0.046)‾1.587‾

10. 0.52)‾3.2318‾

11. 0.658 ÷ 2.8

12. 0.882 ÷ 0.36

13. 3.275 ÷ 0.524

14. 0.6837 ÷ 3.18

Find the quotients to the indicated place.

15. 0.7)‾1.642‾   hundredth

16. 0.6)‾7.695‾   tenth

17. 4.5)‾8.415‾   tenth

18. 5.8)‾16‾   hundredth

19. 3.12)‾4.75‾   hundredth

20. 64.2)‾16.3‾   thousandth

21. 5.38)‾0.205‾   thousandth

22. 0.347)‾0.8193‾   hundredth

1.	18.45
2.	13.55
3.	1.9
4.	4.6
5.	2.235
6.	0.565
7.	6.85
8.	21.4
9.	34.5
10.	6.215
11.	0.235
12.	2.45
13.	6.25
14.	0.215
15.	2.35
16.	12.8
17.	1.9
18.	2.76
19.	1.52
20.	0.254
21.	0.038
22.	2.36

23. $2.42\overline{)1.3}$ hundredth

24. $96.3\overline{)1.753}$ thousandth

25. $0.99 \div 0.624$ thousandth

26. $3.75 \div 1.58$ hundredth

27. $0.125 \div 2.135$ ten thousandth

28. $0.428 \div 1.452$ thousandth

Solve the following applications.

29. Label making. We have 91.25 inches (in.) of plastic labeling tape and wish to make labels that are 1.25 in. long. How many labels can be made?

30. Wages. Alberto worked 32.5 hours (h), earning $306.15. How much did he make per hour?

31. Cost per pound. A roast weighing 5.3 pounds (lb) sold for $14.89. Find the cost per pound to the nearest cent.

32. Weight. One nail weighs 0.025 ounce (oz). How many nails are there in 1 lb? (1 lb is 16 oz.)

33. Mileage. A family drove 1390 miles (mi), stopping for gas 3 times. If they purchased 15.5, 16.2, and 10.8 gallons (gal) of gas, find the number of miles per gallon (the mileage) to the nearest tenth.

34. Mileage. On a trip an odometer changed from 36,213 to 38,319. If 136 gal of gas was used, find the number of miles per gallon (to the nearest tenth).

35. Mileage. When Andrea started on her trip, the odometer read 18,912 mi. When she returned home, it read 19,315 mi. She used 22.9 gal of gas. What was her gas mileage (to the nearest tenth)?

36. Volume. The water in an aquarium weighs 1025 lb. If water weighs 62.5 pounds per cubic foot (lb/ft^3), how many cubic feet of water does the aquarium hold?

37. Conversion. To convert from millimeters (mm) to inches, we can divide by 25.4. If film is 35 mm wide, find the width to the nearest hundredth of an inch.

38. Conversion. To convert from centimeters (cm) to inches, we can divide by 2.54. The rainfall in Paris was 11.8 cm during 1 week. What was that rainfall to the nearest hundredth of an inch?

39. Checking. Sam has written seven checks this week. They are in the amounts

$26.78, $35.18, $89.65, $42.95, $234.65, $356.89, $2565.75

Find the mean amount of the checks he wrote.

40. Selling. During the past 5 weeks, Pam has earned the following commissions on her sales:

$125.98, $178.56, $256.90, $345.55, $396.77

For those 5 weeks, find Pam's mean weekly salary.

41. Weight. Eddie purchased nine bags of sand for use around his house. The individual weights, in pounds, are as follows:

45.25, 43.98, 39.79, 46.25, 39.55, 42.89, 44.89, 38.68, 39.25

Find the mean weight of the bags of sand.

42. Travel. During the past 9 days, Chuck recorded the total number of miles that he drove. Those amounts are

45.67, 56.99, 24.99, 125.98, 34.87, 111.65, 56.12, 350.78, 197.99

Find the mean number of miles driven per day.

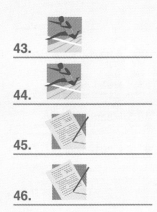
43. The blood alcohol level (BAC) of a person who has been drinking is determined by the formula

$$BAC = \frac{\text{oz of alcohol} \times \text{\% of alcohol} \times 0.075 \text{ of body wt.}}{(\text{hours of drinking} \times 0.015)}$$

A 125-lb person is driving and is stopped by a policewoman on suspicion of driving under the influence (DUI). The driver claims that in the past 2 hours he consumed only six 12-oz bottles of 3.9% beer. If he undergoes a breathalyzer test, what will his BAC be? Will this amount be under legal limit for your state?

44. Four brands of soap are available in a local store.

Brand	Ounces	Total Price	Unit Price
Squeaky Clean	5.5	$0.36	
Smell Fresh	7.5	0.41	
Feel Nice	4.5	0.31	
Look Bright	6.5	0.44	

Compute the unit price, and decide which brand is the best buy.

45. Sophie is a quality control expert. She inspects boxes of #2 pencils. Each pencil weighs 4.4 grams (g). The contents of a box of pencils weigh 66.6 g. If a box is labeled CONTENTS: 16 PENCILS, should Sophie approve the box as meeting specifications? Explain your answer.

46. Write a plan to determine the number of miles per gallon (mpg) your car (or your family car) gets. Use this plan to determine your car's actual mpg.

A N S W E R S

a. 23

b. 4.52

c. 158

d. 24.8

e. 1457.9

f. 427

 Getting Ready for Section 9.3
[Section 8.6]

Multiply.

a. 2.3×10

b. 0.452×10

c. 1.58×100

d. 0.248×100

e. 1.4579×1000

f. $0.0427 \times 10,000$

Answers

1. 18.45 **3.**
$$3.8\overline{)7.2\,2}$$
$$\quad\quad\underline{3\,8}$$
$$\quad\quad3\,4\,2$$
$$\quad\quad\underline{3\,4\,2}$$
$$\quad\quad\quad\;0$$
with quotient 1.9

5. 2.235 **7.** 6.85 **9.** 34.5

11. 0.235 **13.** 6.25 **15.** 2.35 **17.** 1.9 **19.** 1.52 **21.** 0.038

23.
$$2.42\overline{)1.30\,000}$$
quotient .537
$$\quad\underline{1\,21\,0}$$
$$\quad\;\;9\,00$$
$$\quad\;\;\underline{7\,26}$$
$$\quad\;\;1\,740$$
$$\quad\;\;\underline{1\,694}$$
$$\quad\quad\;\;46$$
$1.3 \div 2.42 = 0.54$ (to the nearest hundredth)

25. 1.587 **27.** 0.0585 **29.** 73 labels **31.** $2.81 **33.** 32.7 mi/gal

35. 17.6 mi/gal **37.** 1.38 in. **39.** $478.84 **41.** 42.28 **43.**

45. **a.** 23 **b.** 4.52 **c.** 158 **d.** 24.8 **e.** 1457.9

f. 427

A N S W E R S

1.	5.35
2.	5.2
3.	2.57
4.	3.9
5.	2.15
6.	4.38
7.	4.5
8.	3.17
9.	5.4
10.	5.2
11.	21.14
12.	2.88
13.	0.2419
14.	0.192
15.	38.4 h
16.	128 pieces
17.	34.9 mi/gal
18.	50 mi/h

Divide.

1. $0.8\overline{)4.28}$

2. $2.4\overline{)12.48}$

3. $3.8\overline{)9.766}$

4. $0.58\overline{)2.262}$

5. $0.518\overline{)1.1137}$

6. $1.971 \div 0.45$

7. $2.3625 \div 0.525$

Find the quotients to the indicated place.

8. $0.8\overline{)2.534}$ hundredth

9. $2.8\overline{)15}$ tenth

10. $4.8\overline{)25.07}$ tenth

11. $1.3\overline{)27.48}$ hundredth

12. $0.59\overline{)1.698}$ hundredth

13. $0.1398 \div 0.578$ ten thousandth

14. $5.342 \div 27.8$ thousandth

Solve the following applications.

15. Salary. You received $220.80 in pay during a week. If your hourly rate of pay was $5.75, how many hours did you work?

16. Quantity. A piece of bubble gum weighs 0.25 ounces (oz). How many pieces of gum are there in 2 pounds (lb) (32 oz)?

17. Gas mileage. At the start of a trip, an odometer read 27,458. At the end of the trip, it read 28,808 and 38.7 gallons (gal) of gas had been used. Find the number of miles per gallon (gas mileage) to the nearest tenth.

18. Speed. Carlos drove 224 miles (mi) in 4.5 hours (h). What was his speed to the nearest mile per hour?

Using Your Calculator to Divide Decimals

It would be most surprising if you had reached this point without using your calculator to divide decimals. It is a good way to check your work, and a reasonable way to solve applications. Let's first use the calculator for a straightforward problem.

• Example 1

Dividing Decimals

Use your calculator to find the quotient

$211.56 \div 8.2$

Enter the problem in the calculator to find that the answer is 25.8.

CHECK YOURSELF 1

Use your calculator to find the quotient

$304.32 \div 9.6$

Now that you're convinced that it is easy to divide decimals on your calculator, let's introduce a twist.

• Example 2

An Application Involving the Division of Decimals

Omar drove 256.3 miles on a tank of gas. When he filled up the tank, it took 9.1 gallons. What was his gas mileage?

Solution Here's where students get into trouble when they use a calculator. Entering these values, you may be tempted to answer "28.16483516 miles per gallon." The difficulty is that there is no way you can compute gas mileage to the nearest hundred-millionth mile. How do you decide where to round off the answer that the calculator gives you? A good rule of thumb is to never report more digits than the maximum number of digits in any of the numbers that you are given in the problem. In this case, you were given a number with four digits and another with two digits. Your answer should not have more than four digits. Instead of 28.16483516, the answer could be 28.16 miles per gallon. Think about the question. If you were asked for gas mileage, how precise an answer would you give? The best answer to this question would be to give the nearest whole number of miles per gallon: 28 miles per gallon.

● ● ● **CHECK YOURSELF 2**

Emmet gained a total of 857 yards (yd) in 209 times that he carried the football. How many yards did he average for each time he carried the ball?

_____

● ● ● **CHECK YOURSELF ANSWERS**

1. 31.7.   **2.** 4.1 yd.

_____

Calculator Exercises

Name _____

Section _____ Date _____

A N S W E R S

1.	3.45
2.	58.5
3.	2.89
4.	62.5
5.	0.245
6.	23.48
7.	1.84
8.	4.9
9.	0.468
10.	1.59
11.	0.51
12.	2.835
13.	$4.92
14.	230 lots

Divide and check.

1. $8.901 \div 2.58$ **2.** $16.848 \div 0.288$ **3.** $99.705 \div 34.5$

4. $171.25 \div 2.74$ **5.** $0.01372 \div 0.056$ **6.** $0.200754 \div 0.00855$

Divide and round to the indicated place.

7. $2.546 \div 1.38$ hundredth   **8.** $45.8 \div 9.4$ tenth

9. $0.5782 \div 1.236$ thousandth

10. $1.25 \div 0.785$ hundredth   **11.** $1.34 \div 2.63$ two decimal places

12. $12.364 \div 4.361$ three decimal places

Solve the following applications.

13. Salary. In 1 week, Tom earned $178.30 by working 36.25 hours (h). What was his hourly rate of pay to the nearest cent?

14. Area. An 80.5-acre piece of land is being subdivided into 0.35-acre lots. How many lots are possible in the subdivision?

Answers

1. 3.45   **3.** 2.89   **5.** 0.245   **7.** 1.84   **9.** 0.468   **11.** 0.51   **13.** $4.92

9.3 Dividing Decimals by Powers of 10

9.3 OBJECTIVES

1. Divide a decimal by a power of 10.
2. Use division by powers of 10 to solve an application.

Recall that you can multiply decimals by powers of 10 by simply shifting the decimal point to the right. A similar approach will work for division by powers of 10.

• Example 1

Dividing a Decimal by a Power of 10

(*a*) Divide.

```
      3.53
10)35.30
    30
    ─────
     5 3
     5 0
    ─────
       30
       30
    ─────
        0
```

The dividend is 35.3. The quotient is 3.53. The decimal point has been shifted *one place to the left*. Note also that the divisor, 10, has *one* zero.

(*b*) Divide.

```
        3.785
100)378.500
    300
    ──────
     78 5
     70 0
    ──────
      8 50
      8 00
    ──────
        500
        500
    ──────
          0
```

Here the dividend is 378.5, whereas the quotient is 3.785. The decimal point is now shifted *two places to the left*. In this case the divisor, 100, has *two* zeros.

● ● ● **CHECK YOURSELF 1**

Perform each of the following divisions.

a. $52.6 \div 10$ **b.** $267.9 \div 100$

Examples 2 and 3 suggest the following rule.

To Divide a Decimal by a Power of 10

Move the decimal point *to the left* the same number of places as there are zeros in the power of 10.

●Example 2

Dividing a Decimal by a Power of 10

Divide.

(*a*) $27.3 \div 10 = 2\,7.3$ Shift one place to the left.

(*b*) $57.53 \div 100 = 0\,57.53$ Shift two places to the left.

(*c*) $39.75 \div 1000 = 0\,039.75$ Shift three places to the left.

(*d*) $85 \div 1000 = 0\,085.$ The decimal after the 85 is implied.

(*e*) $235.72 \div 10^4 = 0\,0235.72$ Shift four places to the left.

As you can see, we may have to add zeros to correctly place the decimal point.

Remember: 10^4 is a 1 followed by *four zeros*.

●●● **CHECK YOURSELF 2**

Divide.

a. $3.84 \div 10$ **b.** $27.3 \div 1000$

Let's look at an application of our work in dividing by powers of 10.

●Example 3

Solving an Application Involving a Power of 10

To convert from millimeters (mm) to meters (m), we divide by 1000. How many meters does 3450 mm equal?

Solution

$3450 \text{ mm} = 3\,450. \text{ m}$ Shift three places to the left to divide by 1000.

●●● **CHECK YOURSELF 3**

A shipment of 1000 notebooks cost a stationery store $658. What was the cost per notebook to the nearest cent?

●●● **CHECK YOURSELF ANSWERS**

1. (a) 5.26; **(b)** 2.679.   **2. (a)** 0.384; **(b)** 0.0273.   **3.** 66¢.

9.3 Exercises

A N S W E R S

Divide.

1. $5.8 \div 10$ **2.** $5.1 \div 10$

3. $4.568 \div 100$ **4.** $3.817 \div 100$

5. $24.39 \div 1000$ **6.** $8.41 \div 100$

7. $6.9 \div 1000$ **8.** $7.2 \div 1000$

9. $7.8 \div 10^2$ **10.** $3.6 \div 10^3$

11. $45.2 \div 10^5$ **12.** $57.3 \div 10^4$

Solve the following applications.

13. Construction. The cost of a street-lighting project, $4850, will be shared by 10 homeowners in a neighborhood. What will each homeowner pay?

14. Construction. A road-paving project will cost $23,500. If the cost is to be shared by 100 families, how much will each family pay?

15. Conversion. To convert from milligrams (mg) to grams (g), we divide by 1000. A tablet is 250 mg. What is its weight in grams?

16. Conversion. To convert from milliliters (mL) to liters (L), we divide by 1000. If a bottle of wine holds 750 mL, what is its volume in liters?

17. Unit cost. A shipment of 100 calculators cost a store $593.88. Find the cost per calculator (to the nearest cent).

18. Unit cost. A shipment of 1000 writing tablets cost an office supplier $756.80. Find the cost per tablet (to the nearest cent).

19. Express the width and length of a $1 bill in centimeters (cm). Then express this same length in millimeters (mm).

A N S W E R S

1. 0.58

2. 0.51

3. 0.04568

4. 0.03817

5. 0.02439

6. 0.0841

7. 0.0069

8. 0.0072

9. 0.078

10. 0.0036

11. 0.000452

12. 0.00573

13. $485

14. $235

15. 0.25 g

16. 0.75 L

17. $5.94

18. 76¢

19.

ANSWERS

a. 2 ÷ 3

b. 3 ÷ 8

c. 7 ÷ 10

d. 5 ÷ 7

e. 7 ÷ 9

f. 1 ÷ 4

Getting Ready for Section 9.4 [Section 5.1]

Write each of the following common fractions as a division statement.

a. $\frac{2}{3}$ **b.** $\frac{3}{8}$ **c.** $\frac{7}{10}$

d. $\frac{5}{7}$ **e.** $\frac{7}{9}$ **f.** $\frac{1}{4}$

Answers

1. 0.58 **3.** 0.04568 **5.** 0.02439 **7.** 0.0069 **9.** 0.078 **11.** 0.000452 **13.** \$485 **15.** 0.25 g **17.** \$5.94 **19.** **a.** 2 ÷ 3 **b.** 3 ÷ 8 **c.** 7 ÷ 10 **d.** 5 ÷ 7 **e.** 7 ÷ 9 **f.** 1 ÷ 4

9.3 Supplementary Exercises

Name

Section Date

ANSWERS

1. 0.493

2. 1.579

3. 0.00523

4. 0.0953

5. 0.00271

6. 0.005238

7. \$443.50

8. 0.3 L

9. 44¢

10. 0.75 m

Divide.

1. 4.93 ÷ 10 **2.** 157.9 ÷ 100 **3.** 5.23 ÷ 1000

4. 0.953 ÷ 10 **5.** $27.1 \div 10^4$ **6.** $523.8 \div 10^5$

Solve the following applications.

7. Construction. A sewer project costing \$44,350 will be paid for by 100 families. What will the cost per family be?

8. Conversion. To convert from centiliters (cL) to liters (L), divide by 100. If a glass holds 30 cL, what is its volume in liters?

9. Unit cost. A shipment of 1000 items costs a store \$438.75. What is the cost per item to the nearest cent?

10. Conversion. A desk top is 750 millimeters (mm) wide. You divide by 1000 to convert millimeters to meters (m). What is its width in meters?

9.4 Converting Common Fractions to Decimals

9.4 OBJECTIVES

1. Convert a common fraction to a decimal.
2. Convert a common fraction to a repeating decimal.

Since a common fraction can be interpreted as division, you can divide the numerator of the common fraction by its denominator to convert a common fraction to a decimal. The result is called a **decimal equivalent.**

 Example 1

Converting a Fraction to a Decimal Equivalent

Write $\frac{5}{8}$ as a decimal.

Remember that 5 can be written as 5.0, 5.00, 5.000, and so on. In this case, we continue the division by adding zeros to the dividend until a 0 remainder is reached.

$$\begin{array}{r} 0.625 \\ 8\overline{)5.000} \\ \underline{4\ 8} \\ 20 \\ \underline{16} \\ 40 \\ \underline{40} \\ 0 \end{array}$$

Since $\frac{5}{8}$ means $5 \div 8$, divide 8 into 5.

We see that $\frac{5}{8} = 0.625$; 0.625 is the decimal equivalent of $\frac{5}{8}$.

● ● ● **CHECK YOURSELF 1**

Find the decimal equivalent of $\frac{7}{8}$.

Some fractions are used so often that we have listed their decimal equivalents for your reference.

The division used to find these decimal equivalents stops when a 0 remainder is reached. The equivalents are called **terminating decimals.**

Some Common Decimal Equivalents

$\frac{1}{2} = 0.5$	$\frac{1}{4} = 0.25$	$\frac{1}{5} = 0.2$	$\frac{1}{8} = 0.125$
	$\frac{3}{4} = 0.75$	$\frac{2}{5} = 0.4$	$\frac{3}{8} = 0.375$
		$\frac{3}{5} = 0.6$	$\frac{5}{8} = 0.625$
		$\frac{4}{5} = 0.8$	$\frac{7}{8} = 0.875$

If a decimal equivalent does not terminate, you can round the result to approximate the fraction to some specified number of decimal places. Consider Example 2.

● Example 2

Converting a Fraction to a Decimal Equivalent

Write $\dfrac{3}{7}$ as a decimal. Round the answer to the nearest thousandth.

$$
\begin{array}{r}
0.4285 \\
7\overline{)3.0000} \\
\underline{2\ 8} \\
20 \\
\underline{14} \\
60 \\
\underline{56} \\
40 \\
\underline{35} \\
5
\end{array}
$$

In this example, we are choosing to round to three decimal places, so we must add enough zeros to carry the division to four decimal places.

So $\dfrac{3}{7} = 0.429$ (to the nearest thousandth).

● ● ● CHECK YOURSELF 2

Find the decimal equivalent of $\dfrac{5}{11}$ to the nearest thousandth.

If a decimal equivalent does *not* terminate, it will *repeat* a sequence of digits. These decimals are called **repeating decimals.**

● Example 3

Converting a Fraction to a Repeating Decimal

(*a*) Write $\dfrac{1}{3}$ as a decimal.

$$
\begin{array}{r}
0.333 \\
3\overline{)1.000} \\
\underline{9} \\
10 \\
\underline{9} \\
10 \\
\underline{9}
\end{array}
$$

The digit 3 will just repeat itself indefinitely since each new remainder will be 1.

Adding more zeros and going on will simply lead to more threes in the quotient.

We can say $\dfrac{1}{3} = 0.333\ldots$ The three dots mean "and so on" and tell us that 3 will repeat itself indefinitely.

(*b*) Write $\dfrac{5}{12}$ as a decimal.

$$\begin{array}{r} 0.4166\ldots \\ 12\overline{)5.0000} \\ 4\,8 \\ \hline 20 \\ 12 \\ \hline 80 \\ 72 \\ \hline 80 \\ 72 \\ \hline 8 \end{array}$$

In this example, the digit 6 will just repeat itself since the remainder, 8, will keep occurring if we add more zeros and continue the division.

● ● ●   **CHECK YOURSELF 3**

Find the decimal equivalent of each fraction.

a. $\dfrac{2}{3}$ **b.** $\dfrac{7}{12}$

Another way to write a repeating decimal is with a bar placed over the digit or digits that repeat. For example, we can write

$0.37373737\ldots$

as

$0.\overline{37}$

The bar placed over the digits indicates that "37" repeats indefinitely.

Some important decimal equivalents (rounded to the nearest thousandth) are shown below for reference.

$$\frac{1}{3} = 0.333 \qquad \frac{1}{6} = 0.167 \qquad \frac{2}{3} = 0.667 \qquad \frac{5}{6} = 0.833$$

● Example 4

Converting a Fraction to a Repeating Decimal

Write $\dfrac{5}{11}$ as a decimal.

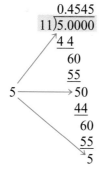

As soon as a remainder repeats itself, as 5 does here, the pattern of digits will repeat in the quotient.

$$\frac{5}{11} = 0.\overline{45}$$

$$= 0.4545\ldots$$

●●● **CHECK YOURSELF 4**

Use the bar notation to write the decimal equivalent of $\dfrac{5}{7}$. (Be patient. You'll have to divide for a while to find the repeating pattern.)

You can find the decimal equivalents for mixed numbers in a similar way. Find the decimal equivalent of the fractional part of the mixed number, and then combine that with the whole-number part. Example 5 illustrates this approach.

•Example 5

Converting a Mixed Number to a Decimal Equivalent

Find the decimal equivalent of $3\dfrac{5}{16}$.

$\dfrac{5}{16} = 0.3125$ First find the equivalent of $\dfrac{5}{16}$ by division.

$3\dfrac{5}{16} = 3.3125$ Add 3 to the result.

●●● **CHECK YOURSELF 5**

Find the decimal equivalent of $2\dfrac{5}{8}$.

We learned something important in this section. To find the decimal equivalent of a fraction, we use long division. Since the remainder must be less than the divisor, the remainder must *either repeat or become 0*. Thus *every common fraction* will have a *repeating* or a *terminating* decimal as its decimal equivalent.

●●● **CHECK YOURSELF ANSWERS**

1. 0.875.   **2.** $\dfrac{5}{11} = 0.455$ (to the nearest thousandth).   **3. (a)** 0.666. . . ;

(b) $\dfrac{7}{12} = 0.583.\ .\ .$ The digit 3 will continue indefinitely.   **4.** $\dfrac{5}{7} = 0.\overline{714285}$.

5. 2.625.

9.4 Exercises

Name

Section   Date

ANSWERS

Find the decimal equivalents for each of the following fractions.

1. $\dfrac{3}{4}$ 2. $\dfrac{4}{5}$ 3. $\dfrac{9}{20}$

4. $\dfrac{3}{10}$ 5. $\dfrac{1}{5}$ 6. $\dfrac{1}{8}$

7. $\dfrac{5}{16}$ 8. $\dfrac{11}{20}$ 9. $\dfrac{7}{10}$

10. $\dfrac{7}{16}$ 11. $\dfrac{27}{40}$ 12. $\dfrac{17}{32}$

Find the decimal equivalents rounded to the indicated place.

13. $\dfrac{5}{6}$ thousandth   14. $\dfrac{7}{12}$ hundredth   15. $\dfrac{4}{15}$ thousandth

Write the decimal equivalents, using the bar notation.

16. $\dfrac{1}{18}$ 17. $\dfrac{4}{9}$ 18. $\dfrac{3}{11}$

Find the decimal equivalents for each of the following mixed numbers.

19. $5\dfrac{3}{5}$ 20. $7\dfrac{3}{4}$ 21. $4\dfrac{7}{16}$

Getting Ready for Section 9.5 [Section 5.4]

Reduce each fraction to simplest form.

a. $\dfrac{8}{10}$ b. $\dfrac{16}{100}$ c. $\dfrac{35}{100}$

d. $\dfrac{225}{1000}$ e. $\dfrac{450}{1000}$ f. $\dfrac{625}{10,000}$

1. 0.75
2. 0.8
3. 0.45
4. 0.3
5. 0.2
6. 0.125
7. 0.3125
8. 0.55
9. 0.7
10. 0.4375
11. 0.675
12. 0.53125
13. 0.833
14. 0.58
15. 0.267
16. $0.0\overline{5}$
17. $0.\overline{4}$
18. $0.\overline{27}$
19. 5.6
20. 7.75
21. 4.4375
a. $\dfrac{4}{5}$
b. $\dfrac{4}{25}$
c. $\dfrac{7}{20}$
d. $\dfrac{9}{40}$
e. $\dfrac{9}{20}$
f. $\dfrac{1}{16}$

9.4 Supplementary Exercises

Name _____

Section _____ Date _____

A N S W E R S

1.	0.6
2.	0.35
3.	0.625
4.	0.1875
5.	0.55
6.	0.34375
7.	0.68
8.	0.325
9.	0.417
10.	0.44
11.	$0.\overline{7}$
12.	$0.\overline{36}$
13.	4.375
14.	5.3125

Find the decimal equivalents for each of the following fractions.

1. $\dfrac{3}{5}$ **2.** $\dfrac{7}{20}$

3. $\dfrac{5}{8}$ **4.** $\dfrac{3}{16}$

5. $\dfrac{11}{20}$ **6.** $\dfrac{11}{32}$

7. $\dfrac{17}{25}$ **8.** $\dfrac{13}{40}$

Find the decimal equivalents rounded to the indicated place.

9. $\dfrac{5}{12}$ thousandth   **10.** $\dfrac{4}{9}$ hundredth

Write the decimal equivalents, using the bar notation.

11. $\dfrac{7}{9}$ **12.** $\dfrac{4}{11}$

Find the decimal equivalents for each of the following mixed numbers.

13. $4\dfrac{3}{8}$ **14.** $5\dfrac{5}{16}$

9.5 Converting Decimals to Common Fractions

© 1998 McGraw-Hill Companies

9.5 OBJECTIVE

Convert a decimal to a common fraction.

Using what we have learned about place values, you can easily write decimals as common fractions. The following rule is used.

To Convert a Terminating Decimal Less Than 1 to a Common Fraction

STEP 1 Write the digits of the decimal without the decimal point. This will be the numerator of the common fraction.

STEP 2 The denominator of the fraction is a 1 followed by as many zeros as there are places in the decimal.

● Example 1

Converting a Decimal to a Common Fraction

$$0.7 = \frac{7}{10} \qquad 0.09 = \frac{9}{100} \qquad 0.257 = \frac{257}{1000}$$

One place   One zero   Two places   Two zeros   Three places   Three zeros

● ● ● CHECK YOURSELF 1

Write as common fractions.

a. 0.3   **b.** 0.311

When a decimal is converted to a common fraction, the common fraction that results should be written in lowest terms.

● Example 2

Converting a Decimal to a Common Fraction

Convert 0.395 to a fraction and reduce the result to lowest terms.

Divide the numerator and denominator by 5.

$$0.395 = \frac{395}{1000} = \frac{79}{200}$$

● ● ● CHECK YOURSELF 2

Write 0.275 as a common fraction.

If the decimal has a whole-number portion, write the digits to the right of the decimal point as a proper fraction and then form a mixed number for your result.

• Example 3

Repeating decimals can also be written as common fractions, although the process is more complicated. We will limit ourselves to the conversion of terminating decimals in this textbook.

Converting a Decimal to a Mixed Number

Write 12.277 as a mixed number.

$$0.277 = \frac{277}{1000}, \text{ so } 12.277 = 12\frac{277}{1000}$$

● ● ● **CHECK YOURSELF 3**

Write 32.433 as a mixed number.

Comparing the sizes of common fractions and decimals requires finding the decimal equivalent of the common fraction and then comparing the resulting decimals.

• Example 4

Comparing the Sizes of Common Fractions and Decimals

Which is larger, $\frac{3}{8}$ or 0.38?

Solution Write the decimal equivalent of $\frac{3}{8}$. That decimal is 0.375. Now comparing 0.375 and 0.38, we see that 0.38 is the larger of the numbers:

$$0.38 > \frac{3}{8}$$

● ● ● **CHECK YOURSELF 4**

Which is larger, $\frac{3}{4}$ or 0.8?

● ● ● **CHECK YOURSELF ANSWERS**

1. (a) $\frac{3}{10}$; (b) $\frac{311}{1000}$.   2. $0.275 = \frac{11}{40}$.   3. $32\frac{433}{1000}$.   4. $0.8 > \frac{3}{4}$.

A N S W E R S

Write each of the following as a common fraction or mixed number.

1. 0.9

2. 0.3

3. 0.8

4. 0.6

5. 0.37

6. 0.97

7. 0.587

8. 0.379

9. 0.48

10. 0.75

11. 0.58

12. 0.65

13. 0.425

14. 0.116

15. 0.375

16. 0.225

17. 0.136

18. 0.575

19. 0.059

20. 0.067

21. 0.0625

22. 0.0425

23. 6.3

24. 5.7

© 1998 McGraw-Hill Companies

1. $\dfrac{9}{10}$

2. $\dfrac{3}{10}$

3. $\dfrac{4}{5}$

4. $\dfrac{3}{5}$

5. $\dfrac{37}{100}$

6. $\dfrac{97}{100}$

7. $\dfrac{587}{1000}$

8. $\dfrac{379}{1000}$

9. $\dfrac{12}{25}$

10. $\dfrac{3}{4}$

11. $\dfrac{29}{50}$

12. $\dfrac{13}{20}$

13. $\dfrac{17}{40}$

14. $\dfrac{29}{250}$

15. $\dfrac{3}{8}$

16. $\dfrac{9}{40}$

17. $\dfrac{17}{125}$

18. $\dfrac{23}{40}$

19. $\dfrac{59}{1000}$

20. $\dfrac{67}{1000}$

21. $\dfrac{1}{16}$

22. $\dfrac{17}{400}$

23. $6\dfrac{3}{10}$

24. $5\dfrac{7}{10}$

ANSWERS

25. $2\dfrac{17}{100}$

26. $3\dfrac{31}{100}$

27. $5\dfrac{7}{25}$

28. $15\dfrac{7}{20}$

29. $>$

30. $<$

31. $<$

---

25. 2.17 **26.** 3.31

27. 5.28 **28.** 15.35

Complete each of the following statements, using the symbol $<$ or $>$.

29. $\dfrac{7}{8}$ _____ 0.87 **30.** $\dfrac{5}{16}$ _____ 0.313 **31.** $\dfrac{9}{25}$ _____ 0.4

Answers

1. $\dfrac{9}{10}$ **3.** $\dfrac{4}{5}$ **5.** $\dfrac{37}{100}$ **7.** $\dfrac{587}{1000}$ **9.** $\dfrac{12}{25}$ **11.** $\dfrac{29}{50}$ **13.** $\dfrac{17}{40}$

15. $\dfrac{3}{8}$ **17.** $\dfrac{17}{125}$ **19.** $\dfrac{59}{1000}$ **21.** $\dfrac{1}{16}$ **23.** $6\dfrac{3}{10}$ **25.** $2\dfrac{17}{100}$

27. $5\dfrac{7}{25}$ **29.** $>$ **31.** $<$

9.5 Supplementary Exercises

Name _____

Section _____ Date _____

ANSWERS

1. $\dfrac{7}{10}$

2. $\dfrac{3}{5}$

3. $\dfrac{41}{100}$

4. $\dfrac{863}{1000}$

5. $\dfrac{7}{20}$

6. $\dfrac{9}{25}$

7. $\dfrac{13}{40}$

8. $\dfrac{31}{125}$

9. $\dfrac{7}{8}$

10. $\dfrac{37}{1000}$

11. $\dfrac{9}{400}$

12. $4\dfrac{7}{10}$

Write each of the following as a common fraction or mixed number.

1. 0.7 **2.** 0.6

3. 0.41 **4.** 0.863

5. 0.35 **6.** 0.36

7. 0.325 **8.** 0.248

9. 0.875 **10.** 0.037

11. 0.0225 **12.** 4.7

Using Your Calculator to Convert Between Decimals and Fractions

A calculator is very useful in converting common fractions to decimals. Just divide the numerator by the denominator, and the decimal equivalent will be in the display.

● Example 1

Converting Fractions to Decimals

Find the decimal equivalent of $\dfrac{7}{16}$.

7 $\div$ 16 $=$

Display | 0.4375 | 0.4375 is the decimal equivalent of $\dfrac{7}{16}$.

● ● ● **CHECK YOURSELF 1**

Find the decimal equivalent of $\dfrac{5}{16}$.

Often, you will want to round the result in the display.

● Example 2

Converting Fractions to Decimals

Some calculators show the 0 to the left of the decimal point and seven digits to the right. Others omit the 0 and show eight digits to the right. Check yours with this example.

Find the decimal equivalent of $\dfrac{5}{24}$ to the nearest hundredth.

5 $\div$ 24 $=$

Display | 0.2083333 |

$\dfrac{5}{24} = 0.21$ (nearest hundredth)

● ● ● **CHECK YOURSELF 2**

Find the decimal equivalent of $\dfrac{7}{29}$ to the nearest hundredth.

To find the decimal equivalent of a mixed number, use the following sequence.

© 1998 McGraw-Hill Companies

533

There are several ways to do this, depending on the calculator you are using. For example,

7 $\boxed{+}$ 5 $\boxed{\div}$ 8 $\boxed{=}$

will work on most scientific calculators. Try it on yours.

•Example 3

Converting Mixed Numbers to Decimals

Change $7\dfrac{5}{8}$ to a decimal.

5 $\boxed{\div}$ 8 $\boxed{+}$ 7 $\boxed{=}$

Display $\boxed{7.625}$

● ● ● **CHECK YOURSELF 3**

Find the decimal equivalent of $3\dfrac{3}{8}$.

Depending on the calculator you are using, the result may be rounded at its last displayed digit.

•Example 4

Converting Fractions to Decimals

For $\dfrac{5}{9}$:

5 $\boxed{\div}$ 9 $\boxed{=}$

Display $\boxed{0.555555555}$ or $\boxed{0.5555555556}$ ⟵ Rounded display

● ● ● **CHECK YOURSELF 4**

Find the decimal equivalent of $\dfrac{7}{11}$.

● ● ● **CHECK YOURSELF ANSWERS**

1. 0.3125.   **2.** 0.24.   **3.** 3.375.   **4.** 0.63636363 or 0.63636364.

Calculator Exercises

Name _____

Section _____ Date _____

A N S W E R S

Find the decimal equivalents.

1. $\dfrac{7}{8}$

2. $\dfrac{11}{16}$

3. $\dfrac{9}{16}$

4. $\dfrac{7}{24}$ hundredth

5. $\dfrac{5}{32}$ thousandth

6. $\dfrac{11}{75}$ thousandth

7. $\dfrac{3}{11}$ use bar notation

8. $\dfrac{7}{11}$ use bar notation

9. $\dfrac{16}{33}$ use bar notation

10. $3\dfrac{4}{5}$

11. $3\dfrac{7}{8}$

12. $8\dfrac{3}{16}$

Convert each decimal to a fraction.

13. 0.3

14. 0.55

15. 0.305

16. 0.1

17. 0.875

18. 0.125

1.	0.875
2.	0.6875
3.	0.5625
4.	0.29
5.	0.156
6.	0.147
7.	$0.\overline{27}$
8.	$0.\overline{63}$
9.	$0.\overline{48}$
10.	3.8
11.	3.875
12.	8.1875
13.	$\dfrac{3}{10}$
14.	$\dfrac{55}{100} = \dfrac{11}{20}$
15.	$\dfrac{305}{1000} = \dfrac{61}{200}$
16.	$\dfrac{1}{10}$
17.	$\dfrac{875}{1000} = \dfrac{7}{8}$
18.	$\dfrac{125}{1000} = \dfrac{1}{8}$

Answers

1. 0.875 **3.** 0.5625 **5.** 0.156 **7.** $0.\overline{27}$ **9.** $0.\overline{48}$

11. 3.875 **13.** $\dfrac{3}{10}$ **15.** $\dfrac{61}{200}$ **17.** $\dfrac{7}{8}$

Special Group Activity. Not all numbers that are used involve place value. We can also use numbers to identify as well as count. When numbers are used in this fashion, place value does not play a role.

In many areas, bar codes are used to identify and code material. The simplest code is the Postnet code the U.S. Postal Service uses to identify the nine-digit ZIP code. This code appears in business reply forms and consists of 52 vertical bars of two lengths: long and short. The long lines at the beginning and end are called *guard bars* and are a frame for the other 50 bars. In groups of 5, the 50 bars within the guard bars represent the ZIP + 4 code and a tenth digit for error correction.

Each block of 5 has 2 long and 3 short bars using the following pattern:

Decimal Digit	Bar Code
1	‖‖‖
2	‖‖‖
3	‖‖‖
4	‖‖‖
5	‖‖‖
6	‖‖‖
7	‖‖‖
8	‖‖‖
9	‖‖‖
0	‖‖‖

The tenth digit of a Postnet code is a check digit chosen so that the digits of a ZIP + 4 plus the check digit is evenly divisible by 10. Thus the ZIP + 4 code 19030-5404 has a check digit of 4 since $1 + 9 + 0 + 3 + 0 + 5 + 4 + 0 + 4 + 4 = 30$.

(a) What ZIP + 4 code is represented by the following? (Lines are inserted for visibility.)

(b) Write a bar code for your home address and check it against a piece of business mail you recently received. Be sure to include the check digit.

(c) Name two other areas in which bar codes are used as identification numbers.

Square Roots and the Pythagorean Theorem

9.6 OBJECTIVES

1. Find the square root of a perfect square.
2. Approximate the square root of a number.

Some numbers can be written as the product of two identical factors, for example,

$$9 = 3 \times 3$$

Either factor is called a **square root** of the number. The symbol $\sqrt{}$ (called a **radical sign**) is used to indicate a square root. Thus $\sqrt{9} = 3$ since $3 \times 3 = 9$.

● Example 1

Finding the Square Root

To use the key on your calculator, first enter the 49, then press the key.

Find the square root of 49 and of 16.

(a) $\sqrt{49} = 7$ Since $7 \times 7 = 49$

(b) $\sqrt{16} = 4$ Since $4 \times 4 = 16$

● ● ● **CHECK YOURSELF 1**

Find the square root of each of the following.

a. $\sqrt{121}$ **b.** $\sqrt{36}$

The most frequently used theorem in geometry is undoubtedly the Pythagorean theorem. In this section you will use that theorem. You will also learn a little about the history of the theorem. It is a theorem that applies only to right triangles.

The side opposite the right angle of a right triangle is called the **hypotenuse.**

● Example 2

Identifying the Hypotenuse

In the following right triangle, the side labeled c is the hypotenuse.

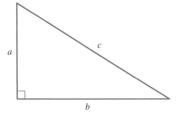

● ● ●   **CHECK YOURSELF 2**

Which side represents the hypotenuse of the given right triangle?

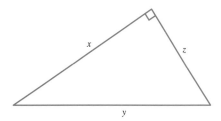

The numbers 3, 4, and 5 have a special relationship. Together they are called a **perfect triple,** which means that when you square all three numbers, the sum of the smaller squares equals the squared value of the larger number.

● Example 3

Identifying Perfect Triples

Show that each of the following is a perfect triple.

(*a*) 3, 4, and 5

$3^2 = 9$, $4^2 = 16$, $5^2 = 25$

and $9 + 16 = 25$, so we can say that $3^2 + 4^2 = 5^2$.

(*b*) 7, 24, and 25

$7^2 = 49$, $24^2 = 576$, $25^2 = 625$

and $49 + 576 = 625$, so we can say that $7^2 + 24^2 = 25^2$.

● ● ●   **CHECK YOURSELF 3**

Show that each of the following is a perfect triple.

a. 5, 12, and 13   **b.** 6, 8, and 10

All the triples that you have seen, and many more, were known by the Babylonians more than 4000 years ago. Stone tablets that had dozens of perfect triples carved into them have been found. The basis of the Pythagorean theorem was understood long before the time of Pythagoras (ca. 540 BC). The Babylonians not only understood perfect triples but also knew how triples related to a right triangle.

The Pythagorean Theorem (Version 1)

If the lengths of the three sides of a right triangle are all integers, they will form a perfect triple, with the hypotenuse as the longest side.

• Example 4

Finding the Length of a Leg of a Right Triangle

Find the missing length for each right triangle.

(*a*)

(*b*)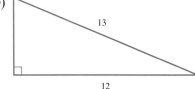

(*a*) A perfect triple will be formed if the hypotenuse is 5 units long, creating the triple 3, 4, 5.

(*b*) The triple must be 5, 12, 13, which makes the missing length 5 units.

● ● ●   CHECK YOURSELF 4

Find the length of the unlabeled side for each right triangle.

a.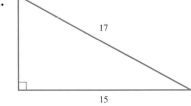

b.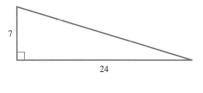

There are two other forms in which the Pythagorean theorem is regularly presented. It is important that you see the connection between the three forms.

The Pythagorean Theorem (Version 2)

The square of the hypotenuse of a right triangle is equal to the sum of the squares of the other two sides.

This is the version that you will refer to in your algebra classes.

The Pythagorean Theorem (Version 3)

Given a right triangle with sides *a* and *b* and hypotenuse *c*, it is always true that

$$c^2 = a^2 + b^2$$

•Example 5

Using the Pythagorean Theorem

The triangle has sides 6, 8, and 10.

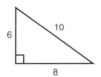

If the lengths of two sides of a right triangle are 6 and 8, find the length of the hypotenuse.

$c^2 = a^2 + b^2$ The value of the hypotenuse is found from the Pythagorean theorem with $a = 6$ and $b = 8$.

$c^2 = (6)^2 + (8)^2 = 36 + 64 = 100$

$c = \sqrt{100} = 10$ The length of the hypotenuse is 10 (since $10^2 = 100$)

● ● ● CHECK YOURSELF 5

Find the hypotenuse of a right triangle whose sides measure 9 and 12.

In some right triangles, the lengths of the hypotenuse and one side are given and we are asked to find the length of the missing side.

•Example 6

Using the Pythagorean Theorem

Find the missing length.

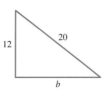

$a^2 + b^2 = c^2$ Use the Pythagorean theorem with $a = 12$ and $c = 20$.

$(12)^2 + b^2 = (20)^2$

$144 + b^2 = 400$

$b^2 = 400 - 144 = 256$

$b = \sqrt{256} = 16$ The missing side is 16.

● ● ● **CHECK YOURSELF 6**

Find the missing length for a right triangle with one leg measuring 8 centimeters (cm) and the hypotenuse measuring 10 cm.

___

Not every square root is a whole number. In fact, there are only 10 whole-number square roots for the numbers from 1 to 100. They are the square roots of 1, 4, 9, 16, 25, 36, 49, 64, 81, and 100. However, we can approximate square roots that are not whole numbers. For example, we know that the square root of 12 is not a whole number. We also know that its value must lie somewhere between the square root of 9 ($\sqrt{9} = 3$) and the square root of 16 ($\sqrt{16} = 4$). That is, $\sqrt{12}$ is between 3 and 4.

● Example 7

Approximating Square Roots

Approximate $\sqrt{29}$.
   The $\sqrt{25} = 5$ and the $\sqrt{36} = 6$, so the $\sqrt{29}$ must be between 5 and 6.

● ● ● **CHECK YOURSELF 7**

$\sqrt{19}$ is between which of the following?

a. 4 and 5   **b.** 5 and 6   **c.** 6 and 7

___

A scientific calculator can be used to evaluate expressions that contain square roots, as Example 8 illustrates.

● Example 8

Evaluating Expressions Using a Calculator

Use a scientific calculator to approximate the value of each expression. Round your answer to the nearest hundredth.

(a) $\sqrt{177}$ Using the calculator, you find $\sqrt{177} = 13.3041\ldots$ To the nearest hundredth, $\sqrt{177} = 13.30$.

(b) $4(\sqrt{93})$ Be certain that you enter the entire expression into the calculator. Then round the answer. Here, $4(\sqrt{93}) = 38.5746\ldots$ to the nearest hundredth, $4(\sqrt{93}) = 38.57$.

● ● ● **CHECK YOURSELF 8**

Use a scientific calculator to approximate the value of each expression. Round your answer to the nearest hundredth.

a. $\sqrt{357}$ **b.** $7(\sqrt{71})$

● ● ● **CHECK YOURSELF ANSWERS**

1. (a) 11; **(b)** 6.   **2.** Side y.   **3. (a)** $5^2 + 12^2 = 25 + 144 = 169$, $13^2 = 169$, so $5^2 + 12^2 = 13^2$; **(b)** $6^2 + 8^2 = 36 + 64 = 100$, $10^2 = 100$ so $6^2 + 8^2 = 10^2$. **4. (a)** 8; **(b)** 25.   **5.** 15.   **6.** 6.   **7. (a)** 4 and 5.   **8. (a)** 18.9; **(b)** 59.0.

9.6 Exercises

In Exercises 1 to 4, find the square root.

1. $\sqrt{64}$

2. $\sqrt{121}$

3. $\sqrt{169}$

4. $\sqrt{196}$

Identify the hypotenuse of the given triangles by giving its letter.

5.

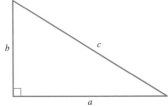

6.

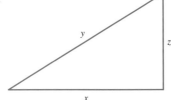

For Exercises 7 to 12, identify which numbers are perfect triples.

7. 3, 4, 5

8. 4, 5, 6

9. 7, 12, 13

10. 5, 12, 13

11. 8, 15, 17

12. 9, 12, 15

For Exercises 13 to 16, find the missing length for each right triangle.

13.

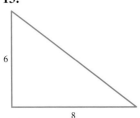

14.

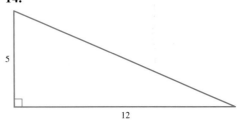

15.

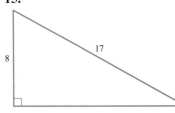

16.

1.	8
2.	11
3.	13
4.	14
5.	c
6.	y
7.	Yes
8.	No
9.	No
10.	Yes
11.	Yes
12.	Yes
13.	10
14.	13
15.	15
16.	24

543

Select the correct approximation for each of the following.

17. Is $\sqrt{23}$ between (*a*) 3 and 4, (*b*) 4 and 5, or (*c*) 5 and 6?

18. Is $\sqrt{15}$ between (*a*) 1 and 2, (*b*) 2 and 3, or (*c*) 3 and 4?

19. Is $\sqrt{44}$ between (*a*) 6 and 7, (*b*) 7 and 8, or (*c*) 8 and 9?

20. Is $\sqrt{31}$ between (*a*) 3 and 4, (*b*) 4 and 5, or (*c*) 5 and 6?

Answers

1. 8   **3.** 13   **5.** *c*   **7.** Yes   **9.** No   **11.** Yes   **13.** 10   **15.** 15
17. b   **19.** a

9.6 Supplementary Exercises

Name

Section   Date

Identify which of the following are right triangles.

1.
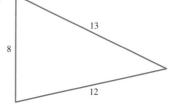

2.
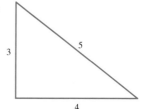

3.
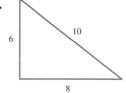

4.
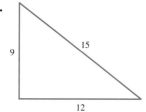

5.
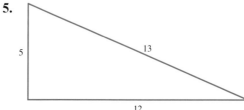

6.
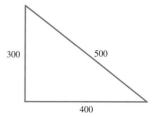

Using Your Calculator to Find Square Roots

To find a square root on your scientific calculator, you use the square root key. On some calculators, you simply enter the number, then press the square root key. With others, you must use the second function on the x^2 (or y^x) key and specify the root you wish to find.

• Example 1

Finding a Square Root Using the Calculator

Find the square root of 256.

256 $\sqrt{}$

Display 16

or

256 2nd $\boxed{\sqrt[x]{y} \atop y^x}$ 2 $=$

Display 16 The "2" is entered for the 2nd (square) root.

●●● CHECK YOURSELF 1

Find the square root of 361.

As we saw in the previous section, not every square root is a whole number. Your calculator can help give you the *approximate* square root of any number.

• Example 2

Finding an Approximate Square Root

Approximate the square root of 29. Round your answer to the nearest tenth.
   Enter

29 $\sqrt{}$

Your calculator display will read something like this:

Display 5.385164807

This is an *approximation* of the square root. It is rounded to the nearest billionth place. The calculator cannot display the exact answer because there is no end to the sequence of digits (and also no pattern.) If the square root of a whole number is not another whole number, then the answer has an infinite number of digits.

To find the approximate square number, we round to the nearest tenth. Our approximation for the square root of 29 is 5.4.

● ● ● CHECK YOURSELF 2

Approximate the square root of 19. Round your answer to the nearest tenth.

● ● ● CHECK YOURSELF ANSWERS

1. 19.   **2.** 4.4.

Calculator Exercises

Name

Section   Date

ANSWERS

Use your calculator to find the square root of each of the following.

1. 64	**2.** 144
3. 289	**4.** 1024
5. 1849	**6.** 784
7. 8649	**8.** 5329
9. 3844	**10.** 3364

1. 8
2. 12
3. 17
4. 32
5. 43
6. 28
7. 93
8. 73
9. 62
10. 58
11. 4.8
12. 5.6
13. 7.1
14. 6.5
15. 11.6
16. 15.8

Use your calculator to approximate the following square roots. Round to the nearest tenth.

11. 23	**12.** 31
13. 51	**14.** 42
15. 134	**16.** 251

Answers

1. 8   **3.** 17   **5.** 43   **7.** 93   **9.** 62   **11.** 4.8   **13.** 7.1   **15.** 11.6

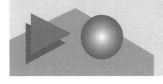

A N S W E R S

1. 2.75

2. 2.385

3. 0.46

4. 0.145

5. $23.28

6. 0.65 in.

7. 6.7

8. 3.225

9. 2.84

10. 5.53

11. 2.02

12. 0.541

13. 30.3 mi/gal

14. 29 shirts

The purpose of the Self-Test is to help you check your progress and review for a chapter test in class. Allow yourself about 1 hour to take the test. When you are done, check your answers in the back of the book. If you missed any answers, be sure to go back and review the appropriate sections in the chapter and do the supplementary exercises provided there.

[9.1] In Exercises 1 to 4, divide.

1. $9\overline{)24.75}$

2. $138.33 \div 58$

3. $28\overline{)12.85}$ hundredth

4. $8.97 \div 62$ thousandth

In Exercises 5 and 6, solve the applications.

5. Alice buys a television set and, with interest charges included, owes $558.72. If she agrees to make equal monthly payments for 2 years, what is the amount of each monthly payment?

6. Rainfall amounts at an airport weather station were measured at 1.12, 0.68, 0.04, 0.2, 1.31, 0.5, and 0.72 inches (in.) for a 7-day period. What was the average rainfall (to the nearest hundredth of an inch) for the 7 days?

[9.2] In Exercises 7 to 12, divide.

7. $5.8\overline{)38.86}$

8. $0.24\overline{)0.774}$

9. $0.075\overline{)0.213}$

10. $5.3\overline{)29.284}$ hundredth

11. $0.258\overline{)0.5218}$ hundredth

12. $2 \div 3.7$ thousandth

In Exercises 13 to 15, solve the following applications.

13. Gas mileage. An automobile travels 627 miles (mi) on 20.7 gallons (gal) of gas. What is the average gas mileage for that period (to the nearest tenth of a mile per gallon)?

14. Shirts produced. A shirt pattern requires 2.25 yards (yd) of fabric. If a tailor has 65.25 yd of the fabric available, how many shirts can be made?

547

15.	45 lots
16.	3.857
17.	0.02847
18.	0.003795
19.	$5.37
20.	0.828 L
21.	0.875
22.	2.5625
23.	$0.\overline{63}$
24.	$\frac{29}{100}$
25.	$\frac{14}{25}$
26.	$\frac{313}{400}$
27.	$\frac{49}{100}$
28.	$\frac{3}{8}$
29.	$7\frac{1}{5}$
30.	$23\frac{39}{50}$
31.	13
32.	16
33.	20
34.	10
35.	24

15. Land division. A 21-acre piece of land is being divided for a housing project. Roads will use 4.8 acres, and the remainder will be divided into 0.36-acre lots. How many lots can be formed?

[9.3] In Exercises 16 to 18, divide.

16. $385.7 \div 100$ **17.** $28.47 \div 1000$

18. $37.95 \div 10^4$

Solve the following applications.

19. Cost per item. A shipment of 100 items costs a store $537. What is the cost of an individual item?

20. Conversion. To convert from milliliters (mL) to liters (L), we divide by 1000. A bottle of soft drink contains 828 mL. What is its volume in liters?

[9.4] In Exercises 21 to 23, find the decimal equivalents of the given fractions.

21. $\dfrac{7}{8}$ **22.** $2\dfrac{9}{16}$ **23.** $\dfrac{7}{11}$ use bar notation

In Exercises 24 to 26, write the decimals as common fractions or mixed numbers.

24. 0.29   **25.** 0.56   **26.** 0.7825

[9.5] Write each of the following as a common fraction or mixed number.

27. 0.49   **28.** 0.375   **29.** 7.2   **30.** 23.78

[9.6] Find the square root of each of the following.

31. $\sqrt{169}$ **32.** $\sqrt{256}$ **33.** $\sqrt{400}$

34. Find the hypotenuse of the right triangle whose sides measure 6 and 8.

35. Find the missing side in the given triangle.

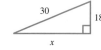

Summary for Chapters 8 and 9

The Language of Decimals

$\frac{7}{10}$ and $\frac{47}{100}$ are decimal fractions.

Decimal Fraction A fraction whose denominator is a power of 10. We call decimal fractions *decimals*.

2.3456
— Ten thousandths
— Thousandths
— Hundredths
— Tenths

Decimal Place Each position for a digit to the right of the decimal point. Each decimal place has a place value that is one-tenth the value of the place to its left.

Reading and Writing Decimals in Words

Hundredths

8.15 is read "eight and fifteen hundredths."

1. Read the digits *to the left* of the decimal point as a whole number.
2. Read the decimal point as the word "and."
3. Read the digits *to the right* of the decimal point as a whole number followed by the place value of the rightmost digit.

Adding Decimals

To add 2.7, 3.15, and 0.48:

```
  2.7
  3.15
+0.48
  6.33
```

1. Write the numbers being added in column form with their decimal points in a vertical line.
2. Add just as you would with whole numbers.
3. Place the decimal point of the sum in line with the decimal points of the addends.

Subtracting Decimals

To subtract 5.875 from 8.5:

```
 8.500
-5.875
 2.625
```

1. Write the numbers being subtracted in column form with their decimal points in a vertical line. You may have to place zeros to the right of the existing digits.
2. Subtract just as you would with whole numbers.
3. Place the decimal point of the difference in line with the decimal points of the numbers being subtracted.

Multiplying Decimals

To multiply 2.85×0.045:

```
  2.85   ← Two places
×0.045   ← Three places
  1425
  1140
0.12825  ← Five places
```

1. Multiply the decimals as though they were whole numbers.
2. Add the number of decimal places in the numbers being multiplied.
3. Place the decimal point in the product so that the number of decimal places in the product is the sum of the number of decimal places in the factors.

Multiplying by Powers of 10

$2.37 \times 10 = 23.7$

$0.567 \times 1000 = 567$

Move the decimal point to the right the same number of places as there are zeros in the power of 10.

Rounding Decimals

To round 5.87 to the nearest tenth:

5.87 is rounded to 5.9

To round 12.3454 to the nearest thousandth:

12.3454 is rounded to 12.345.

1. Find the place to which the decimal is to be rounded.
2. If the next digit to the right is 5 or more, increase the digit in the place you are rounding to by 1. Discard any remaining digits to the right.
3. If the next digit to the right is less than 5, just discard that digit and any remaining digits to the right.

Dividing Decimals

To Divide by a Decimal

To divide 2.3147 by 0.395, move the decimal points:

$$
\begin{array}{r}
5.86 \\
0.395\overline{)2.314\,70} \\
\underline{1\,975} \\
339\;7 \\
\underline{316\;0} \\
23\;70 \\
\underline{23\;70} \\
0
\end{array}
$$

1. Move the decimal point to the right, making the divisor a whole number.
2. Move the decimal point in the dividend to the right the same number of places. Add zeros if necessary.
3. Place the decimal point in the quotient directly above the decimal point of the dividend.
4. Divide as you would with whole numbers.

To Divide by a Power of 10

$25.8 \div 10 = 2\,5.8 = 2.58$

$34.789 \div 1000 = 0\,034.789$

$= 0.034789$

Move the decimal point to the left the same number of places as there are zeros in the power of 10.

Converting Common Fractions to Decimals

To Convert a Common Fraction to a Decimal

To convert $\frac{5}{8}$ to a decimal:

$$
\begin{array}{r}
0.625 \\
8\overline{)5.000} \\
\underline{4\,8} \\
20 \\
\underline{16} \\
40 \\
\underline{40} \\
0
\end{array}
\qquad \frac{5}{8} = 0.625
$$

1. Divide the numerator of the common fraction by its denominator.
2. The quotient is the decimal equivalent of the common fraction.

To Convert a Terminating Decimal Less Than 1 to a Common Fraction

To convert 0.275 to a common fraction:

$0.275 = \frac{275}{1000} = \frac{11}{40}$

1. Write the digits of the decimal without the decimal point. This will be the numerator of the common fraction.
2. The denominator of the fraction is a 1 followed by as many zeros as there are places in the decimal.

Square Roots and the Pythagorean Theorem

$\sqrt{25} = 5$ because $5^2 = 25$

$3^2 + 4^2 = 5^2$

The square root of a number is a value that, when squared, gives us that number.

The length of the three sides of a right triangle will form a perfect triple.

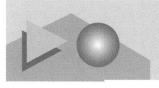

Summary Exercises

Name

Section   Date

You should now be reviewing the material in Chapters 8 and 9 of the text. The following exercises will help in that process. Work all the exercises carefully. Then check your answers in the back of the book. References are provided to the chapter and section for each exercise. If you made an error, go back and review the related material and do the supplementary exercises for that section.

[8.1]   In Exercises 1 and 2, find the indicated place values.

  1. 7 in 3.5742   **2.** 3 in 0.5273

In Exercises 3 and 4, write the fractions in decimal form.

  3. $\dfrac{37}{100}$ **4.** $\dfrac{307}{10,000}$

In Exercises 5 and 6, write the decimals in words.

  5. 0.071   **6.** 12.39

In Exercises 7 and 8, write the fractions in decimal form.

  7. Four and five tenths

  8. Four hundred and thirty-seven thousandths

In Exercises 9 to 12, complete each statement using the symbol $<$, $=$, or $>$.

  9. 0.79 _____ 0.785   **10.** 1.25 _____ 1.250

  11. 12.8 _____ 13   **12.** 0.832 _____ 0.83

[8.2]   In Exercises 13 to 15, round to the indicated place.

  13. 5.837   hundredth   **14.** 9.5723   thousandth

  15. 4.87625   three decimal places

[8.2]   In Exercise 16, solve the application.

  16. Find the area (to the nearest hundredth of a square centimeter) of a rectangle that has dimensions 5.25 by 8.75 cm.

ANSWERS

1. Hundredths

2. Ten thousandths

3. 0.37

4. 0.0307

5. Seventy-one thousandths

6. Twelve and thirty-nine hundredths

7. 4.5

8. 400.037

9. $>$

10. $=$

11. $<$

12. $>$

13. 5.84

14. 9.572

15. 4.876

16. 45.94 cm^2

551

[8.3]   In Exercises 17 to 20, add.

17.   2.58
          +0.89

18.   3.14
           0.8
           2.912
          +12.

19. 1.3, 25, 5.27, and 6.158

20. Add eight, forty-three thousandths, five and nineteen hundredths, and seven and three tenths.

In Exercises 21 and 22, solve the applications.

21. Distance. Janice ran 4.8 miles (mi) on Sunday, 5.3 mi on Tuesday, 3.9 mi on Thursday, and 8.2 mi on Saturday. How far did she run during the week?

22. Dimensions. Find dimension *a* in the following figure.

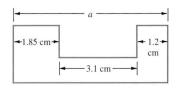

[8.4]   In Exercises 23 to 26, subtract.

23.   29.21
          − 5.89

24.   6.73
          −2.485

25. 1.735 from 2.81

26. 12.38 from 19

In Exercises 27 and 28, solve the applications.

27. Savings. A stereo system that normally sells for $499.50 is discounted (or marked down) to $437.75 for a sale. Find the savings.

28. Cash remaining. If you cash a $50 check and make purchases of $8.71, $12.53, and $9.83, how much money do you have left?

[8.5] In Exercises 29 to 32, multiply.

29. 22.8
       ×0.72

30. 0.0045
       × 0.058

31. 1.24 × 56

32. 0.0025 × 0.491

In Exercises 33 to 35, solve the applications.

33. Earnings. Neal worked for 37.4 hours (h) during a week. If his hourly rate of pay was $7.25, how much did he earn?

34. Interest. To find the interest on a loan at $11\frac{1}{2}$ percent for 1 year, we must multiply the amount of the loan by 0.115. Find the interest on a $2500 loan at $11\frac{1}{2}$ percent for 1 year.

35. Installment plan costs. A television set has an advertised price of $499.50. You buy the set and agree to make payments of $27.15 per month for 2 years. How much extra are you paying by buying on this installment plan?

[8.6] Find the perimeter or circumference of each figure.

36.

12 ft

37.

5 in.

5 in.

38. Find the area of a circle with radius 10 ft.

[8.7] In Exercises 39 and 40, find the area of each figure.

39.

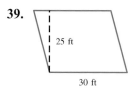

25 ft

30 ft

40.

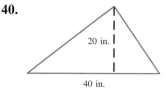

20 in.

40 in.

29. 16.416

30. 0.000261

31. 69.44

32. 0.0012275

33. $271.15

34. $287.50

35. $152.10

36. 37.7 ft

37. 22.9 in.

38. 314 ft²

39. 750 ft²

40. 400 in.²

In Exercises 41 and 42, solve the applications.

41. Area. How many square feet of vinyl floor covering will be needed to cover the floor of a room that is 10 by 18 feet (ft)? How many square yards (yd²) will be needed?

42. Cost. A rectangular roof for a house addition measures 15 by 30 ft. A roofer will charge $175 per "square" (100 ft²). Find the cost of the roofing for the addition.

In Exercises 43 and 44, multiply.

43. 0.052×1000

44. 0.045×10^4

In Exercise 45, solve the application.

45. Total cost. A stereo dealer buys 100 portable radios for a promotion sale. If she pays $57.42 per radio, what is her total cost?

[9.1] In Exercises 46 to 48, divide.

46. $8\overline{)3.08}$

47. $58\overline{)269.7}$

48. $55\overline{)17.69}$ thousandth

In Exercises 49 and 50, solve the applications.

49. Employee donation. During a charity fund-raising drive 37 employees of a company donated a total of $867.65. What was the donation per employee?

50. Mileage. In six readings, Faith's gas mileage was 38.9, 35.3, 39.0, 41.2, 40.5, and 40.8 miles per gallon (mi/gal). What was the mean mileage to the nearest tenth of a mile per gallon?

[9.2] In Exercises 51 to 54, divide.

51. $0.7\overline{)1.862}$

52. $3.042 \div 0.36$

53. $5.3\overline{)6.748}$ tenth

54. $0.2549 \div 2.87$ three decimal places

In Exercises 55 and 56, solve the applications.

55. Lot quantity. A developer is planning to subdivide an 18.5-acre piece of land. She estimates that 5 acres will be used for roads and wants individual lots of 0.25 acre. How many lots are possible?

56. Mileage. Paul drives 949 mi using 31.8 gal of gas. What is his mileage for the trip (to the nearest tenth of a mile per gallon)?

[9.3] In Exercises 57 to 59, divide.

57. $7.6 \div 10$ **58.** $80.7 \div 1000$ **59.** $457 \div 10^4$

In Exercise 60, solve the application.

60. Tape cost. A shipment of 1000 videotapes cost a dealer $7090. What was the cost per tape to the dealer?

[9.4] In Exercises 61 to 64, find the decimal equivalents.

61. $\dfrac{7}{16}$ **62.** $\dfrac{3}{7}$ thousandth

63. $\dfrac{4}{15}$ (use bar notation)   **64.** $3\dfrac{3}{4}$

[9.5] In Exercises 65 to 67, write as common fractions or mixed numbers.

65. 0.21 **66.** 0.084 **67.** 5.28

[9.6] Find the square root of each of the following.

68. $\sqrt{324}$ **69.** $\sqrt{784}$

70. Find the hypotenuse of the triangle whose sides are 33 and 44.

ANSWERS

55. 54 lots

56. 29.8 mi/gal

57. 0.76

58. 0.0807

59. 0.0457

60. $7.09

61. 0.4375

62. 0.429

63. $0.2\overline{6}$

64. 3.75

65. $\dfrac{21}{100}$

66. $\dfrac{21}{250}$

67. $5\dfrac{7}{25}$

68. 18

69. 28

70. 55

© 1998 McGraw-Hill Companies

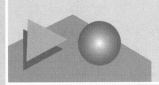

Cumulative Test for Chapters 8 and 9

Name

Section   Date

ANSWERS

1. Ten thousandths

2. 0.049

3. Two and fifty-three hundredths

4. 12.017

5. <

6. >

7. 16.64

8. 47.253

9. 12.803

10. 50.2 gal

11. 10.54

12. 24.375

13. 3.888

14. $3.06

This test is provided to help you in the process of reviewing Chapters 8 and 9. Answers are provided in the back of the book. If you missed any answers, be sure to go back and review the appropriate chapter sections.

1. Find the place value of 8 in 0.5248.

2. Write $\dfrac{49}{1000}$ in decimal form.

3. Write 2.53 in words.

4. Write twelve and seventeen thousandths in decimal form.

In Exercises 5 and 6, complete the statement, using the symbol < or >.

5. 0.889 _____ 0.89

6. 0.531 _____ 0.53

In Exercises 7 to 9, add.

7.   3.45
       0.6
   +12.59

8. 2.4, 35, 4.73, and 5.123.

9. Seven, seventy-nine hundredths, and five and thirteen thousandths.

In Exercise 10, solve the application.

10. Gasoline purchased. On a business trip, Martin bought the following amounts of gasoline: 14.4, 12, 13.8, and 10 gallons (gal). How much gasoline did he purchase on the trip?

In Exercises 11 to 13, subtract.

11.   18.32
    − 7.78

12.   40
  −15.625

13. 1.742 from 5.63

In Exercise 14, solve the application.

14. Cash remaining. You pay for purchases of $13.99, $18.75, $9.20, and $5 with a $50 check. How much cash will you have left?

In Exercises 15 to 17, multiply.

15. 32.9
  $\times 0.53$

16. 0.049
  $\times\ 0.57$

17. 2.75×0.53

In Exercise 18, solve the application.

18. Find the area of a rectangle with length 3.5 inches (in.) and width 2.15 in.

In Exercises 19 and 20, multiply.

19. 0.735×1000

20. 1.257×10^4

In Exercise 21, solve the application.

21. Total costs. A college bookstore purchases 1000 pens at a cost of 54.3 cents per pen. Find the total cost of the order in dollars.

In Exercises 22 and 23, round to the indicated place.

22. 0.5977   thousandth

23. 23.5724   two decimal places

In Exercise 24, solve the application.

24. Circumference. We find the circumference of a circle by multiplying the diameter of the circle by 3.14. If a circle has a diameter of 3.2 ft, find its circumference, to the nearest hundredth of a foot.

In Exercises 25 to 31, divide.

25. $8\overline{)3.72}$

26. $27\overline{)63.45}$

27. $2.72 \div 53$ thousandth

28. $4.1\overline{)10.455}$

29. $0.6\overline{)1.431}$

30. $3.969 \div 0.54$

31. $0.263 \div 3.91$ three decimal places

In Exercise 32, solve the application.

32. Number of lots. A 14-acre piece of land is being developed into home lots. If 2.8 acres of land will be used for roads and each home site is to be 0.35 acre, how many lots can be formed?

In Exercises 33 and 34, divide.

33. $4.983 \div 1000$

34. $523 \div 10^5$

ANSWERS	
15.	17.437
16.	0.02793
17.	1.4575
18.	7.525 in.^{2}
19.	735
20.	12,570
21.	$543
22.	0.598
23.	23.57
24.	10.05 ft
25.	0.465
26.	2.35
27.	0.051
28.	2.55
29.	2.385
30.	7.35
31.	0.067
32.	32 lots
33.	0.004983
34.	0.00523

In Exercise 35, solve the application.

35. Costs to families. A street improvement project will cost $57,340 and that cost is to be divided among the 100 families in the area. What will be the cost to each individual family?

In Exercises 36 to 38, find the decimal equivalents of the common fractions.

36. $\dfrac{7}{16}$ **37.** $\dfrac{3}{7}$ thousandth   **38.** $\dfrac{7}{11}$ (use bar notation)

In Exercises 39 and 40, write the decimals as common fractions or mixed numbers.

39. 0.072   **40.** 4.44

INTRODUCTION

Many people enjoy building model airplanes and cars. Some hobbyists enjoy it enough to pursue it as a career. Tranh is one who did just that.

When Tranh looked for a career, he knew that he had a head start in pursuing a degree in drafting. His interest in model building had led him to working with many of the tools draftspeople use. Once he got his degree, he turned his attention back to model building. For several years, Tranh worked with an architectural firm. He helped construct building models that were given to prospective customers. In the past 2 years, Tranh has combined his interest in theater with his skills: He now builds models of sets to be used in theater productions.

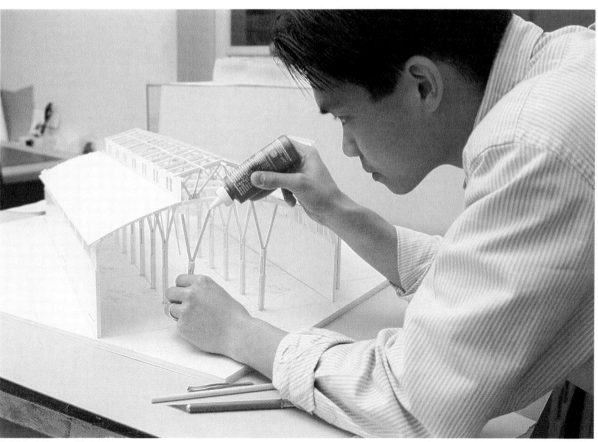

© Michael Newman/PhotoEdit

ANSWERS

1. $\dfrac{7}{10}$

2. $\dfrac{4}{3}$

3. Yes

4. No

5. 10

6. 15

7. 12

8. $6.30

9. 32

10. 11 gal

Ratio and Proportion

This pretest will point out any difficulties you may be having with ratios and proportions. Do all the problems. Then check your answers with those in the back of the book.

1. Write the ratio of 7 to 10.

2. Write the ratio of 20 to 15 in lowest terms.

3. Is $\dfrac{4}{7} = \dfrac{12}{21}$ a true proportion?

4. Is $\dfrac{5}{9} = \dfrac{9}{16}$ a true proportion?

5. Solve for x: $\dfrac{x}{4} = \dfrac{5}{2}$

6. Solve for a: $\dfrac{5}{a} = \dfrac{7}{21}$

7. Solve for n: $\dfrac{\frac{1}{2}}{2} = \dfrac{3}{n}$

8. **Pricing.** Cans of tomato juice are marked 2 for $1.05. At this price, what will 12 cans cost?

9. **Car sales.** The ratio of compact cars to larger model cars sold during a month was 9 to 4. If 72 compact cars were sold during that period, how many of larger cars were sold?

10. **Cost.** If 2 gal of paint will cover 450 ft^2, how many gallons will be needed to paint a room with 2475 ft^2 of wall surface?

Using Ratios

10.1 OBJECTIVES

1. Write the ratio of two numbers in simplest form.
2. Write the ratio of two quantities in simplest form.

In earlier chapters, you saw two meanings for a fraction:

First Meaning   A fraction can name a certain number of parts of a whole.

$\dfrac{3}{5}$ names 3 parts of a whole that has been divided into 5 equal parts.

Second Meaning   A fraction can indicate division.

$\dfrac{3}{5}$ can be thought of as $3 \div 5$.

We now want to turn to a third meaning for a fraction:

Third Meaning   A fraction can be a ratio. A **ratio** is a means of comparing two numbers or quantities.

● Example 1

Writing a Ratio as a Fraction

Write the ratio 3 to 5 as a fraction.

   To compare 3 to 5, we write the ratio of 3 to 5 as $\dfrac{3}{5}$. So $\dfrac{3}{5}$ also means "the ratio of 3 to 5."

Note: Another way of writing the ratio of 3 to 5 is 3:5. We have chosen to stay with the fraction notation for a ratio in this textbook.

● ● ● CHECK YOURSELF 1

Write the ratio of 7 to 12 as a fraction.

   Example 2 illustrates the use of a ratio in comparing *like quantities,* which means we're comparing inches to inches, cm to cm, apples to apples, etc.

● Example 2

Applying the Concept of Ratio

The width of a rectangle is 7 cm and its length is 19 cm. Write the ratio of its width to its length as a fraction.

$$\dfrac{7 \text{ cm}}{19 \text{ cm}} = \dfrac{7}{19}$$

We are comparing centimeters to centimeters, so we don't need to write the units.

A ratio fraction can be greater than 1.

Note: In this case the ratio is *never* written as a mixed number. It is left as an improper fraction.

The ratio of its length to its width is

$$\frac{19 \text{ cm}}{7 \text{ cm}} = \frac{19}{7}$$

● ● ● **CHECK YOURSELF 2**

A basketball team wins 17 of its 29 games in a season.

a. Write the ratio of wins to games played.
b. Write the ratio of wins to losses.

Since a ratio is a fraction, we can reduce it to simplest form. Consider Example 3.

● Example 3

Writing a Ratio in Simplest Form

Write the ratio of 20 to 30 in lowest terms.

$$\frac{20}{30} = \frac{2}{3}$$ Divide the numerator and denominator by the common factor of 10.

● ● ● **CHECK YOURSELF 3**

Write the ratio of 24 to 32 in lowest terms.

Ratios can be used to compare similar quantities for different items. Example 4 illustrates how this can be done.

● Example 4

An Application Comparing Gas Mileage

A medium-sized car has a mileage rating of 18 miles per gallon (mi/gal). A compact car has a rating of 30 mi/gal. The ratio of the mileage rating of the medium-sized car to that of the compact car is

$$\frac{18}{30} = \frac{3}{5}$$ Divide the numerator and denominator by 6.

● ● ● **CHECK YOURSELF 4**

A station wagon has a mileage rating of 16 mi/gal. A subcompact car has a rating of 36 mi/gal. Write the ratio of the mileage rating of the station wagon to that of the subcompact car.

In some cases, the ratio of two unlike quantities must be found, as shown in Example 5.

● Example 5

An Application Comparing Unlike Quantities

The cost of 32 ounces (oz) of orange juice is $1.20. Write the ratio of ounces to cost (in cents).

$$\frac{32 \text{ oz}}{120\cent} = \frac{4 \text{ oz}}{15\cent}$$ Divide the numerator and denominator by 8. Because the units are different, we write them in the fraction.

● ● ● **CHECK YOURSELF 5**

A baseball player has 12 hits in 42 times at bat. Write the ratio of hits to times at bat.

Sometimes we must convert the identifying units to simplify a ratio, as shown in Example 6.

● Example 6

Converting Units to Simplify Ratios

(*a*) Write the ratio of 3 minutes (min) to 2 hours (h).

1 h is 60 min, so 2 h is 120 min.

$$\frac{3 \text{ min}}{2 \text{ h}} = \frac{3 \text{ min}}{120 \text{ min}} = \frac{3}{120} = \frac{1}{40}$$

Write 2 h as 120 min. Then divide the like units and reduce the ratio fraction to lowest terms.

(b) Write the ratio of 8 oz to 3 pounds (lb).

1 lb is 16 oz, so write 3 lb as 48 oz.

$$\frac{8 \text{ oz}}{3 \text{ lb}} = \frac{8 \text{ oz}}{48 \text{ oz}} = \frac{1}{6}$$

(c) Write the ratio of 5 quarters to 3 dollars.

$$\frac{5 \text{ quarters}}{3 \text{ dollars}} = \frac{125\cent}{300\cent} = \frac{5}{12}$$

Convert both terms of the ratio to cents. Then reduce the fraction.

● ● ● **CHECK YOURSELF 6**

Write the ratio of

a. 3 inches (in.) to 2 feet (ft)   **b.** 3 days to 5 hours (h)   **c.** 5 dimes to 3 quarters

---

● ● ● **CHECK YOURSELF ANSWERS**

1. $\dfrac{7}{12}$.   **2. (a)** $\dfrac{17}{29}$; **(b)** $\dfrac{17}{12}$ (the team lost 12 games).   **3.** $\dfrac{3}{4}$.

4. $\dfrac{4}{9}$. Did you reduce to lowest terms?   **5.** $\dfrac{2}{7}$.   **6. (a)** $\dfrac{1}{8}$; **(b)** $\dfrac{72}{5}$; **(c)** $\dfrac{2}{3}$.

Write each of the following ratios in simplest form.

1. The ratio of 9 to 13

2. The ratio of 5 to 4

3. The ratio of 9 to 4

4. The ratio of 5 to 12

5. The ratio of 10 to 15

6. The ratio of 16 to 12

7. The ratio of 21 to 14

8. The ratio of 25 to 40

9. The ratio of 17 in. to 30 in.

10. The ratio of 23 lb to 36 lb

11. The ratio of 12 miles (mi) to 18 mi

12. The ratio of 100 centimeters (cm) to 90 cm

13. The ratio of 40 ft to 65 ft

14. The ratio of 12 oz to 18 oz

15. The ratio of $48 to $42

16. The ratio of 20 ft to 24 ft

17. The ratio of 75 seconds (s) to 3 minutes (min)

18. The ratio of 7 oz to 3 lb

19. The ratio of 4 nickels to 5 dimes

20. The ratio of 8 in. to 3 ft

21. The ratio of 2 days to 10 h

22. The ratio of 4 ft to 4 yd

23. The ratio of 5 gallons (gal) to 12 quarts (qt)

24. The ratio of 7 dimes to 3 quarters

A N S W E R S

1. $\dfrac{9}{13}$

2. $\dfrac{5}{4}$

3. $\dfrac{9}{4}$

4. $\dfrac{5}{12}$

5. $\dfrac{2}{3}$

6. $\dfrac{4}{3}$

7. $\dfrac{3}{2}$

8. $\dfrac{5}{8}$

9. $\dfrac{17}{30}$

10. $\dfrac{23}{36}$

11. $\dfrac{2}{3}$

12. $\dfrac{10}{9}$

13. $\dfrac{8}{13}$

14. $\dfrac{2}{3}$

15. $\dfrac{8}{7}$

16. $\dfrac{5}{6}$

17. $\dfrac{5}{12}$

18. $\dfrac{7}{48}$

19. $\dfrac{2}{5}$

20. $\dfrac{2}{9}$

21. $\dfrac{24}{5}$

22. $\dfrac{1}{3}$

23. $\dfrac{5}{3}$

24. $\dfrac{14}{15}$

Solve the following applications.

25. **Class make-up ratio.** An algebra class has 7 men and 13 women. Write the ratio of men to women. Write the ratio of women to men.

26. **Recipe ratio.** A French bread recipe calls for 7 cups of flour for 4 loaves of bread. Write the ratio of cups of flour to loaves of bread.

27. **Football ratio.** A football team wins 9 of its 16 games with no ties. Write the ratio of wins to games played. Write the ratio of wins to losses.

28. **Wage ratio.** Rick makes $53 in an 8-h day. Write the ratio of dollars to hours.

29. **Election ratio.** In a school election 4500 yes votes were cast, and 3000 no votes were cast. Write the ratio of yes to no votes.

30. **Basketball ratio.** A basketball player made 42 of the 70 shots taken in a tournament. Write the ratio of shots made to shots taken.

31. **Unit cost ratio.** A 32-oz bottle of dishwashing liquid costs $1.92. Write the ratio of cents to ounces.

32. **Fuel consumption ratio.** A new compact automobile travels 252 mi on 7 gal of gasoline. Write the ratio of miles to gallons.

Write each ratio as a fraction.

33. $6\dfrac{1}{2}$ oz to $2\dfrac{3}{4}$ lb

34. $7\dfrac{1}{4}$ days to $3\dfrac{1}{2}$ weeks

35. $5\dfrac{3}{16}$ ft to $2\dfrac{1}{8}$ yards (yd)

36. $3\dfrac{3}{8}$ in. to $2\dfrac{1}{4}$ ft

Solve the following applications.

37. **Price increase ratio.** The price of a barbecue grill went from $125 to $160. Find the ratio of the increase in price to the original price.

38. **Price decrease ratio.** The price of a case of dishwashing detergent decreased from $12.50 to $11.75. Find the ratio of the decrease in price to the original price.

39. Employment ratio. A company employs 24 women and 18 men. Write the ratio of men to women employed by the company.

40. Measurement ratio. If a room is 24 ft long by 18 ft wide, write the ratio of the length to the width of the room.

41. Triangle sides ratio. In the triangle pictured, find the ratio of the length of the longest side to the hypotenuse.

42. Money ratio. Sean has 15 dimes and 25 nickels in his pocket. Write the ratio of nickels to dimes.

43. (a) Buy a 1.69-oz (medium size) bag of M&Ms. Determine the ratio of the different colors (yellow, red, blue, orange, brown, and green) to the total number of M&Ms in the bag.

(b) Compare your ratios to those your classmates obtain.
(c) Use the information from parts a and b to determine a ratio for all the different colors in a bag of M&Ms.
(d) Call the manufacturer of M&Ms (Mars, Inc.) and see if they have a fixed ratio they use to determine the distribution of the colors in a bag. If they do, compare this ratio to yours.

⬤ **Getting Ready for Section 10.2 [Section 5.3]**

Determine whether the following pairs of fractions are equivalent.

a. $\dfrac{1}{3}, \dfrac{3}{9}$ **b.** $\dfrac{2}{5}, \dfrac{3}{7}$ **c.** $\dfrac{5}{6}, \dfrac{6}{7}$

d. $\dfrac{5}{6}, \dfrac{25}{30}$ **e.** $\dfrac{3}{10}, \dfrac{2}{7}$ **f.** $\dfrac{4}{15}, \dfrac{12}{45}$

Answers

1. $\dfrac{9}{13}$ **3.** $\dfrac{9}{4}$ **5.** $\dfrac{2}{3}$ **7.** $\dfrac{3}{2}$ **9.** $\dfrac{17}{30}$ **11.** $\dfrac{2}{3}$ **13.** $\dfrac{8}{13}$ **15.** $\dfrac{8}{7}$

17. $\dfrac{5}{12}$ **19.** $\dfrac{2}{5}$ **21.** $\dfrac{24}{5}$ **23.** $\dfrac{5}{3}$ **25.** $\dfrac{7}{13}, \dfrac{13}{7}$ **27.** $\dfrac{9}{16}, \dfrac{9}{7}$

29. $\dfrac{3}{2}$ **31.** $\dfrac{6¢}{1\ oz}$ **33.** $\dfrac{13}{88}$ **35.** $\dfrac{83}{102}$ **37.** $\dfrac{7}{25}$ **39.** $\dfrac{3}{4}$ **41.** $\dfrac{12}{13}$

43.   **a.** Yes **b.** No **c.** No **d.** Yes **e.** No **f.** Yes

567

ANSWERS

39. $\dfrac{3}{4}$

40. $\dfrac{4}{3}$

41. $\dfrac{12}{13}$

42. $\dfrac{5}{3}$

43.

a. Yes

b. No

c. No

d. Yes

e. No

f. Yes

10.1 Supplementary Exercises

Name

Section   Date

ANSWERS

1. $\dfrac{8}{13}$

2. $\dfrac{8}{5}$

3. $\dfrac{4}{3}$

4. $\dfrac{2}{3}$

5. $\dfrac{17}{25}$

6. $\dfrac{12}{7}$

7. $\dfrac{4}{3}$

8. $\dfrac{2}{3}$

9. $\dfrac{3}{40}$

10. $\dfrac{5}{24}$

11. $\dfrac{5}{8}$

12. $\dfrac{5}{36}$

13. $\dfrac{600 \text{ mi}}{17 \text{ gal}}$

14. $\dfrac{2}{3}$, $\dfrac{2}{1}$

15. $\dfrac{2}{3}$

16. $\dfrac{7}{3}$

Write each of the following ratios in simplest form.

1. The ratio of 8 to 13

2. The ratio of 8 to 5

3. The ratio of 20 to 15

4. The ratio of 10 to 15

5. The ratio of 17 lb to 25 lb

6. The ratio of $12 to $7

7. The ratio of 28 oz to 21 oz

8. The ratio of 50 mi to 75 mi

9. The ratio of 6 oz to 5 lb

10. The ratio of 5 in. to 2 ft

11. The ratio of 5 nickels to 4 dimes

12. The ratio of 25 min to 3 h

Solve the following applications.

13. Gas mileage ratio. A Toyota travels 600 mi on 17 gal of gas. Write the ratio of miles to gallons.

14. Basketball ratio. A basketball team wins 30 of its 45 games. Write the ratio of wins to games played. Write the ratio of wins to losses.

15. Weight ratio. A compact car weighs 2400 lb, and a medium-sized car weighs 3600 lb. Write the ratio of the weight of the smaller car to that of the larger.

16. Election ratio. An election on a city budget resulted in 7700 no votes and 3300 yes votes. Write the ratio of no votes to yes votes.

568

© 1998 McGraw-Hill Companies

The Language of Proportions

10.2 OBJECTIVES

1. Use the language of proportions.
2. Determine whether a proportion is a true statement.

This is the same as saying the fractions are equivalent. They name the same number.

Remember that we call a letter representing an unknown value a *variable*. Here a, b, c, and d variables. We could have chosen any other letter.

Proportion

A statement that two ratios are equal is called a **proportion**.

Since the ratio of 1 to 3 is equal to the ratio of 2 to 6, we can write the proportion

$$\frac{1}{3} = \frac{2}{6}$$

The proportion $\frac{a}{b} = \frac{c}{d}$ is read "a is to b as c is to d." We read the proportion $\frac{1}{3} = \frac{2}{6}$ as "one is to three as two is to six."

We can number the terms of a proportion and give them special names. The terms are numbered as shown.

First term   Third term

$$\frac{a}{b} = \frac{c}{d}$$

Second term   Fourth term

The second and third terms are called the **means** of the proportion. The first and fourth terms are the **extremes** of the proportion.

Extreme   Mean

$$\frac{a}{b} = \frac{c}{d}$$

Mean   Extreme

● Example 1

Identifying the Means and the Extremes

For each proportion, identify the means and the extremes.

(a) $\dfrac{1}{3} = \dfrac{2}{6}$

3 and 2 are the means, and 1 and 6 are the extremes.

(b) $\dfrac{a}{5} = \dfrac{3}{15}$

5 and 3 are the means, and a and 15 are the extremes.

● ● ● **CHECK YOURSELF 1**

List the means and extremes in the proportion $\dfrac{5}{8} = \dfrac{20}{32}$.

A useful property holds for the terms of a true proportion. Let's look at one and calculate the product of the means and the product of the extremes.

In the proportion

$$\frac{3}{8} = \frac{9}{24}$$

Since $8 \times 9 = 3 \times 24 = 72$

the means are 8 and 9, so the product of the means is $8 \times 9 = 72$. The extremes are 3 and 24, so the product of the extremes is $3 \times 24 = 72$. The product of the means is equal to the product of the extremes. This property holds for any true proportion.

The Proportion Rule

In a true proportion, the product of the means is equal to the product of the extremes.

Using symbols in the proportion, we get

This is sometimes called **cross multiplication.**

$$\frac{a}{b} = \frac{c}{d}$$

$$\underset{\underset{\text{The product}}{\uparrow}}{b \times c} = \underset{\underset{\text{The product}}{\nwarrow}}{a \times d}$$
of the means   of the extremes

● **Example 2**

Verifying a Proportion

In the following proportions, determine if the product of the means equals the product of the extremes.

(a) $\dfrac{5}{6} \overset{?}{=} \dfrac{10}{12}$

Multiply:

Means: $6 \times 10 = 60$
Extremes: $5 \times 12 = 60$ } Equal

The product of the means equals the product of the extremes, so $\dfrac{5}{6} = \dfrac{10}{12}$ is a true proportion.

(b) $\dfrac{3}{7} \overset{?}{=} \dfrac{4}{9}$

Multiply:

Means: $7 \times 4 = 28$
Extremes: $3 \times 9 = 27$ } Not equal

The products are not equal, so $\dfrac{3}{7} = \dfrac{4}{9}$ is not a true proportion.

● ● ● **CHECK YOURSELF 2**

Determine if the product of the means is equal to the product of the extremes in the following proportions.

a. $\dfrac{5}{8} = \dfrac{20}{32}$ 　　　　　　　　　**b.** $\dfrac{7}{9} = \dfrac{3}{4}$

Do you see that the proportion rule is the same as our earlier rule for testing equivalent fractions? If the cross products are equal, the two fractions are equivalent and a true proportion is formed.

The proportion rule can be used to find out whether a given proportion is a true statement. Look at Example 3.

●Example 3

Verifying a Proportion

(a) Is $\dfrac{4}{5} = \dfrac{20}{25}$ a true proportion?

$5 \times 20 = 100$　　　The product of the means is 100.
$4 \times 25 = 100$　　　The product of the extremes is also 100.

Since the products are equal, the statement is a true proportion.

(b) Is $\dfrac{2}{3} = \dfrac{15}{20}$ a true proportion?

$3 \times 15 = 45$　　　The product of the means is 45.
$2 \times 20 = 40$　　　The product of the extremes is 40.

Since the products are not equal, the statement is not a true proportion.

● ● ● **CHECK YOURSELF 3**

Are the following true proportions?

a. $\dfrac{3}{8} = \dfrac{12}{30}$ **b.** $\dfrac{5}{9} = \dfrac{15}{27}$

Later we will use the Proportion Rule with terms that are fractions or decimals. Example 4 illustrates this rule.

● Example 4

Verifying Proportions

(a) Is $\dfrac{3}{\frac{1}{2}} = \dfrac{30}{5}$ a true proportion?

$\dfrac{1}{2} \times 30 = 15$ The product of the means is 15.

$3 \times 5 = 15$ The product of the extremes is 15.

Since the products are equal, the statement is a true proportion.

(b) Is $\dfrac{0.4}{20} = \dfrac{3}{100}$ a true proportion?

$20 \times 3 = 60$ The product of the means is 60.
$0.4 \times 100 = 40$ The product of the extremes is 40.

Since the products are *not* equal, the statement is not a true proportion.

● ● ● **CHECK YOURSELF 4**

a. Is $\dfrac{0.5}{8} = \dfrac{3}{48}$ a true proportion?   **b.** Is $\dfrac{\frac{1}{4}}{6} = \dfrac{3}{80}$ a true proportion?

● ● ● **CHECK YOURSELF ANSWERS**

1. The means are 8 and 20. The extremes are 5 and 32.
2. **(a)** $8 \times 20 = 160$
    $5 \times 32 = 160$ The products are equal.
    (b) $9 \times 3 = 27$
    $7 \times 4 = 28$ The products are not equal.
3. **(a)** No; **(b)** yes.   **4.** **(a)** Yes; **(b)** no.

List the means and the extremes for each of the following proportions.

1. $\dfrac{2}{3} = \dfrac{6}{9}$

2. $\dfrac{3}{4} = \dfrac{6}{8}$

3. $\dfrac{8}{11} = \dfrac{16}{22}$

4. $\dfrac{5}{x} = \dfrac{20}{24}$

5. $\dfrac{3}{8} = \dfrac{a}{32}$

6. $\dfrac{3}{8} = \dfrac{15}{40}$

7. $\dfrac{x}{6} = \dfrac{5}{30}$

8. $\dfrac{4}{7} = \dfrac{n}{28}$

9. $\dfrac{1}{4} = \dfrac{a}{b}$

Which of the following are true proportions?

10. $\dfrac{3}{4} = \dfrac{9}{12}$

11. $\dfrac{6}{7} = \dfrac{18}{21}$

12. $\dfrac{3}{4} = \dfrac{15}{20}$

13. $\dfrac{3}{5} = \dfrac{6}{10}$

14. $\dfrac{11}{15} = \dfrac{9}{13}$

15. $\dfrac{9}{10} = \dfrac{2}{7}$

16. $\dfrac{8}{3} = \dfrac{24}{9}$

17. $\dfrac{5}{8} = \dfrac{15}{24}$

18. $\dfrac{6}{17} = \dfrac{9}{11}$

19. $\dfrac{5}{12} = \dfrac{8}{20}$

20. $\dfrac{7}{16} = \dfrac{21}{48}$

21. $\dfrac{2}{5} = \dfrac{7}{9}$

ANSWERS

1. Means: 3, 6; extremes: 2, 9

2. Means: 4, 6; extremes: 3, 8

3. Means: 11, 16; extremes: 8, 22

4. Means: x, 20; extremes: 5, 24

5. Means: 8, a; extremes: 3, 32

6. Means: 8, 15; extremes: 3, 40

7. Means: 6, 5; extremes: x, 30

8. Means: 7, n; extremes: 4, 28

9. Means: 4, a; extremes: 1, b

10. True

11. True

12. True

13. True

14. False

15. False

16. True

17. True

18. False

19. False

20. True

21. False

22. $\dfrac{10}{3} = \dfrac{150}{50}$

23. $\dfrac{5}{8} = \dfrac{75}{120}$

24. $\dfrac{3}{7} = \dfrac{18}{42}$

25. $\dfrac{12}{7} = \dfrac{96}{50}$

26. $\dfrac{7}{15} = \dfrac{84}{180}$

27. $\dfrac{76}{24} = \dfrac{19}{6}$

28. $\dfrac{60}{36} = \dfrac{25}{15}$

29. $\dfrac{\frac{1}{2}}{4} = \dfrac{5}{40}$

30. $\dfrac{3}{\frac{1}{5}} = \dfrac{30}{6}$

31. $\dfrac{\frac{2}{3}}{6} = \dfrac{1}{12}$

32. $\dfrac{\frac{3}{4}}{12} = \dfrac{1}{16}$

33. $\dfrac{0.3}{4} = \dfrac{1}{20}$

34. $\dfrac{3}{60} = \dfrac{0.3}{6}$

35. $\dfrac{0.6}{0.12} = \dfrac{2}{0.4}$

36. $\dfrac{0.6}{15} = \dfrac{2}{75}$

Use your calculator to determine which of the following are true proportions.

37. $\dfrac{21.2}{11.5} = \dfrac{3.6}{14.5}$

38. $\dfrac{18.1}{6.3} = \dfrac{2.7}{5.3}$

39. $\dfrac{3.8}{4.75} = \dfrac{2}{2.5}$

40. $\dfrac{18.3}{2.4} = \dfrac{57.95}{7.6}$

41. $\dfrac{61.34}{11.97} = \dfrac{24.86}{13.12}$

42. $\dfrac{18.36}{73.44} = \dfrac{188.1}{752.4}$

43. A pitcher's earned run average (ERA) is determined by finding the number of earned runs given up every nine innings. We find the number by using the proportion.

$$\dfrac{\text{Earned runs}}{\text{Total innings}} = \dfrac{\text{ERA}}{9}$$

(a) If a pitcher gives up 60 earned runs in 180 innings, what is his ERA?
(b) During his next outing, the pitcher gives up 6 runs and lasts 4 innings. What is his new ERA?
(c) How many more innings of shutout ball would the pitcher have to pitch to drop his ERA to under 3.00?

44. You want to estimate the number of fish in a lake. You catch 40 fish and tag them all and return them to the lake. About 2 weeks later, you return and catch 24 fish, of which 6 are tagged.

(a) Using the above information, devise a plan to estimate the number of fish in the lake.
(b) Using this plan, estimate the number of fish in the lake.

44. _____
a. 48 _____
b. 96 _____
c. 30 _____
d. 24 _____
e. 40 _____
f. 60 _____
g. 60 _____
h. 50 _____
i. 24 _____
j. $\dfrac{2}{3}$ _____

● Getting Ready for Section 10.3 [Sections 6.4 and 9.2]

Divide.

a. $16 \div \dfrac{1}{3}$ **b.** $24 \div \dfrac{1}{4}$

c. $25 \div \dfrac{5}{6}$ **d.** $32 \div \dfrac{4}{3}$

e. $20 \div 0.5$ **f.** $18 \div 0.3$

g. $48 \div 0.8$ **h.** $60 \div 1.2$

i. $\dfrac{\dfrac{6}{1}}{\dfrac{1}{4}}$ **j.** $\dfrac{\dfrac{3}{4}}{\dfrac{9}{8}}$

Answers

1. 3 and 6 are the means, 2 and 9 are the extremes.   **3.** 11 and 16 are the means, 8 and 22 are the extremes.   **5.** 8 and *a* are the means, 3 and 32 are the extremes.   **7.** 6 and 5 are the means, *x* and 30 are the extremes.   **9.** 4 and *a* are the means, 1 and *b* are the extremes.   **11.** True   **13.** True; $5 \times 6 = 3 \times 10$ **15.** False   **17.** True   **19.** False; $12 \times 8 \neq 5 \times 20$ **21.** False

23. True   **25.** False   **27.** True   **29.** True, $4 \times 5 = 20$; $\dfrac{1}{2} \times 40 = 20$.

The products are equal, and so the proportion is a true statement.   **31.** False
33. False   **35.** True   **37.** False   **39.** True   **41.** False   **43. a.** 3.00;
b. 3.23; **c.** 14   **a.** 48   **b.** 96   **c.** 30   **d.** 24   **e.** 40   **f.** 60

g. 60   **h.** 50   **i.** 24   **j.** $\dfrac{2}{3}$

Supplementary Exercises

A N S W E R S

1. Means: 5, 8; extremes: 4, 10

2. Means: 3, 16; extremes: a, 24

3. Means: x, 28; extremes: 7, 20

4. Means: 5, 36; extremes: 9, 20

5. False

6. True

7. True

8. False

9. True

10. True

11. False

12. False

13. True

14. True

15. True

16. False

17. False

18. True

List the means and the extremes for the following proportions.

1. $\dfrac{4}{5} = \dfrac{8}{10}$

2. $\dfrac{a}{3} = \dfrac{16}{24}$

3. $\dfrac{7}{x} = \dfrac{28}{20}$

4. $\dfrac{9}{5} = \dfrac{36}{20}$

Which of the following are true proportions?

5. $\dfrac{3}{4} = \dfrac{6}{9}$

6. $\dfrac{2}{7} = \dfrac{8}{28}$

7. $\dfrac{4}{5} = \dfrac{8}{10}$

8. $\dfrac{5}{11} = \dfrac{2}{9}$

9. $\dfrac{12}{7} = \dfrac{60}{35}$

10. $\dfrac{5}{8} = \dfrac{15}{24}$

11. $\dfrac{2}{7} = \dfrac{13}{9}$

12. $\dfrac{7}{9} = \dfrac{35}{50}$

13. $\dfrac{5}{25} = \dfrac{50}{250}$

14. $\dfrac{28}{12} = \dfrac{35}{15}$

15. $\dfrac{\frac{1}{3}}{8} = \dfrac{2}{48}$

16. $\dfrac{6}{\frac{1}{5}} = \dfrac{24}{6}$

17. $\dfrac{0.4}{6} = \dfrac{3}{18}$

18. $\dfrac{5}{0.2} = \dfrac{100}{4}$

10.3 Solving Proportions

10.3 OBJECTIVES

1. Solve a proportion for an unknown value.
2. Use proportions to solve an application.

$\dfrac{?}{3} = \dfrac{10}{15}$ is a proportion in which the first term is unknown. Our work in this section will be learning how to find that unknown value.

Use the raised dot ($\cdot$) for multiplication rather than the cross ($\times$), so that the cross won't be confused with the letter x.

In solving applied problems later in this chapter, we will use proportions in which one of the four terms is *missing* or *unknown*. If three of the four terms of a proportion are known, you can always find the missing or unknown term.

In the proportion $\dfrac{a}{3} = \dfrac{10}{15}$, the first term is unknown. We have chosen to represent the unknown value with the letter a. Since the product of the means is equal to the product of the extremes, we can proceed as follows.

$$\frac{a}{3} = \frac{10}{15}$$

Product of the extremes: $15 \cdot a$
Product of the means: $3 \cdot 10$

Since the products of the means and extremes must be equal, we have

$$15 \cdot a = 3 \cdot 10 \qquad \text{or} \qquad 15 \cdot a = 30$$

The equals sign tells us that $15 \cdot a$ and 30 are just different names for the same number. This type of statement is called an **equation.**

Definition of an Equation

An *equation* is a statement that two expressions are equal.

One important property of an equation is that we can divide both sides by the same nonzero number. Here let's divide by 15.

$$15 \cdot a = 30$$

$$\frac{15 \cdot a}{15} = \frac{30}{15}$$

We will always divide by the number multiplying the variable. This is called the *coefficient*.

$$\frac{\cancel{15} \cdot a}{\cancel{15}} = \frac{\overset{2}{\cancel{30}}}{\underset{1}{\cancel{15}}}$$

Divide by the like factors. Do you see why we divided by 15? It leaves our unknown a by itself in the left term.

$$a = 2$$

We have found a value of 2 for a, the missing term.

You should always check your result. It is easy in this case. We found a value of 2 for a. Replace the unknown a with that value. Then cross multiply to verify that the proportion is true. We started with $\dfrac{a}{3} = \dfrac{10}{15}$ and found a value of 2 for a. So we write

Replace a with 2 and multiply.

$$\frac{2}{3} = \frac{10}{15}$$

$$3 \cdot 10 = 2 \cdot 15$$

$$30 = 30$$

The value of 2 for a is correct.

The procedure for solving a proportion is summarized as follows.

To Solve a Proportion

STEP 1 Cross multiply the means and extremes of the proportion. Write the product of the means equal to the product of the extremes.

STEP 2 Divide both terms of the resulting equation by the coefficient of the variable.

STEP 3 Use the value found to replace the unknown in the original proportion. Cross multiply to check that the proportion is true.

This gives us the unknown value.

Now check the result.

● Example 1

Solving Proportions for Unknown Values

Find the unknown value.

(a) $\dfrac{8}{x} = \dfrac{6}{9}$

You are really using algebra to solve these proportions. In algebra, we write the product $6\{\cdot\}x$ as $6x$, omitting the dot. Multiplication of the number and the variable is understood and doesn't need to be written.

Step 1 Cross-multiply and write the product of the means equal to the product of the extremes.

$$6 \cdot x = 8 \cdot 9$$

or $\quad 6x = 72$

Step 2 Locate the coefficient of the variable, 6, and divide both sides of the equation by that coefficient.

$$\frac{\overset{1}{\cancel{6}}x}{\underset{1}{\cancel{6}}} = \frac{\overset{12}{\cancel{72}}}{\underset{1}{\cancel{6}}}$$

$$x = 12$$

Step 3 To check, replace x with 12 in the original proportion.

$$\frac{8}{12} = \frac{6}{9}$$

Multiply:

$$12 \cdot 6 = 8 \cdot 9$$

$$72 = 72 \qquad \text{The value of 12 for } x \text{ checks.}$$

(b) $\dfrac{3}{5} = \dfrac{b}{25}$

Step 1 Set the product of the means equal to the product of the extremes.

$$5 \cdot b = 3 \cdot 25$$

or $5 \cdot b = 75$

Step 2 Locate the coefficient of the variable, 5, and divide both sides of the equation by that coefficient.

$$\dfrac{\overset{1}{\cancel{5b}}}{\underset{1}{\cancel{5}}} = \dfrac{\overset{15}{\cancel{75}}}{\underset{1}{\cancel{5}}}$$

$$b = 15$$

Step 3 To check, replace b with 15 in the original proportion.

$$\dfrac{3}{5} = \dfrac{15}{25}$$

Multiply:

$$5 \cdot 15 = 3 \cdot 25$$

$$75 = 75 \qquad \text{The value of 15 checks for } b.$$

● ● ● **CHECK YOURSELF 1**

Solve the proportion for n. Check your result.

a. $\dfrac{4}{5} = \dfrac{n}{25}$
 b. $\dfrac{7}{9} = \dfrac{42}{n}$

In solving for a missing term in a proportion, we may find an equation involving fractions or decimals. Example 2 involves finding the unknown value in such cases.

● Example 2

Solving Proportions for Unknown Values

(a) Solve the proportion for x.

$$\dfrac{\frac{1}{4}}{3} = \dfrac{4}{x}$$

$$\dfrac{1}{4}x = 12 \qquad \text{The product of the extremes equals the product of the means.}$$

$$\frac{\frac{1}{4}x}{\frac{1}{4}} = \frac{12}{\frac{1}{4}}$$ We divide by the coefficient of x.

In this case it is $\frac{1}{4}$.

$$x = \frac{12}{\frac{1}{4}}$$ Remember: $\frac{12}{\frac{1}{4}}$ is $12 \div \frac{1}{4}$.

$x = 12 \cdot \frac{4}{1} = 48$

$x = \boxed{48}$ Invert the divisor and multiply.

To check, replace x with 48 in the original proportion.

$$\frac{\frac{1}{4}}{3} = \frac{4}{48}$$

$$3 \cdot 4 = \frac{1}{4} \cdot 48$$

$$12 = 12$$

(*b*) Solve the proportion for *a*.

$$\frac{0.5}{2} = \frac{3}{a}$$

$$0.5a = 6$$

Here we must divide 6 by 0.5 to find the unknown value. The steps of that division are shown below for review.

$$\frac{0.5a}{0.5} = \frac{6}{0.5}$$ Divide by the coefficient, 0.5.

$$a = 12$$

```
        1 2.
0.5∧)6.0∧
      5
      1 0
      1 0
        0
```

● ● ● CHECK YOURSELF 2

a. Solve for *a*.

$$\frac{\frac{1}{2}}{5} = \frac{3}{a}$$

b. Solve for *x*.

$$\frac{0.4}{x} = \frac{2}{30}$$

Now that you have learned how to find an unknown value in a proportion, let's see how this can be used in the solution of applications.

Solving Applications of Proportions

STEP 1 Read the problem carefully to determine the given information.
STEP 2 Write the proportion necessary to solve the problem. Use a letter to represent the unknown quantity. Be sure to include the units in writing the proportion.
STEP 3 Solve, answer the question of the original problem, and check the proportion as before.

● Example 3

Using Proportions to Find an Unknown Value

If 6 tickets to a play cost $15, how much will 8 tickets cost?

Solution Several proportions can be written for this solution. For example, since the ratio of the number of tickets to the cost must remain the same, we can write

Note that each ratio must have the same units in the numerators and in the denominators (here tickets and dollars). This helps check that the proportion is written correctly.

$$\frac{6 \text{ tickets}}{\$15} = \frac{8 \text{ tickets}}{\$x}$$

We have decided to let x be the cost (in dollars) of 8 tickets.

Multiply as before.

$$6x = 120$$

$$\frac{6x}{6} = \frac{120}{6}$$

You can drop the labels in solving the proportion.
Divide by the coefficient, 6.

$$x = 20 \text{ (dollars)}$$

The 8 tickets will cost $20.

● ● ● CHECK YOURSELF 3

If the ratio of women to men in a class is $\dfrac{4}{5}$, how many women will there be in a class with 25 men?

● Example 4

Using Proportions to Find an Unknown Value

In a shipment of 400 parts, 14 are found to be defective. How many defective parts should be expected in a shipment of 1000?

Solution Assume that the ratio of defective parts to the total number remains the same.

$$\frac{14 \text{ defective}}{400 \text{ total}} = \frac{x \text{ defective}}{1000 \text{ total}}$$

We have decided to let x be the unknown number of defective parts.

Multiply:

$$400x = 14{,}000$$

Divide by the coefficient, 400.

$$x = 35$$

35 defective parts should be expected in the shipment.

● ● ● **CHECK YOURSELF 4**

An investment of $3000 earned $330 for 1 year. How much will an investment of $10,000 earn at the same rate for 1 year?

Let's look at an application involving fractions in the proportion.

● Example 5

Using Proportions to Find an Unknown Value

The scale on a map is given as $\frac{1}{4}$ inches (in.) = 3 miles (mi). The distance between two towns is 4 in. on the map. How far apart are they in miles?

Solution   For this solution we use the fact that the ratio of inches (on the map) to miles remains the same.

$$\frac{\frac{1}{4}\text{ in.}}{3\text{ mi}} = \frac{4\text{ in.}}{x\text{ mi}}$$
$$x = 48\text{ (mi)}$$

● ● ● **CHECK YOURSELF 5**

Jack drives 125 mi in $2\frac{1}{2}$ hours (h). At the same rate, how far will he be able to travel in 4 h? (*Hint:* Write $2\frac{1}{2}$ as an improper fraction.)

We may also find decimals in the solution of an application.

● Example 6

Using Proportions to Find an Unknown Value

Jill works 4.2 h and receives $21. How much will she get if she works 10 h?

Solution   The ratio of hours worked to the amount of pay remains the same.

$$\frac{4.2\text{ h}}{\$21} = \frac{10\text{ h}}{\$a}$$ Let *a* be the unknown amount of pay.

$$4.2a = 210$$

$$\frac{4.2a}{4.2} = \frac{210}{4.2}$$ Divide both terms by 4.2.

$$a = \$50$$

● ● ● **CHECK YOURSELF 6**

A piece of cable 8.5 centimeters (cm) long weighs 68 grams (g). What will a 10-cm length of the same cable weigh?

Be careful that the ratios used in a proportion are comparing the *same units*. In Example 7 we must convert the units stated in the problem.

● Example 7

Using Proportions to Find an Unknown Value

A machine can produce 15 tin cans in 2 minutes (min). At this rate how many cans can it make in an 8-h period?

Solution In writing a proportion for this problem, we must write the times involved in terms of the same units.

$$\frac{15 \text{ cans}}{2 \text{ min}} = \frac{x \text{ cans}}{480 \text{ min}}$$ Since 1 h is 60 min, convert 8 h to 480 min.

$$2x = 15 \cdot 480$$

or $\quad 2x = 7200$

$$x = 3600 \text{ cans}$$

● ● ● **CHECK YOURSELF 7**

Instructions on a can of film developer call for 2 ounces (oz) of concentrate to 1 quart (qt) of water. How much of the concentrate is needed to mix with 1 gallon (gal) of water? (4 qt = 1 gal.)

An important use of proportions is in solving problems involving *similar* geometric figures. These are figures that have the same shape and their corresponding sides are proportional. For instance, in the similar triangles shown below,

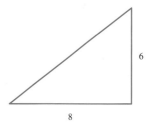

a proportion involving corresponding sides is

$$\frac{3}{4} = \frac{6}{8}$$

• Example 8

Using Proportions to Find Unknown Sides in Similar Geometric Figures

Use a proportion to find the unknown side, labeled x, in the following pair of similar triangles:

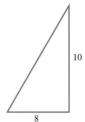

$$\frac{x}{4} = \frac{10}{8}$$

$$x = 5$$

● ● ● CHECK YOURSELF ANSWERS

1. (a) $5n = 100$ To check:

$$\frac{5n}{5} = \frac{100}{5} \qquad \frac{4}{5} = \frac{20}{25}$$

$$n = 20. \qquad 5 \cdot 20 = 4 \cdot 25$$

$$100 = 100.$$

(b) $7n = 42 \cdot 9$ To check:

$$\frac{7n}{7} = \frac{42 \cdot 9}{7} \qquad \frac{7}{9} = \frac{42}{54}$$

$$n = 6 \cdot 9 \qquad 7 \cdot 54 = 9 \cdot 42$$

$$n = 54. \qquad 378 = 378.$$

2. (a) 30;   **(b)** 6.

3. The ratio of women to men remains the same, and so you can write

$$\frac{4 \text{ women}}{5 \text{ men}} = \frac{n \text{ women}}{25 \text{ men}}$$

You solved this proportion in the last section and found that n has the value of 20, the number of women.

4. $1100.   **5.** $\dfrac{125 \text{ mi}}{\dfrac{5}{2} \text{ h}} = \dfrac{x \text{ mi}}{4 \text{ h}}$

$$\frac{5}{2} x = 500$$

Divide both members by $\dfrac{5}{2}$.

$x = 200 \text{ mi}.$ **6.** 80 g.   **7.** 8 oz.

Name

Section   Date

Solve for the unknown in each of the following proportions.

1. $\dfrac{x}{3} = \dfrac{6}{9}$

2. $\dfrac{x}{6} = \dfrac{3}{9}$

3. $\dfrac{10}{n} = \dfrac{15}{6}$

4. $\dfrac{4}{3} = \dfrac{8}{n}$

5. $\dfrac{4}{7} = \dfrac{y}{14}$

6. $\dfrac{7}{m} = \dfrac{14}{8}$

7. $\dfrac{5}{8} = \dfrac{a}{16}$

8. $\dfrac{5}{7} = \dfrac{x}{35}$

9. $\dfrac{8}{p} = \dfrac{6}{3}$

10. $\dfrac{4}{15} = \dfrac{8}{n}$

11. $\dfrac{11}{a} = \dfrac{2}{44}$

12. $\dfrac{5}{x} = \dfrac{15}{9}$

13. $\dfrac{35}{40} = \dfrac{7}{n}$

14. $\dfrac{x}{8} = \dfrac{15}{24}$

15. $\dfrac{a}{42} = \dfrac{5}{7}$

16. $\dfrac{7}{12} = \dfrac{m}{24}$

17. $\dfrac{18}{12} = \dfrac{12}{p}$

18. $\dfrac{x}{32} = \dfrac{7}{8}$

19. $\dfrac{x}{18} = \dfrac{64}{72}$

20. $\dfrac{20}{15} = \dfrac{100}{a}$

21. $\dfrac{6}{n} = \dfrac{75}{100}$

22. $\dfrac{36}{x} = \dfrac{8}{6}$

23. $\dfrac{5}{35} = \dfrac{a}{28}$

24. $\dfrac{20}{24} = \dfrac{p}{18}$

A N S W E R S

1. 2
2. 2
3. 4
4. 6
5. 8
6. 4
7. 10
8. 25
9. 4
10. 30
11. 242
12. 3
13. 8
14. 5
15. 30
16. 14
17. 8
18. 28
19. 16
20. 75
21. 8
22. 27
23. 4
24. 15

25. 25

26. 3

27. 5

28. 22

29. 12

30. 30

31. 2

32. 3

33. 24

34. 100

35. 12

36. 36

37. 80

38. 100

39. 0.55

40. 2

41. $60

42. $1.44

43. $2.40

25. $\dfrac{12}{100} = \dfrac{3}{x}$

26. $\dfrac{b}{7} = \dfrac{21}{49}$

27. $\dfrac{p}{24} = \dfrac{25}{120}$

28. $\dfrac{5}{x} = \dfrac{20}{88}$

29. $\dfrac{\frac{1}{2}}{2} = \dfrac{3}{a}$

30. $\dfrac{x}{5} = \dfrac{2}{\frac{1}{3}}$

31. $\dfrac{\frac{1}{4}}{12} = \dfrac{m}{96}$

32. $\dfrac{12}{\frac{1}{3}} = \dfrac{108}{y}$

33. $\dfrac{\frac{2}{5}}{8} = \dfrac{1.2}{n}$

34. $\dfrac{4}{a} = \dfrac{\frac{2}{5}}{10}$

35. $\dfrac{0.2}{2} = \dfrac{1.2}{a}$

36. $\dfrac{n}{3} = \dfrac{6}{0.5}$

37. $\dfrac{p}{7} = \dfrac{8}{0.7}$

38. $\dfrac{y}{12} = \dfrac{5}{0.6}$

39. $\dfrac{x}{3.3} = \dfrac{1.1}{6.6}$

40. $\dfrac{0.5}{a} = \dfrac{1.25}{5}$

Solve the following applications.

41. Book purchases. If 12 books are purchased for $40, how much will you pay for 18 books at the same rate?

42. Construction. If an 8-foot (ft) two-by-four costs 96¢, what should a 12-ft two-by-four cost?

43. Consumer affairs. A box of 18 tea bags is marked 90¢. At that price, what should a box of 48 tea bags cost?

44. **Consumer affairs.** Cans of orange juice are marked 2 for 93¢. What would the price of a case of 24 cans be?

45. **Workload.** A worker can complete the assembly of 15 tape players in 6 hours (h). At this rate, how many can the worker complete in a 40-h workweek?

46. **Consumer affairs.** If 3 pounds (lb) of apples costs 90¢, what will 10 lb cost?

47. **Elections.** The ratio of yes to no votes in an election was 3 to 2. How many no votes were cast if there were 2880 yes votes?

48. **College enrollment.** The ratio of men to women at a college is 7 to 5. How many women students are there if there are 3500 men?

49. **Photography.** A photograph 5 inches (in.) wide by 6 in. high is to be enlarged so that the new width is 15 in. What will be the height of the enlargement?

50. **Shift work.** Meg's job is assembling lawn chairs. She can put together 55 chairs in 4 h. At this rate, how many chairs can she assemble in an 8-h shift?

51. **Distance.** Christy can travel 110 miles (mi) in her new car on 5 gallons (gal) of gas. How far can she travel on a full tank, which has 12 usable gal?

52. **Property taxes.** The Changs purchased an $80,000 home, and the property taxes were $1400. If they make improvements and the house is now valued at $120,000, what will the new property tax be?

53. **Distance.** A car travels 165 mi in 3 h. How far will it travel in 8 h if it continues at the same speed?

54. **Distance.** If two cities on a map are 7 in. apart and the actual distance between the cities is 420 mi, find the distance between two cities that are 4 in. apart on the map.

55. **Manufacturing.** The ratio of teeth on a smaller gear to those on a larger gear is 3 to 7. If the smaller gear has 15 teeth, how many teeth does the larger gear have?

56. **Consumer affairs.** A store has T-shirts on sale at 2 for $5.50. At this rate, what will five shirts cost?

57. **Manufacturing.** An inspection reveals 30 defective parts in a shipment of 500. How many defective parts should be expected in a shipment of 1200?

58. **Investments.** You invest $4000 in a stock that pays a $180 dividend in 1 year. At the same rate, how much will you need to invest to earn $270?

59. **Football.** A football back ran 212 yards (yd) in the first two games of the season. If he continues at the same pace, how many yards should he gain in the 11-game season?

60. **Cooking.** A 6-lb roast will serve 14 people. What size roast is needed to serve 21 people?

61. **Lawn care.** A 2-lb box of grass seed is supposed to cover 2500 square feet (ft^2) of lawn. How much seed will you need for 8750 ft^2 of lawn?

62. **Fencing.** A 6-ft fence post casts a 9-ft shadow. How tall is a nearby pole that casts a 15-ft shadow?

63. **Lighting.** A 9-ft light pole casts a 15-ft shadow. Find the height of a nearby tree that is casting a 40-ft shadow.

64. **Construction.** On the blueprint of the Wilsons' new home, the scale is 5 in. equals 7 ft. What will be the actual length of a bedroom if it measures 10 in. long on the blueprint?

65. **Distance.** The scale on a map is $\frac{1}{2}$ in. = 50 mi. If the distance between towns on the map is 6 in., how far apart are they in miles?

66. **Science.** A metal bar expands $\frac{1}{4}$ in. for each 12°F rise in temperature. How much will it expand if the temperature rises 48°F?

67. **Car maintenance.** Your car burns $2\frac{1}{2}$ quarts (qt) of oil on a trip of 5000 mi. How many quarts should you expect to use when driving 7200 mi?

68. **Lighting.** A 6-ft person casts a $7\frac{1}{2}$-ft shadow. If the shadow of a nearby pole is 30 ft long, how tall is the pole?

69. **Manufacturing.** A piece of tubing 10.5 centimeters (cm) long weighs 35 grams (g). What is the weight of a piece of the same tubing that is 15 cm long?

70. **Salary.** Jane works 7.75 h and receives $38.75 pay. What will she receive at the same rate if she works 12 h?

71. **Sales tax.** The sales tax on an item costing $80 is $5.20. What will the tax be for an item costing $150?

72. **Conversion.** If 8 kilometers (km) is approximately 4.8 mi, how many kilometers will equal 12 mi?

73. **Timing.** You find that your watch gains 2 minutes (min) in 6 h. How much will it gain in 3 days?

74. **Painting.** If 2 qt of paint will cover 225 ft^2, how many square feet will 2 gal cover? (1 gal = 4 qt.)

75. **Construction.** Directions on a box of 4 cups of wallpaper paste are to mix the contents with 5 qt of water. To mix a smaller batch using 1 cup of paste, how much water (in ounces) should be added? (1 qt = 32 oz.)

76. **Film processing.** A film processing machine can develop three rolls of film every 8 min. At this rate, how many rolls can be developed in a 4-h period?

77. **Carpooling.** Approximately 7 out of every 10 people in the U.S. workforce drive to work alone. During morning rush hour there are 115,000 cars on the streets of a medium-sized city. How many of these cars have one person in them?

78. **Carpooling.** Approximately 15 out of every 100 people in the U.S. workforce carpool to work. There are an estimated 320,000 people in the workforce of a given city. How many of these people are in car pools?

Use a proportion to find the unknown side, labeled x, in each of the following pairs of similar figures.

79.
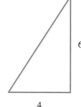

80.
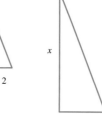

69. 50 g

70. $60

71. $9.75

72. 20 km

73. 24 min

74. 900 ft^2

75. 40 oz

76. 90 rolls

77. 80,500 cars with one person

78. 48,000 people

79. 3

80. 15

589

81. 6

82. 9

83.

81.

82.

83. A recipe for 12 servings lists the following ingredients:

12 cups ziti	7 cups spaghetti sauce	4 cups ricotta cheese
1/2 cup parsley	1 teaspoon garlic powder	1/2 teaspoon pepper
4 cups mozzarella cheese	2 tablespoons parmesan cheese	

Determine the amount of ingredients necessary to serve 5 people.

Answers

1. $9x = 18$; $\dfrac{9x}{9} = \dfrac{18}{9}$; $x = 2$ **3.** 4   **5.** 8   **7.** 10   **9.** 4   **11.** 242

13. $35n = 280$; $\dfrac{35n}{35} = \dfrac{280}{35}$; $n = 8$ **15.** 30   **17.** $18p = 144$; $\dfrac{18p}{18} = \dfrac{144}{18}$;

$p = 8$ **19.** 16   **21.** 8   **23.** $35a = 140$; $\dfrac{35a}{35} = \dfrac{140}{35}$; $a = 4$ **25.** 25

27. 5   **29.** 12   **31.** 2   **33.** 24   **35.** 12   **37.** 80   **39.** 0.55

41. $60   **43.** $\dfrac{18 \text{ tea bags}}{90¢} = \dfrac{48 \text{ tea bags}}{x¢}$; $18x = 4320$; $x = 240¢$, or $2.40

45. 100 players   **47.** 1920 no votes   **49.** 18 in.   **51.** $\dfrac{5 \text{ gal}}{110 \text{ mi}} = \dfrac{12 \text{ gal}}{x \text{ mi}}$;

$5x = 1320$; $x = 264$ mi   **53.** 440 mi   **55.** 35 teeth

57. 72 defective parts   **59.** $\dfrac{212 \text{ yd}}{2 \text{ games}} = \dfrac{x \text{ yd}}{11 \text{ games}}$; $x = 1166$ yd   **61.** 7 lb

63. 24 ft   **65.** 600 mi   **67.** (Write $2\dfrac{1}{2}$ as $\dfrac{5}{2}$) $\dfrac{\frac{5}{2} \text{ qt}}{5000 \text{ mi}} = \dfrac{x \text{ qt}}{7200 \text{ mi}}$; $5000x =$

$18,000$; $x = 3.6$ qt   **69.** 50 g   **71.** $9.75   **73.** (Write 3 days as 72 h)

$\dfrac{2 \text{ min}}{6 \text{ h}} = \dfrac{x \text{ min}}{72 \text{ h}}$; $x = 24$ min   **75.** 40 oz   **77.** 80,500 cars with one person

79. 3   **81.** 6   **83.**

Name

Section   Date

Solve for the unknown in each of the following proportions.

1. $\dfrac{x}{4} = \dfrac{15}{20}$

2. $\dfrac{4}{5} = \dfrac{n}{20}$

3. $\dfrac{7}{n} = \dfrac{56}{72}$

4. $\dfrac{8}{10} = \dfrac{x}{5}$

5. $\dfrac{3}{10} = \dfrac{9}{y}$

6. $\dfrac{4}{24} = \dfrac{6}{p}$

7. $\dfrac{12}{s} = \dfrac{6}{9}$

8. $\dfrac{5}{7} = \dfrac{x}{35}$

9. $\dfrac{5}{12} = \dfrac{15}{t}$

10. $\dfrac{3}{s} = \dfrac{15}{25}$

11. $\dfrac{7}{b} = \dfrac{28}{48}$

12. $\dfrac{x}{15} = \dfrac{12}{60}$

13. $\dfrac{p}{40} = \dfrac{9}{12}$

14. $\dfrac{45}{a} = \dfrac{27}{21}$

15. $\dfrac{\frac{1}{2}}{5} = \dfrac{3}{x}$

16. $\dfrac{a}{6} = \dfrac{3}{\frac{1}{4}}$

17. $\dfrac{\frac{3}{4}}{3} = \dfrac{6}{y}$

18. $\dfrac{0.3}{6} = \dfrac{2}{x}$

19. $\dfrac{n}{7} = \dfrac{10}{0.7}$

20. $\dfrac{3}{x} = \dfrac{7.5}{1.5}$

21. **Recipe.** A bread recipe calls for 3 cups of flour for 2 loaves of bread. How many cups of flour will be needed for 6 loaves of bread?

22. **Painting.** A painter uses 2 gallons (gal) of paint in doing 3 rooms. At this rate, how many gallons will be needed to paint 15 similar rooms?

23. **Cost.** Cans of tomato soup are marked 2 for 79¢. How much will a case of 24 cans cost at the same rate?

24. **Distance.** Dick drives 256 miles (mi), using 8 gal of gasoline. If the gas tank holds 11 gal, how far can he travel on a full tank?

#	Answer
1.	3
2.	16
3.	9
4.	4
5.	30
6.	36
7.	18
8.	25
9.	36
10.	5
11.	12
12.	3
13.	30
14.	35
15.	30
16.	72
17.	24
18.	40
19.	100
20.	0.6
21.	9 cups
22.	10 gal
23.	$9.48
24.	352 mi

25. 13.5 in.

26. 140 games

27. 72 parts

28. $600

29. 25 ft

30. 350 mi

31. 11 lb

32. 5 h

33. $\frac{1}{2}$ oz

34. $38.50

25. Photography. You want to enlarge a color print that is 4 inches (in.) wide by 6 in. long. If the new width is to be 9 in., find the length of the enlargement.

26. Baseball. A baseball team won 35 of its first 41 games. At this rate, how many games should it win in the 164-game season?

27. Manufacturing. In a sample of 250 parts taken from a shipment, 12 are found to be improperly assembled. At this rate, how many faulty parts can be expected in a shipment of 1500?

28. Investment. An investment of $5000 earns $375 in 1 year. What will an investment of $8000 earn at the same rate?

29. Geometry. The shadow cast by a 30-feet (ft) telephone pole is 18 ft long. How tall is a nearby tree that is casting a 15-ft shadow?

30. Map scales. The scale on a map is 2 in. to 100 mi. How many miles is it between two cities that are 7 in. apart on the map?

31. Farming. If 7 lb of fertilizer will cover 1400 ft^2, what amount will be needed to cover 2200 ft^2?

32. Distance. A plane can fly 480 mi in 3 hours (h). At this rate, how long will it take to travel 800 mi?

33. Pharmacy. A medicine label calls for $\frac{1}{16}$ ounces (oz) of medicine for 20 pounds (lb) of body weight. If a patient weighs 160 lb, how much of the medicine should be prescribed?

34. Cost. A $3\frac{1}{2}$-lb roast costs $24.50. At the same price, what will a $5\frac{1}{2}$-lb roast cost?

Using Your Calculator to Solve Proportions

When the numbers involved in a proportion are large, your calculator can be very useful. Consider Example 1.

•Example 1

Solving a Proportion for an Unknown Value

Solve the proportion for n.

$$\frac{n}{43} = \frac{105}{1505}$$

Start the solution as before.

$1505n = 43 \cdot 105$ Set the product of the means equal to the product of the extremes.

$\dfrac{\cancel{1505}n}{\cancel{1505}} = \dfrac{43 \cdot 105}{1505}$ Divide both terms by the coefficient, 1505.

Find n, using the calculator.

$43 \boxed{\times} 105 \boxed{\div} 1505 \boxed{=}$

Display $\boxed{3}$ n has the value 3. To check, replace n with 3 and use your calculator to multiply.

● ● ● CHECK YOURSELF 1

Solve the proportion for n.

$$\frac{n}{27} = \frac{35}{189}$$

In practical applications, you may have to round the result after using your calculator in the solution of a proportion. Example 2 shows such a situation.

•Example 2

Using Proportions to Find an Unknown Value

Micki drives 278 miles (mi) on 13.6 gallons (gal) of gas. If the gas tank of her car holds 21 gal, how far can she travel on a full tank of gas?

Solution   We can write the proportion

$$\frac{278 \text{ mi}}{13.6 \text{ gal}} = \frac{x \text{ mi}}{21 \text{ gal}}$$

Multiply.

$$13.6x = 278 \cdot 21$$

Now divide both terms by 13.6.

$$\frac{\cancel{13.6}x}{\cancel{13.6}} = \frac{278 \cdot 21}{13.6}$$

Now to find x, we must multiply 278 by 21 and then divide by 13.6.
On the calculator,

$$278 \boxed{\times} 21 \boxed{\div} 13.6 \boxed{=} \boxed{429.26471}$$

Let's round the result to the nearest mile; Micki can drive 429 mi on a full tank of gas.

● ● ●   **CHECK YOURSELF 2**

Life insurance costs $4.37 for each $1000 of insurance. How much does a $25,000 policy cost?

Unit Pricing

Your calculator can be very handy for comparing prices at the grocery store. When items are sold in different-sized containers, it is often difficult to determine which is the better buy. Using your calculator to find your own unit prices may help you save some money.

> **Definition of Unit Price**
>
> A **unit price** is a ratio of price to some unit.

A unit price is a price *per* unit.

To find the unit price, just divide the cost of the item by the number of units.

The unit used may be ounces, pints, pounds, or some other measure.

● **Example 3**

Using Proportions to Find an Unknown Value

A dishwashing liquid comes in three sizes:

(*a*) 12 ounces (oz) for 77¢
(*b*) 22 oz for $1.33
(*c*) 32 oz for $1.85

Which is the best buy?

Solution For each size, let's find the unit price in cents per ounce.

(*a*) For the first size (77¢ for 12 oz), using your calculator, divide.

77 ÷ 12 = | 6.4166667 |

$$\frac{77¢}{12 \text{ oz}} \approx 6.4¢ \text{ per ounce}$$ We have chosen to round to the nearest tenth of a cent.

(*b*) To find the unit price for the second size, divide again:

Note that we consider $1.33 as 133¢ to find the ratio "cents per ounce."

133 ÷ 22 = | 6.0454545 | Treat $1.33 as 133¢.

$$\frac{\$1.33}{22 \text{ oz}} \approx 6¢ \text{ per ounce}$$ Again round to the nearest tenth of a cent.

(*c*) For the third size,

185 ÷ 32 = | 5.78125 |

$$\frac{\$1.85}{32 \text{ oz}} \approx 5.8¢ \text{ per ounce}$$

Comparing the three unit prices, we see that the 32-oz size of dishwashing liquid, at 5.8¢ per ounce, is the best buy.

Note: All the ratios used must be in terms of the same units. If quantities involve different units, they must be converted.

● ● ● **CHECK YOURSELF 3**

A floor cleaner comes in three sizes:

32 oz for $2.89
48 oz for $3.43
70 oz for $4.96

Which is the best buy?

●**Example 4**

Applying Proportions to Find an Unknown Value

Vegetable oil is sold in the following quantities:

(*a*) 16 oz for $1.27
(*b*) 1 pint (pt) 8 oz for $1.79
(*c*) 1 quart (qt) 6 oz for $2.89

Which is the best buy?

Solution

(a) $\dfrac{\$1.27}{16 \text{ oz}} \approx 7.9¢$ per ounce

(b) Since 1 pt is 16 oz, 1 pt 8 oz is (16 + 8) oz, or 24 oz. So we write 1 pt 8 oz as 24 oz.

$\dfrac{\$1.79}{24 \text{ oz}} \approx 7.5¢$ per ounce

(c) Since 1 qt is 32 oz, 1 qt 6 oz is (32 + 6) oz, or 38 oz. Write 1 qt 6 oz as 38 oz.

$\dfrac{\$2.89}{38 \text{ oz}} \approx 7.6¢$ per ounce

In this case, by comparing the unit prices we see that the 1-pt 8-oz size is the best buy.

● ● ● **CHECK YOURSELF 4**

Ketchup is sold in the following quantities:

12 oz for $0.68
1 pt 5 oz for $1.05
1 qt 7 oz for $1.89

Which is the best buy?

● ● ● **CHECK YOURSELF ANSWERS**

1. 5.   **2.** $109.25.   **3.** 70-oz size at 7.09¢ per oz.   **4.** 1 qt 7 oz for 4.8¢ per oz.

A N S W E R S

Solve for the unknowns.

1. $\dfrac{630}{1365} = \dfrac{15}{a}$

2. $\dfrac{770}{1988} = \dfrac{n}{71}$

3. $\dfrac{x}{4.7} = \dfrac{11.8}{16.9}$ (to nearest tenth)

4. $\dfrac{13.9}{8.4} = \dfrac{n}{9.2}$ (to nearest hundredth)

5. $\dfrac{2.7}{3.8} = \dfrac{5.9}{n}$ (to nearest tenth)

6. $\dfrac{12.2}{0.042} = \dfrac{x}{0.08}$ (to nearest hundredth)

Solve the following applications.

7. **Salary.** Bill earns $248.40 for working 34.5 hours (h). How much will he receive if he works at the same pay rate for 31.75 h?

8. **Construction.** Construction-grade lumber costs $384.50 per 1000 board-feet. What will be the cost of 686 board-feet?

9. **Speed of sound.** A speed of 88 feet per second (ft/s) is equal to a speed of 60 miles per hour (mi/h). If the speed of sound is 750 mi/h, what is the speed of sound in feet per second?

10. **Manufacturing.** A shipment of 75 parts is inspected, and 6 are found to be faulty. At the same rate, how many defective parts should be found in a shipment of 139? Round your result to the nearest whole number.

11. **Taxes.** The property tax on a $67,250 home is $2315. At the same rate, what will be the tax on a home valued at $87,625? Round your result to the nearest dollar.

12. **Production.** A machine produces 158 items in 12 minutes (min). At the same rate, how many items will it produce in 8 h?

13. **Gas mileage.** Sally travels 510 mi, using 11.6 gallons (gal) of gas. How many gallons of gas will she need for a trip of 1800 mi? Give your answer to the nearest tenth of a gallon.

1.	32.5
2.	27.5
3.	3.3
4.	15.22
5.	8.3
6.	23.24
7.	$228.60
8.	$263.77
9.	1100 ft/s
10.	11 parts
11.	$3016
12.	6320 items
13.	40.9 gal

14.

15. b

16. a

17. c

18. b

19. b

20. c

21. c

22. d

14. Tom and Jerry operate a food concession stand at a local amusement park. They sell the most food when the attendance at the park is at maximum capacity. When this happens, they sell an average of 450 pork roll sandwiches and 550 cheese steak sandwiches. The company that owns the park is going to expand; they will increase the capacity of the park from 6000 to 9000 people next season. Tom and Jerry plan to expand their concession stand so they can sell more sandwiches.

(a) Using the same ratio of attendance to sandwiches, how many additional sandwiches of each kind would Tom and Jerry expect to sell?

(b) The following costs are associated with the anticipated expansion:

Item	Cost per Unit
Construction	$70/square foot
Supplies	$0.65/pork roll sandwich
	$1.25/steak sandwich
Employee costs	$8/hour

Currently, pork roll sandwiches sell for $1.50, and steak sandwiches sell for $2.75. Based on these prices and the information in part a how would you plan the expansion, and what would you charge for a sandwich?

Find the best buy in each of the following exercises.

15. Dishwashing liquid:
(a) 12 oz for 79¢
(b) 22 oz for $1.29

16. Canned corn:
(a) 10 oz for 21¢
(b) 17 oz for 39¢

17. Syrup:
(a) 12 oz for 99¢
(b) 24 oz for $1.59
(c) 36 oz for $2.19

18. Shampoo:
(a) 4 oz for $1.16
(b) 7 oz for $1.52
(c) 15 oz for $3.39

19. Salad oil (1 qt is 32 oz):
(a) 18 oz for 89¢
(b) 1 qt for $1.39
(c) 1 qt 16 oz for $2.19

20. Tomato juice (1 pt is 16 oz):
(a) 8 oz for 37¢
(b) 1 pt 10 oz for $1.19
(c) 1 qt 14 oz for $1.99

21. Peanut butter (1 lb is 16 oz):
(a) 12 oz for $1.25
(b) 18 oz for $1.72
(c) 1 lb 12 oz for $2.59
(d) 2 lb 8 oz for $3.76

22. Laundry detergent:
(a) 1 lb 2 oz for $1.99
(b) 1 lb 12 oz for $2.89
(c) 2 lb 8 oz for $4.19
(d) 5 lb for $7.99

Answers

1. 32.5   **3.** 3.3   **5.** 8.3   **7.** $228.60   **9.** 1100 ft/s   **11.** $3016
13. 40.9 gal   **15. b**   **17. c**   **19. b**   **21. c**

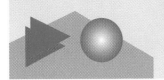
The purpose of the Self-Test is to help you check your progress and review for the class-held Chapter Test. Allow yourself about 1 hour to take the test. When you are done, check your answers in the back of the book. If you missed any answers, be sure to go back and review the appropriate sections in the chapter and do the supplementary exercises provided there.

[10.1]   In Exercises 1 to 5, write the ratios in lowest terms.

1. The ratio of 7 to 19

2. The ratio of 75 to 45

3. The ratio of 8 ft to 4 yd

4. The ratio of 6 h to 3 days

Solve the following application.

5. Basketball ratio. A basketball team wins 26 of its 33 games during a season. What is the ratio of wins to games played? What is the ratio of wins to losses?

[10.2]   In Exercises 6 and 7, list the means and extremes for the proportions.

6. $\dfrac{6}{13} = \dfrac{18}{39}$

7. $\dfrac{5}{a} = \dfrac{15}{21}$

In Exercises 8 to 11, which are true proportions?

8. $\dfrac{3}{9} = \dfrac{27}{81}$

9. $\dfrac{6}{7} = \dfrac{9}{11}$

10. $\dfrac{9}{10} = \dfrac{27}{30}$

11. $\dfrac{\frac{1}{2}}{5} = \dfrac{2}{18}$

[10.3]   In Exercises 12 to 21, solve for the unknowns.

12. $\dfrac{a}{9} = \dfrac{6}{27}$

13. $\dfrac{8}{x} = \dfrac{32}{48}$

14. $\dfrac{10}{14} = \dfrac{n}{35}$

15. $\dfrac{8}{15} = \dfrac{24}{t}$

16. $\dfrac{45}{75} = \dfrac{12}{x}$

17. $\dfrac{a}{24} = \dfrac{45}{60}$

18. $\dfrac{\frac{1}{2}}{p} = \dfrac{5}{30}$

19. $\dfrac{\frac{5}{6}}{8} = \dfrac{5}{a}$

20. $\dfrac{x}{0.3} = \dfrac{60}{3}$

21. $\dfrac{3}{m} = \dfrac{0.9}{4.8}$

In Exercises 22 to 30, solve each application, using a proportion.

22. **Consumer affairs.** If ballpoint pens are marked 5 for 95¢, how much will a dozen cost?

23. **Basketball.** A basketball player scores 207 points in her first 9 games. At the same rate, how many points will she score in the 28-game season?

24. **Distance.** Your new compact car travels 324 mi on 9 gal of gas. If the tank holds 16 usable gal, how far can you drive on a tankful of gas?

25. **Elections.** The ratio of yes to no votes in an election was 6 to 5. How many no votes were cast if 3600 people voted yes?

26. **Distance.** Two towns that are 120 mi apart are 2 in. apart on a map. What is the actual distance between two towns that are 7 in. apart on the map?

27. **Reduction.** You are using a photocopier to reduce an article that was 15 in. long by 9 in. wide. If the new width is 6 in., find the length of the reduced article.

28. **Length.** If the scale of a drawing is $\dfrac{1}{4}$ in. equals 1 ft, what actual length is represented by 3 in. on the drawing?

29. **Mufflers installed.** An assembly line can install 5 car mufflers in 4 min. At this rate, how many mufflers can be installed in an 8-h shift?

30. **Mixing.** Instructions on a package of concentrated plant food call for 2 teaspoons (tsp) to 1 qt of water. We have a large job and wish to use 3 gal of water. How much of the plant food concentrate should be added to the 3 gal of water?

CHAPTER 11

PERCENT

INTRODUCTION

With a degree in literature and a love of books, Don set out looking for a career in publishing. He was surprised to find that he ended up selling books to bookstores. He was more surprised to find that he enjoyed the selling. He finds having to talk about all types of books (even math books!) with bookstore managers very challenging.

Don does admit that he wishes he had paid a little more attention in his math class. Thinking that math was not relevant to his field, he took it only because it was required. Now he has to do numerous calculations every day. Figuring discounts, quotas, bonuses, and commissions occupies much of Don's office time. ◼

© Brownie Harris/The Stock Market

Pretest
for Chapter 11

Name

Section   Date

A N S W E R S

1. $\dfrac{7}{100}$

2. 0.23

3. 3.5% or $3\dfrac{1}{2}$%

4. 80%

5. 63

6. 9%

7. 600

8. $560

9. 5%

10. $1200

Percent

This pretest will point out any difficulties you may be having with percents. Do all the problems. Then check your answers with those in the back of the book.

1. Write 7% as a fraction.

2. Write 23% as a decimal.

3. Write 0.035 as a percent.

4. Write $\dfrac{4}{5}$ as a percent.

5. What is 25% of 252?

6. What percent of 500 is 45?

7. 35% of a number is 210. What is the number?

8. Interest. How much interest will you pay on a $4000 loan for 1 year if the interest rate is 14%?

9. Commission. A salesperson earns a $400 commission on sales of $8000. What is the commission rate?

10. Salary. A salary increase of 5% amounts to a $60 monthly raise. What was the monthly salary before the increase?

© 1998 McGraw-Hill Companies

602

Changing a Percent to a Fraction or Decimal

© 1998 McGraw-Hill Companies

11.1 OBJECTIVES

1. Use the percent notation.
2. Change a percent to a fraction.
3. Change a percent to a mixed number.
4. Change a percent to a decimal.

Since

$$\frac{1}{4} = \frac{25}{100} = 0.25$$

The symbol % comes from an arrangement of the digits 1, 0, 0.

When we considered parts of a whole in earlier chapters, we used fractions and decimals. The idea of *percent* is another useful way of naming parts of a whole. We can think of percents as ratios whose denominators are 100. In fact, the word **percent** means "for each hundred." Look at the following drawing:

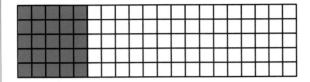

How can we describe how much of the drawing is shaded? We can say that

1. $\frac{1}{4}$ of the drawing is shaded.

   or
2. 0.25 of the drawing is shaded.

   or
3. 25% (read "twenty-five percent") of the drawing is shaded.

•Example 1

Using Percent Notation

(*a*) Four out of five geography students passed their midterm exams. Write this statement, using the percent notation.

The ratio of students passing to students taking the class is $\frac{4}{5}$.

$$\frac{4}{5} = \frac{80}{100} = 80\%$$

Percent means for each hundred. To obtain a denominator of 100, multiply the numerator and denominator of the original fraction by 20.

So we can say that 80% of the geography students passed.

(*b*) Of 50 automobiles sold by a dealer in one month 35 were compact cars. Write this statement, using the percent notation.

The ratio of compact cars to all cars is $\frac{35}{50}$.

$$\frac{35}{50} = \frac{70}{100} = 70\%$$

We can say that 70% of the cars sold were compact cars.

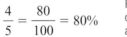 **CHECK YOURSELF 1**

Rewrite the following statement, using the percent notation: 4 of the 50 parts in a shipment were defective.

Since there are different ways of naming the parts of a whole, you need to know how to change from one of these ways to another. First let's look at changing a percent to a fraction. Since a percent is a fraction or a ratio with denominator 100, we can use the following rule.

Changing a Percent to a Fraction

To change a percent to a common fraction, remove the percent symbol and write the number over 100.

The use of this rule is shown in our first example.

●Example 2

Changing a Percent to a Fraction

Change each percent to a fraction.

(a) $7\% = \dfrac{7}{100}$

You can choose to reduce $\dfrac{25}{100}$ to simplest form.

(b) $25\% = \dfrac{25}{100} = \dfrac{1}{4}$

● ● ● **CHECK YOURSELF 2**

Write 12% as a fraction.

If a percent is *greater than 100,* the resulting fraction will be *greater than 1.* This is shown in Example 3.

●Example 3

Changing a Percent to a Mixed Number

Change 150% to a mixed number.

$$150\% = \frac{150}{100} = 1\frac{50}{100} = 1\frac{1}{2}$$

● ● ● **CHECK YOURSELF 3**

Write 125% as a fraction.

The fractional equivalents of certain percents should be memorized.

$$33\frac{1}{3}\% = \frac{1}{3} \qquad 66\frac{2}{3}\% = \frac{2}{3}$$

To see how these values are formed, look at $33\frac{1}{3}\%$:

$$33\frac{1}{3}\% = \frac{33\frac{1}{3}}{100} = \frac{\frac{100}{3}}{\frac{100}{1}} \qquad \text{Convert to improper fractions.}$$

$$= \frac{100}{3} \div \frac{100}{1} \qquad \text{Rewrite, using the division symbol.}$$

$$= \frac{100}{3} \times \frac{1}{100} = \frac{1}{3}$$

Invert the divisor and multiply.

Instead of working this hard, it's better just to try to remember these fractional equivalents.

In Example 2, we wrote percents as fractions by removing the percent sign and dividing by 100. How do we divide by 100 when we are working with decimals? Just move the decimal point two places to the left. This gives us a second rule for converting percents.

Changing a Percent to a Decimal

To change a percent to a decimal, remove the percent symbol and move the decimal point *two* places to the *left*.

• Example 4

Changing a Percent to a Decimal

Change each percent to a decimal equivalent.

$25\% = \dfrac{25}{100} = 0.25$

Note: A percent greater than 100 gives a decimal greater than 1.

(*a*) $25\% = 0.25$ The decimal point in 25% is understood to be after the 5.

(*b*) $8\% = 0.08$ We must add a zero to move the decimal point.

(*c*) $130\% = 1.30$

● ● ● CHECK YOURSELF 4

Write as decimals.

a. 5%   **b.** 32%   **c.** 115%

Look at Example 5, which involves fractions of a percent. In this case, decimal fractions are involved.

• Example 5

Changing a Percent to a Decimal

Write as decimals.

(*a*) 4.5% = 0.045

(*b*) 0.5% = 0.005

● ● ● CHECK YOURSELF 5

Write as decimals.

a. 8.5%   **b.** 0.3%

You will also find examples in which common fractions are involved in a percent. Example 6 illustrates this approach.

• Example 6

Changing a Percent to a Decimal

Write as decimals.

Write the common fractions as decimals. Then remove the percent symbol by our earlier rule.

$$9\frac{1}{2}\% = 9.5\% = 0.095$$

$$\frac{3}{4}\% = 0.75\% = 0.0075$$

● ● ● CHECK YOURSELF 6

Write as decimals.

a. $7\frac{1}{2}\%$ **b.** $\frac{1}{2}\%$

● ● ● CHECK YOURSELF ANSWERS

1. 8% were defective.   **2.** $12\% = \frac{12}{100} = \frac{3}{25}$.   **3.** $1\frac{1}{4}$.   **4. (a)** 0.05;

(b) 0.32; **(c)** 1.15.   **5. (a)** 0.085; **(b)** 0.003.   **6. (a)** 0.075; **(b)** 0.005.

Name

Section   Date

Use percents to name the shaded portion of each drawing.

1.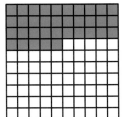

2.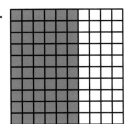

3.

4.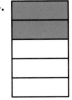

Rewrite each statement, using the percent notation.

5. Out of every 100 eligible people, 53 voted in a recent election.

6. You receive $5 in interest for every $100 saved for 1 year.

7. Out of every 100 entering students, 74 register for English composition.

8. Of 100 people surveyed, 29 watched a particular sports event on television.

9. Out of 10 voters in a state, 3 are registered as independents.

10. A dealer sold 9 of the 20 cars available during a 1-day sale.

11. Of 50 houses in a development, 27 are sold.

12. Of the 25 employees of a company, 9 are part-time.

13. Out of 50 people surveyed, 23 prefer decaffeinated coffee.

14. 17 out of 20 college students work at part-time jobs.

15. Of the 20 students in an algebra class, 5 receive a grade of A.

16. Of the 50 families in a neighborhood, 31 have children in public schools.

ANSWERS

1. 35%
2. 60%
3. 75%
4. 40%
5. 53%
6. 5%
7. 74%
8. 29%
9. 30%
10. 45%
11. 54%
12. 36%
13. 46%
14. 85%
15. 25%
16. 62%

17. $\dfrac{3}{50}$

18. $\dfrac{17}{100}$

19. $\dfrac{3}{4}$

20. $\dfrac{1}{5}$

21. $\dfrac{13}{20}$

22. $\dfrac{12}{25}$

23. $\dfrac{1}{2}$

24. $\dfrac{13}{25}$

25. $\dfrac{23}{50}$

26. $\dfrac{7}{20}$

27. $\dfrac{33}{50}$

28. $\dfrac{12}{25}$

29. $1\dfrac{1}{2}$

30. $1\dfrac{2}{5}$

31. $1\dfrac{2}{3}$

32. $1\dfrac{1}{3}$

33. 0.2

34. 0.7

35. 0.35

36. 0.75

37. 0.39

38. 0.27

39. 0.05

40. 0.07

41. 1.35

42. 2.5

43. 2.4

44. 1.6

Write as fractions or mixed numbers.

17. 6%　　　　**18.** 17%

19. 75%　　　　**20.** 20%

21. 65%　　　　**22.** 48%

23. 50%　　　　**24.** 52%

25. 46%　　　　**26.** 35%

27. 66%　　　　**28.** 48%

29. 150%　　　　**30.** 140%

31. $166\dfrac{2}{3}\%$　　　　**32.** $133\dfrac{1}{3}\%$

Write as decimals.

33. 20%　　　　**34.** 70%

35. 35%　　　　**36.** 75%

37. 39%　　　　**38.** 27%

39. 5%　　　　**40.** 7%

41. 135%　　　　**42.** 250%

43. 240%　　　　**44.** 160%

45. 23.6%

46. 10.5%

47. 6.4%

48. 3.5%

49. 0.2%

50. 0.5%

51. $7\frac{1}{2}$%

52. $8\frac{1}{4}$%

Solve the following applications.

53. Travel. Automobiles account for 85% of the travel between cities in the United States. What fraction does this percent represent?

54. Travel. Automobiles and small trucks account for 84% of the travel to and from work in the United States. What fraction does this percent represent?

55. Explain the difference between $\frac{1}{4}$ of a quantity and $\frac{1}{4}$% of a quantity.

56. Match the percents in column A with their equivalent fractions in column B.

Column A	Column B
(a) $37\frac{1}{2}$%	**(1)** $\frac{3}{5}$
(b) 5%	**(2)** $\frac{5}{8}$
(c) $33\frac{1}{3}$%	**(3)** $\frac{1}{20}$
(d) $83\frac{1}{3}$%	**(4)** $\frac{3}{8}$
(e) 60%	**(5)** $\frac{5}{6}$
(f) $62\frac{1}{2}$%	**(6)** $\frac{1}{3}$

Getting Ready for Section 11.2
[Sections 8.7 and 9.4]

Multiply.

a. 0.05×100 **b.** 0.15×100

c. 0.45×100 **d.** 1.40×100

Find the decimal equivalents for each of the following.

e. $\dfrac{2}{5}$ **f.** $\dfrac{3}{4}$

g. $1\dfrac{1}{2}$ **h.** $2\dfrac{4}{5}$

Answers

1. 35%   **3.** 75%   **5.** 53% of the eligible people voted.   **7.** 74% registered for English composition.   **9.** 30% are registered as independents.

11. 54% of the houses are sold.   **13.** 46% prefer decaffeinated coffee.

15. $\dfrac{5}{20} = \dfrac{25}{100} = 25\%$; 25% of the students received A's.   **17.** $\dfrac{3}{50}$ **19.** $\dfrac{3}{4}$

21. $\dfrac{13}{20}$ **23.** $\dfrac{1}{2}$ **25.** $\dfrac{23}{50}$ **27.** $\dfrac{33}{50}$ **29.** $1\dfrac{1}{2}$ **31.** $1\dfrac{2}{3}$ **33.** 0.2

35. 0.35   **37.** 0.39   **39.** 0.05   **41.** 1.35   **43.** 2.4   **45.** 0.236

47. 0.064   **49.** 0.002   **51.** 0.075   **53.** $\dfrac{17}{20}$ **55.**   **a.** 5

b. 15   **c.** 45   **d.** 140   **e.** 0.4   **f.** 0.75   **g.** 1.5   **h.** 2.8

11.1 Supplementary Exercises

Name

Section   Date

ANSWERS

1. 40%

2. 80%

3. 63%

4. 38%

5. 40%

6. 35%

7. 6%

8. 32%

9. $\frac{2}{25}$

10. $\frac{23}{100}$

11. $\frac{3}{5}$

12. $\frac{9}{20}$

Use percents to name the shaded portions of each drawing.

1. 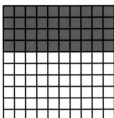   2.

Rewrite each statement, using the percent notation.

3. Of 100 people surveyed, 63 prefer brand A of a product.

4. Of the 100 cars in a parking lot, 38 are imported cars.

5. Out of 10 students at a college, 4 are first-year students.

6. Out of 20 people working downtown, 7 ride a bus to work.

7. Out of 50 parts in a shipment, 3 were defective.

8. Of the 25 students in a psychology class, 8 receive a grade of B.

Write as fractions or as mixed numbers.

9. 8%   **10.** 23%

11. 60%   **12.** 45%

© 1998 McGraw-Hill Companies

611

13. $\dfrac{9}{25}$

14. $\dfrac{12}{25}$

15. $\dfrac{8}{25}$

16. $1\dfrac{1}{4}$

17. 0.6

18. 0.37

19. 0.85

20. 0.04

21. 1.2

22. 4.5

23. 0.155

24. 0.045

13. 36%   **14.** 48%

15. 32%   **16.** 125%

Write as decimals.

17. 60%   **18.** 37%

19. 85%   **20.** 4%

21. 120%   **22.** 450%

23. 15.5%   **24.** 4.5%

 **11.2**

Changing a Decimal or a Fraction to a Percent

11.2 OBJECTIVES

1. Change a decimal to a percent.
2. Change a fraction to a percent.
3. Change a mixed number to a percent.

Changing a decimal to a percent is the opposite of changing from a percent to a decimal. We reverse the process of Section 11.1. Here is the rule:

Changing a Decimal to a Percent

To change a decimal to a percent, move the decimal point *two* places to the right and attach the percent symbol.

● Example 1

Changing a Decimal to a Percent

Write 0.18 as a percent.

To change 0.18 to a percent, we move the decimal point two places to the right. The decimal point is no longer necessary.

0.18 = 18%

● ● ● **CHECK YOURSELF 1**

Write 0.27 as a percent.

● Example 2

Changing a Decimal to a Percent

Write 0.03 as a percent.

0.03 = 3%

● ● ● **CHECK YOURSELF 2**

Write 0.05 as a percent.

● Example 3

Changing a Decimal to a Percent

Write 1.25 as a percent.

A decimal greater than 1 always gives a percent greater than 100.

1.25 = 125%

613

● ● ● **CHECK YOURSELF 3**

Write 1.3 as a percent.

If the percent still includes a decimal after the decimal point is moved two places to the right, the fractional portion can be written as a decimal or as a fraction.

● Example 4

Changing a Decimal to a Percent

Write as a percent.

(*a*) $0.045 = 4.5\%$ or $4\frac{1}{2}\%$

(*b*) $0.003 = 0.3\%$ or $\frac{3}{10}\%$

● ● ● **CHECK YOURSELF 4**

Write 0.075 as a percent.

The following rule allows us to change fractions to percents.

> You may want to review Section 9.4 on writing decimal equivalents.

Changing a Fraction to a Percent

To change a fraction to a percent, write the decimal equivalent of the fraction. Then use the previous rule to change the decimal to a percent.

● Example 5

Changing a Fraction to a Percent

Write $\frac{3}{5}$ as a percent.

> **Remember:** Move the decimal point two places to the right and attach the percent symbol.

Solution First write the decimal equivalent.

$\frac{3}{5} = 0.60$ To find the decimal equivalent, just divide the denominator into the numerator.

Now write the percent.

$\frac{3}{5} = 0.60 = 60\%$

••• **CHECK YOURSELF 5**

Write $\dfrac{3}{4}$ as a percent.

Again, you will find both decimals and fractions used in writing percents. Consider the following.

•Example 6

Changing a Fraction to a Percent

Write $\dfrac{1}{8}$ as a percent.

$$\dfrac{1}{8} = 0.125 = 12.5\% \text{ or } 12\dfrac{1}{2}\%$$

••• **CHECK YOURSELF 6**

Write $\dfrac{3}{8}$ as a percent.

To write a mixed number as a percent, we use exactly the same steps.

•Example 7

Changing a Mixed Number to a Percent

Write $1\dfrac{1}{4}$ as a percent.

$$1\dfrac{1}{4} = 1.25 = 125\%$$

Note that the resulting percent must be greater than 100 because the original mixed number was greater than 1.

••• **CHECK YOURSELF 7**

Write $1\dfrac{2}{5}$ as a percent.

Some fractions have repeating-decimal equivalents. In writing these as percents, we will either round to some indicated place or use a fractional remainder form.

• Example 8

Changing a Fraction to a Percent

Write $\dfrac{1}{3}$ as a percent.

Here we use the fractional remainder in writing the decimal equivalent.

$$\frac{1}{3} = 0.33\frac{1}{3} = 33\frac{1}{3}\%$$

● ● ● CHECK YOURSELF 8

Write $\dfrac{2}{3}$ as a percent.

• Example 9

Changing a Fraction to a Percent

Write $\dfrac{5}{7}$ as a percent.

In this case, we round the decimal equivalent. Then we write the percent.

$$\frac{5}{7} = 0.714 \qquad \text{(to the nearest thousandth)}$$

$$= 71.4\% \qquad \text{(to the nearest tenth of a percent)}$$

● ● ● CHECK YOURSELF 9

Write $\dfrac{2}{9}$ to the nearest tenth of a percent.

● ● ● CHECK YOURSELF ANSWERS

1. 27%.   **2.** $0.05 = 5\%$.   **3.** 130%.

4. 7.5% or $7\dfrac{1}{2}\%$.   **5.** $\dfrac{3}{4} = 0.75 = 75\%$.   **6.** 37.5% or $37\dfrac{1}{2}\%$.

7. $1\dfrac{2}{5} = 1.4 = 140\%$.   **8.** $\dfrac{2}{3} = 66\dfrac{2}{3}\%$.   **9.** 22.2%.

11.2 Exercises

Write each decimal as a percent.

1. 0.08

2. 0.09

3. 0.05

4. 0.13

5. 0.18

6. 0.63

7. 0.86

8. 0.45

9. 0.4

10. 0.3

11. 0.7

12. 0.6

13. 1.10

14. 2.50

15. 4.40

16. 5

17. 0.065

18. 0.095

19. 0.025

20. 0.085

21. 0.002

22. 0.008

23. 0.004

24. 0.001

A N S W E R S

1. 8%

2. 9%

3. 5%

4. 13%

5. 18%

6. 63%

7. 86%

8. 45%

9. 40%

10. 30%

11. 70%

12. 60%

13. 110%

14. 250%

15. 440%

16. 500%

17. 6.5% or $6\frac{1}{2}$%

18. 9.5% or $9\frac{1}{2}$%

19. 2.5% or $2\frac{1}{2}$%

20. 8.5% or $8\frac{1}{2}$%

21. 0.2% or $\frac{1}{5}$%

22. 0.8% or $\frac{4}{5}$%

23. 0.4% or $\frac{2}{5}$%

24. 0.1% or $\frac{1}{10}$%

25. 25%

26. 80%

27. 40%

28. 50%

29. 20%

30. 75%

31. 62.5% or $62\frac{1}{2}$%

32. 87.5% or $87\frac{1}{2}$%

33. 31.25% or $31\frac{1}{4}$%

34. 120%

35. 350%

36. $66\frac{2}{3}$%

37. $16\frac{2}{3}$%

38. 18.75% or $18\frac{3}{4}$%

39. 77.8%

40. 45.5%

41. 85.9%

42. 56.6%

Write each fraction as a percent.

25. $\frac{1}{4}$ **26.** $\frac{4}{5}$

27. $\frac{2}{5}$ **28.** $\frac{1}{2}$

29. $\frac{1}{5}$ **30.** $\frac{3}{4}$

31. $\frac{5}{8}$ **32.** $\frac{7}{8}$

33. $\frac{5}{16}$ **34.** $1\frac{1}{5}$

35. $3\frac{1}{2}$ **36.** $\frac{2}{3}$

37. $\frac{1}{6}$ **38.** $\frac{3}{16}$

39. $\frac{7}{9}$ (to nearest tenth of a percent)   **40.** $\frac{5}{11}$ (to nearest tenth of a percent)

Between 1973 and 1990, the average fuel efficiency of new U.S. cars increased from 6.4 to 11.9 kilometers per liter (km/L). During this same time, the average fuel efficiency for the entire fleet of U.S. cars rose from 5.3 to 8.3 km/L. For Exercises 41 and 42, solve the applications involving fuel efficiency.

41. Fuel efficiency. The percentage increase in fuel efficiency for new cars is given by the fraction $\frac{(11.9 - 6.4)}{6.4}$. Change this fraction to a percent.

42. Fuel efficiency. The percentage increase in fuel efficiency for the fleet of U.S. cars is given by the fraction $\frac{(8.3 - 5.3)}{5.3}$. Change this fraction to a percent.

Getting Ready for Section 11.3
[Section 11.1]

Change the following percents to decimals.

a. 11%   **b.** 15%   **c.** 27%

d. 112%   **e.** 123%   **f.** 8%

Answers

1. 8%   **3.** 5%   **5.** 18%   **7.** 86%   **9.** 40%   **11.** 70%   **13.** 110%

15. 440%   **17.** 6.5% or $6\frac{1}{2}$%   **19.** 2.5% or $2\frac{1}{2}$%   **21.** 0.2% or $\frac{1}{5}$%

23. 0.4% or $\frac{2}{5}$%   **25.** $\frac{1}{4}$ = 0.25 = 25%   **27.** 40%   **29.** 20%

31. $\frac{5}{8}$ = 0.625 = 62.5%   **33.** 31.25%   **35.** 350%   **37.** $16\frac{2}{3}$%

39. 77.8%   **41.** 85.9%

11.2 Supplementary Exercises

Name _____

Section _____ Date _____

Write as percents:

1. 0.06   **2.** 0.02

3. 0.24   **4.** 0.53

5. 0.4   **6.** 0.6

7. 2.75   **8.** 2.65

9. 8.5% or $8\frac{1}{2}\%$

10. 3.5% or $3\frac{1}{2}\%$

11. 0.8% or $\frac{4}{5}\%$

12. 0.5% or $\frac{1}{2}\%$

13. 75%

14. 60%

15. 80%

16. 12.5% or $12\frac{1}{2}\%$

17. 43.75% or $43\frac{3}{4}\%$

18. 275%

19. $83\frac{1}{3}\%$

20. 38.5%

9. 0.085

10. 0.035

11. 0.008

12. 0.005

13. $\dfrac{3}{4}$

14. $\dfrac{3}{5}$

15. $\dfrac{4}{5}$

16. $\dfrac{1}{8}$

17. $\dfrac{7}{16}$

18. $2\dfrac{3}{4}$

19. $\dfrac{5}{6}$

20. $\dfrac{5}{13}$ (to nearest tenth of a percent)

Identifying the Rate, Base, and Amount

11.3 OBJECTIVES

1. Identify the rate in an application problem.
2. Identify the base in an application problem.
3. Identify the amount in an application problem.

There are many practical applications of our work with percents. All these problems have three basic parts that need to be identified. Let's look at some definitions that will help with that process.

Definitions of Base, Amount, and Rate

The **base** is the whole in a problem. It is the standard used for comparison.

The **amount** is the part of the whole being compared to the base.

The **rate** is the ratio of the amount to the base. It is written as a percent.

Let's look at some examples of determining the parts of a percent problem.

● Example 1

Identifying Rates

Identify each rate.

The *rate* (which we will label R%) is the easiest of the terms to identify. The rate is written with the percent symbol (%) or the word "percent."

(*a*) What is 15% of 200?

      ↑
     R%

Here 15% is the rate because it has the percent symbol attached.

(*b*) 25% of what number is 50?

↑
R%

25% is the rate.

(*c*) 20 is what percent of 40?

       ↑
R%

Here the rate is unknown.

● ● ● CHECK YOURSELF 1

Identify the rate.

a. 15% of what number is 75?
b. What is 8.5% of 200?
c. 200 is what percent of 500?

The *base* (*B*) is the whole, or 100%, in the problem. The base will often follow the word "of." Look at our next example.

● Example 2

Identifying Bases

Identify each base.

(*a*) What is 15% of 200?

  B   200 is the base. It follows the word "of."

(*b*) 25% of what number is 50?

  B   Here the base is the unknown.

(*c*) 20 is what percent of 40?

  B   40 is the base.

● ● ●   **CHECK YOURSELF 2**

Identify the base.

a. 70 is what percent of 350?   **b.** What is 25% of 300?
c. 14% of what number is 280?

The *amount* (*A*) will be the part of the problem remaining once the rate and the base have been identified.

In many applications, the amount is found with the word "is."

● Example 3

Identifying Amounts

Identify the amount.

(*a*) What is 15% of 200?   Here the amount is the
               unknown part of the problem.
  A   Note that the word "is" follows.

(*b*) 25% of what number is 50?   Here the amount, 50, follows
               the word "is."
                        A

(*c*) 20 is what percent of 40?

  A   Again the amount, here 20,
                                   can be found with the word "is."

●●● **CHECK YOURSELF 3**

Identify the amount.

a. 30 is what percent of 600?   **b.** What is 12% of 5000?
c. 24% of what number is 96?

Let's look at another example of identifying the three parts in a percent problem.

● Example 4

Identifying the Rate, Base, and Amount

Determine the rate, base, and amount in this problem:

12% of 800 is what number?

Solution Finding the *rate* is not difficult. Just look for the percent symbol or the word "percent." In this exercise, 12% is the rate.

The *base* is the whole. Here it follows the word "of." 800 is the whole or the base.

The *amount* remains when the rate and the base have been found. Here the amount is the unknown. It follows the word "is." "What number" asks for the unknown amount.

●●● **CHECK YOURSELF 4**

Find the rate, base, and amount in the following statements or questions.

a. 75 is 25% of 300.   **b.** 20% of what number is 50?

We will use percents to solve a variety of applied problems. In all these situations, you will have to identify the three parts of the problem. Let's work through some examples intended to help you build that skill.

● Example 5

Identifying the Rate, Base, and Amount

Determine the rate, base, and amount in the following application.

In an algebra class of 35 students, 7 received a grade of A. What percent of the class received an A?

Solution The *base* is the whole in the problem, or the number of students in the class. 35 is the base.

The *amount* is the portion of the base, here the number of students that receive the A grade. 7 is the amount.

The *rate* is the unknown in this example. "What percent" asks for the unknown rate.

● ● ● **CHECK YOURSELF 5**

Determine the rate, base, and amount in the following application: In a shipment of 150 parts, 9 of the parts were defective. What percent were defective?

In Example 6, we look at a practical (business) application.

● Example 6

Identifying the Rate, Base, and Amount in a Business Application

Determine the rate, base, and amount in the following application:

Doyle borrows $2000 for 1 year. If the interest rate is 12%, how much interest will he pay?

Solution The *base* is again the whole, the size of the loan in this example. $2000 is the base.

The *rate* is, of course, the interest rate. 12% is the rate.

The *amount* is the quantity left once the base and rate have been identified. Here the amount is the amount of interest that Doyle must pay. The amount is the unknown in this example.

● ● ● **CHECK YOURSELF 6**

Determine the rate, base, and amount in the following application: Robert earned $120 interest from a savings account paying 8% interest. What amount did he have invested?

● ● ● **CHECK YOURSELF ANSWERS**

1. **(a)** 15%; **(b)** 8.5%; **(c)** what percent (the unknown).
2. **(a)** 350; **(b)** 300; **(c)** what number (the unknown).
3. **(a)** 30; **(b)** "what is" (the unknown); **(c)** 96.
4. **(a)** $R\% = 25\%$, $B = 300$, $A = 75$; **(b)** $R\% = 20\%$, $B =$ "what number," $A = 50$.
5. $B = 150$, $A = 9$, $R\% =$ "what percent" (the unknown).
6. $R\% = 8\%$, $A = \$120$, $B =$ "what amount" (the unknown).

A N S W E R S

Identify the rate, base, and amount in each statement or question. *Do not solve* the exercise at this point.

1. 23% of 400 is 92.
   R%   B   A

2. 150 is 20% of 750.
   A   R%   B

3. 40% of 600 is 240.
   R%   B   A

4. 200 is 40% of 500.
   A   R%   B

5. What is 7% of 325?
   A   R%   B

6. 80 is what percent of 400?
   A   R%   B

7. 16% of what number is 56?
   R%   B   A

8. What percent of 150 is 30?
   R%   B   A

9. 480 is 60% of what number?
   A   R%   B

10. What is 60% of 250?
   A   R%   B

11. What percent of 120 is 40?
   R%   B   A

12. 150 is 75% of what number?
   A   R%   B

1. See exercise
2. See exercise
3. See exercise
4. See exercise
5. See exercise
6. See exercise
7. See exercise
8. See exercise
9. See exercise
10. See exercise
11. See exercise
12. See exercise
13. See exercise
14. See exercise
15. See exercise

Identify the rate, base, and amount in the following applications. *Do not solve* the applications at this point.

13. Commission. Jan has a 5% commission rate on all her sales. If she sells
                                                R%
$40,000 worth of merchandise in 1 month, what commission will she earn?
   B   A

14. Salary. 22% of Shirley's monthly salary is deducted for withholding. If those
                R%
deductions total $209, what is her salary?
              A   B

15. Chemistry. In a chemistry class of 30 students, 5 received a grade of A.
                                            B   A
What percent of the students received A's?
   R%

A N S W E R S

16. See exercise

17. See exercise

18. See exercise

19. See exercise

20. See exercise

21.

22.

23.

16. Mixtures. A can of mixed nuts contains 80% peanuts. If the can holds 16 ounces (oz), how many ounces of peanuts does it contain?

17. Selling price. The sales tax rate in a state is 5.5%. If you pay a tax of $3.30 on an item that you purchase, what is its selling price?

18. Manufacturing. In a shipment of 750 parts, 75 were found to be defective. What percent of the parts were faulty?

19. Enrollments. A college had 9000 students at the start of a school year. If there is an enrollment increase of 6% by the beginning of the next year, how many additional students were there?

20. Investments. Paul invested $5000 in a time deposit. What interest will he earn for 1 year if the interest rate is 6.5%?

21. Using the latest census figures for your state, determine the percent of the following minorities: African-American, Hispanic, and Asian. These figures can be found on the World Wide Web at **www.census.gov/.**

22. An advertisement for a local supermarket proclaims "Chicken: only $1.97 per pound. Save 40%." Last week the same brand of chicken was on sale for $2.75. Is the ad accurate? If not, what should it say?

23. At True Grip hardware, you pay $10 in tax for a barbecue grill that is 6% of the purchase price. At Loose Fit hardware, you pay $10 in tax for the same grill, but it is 8% of the purchase price. At which store do you get the better buy? Why?

Getting Ready for Section 11.4
[Section 10.3]

Solve each of the following proportions.

a. $\dfrac{x}{150} = \dfrac{20}{100}$

b. $\dfrac{36}{y} = \dfrac{30}{100}$

c. $\dfrac{45}{180} = \dfrac{m}{100}$

d. $\dfrac{21}{p} = \dfrac{35}{100}$

e. $\dfrac{a}{80} = \dfrac{45}{100}$

f. $\dfrac{150}{120} = \dfrac{r}{100}$

Answers

1. 23% of 400 is 92.
   ↑   ↑   ↑
  R%   *B*   *A*

3. 40% of 600 is 240.
   ↑   ↑   ↑
  R%   *B*   *A*

5. What is 7% of 325?
   ↑   ↑   ↑
  A   *R%*   *B*

7. 16% of what number is 56?
 ↑   ↑   ↑
R%   *B*   *A*

9. 480 is 60% of what number?
 ↑   ↑   ↑
A   *R%*   *B*

11. What percent of 120 is 40?
      ↑   ↑   ↑
     R%   *B*   *A*

13. $40,000 is the base. 5% is the rate. Her commission, the unknown, is the amount.

15. 30 is the base. 5 is the amount. The unknown percent is the rate.

17. 5.5% is the rate. The tax, $3.30, is the amount. The unknown selling price is the base.

19. The base is 9000. The rate is 6%. The unknown number of additional students is the amount.   **21.**   **23.**

   a. 30   **b.** 120   **c.** 25   **d.** 60   **e.** 36   **f.** 125

11.3 Supplementary Exercises

Name

Section   Date

Identify the rate, base, and amount in each statement or question. *Do not solve* the exercises at this point.

1. 60 is 30% of 200.
  A $R\%$ B

2. 45% of 200 is 90.
  $R\%$ B A

3. What is 9% of 300?
  A $R\%$ B

4. 80 is 16% of what number?
  A $R\%$ B

5. 800 is what percent of 1200?
  A $R\%$ B

6. What is 150% of 1200?
  A $R\%$ B

Identify the rate, base, and amount in the following applications. *Do not solve* the applications at this point.

7. Test scores. On a test, Alice had 80% of the problems right. If she did 40
   $R\%$ A
problems correctly, how many questions were on the test?
                        B

8. Manufacturing. In a shipment of 250 parts, 40 are found to be defective.
                                      B A
What percent of the parts are faulty?
  $R\%$

9. Chemistry. There are 99 grams (g) of acid in 900 g of a solution.
                        A B
What percent of the solution is acid?
  $R\%$

10. Enrollments. 76% of the students in a psychology class passed their midterm
       $R\%$
examination. If 114 students passed, how many students were in the class?
                        A B

Three Types of Percent Problems

11.4 OBJECTIVES

1. Find the unknown amount in a percent problem.
2. Find the unknown rate in a percent problem.
3. Find the unknown base in a percent problem.

From your work in Section 11.3, you may have observed that there are three basic types of percent problems. These depend on which of the three parts—the amount, the rate, or the base—is missing in the problem statement. The solution for each type of problem depends on the **percent relationship.**

> ### Formula for an Unknown Amount
>
> Amount = rate × base

We will illustrate the solution of each type of problem in the following examples. Let's start with a problem in which we want to find the amount.

• Example 1

Finding an Unknown Amount

What is 18% of 300?

Type 1: Finding an unknown amount.

Solution We know the rate, 18%; and the base, 300; the amount is the unknown. Using the percent relationship, we can translate the problem to an equation.

$$\text{Amount} = \underset{\text{Rate}}{0.18} \times \underset{\text{Base}}{300}$$
$$= 54$$

Write 18% as the decimal 0.18 by the rule of Section 11.2. Then multiply to find the amount.

So 54 is 18% of 300.

● ● ● **CHECK YOURSELF 1**

Find 65% of 200.

Note

1. If the rate is *less than* 100%, the amount will be *less than* the base.

   25 is 50% of 50   and   25 < 50

2. If the rate is *greater than* 100%, the amount will be *greater than* the base.

   75 is 150% of 50   and   75 > 50

Let's consider a second type of percent problem involving an unknown rate.

● Example 2

Finding an Unknown Percent

30 is what percent of 150?

Type 2: Finding an unknown percent.

Solution We know the amount, 30, and the base, 150; the rate (what percent) is the unknown. Again using the percent relationship to translate to an equation, we have

$$\overset{\text{Base}}{\underset{\searrow}{}}\ \overset{\text{Amount}}{\underset{\searrow}{}}$$
$$\text{Rate} \times 150 = 30$$

This will leave the rate *alone* on the left.

We *divide* both sides by 150 to find the rate.

$$\text{Rate} = \frac{30}{150} = \frac{1}{5} = 0.20 = 20\%$$

Remember that 0.20 = 20% by the rule of Section 11.3.

30 is 20% of 150.

Formula for an Unknown Rate

$$\text{Rate} = \frac{\text{amount}}{\text{base}}$$

● ● ● **CHECK YOURSELF 2**

75 is what percent of 300?

Note

1. If the amount is *less than* the base, the rate will be *less than* 100%.
2. If the amount is *greater than* the base, the rate will be greater than 100%.

Let's look at another percent problem involving an unknown base in Example 3.

● Example 3

Finding an Unknown Base

28 is 40% of what number?

Type 3: Finding an unknown base.

Solution We know the amount, 28, and the rate, 40%. The base (what number) is the unknown. From the percent relationship we have

Rate   Amount

Note that 40% is written as 0.40.

$$0.40 \times \text{base} = 28$$

We *divide* both sides by 0.40 to find the base.

$$\text{Base} = \frac{28}{0.40} = 70$$

So 28 is 40% of 70.

Formula for an Unknown Base

$$\text{Base} = \frac{\text{amount}}{\text{rate}}$$

● ● ● CHECK YOURSELF 3

70 is 35% of what number?

We have now seen solution methods for the three basic types of percent problems: finding the amount, the rate, and the base. As you will see in the remainder of this section, our work in Chapter 10 with proportions will allow us to solve each type of problem in an identical fashion. In fact, many students find percent problems easier to approach with the proportion method.

First, we will write what is called the **percent proportion.**

The Percent Proportion

$$\frac{\text{Amount}}{\text{Base}} = \frac{R}{100}$$

In symbols,

$$\frac{A}{B} = \frac{R}{100}$$

On the right, $\frac{R}{100}$ is the rate, and this proportion is equivalent to our earlier percent relationship.

Since in any percent problem we know two of the three quantities (A, B, or R), we can always solve for the unknown term. Consider in Example 4 the use of the percent proportion.

● Example 4

Solving a Problem Involving an Unknown Amount

This is an **unknown-amount problem.**

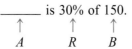

_____ is 30% of 150.

$\quad\quad A \quad\quad\quad\quad R \quad\quad\quad B$

Solution Substitute the values into the percent proportion.

$$\underset{B}{\nearrow}\frac{A}{150} = \frac{\overset{R}{\swarrow}30}{100} \qquad \text{The amount } A \text{ is the unknown term of the proportion.}$$

We solve the proportion with the methods of Section 10.3.

$100A = 150 \cdot 30$
$100A = 4500$

Divide by the coefficient, 100.

$$\frac{\cancel{100}A}{\cancel{100}} = \frac{4500}{100}$$
$$A = 45$$

The amount is 45. This means that 45 is 30% of 150.

● ● ● **CHECK YOURSELF 4**

Use the percent proportion to answer this question: What is 24% of 300?

The same percent proportion will work if you want to find the rate.

●Example 5

Solving a Problem Involving an Unknown Rate

This is an **unknown-rate problem.**

$$\underset{\underset{R}{\uparrow}}{\rule{2cm}{0.4pt}}\% \text{ of } \underset{\underset{B}{\uparrow}}{400} \text{ is } \underset{\underset{A}{\uparrow}}{72}.$$

Solution Substitute the known values into the percent proportion.

$$\overset{A}{\searrow}\frac{72}{400} = \frac{R}{100} \qquad R, \text{ the rate, is the unknown term in this case.}$$
$$\underset{B}{\nearrow}$$

Solving, we get

$$400R = 7200$$
$$\frac{\cancel{400}R}{\cancel{400}} = \frac{7200}{400}$$
$$R = 18$$

The rate is 18%. So 18% of 400 is 72.

● ● ● **CHECK YOURSELF 5**

Use the percent proportion to answer this question: What percent of 50 is 12.5?

___

Finally, we use the same proportion to find an unknown base.

● Example 6

Solving a Problem Involving an Unknown Base

This is an **unknown-base problem.**

40% of _____ is 200.
 ↑   ↑   ↑
 R   B   A

Solution   Substitute the known values into the percent proportion.

$$\frac{200}{B} = \frac{40}{100}$$

A points to 200, *R* points to 40.

In this case *B*, the base, is the unknown term of the proportion.

Solving gives

$$40B = 200 \cdot 100$$

$$\frac{\cancel{40}B}{\cancel{40}} = \frac{20,000}{40}$$

$$B = 500$$

The base is 500, and 40% of 500 is 200.

● ● ● **CHECK YOURSELF 6**

288 is 60% of what number?

___

Remember that a percent (the rate) can be greater than 100.

● Example 7

Solving a Percent Problem

The rate is 125%. The base is 300.

What is 125% of 300?

When the rate is greater than 100%, the amount will be *greater than* the base.

Solution In the percent proportion, we have

$$\frac{A}{300} = \frac{125}{100}$$

So $100A = 300 \cdot 125$.

Dividing by 100 yields

$$A = \frac{37,500}{100} = 375$$

So 375 is 125% of 300.

● ● ● **CHECK YOURSELF 7**

Find 150% of 500.

In Example 8, we want to find a rate where the amount is greater than the base.

● Example 8

Solving a Percent Problem

The amount is 92, the base is 80.

92 is what percent of 80?

Solution In the percent proportion we have

$$\frac{92}{80} = \frac{R}{100}$$

So $80R = 92 \cdot 100$.

Now, solving for R, we divide by 80:

$$R = \frac{9200}{80} = 115$$

So 92 is 115% of 80.

● ● ● **CHECK YOURSELF 8**

What percent of 120 is 156?

We will conclude this section by looking at two examples of solving percent problems involving fractions of a percent.

• Example 9

Solving a Percent Problem

34 is 8.5% of what number?

Solution   Using the percent proportion yields

The amount is 34, the rate is 8.5%. We want to find the base.

$$\frac{34}{B} = \frac{8.5}{100}$$

Solving, we have

$$8.5B = 34 \cdot 100$$

or

Divide by 8.5.

$$B = \frac{3400}{8.5} = 400$$

So 34 is 8.5% of 400.

● ● ●   CHECK YOURSELF 9

12.5% of what number is 75?

• Example 10

Solving a Percent Problem

40 is what percent of 120?

Solution   With the percent proportion we have

Here we know the amount, 40, and the base, 120.

$$\frac{40}{120} = \frac{R}{100}$$

or

$$120R = 40 \cdot 100$$

Now solving for R, we get

We have chosen to solve for R by dividing numerator and denominator by 40. We then write the result as a mixed number.

$$R = \frac{4000}{120} = \frac{100}{3} = 33\frac{1}{3}$$

and 40 is $33\frac{1}{3}$% of 120.

● ● ● **CHECK YOURSELF 10**

80 is what percent of 120?

Estimation is a very useful skill in working with percents. For example, to see whether an amount seems reasonable, one approach is to round the rate to a "convenient" value and then use that rounded rate to estimate the amount.

● Example 11

Estimating Percentages

Find 19.3% of 500.

Solution Round the rate to 20% $\left(\text{as a fraction, } \dfrac{1}{5}\right)$. An estimate of the amount is then

$$\frac{1}{5} \times 500 = 100$$

Rounded rate   Base   Estimate of amount

● ● ● **CHECK YOURSELF 11**

Estimate the amount.

20.2% of 800

● ● ● **CHECK YOURSELF ANSWERS**

1. 130.   **2.** 25%.   **3.** 200.

4.

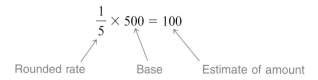

$$\frac{A}{300} = \frac{24}{100} \quad R$$

$100A = 7200; A = 72.$

5. A

$$\frac{12.5}{50} = \frac{R}{100}$$

B

$50R = 1250; R = 25\%.$

6. A

$$\frac{288}{B} = \frac{60}{100} \quad R$$

$60B = 28{,}800; B = 480.$

7. 750.   **8.** 130%.

9. 600.   **10.** $66\dfrac{2}{3}\%$.   **11.** 160.

Solve each of the following problems involving percent.

1. What is 35% of 600?

2. 20% of 400 is what number?

3. 45% of 200 is what number?

4. What is 40% of 1200?

5. Find 40% of 2500.

6. What is 75% of 120?

7. What percent of 50 is 4?

8. 51 is what percent of 850?

9. What percent of 500 is 45?

10. 14 is what percent of 200?

11. What percent of 200 is 340?

12. 392 is what percent of 2800?

13. 46 is 8% of what number?

14. 7% of what number is 42?

15. Find the base if 11% of the base is 55.

16. 16% of what number is 192?

17. 58.5 is 13% of what number?

18. 21% of what number is 73.5?

19. Find 110% of 800.

20. What is 115% of 600?

21. What is 108% of 4000?

22. Find 160% of 2000.

23. 210 is what percent of 120?

24. What percent of 40 is 52?

A N S W E R S

1. 210
2. 80
3. 90
4. 480
5. 1000
6. 90
7. 8%
8. 6%
9. 9%
10. 7%
11. 170%
12. 14%
13. 575
14. 600
15. 500
16. 1200
17. 450
18. 350
19. 880
20. 690
21. 4320
22. 3200
23. 175%
24. 130%

25. 400%

26. 120%

27. 500

28. 250

29. 850

30. 900

31. 25.5

32. 66

33. 705

34. 17.5

35. 157.5

36. 551

37. 7.5%

38. $33\frac{1}{3}$%

39. $66\frac{2}{3}$%

40. 9.75%

41. 62.5%

42. 8.5%

43. 4000

44. 400

45. 450

46. 5000

47. 2600

48. 350

25. 360 is what percent of 90?

26. What percent of 15,000 is 18,000?

27. 625 is 125% of what number?

28. 140% of what number is 350?

29. Find the base if 110% of the base is 935.

30. 130% of what number is 1170?

31. Find 8.5% of 300.

32. $8\frac{1}{4}$% of 800 is what number?

33. Find $11\frac{3}{4}$% of 6000.

34. What is 3.5% of 500?

35. What is 5.25% of 3000?

36. What is 7.25% of 7600?

37. 60 is what percent of 800?

38. 500 is what percent of 1500?

39. What percent of 180 is 120?

40. What percent of 800 is 78?

41. What percent of 1200 is 750?

42. 68 is what percent of 800?

43. 10.5% of what number is 420?

44. Find the base if $11\frac{1}{2}$% of the base is 46.

45. 58.5 is 13% of what number?

46. 6.5% of what number is 325?

47. 195 is 7.5% of what number?

48. 21% of what number is 73.5?

Estimate the amount in each of the following problems.

49. Find 25.8% of 4000. **50.** What is 48.3% of 1500?

51. 74.7% of 600 is what number? **52.** 9.8% of 1200 is what number?

53. Find 152% of 400. **54.** What is 118% of 5000?

55. It is customary when eating in a restaurant to leave a 15% tip.

    (a) Outline a method to do a quick approximation for the amount of tip to leave.

    (b) Use this method to figure a 15% tip on a bill of $47.76.

56. The dean of Enrollment Management at a college states, "Last year was not a good year. Our enrollments were down 25%. But this year we increased our enrollment by 30%. I think we have turned the corner." Evaluate the dean's analysis.

● **Getting Ready for Section 11.5**
    [Section 3.6]

Evaluate the following.

a. $\dfrac{270 \cdot 10}{90}$ **b.** $\dfrac{660 \cdot 100}{11}$ **c.** $\dfrac{120 \cdot 100}{4000}$

d. $\dfrac{320 \cdot 100}{2.5}$ **e.** $\dfrac{23 \cdot 4.5}{10}$ **f.** $\dfrac{46 \cdot 15}{100}$

Answers

1. 210   **3.** 90   **5.** 1000   **7.** 8%   **9.** 9%   **11.** 170%   **13.** 575
15. 500   **17.** 450   **19.** 880   **21.** 4320   **23.** 175%   **25.** 400%
27. 500   **29.** 850   **31.** 25.5   **33.** 705   **35.** 157.5   **37.** 7.5%

39. $66\frac{2}{3}$%   **41.** 62.5%   **43.** 4000   **45.** 450   **47.** 2600   **49.** 1000

51. 450   **53.** 600   **55.**   **a.** 30   **b.** 6000   **c.** 3   **d.** 1280

e. 10.35   **f.** 6.9

A N S W E R S

49. 1000

50. 750

51. 450

52. 120

53. 600

54. 6000

55.

56.

a. 30

b. 6000

c. 3

d. 1280

e. 10.35

f. 6.9

Supplementary Exercises

A N S W E R S

1.	150
2.	45
3.	900
4.	21%
5.	12%
6.	15%
7.	400
8.	800
9.	350
10.	2040
11.	4480
12.	125%
13.	140%
14.	600
15.	1200
16.	255
17.	33.25
18.	51
19.	37.5%
20.	76.5%
21.	$66\frac{2}{3}$%
22.	1200
23.	400
24.	660

Solve each of the following problems involving percent.

1. What is 30% of 500?

2. Find 15% of 300.

3. 45% of 2000 is what number?

4. What percent of 400 is 84?

5. 120 is what percent of 1000?

6. What percent of 800 is 120?

7. 7% of what number is 28?

8. Find the base if 12% of the base is 96.

9. 59.5 is 17% of what number?

10. What is 170% of 1200?

11. Find 112% of 4000.

12. What percent of 480 is 600?

13. 840 is what percent of 600?

14. Find the base if 125% of the base is 750.

15. 1560 is 130% of what number?

16. Find 8.5% of 3000.

17. What is 4.75% of 700?

18. Find $8\frac{1}{2}$% of 600.

19. 225 is what percent of 600?

20. 1071 is what percent of 1400?

21. What percent of 2400 is 1600?

22. 102 is 8.5% of what number?

23. Find the base if 10.5% of the base is 42.

24. 178.2 is 27% of what number?

A *rate, base,* and *amount* will appear in *all* problems involving percents.

The concept of percent is perhaps the most frequently encountered arithmetic idea that we will consider in this book. In this section, we will show some of the many applications of percent and the special terms that are used in these applications.

To use percents in problem solving, you should always read the problem carefully to determine the rate, base, and amount in the problem. This is illustrated in our first example.

● Example 1

Solving a Problem Involving an Unknown Amount

A student needs 70% to pass an examination containing 50 questions. How many questions must she get right?

Solution The *rate* is 70%. The *base* is the number of questions on the test, here 50. The *amount* is the number of questions that must be correct.

To find the amount, we will use the percent proportion from Section 11.4.

Substitute 50 for *B* and 70 for *R*.

$$B \longrightarrow \frac{A}{50} = \frac{70}{100} \longleftarrow R$$

so

$$100A = 50 \cdot 70$$

Dividing by 100 gives

$$A = \frac{3500}{100} = 35$$

She must answer 35 questions correctly to pass.

● ● ● CHECK YOURSELF 1

Generally, 72% of the students in a chemistry class pass the course. If there are 150 students in the class, how many can you expect to pass?

The money borrowed or saved is called the **principal.**

As we said earlier, there are many applications of percent to daily life. One that almost all of us encounter involves **interest.** When you borrow money, you pay interest. When you place money in a savings account, you earn interest. Interest is a percent of the whole (in this case, the *principal*), and the percent is called the **interest rate.**

•Example 2

Solving a Problem Involving an Unknown Amount

Find the interest you must pay if you borrow $2000 for 1 year with an interest rate of $9\frac{1}{2}\%$.

Remember:

$9\frac{1}{2}\% = 9.5\%$

Solution The base (the principal) is $2000, the rate is $9\frac{1}{2}\%$, and we want to find the interest (the amount). Using the percent proportion gives

$$\frac{A}{2000} = \frac{9.5}{100}$$

so

$$100A = 2000 \cdot 9.5$$

or

$$A = \frac{19,000}{100} = 190$$

The interest (amount) is $190.

● ● ● CHECK YOURSELF 2

You invest $5000 for 1 year at $8\frac{1}{2}\%$. How much interest will you earn?

Let's look at an application that requires finding the rate.

•Example 3

Solving a Problem Involving an Unknown Rate

You borrow $2000 from a bank for 1 year and are charged $150 interest. What is the interest rate?

Solution The base is the amount of the loan (the principal). The amount is the interest paid. To find the interest rate, we again use the percent proportion.

$$\frac{150}{2000} = \frac{R}{100}$$

Then

$$2000R = 150 \cdot 100$$

$$R = \frac{15,000}{2000} = 7.5$$

The interest rate is 7.5%.

● ● ● **CHECK YOURSELF 3**

Xian borrowed $3200 and was charged $352 in interest for 1 year. What was the interest rate?

Now let's look at an application that requires finding the base.

● Example 4

Solving a Problem Involving an Unknown Base

Ms. Hobson agrees to pay 11% interest on a loan for her new automobile. She is charged $550 interest on a loan for 1 year. How much did she borrow?

Solution The rate is 11%. The amount, or interest, is $550. We want to find the base, which is the principal, or the size of the loan. To solve the problem, we have

$$\frac{550}{B} = \frac{11}{100}$$

The product of the means equals the product of the extremes.

$$11B = 550 \cdot 100$$

$$B = \frac{55,000}{11} = 5000$$

She borrowed $5000.

● ● ● **CHECK YOURSELF 4**

Sue pays $210 interest for a 1-year loan at 10.5%. What was the size of her loan?

Percents are used in too many ways for us to list. Look at the variety in the following examples, which illustrate some additional situations in which you will find percents.

•Example 5

Solving a Percent Problem

A **commission** is the amount that a person is paid for a sale. That commission is the amount of the total sold.

A salesman sells a car for $9500. His commission rate is 4%. What will be his commission for the sale?

Solution   The base is the total of the sale, in this problem, $9500. The rate is 4%, and we want to find the commission. This is the amount. By the percent proportion

$$\frac{A}{9500} = \frac{4}{100}$$

$$100A = 4 \cdot 9500$$

$$A = \frac{38,000}{100} = 380$$

The salesman's commission is $380.

● ● ●   **CHECK YOURSELF 5**

Jenny sells a $12,000 building lot. If her real estate commission rate is 5%, what commission will she receive for the sale?

•Example 6

Solving a Percent Problem

A clerk sold $3500 in merchandise during 1 week. If he received a commission of $140, what was the commission rate?

Solution   The base is $3500, and the amount is the commission of $140. Using the percent proportion we have

$$\frac{140}{3500} = \frac{R}{100}$$

Again, the product of the means equals the product of the extremes.

$$3500R = 140 \cdot 100$$

$$R = \frac{14,000}{3500}$$

$$= 4$$

The commission rate is 4%.

● ● ●   **CHECK YOURSELF 6**

On a purchase of $500 you pay a sales tax of $21. What is the tax rate?

Example 7, involving a commission, shows how to find the total sold.

● Example 7

Solving a Percent Problem

A saleswoman has a commission rate of 3.5%. To earn $280, how much must she sell?

Solution   The rate is 3.5%. The amount is the commission, $280. We want to find the base. In this case, this is the amount that the saleswoman needs to sell.

By the percent proportion

$$\frac{280}{B} = \frac{3.5}{100} \text{ or } 3.5B = 280 \cdot 100$$

$$B = \frac{28{,}000}{3.5} = 8000$$

The saleswoman must sell $8000 to earn $280 in commissions.

● ● ●   **CHECK YOURSELF 7**

Kerri works with a commission rate of 5.5%. If she wants to earn $825 in commissions, find the total sales that she must make.

Another common application of percents involves tax rates.

● Example 8

Solving a Percent Problem

A state taxes sales at 5.5%. How much sales tax will you pay on a purchase of $48?

Solution   The tax you pay is the amount (the part of the whole). Here the base is the purchase price, $48, and the rate is the tax rate, 5.5%.

In an application involving taxes, the tax paid is always the amount.

$$\frac{A}{48} = \frac{5.5}{100} \qquad \text{or} \qquad 100A = 48 \cdot 5.5$$

Now

$48 \times 5.5 = 264$

$$A = \frac{264}{100} = 2.64$$

The sales tax paid is $2.64.

● ● ● **CHECK YOURSELF 8**

Suppose that a state has a sales tax rate of $6\frac{1}{2}$%. If you buy a used car for $1200, how much sales tax must you pay?

Percents are also used to deal with store markups or discounts. Consider Example 9.

● Example 9

Solving a Percent Problem

A store marks up items to make a 30% profit. If an item cost $2.50 from the supplier, what will the selling price be?

This problem can be done in one step. We'll look at that method in the calculator section later in this chapter.

Solution The base is the cost of the item, $2.50, and the rate is 30%. In the percent proportion, the markup is the amount in this application.

$$\frac{A}{2.50} = \frac{30}{100} \qquad \text{or} \qquad 100A = 30 \cdot 2.50$$

Then

$$A = \frac{75}{100} = 0.75$$

The markup is $0.75. Finally we have

Selling price = original cost + markup

Selling price = $2.50 + $0.75 = $3.25   Add the cost and the markup to find the selling price.

● ● ● **CHECK YOURSELF 9**

A store wants to discount (or mark down) an item by 25% for a sale. If the original price of the item was $45, find the sale price. [*Hint:* Find the discount (the amount the item will be marked down), and subtract that from the original price.]

Increases and decreases are often stated in terms of percents, as our final examples illustrate.

● Example 10

Solving a Percent Problem

The population of a town increased 15% in a 3-year period. If the original population was 12,000, what was the population at the end of the period?

Solution   First we find the increase in the population. That increase is the amount in the problem.

$$\frac{A}{12,000} = \frac{15}{100} \text{ so } 100A = 15 \cdot 12,000$$

$$A = \frac{180,000}{100}$$

$$= 1800$$

To find the population at the end of the period, we add

$$12,000 + 1800 = 13,800$$

Original population   Increase   New population

● ● ●   **CHECK YOURSELF 10**

A school's enrollment decreased by 8% from a given year to the next. If the enrollment was 550 students the first year, how many students were enrolled the second year?

● Example 11

Solving a Percent Problem

Enrollment at a school increased from 800 to 888 students from a given year to the next. What was the rate of increase?

Solution   First we must subtract to find the actual increase.

Increase: $888 - 800 = 88$ students

Now to find the rate, we have

Note: We use the *original* enrollment, 800, as our base.

$$\frac{88}{800} = \frac{R}{100} \text{ so } 800R = 88 \cdot 100$$

$$R = \frac{8800}{800} = 11$$

The enrollment has increased at a rate of 11%.

● ● ● **CHECK YOURSELF 11**

Car sales at a dealership decreased from 350 units one year to 322 units the next. What was the rate of decrease?

● **Example 12**

Solving a Percent Problem

A company hired 18 new employees in 1 year. If this was a 15% increase, how many employees did the company have before the increase?

Solution The rate is 15%. The amount is 18, the number of new employees. The base in this problem is the number of employees *before the increase.* So

$$\frac{18}{B} = \frac{15}{100}$$

$$15B = 18 \cdot 100 \qquad \text{or} \qquad B = \frac{1800}{15} = 120$$

The company had 120 employees before the increase.

● ● ● **CHECK YOURSELF 12**

A school had 54 new students in one term. If this was a 12% increase over the previous term, how many students were there before the increase?

● ● ● **CHECK YOURSELF ANSWERS**

1. 108. 2. $425. 3. 11%. 4. $2000.
5. $600. 6. 4.2%. 7. $15,000. 8. $78.
9. $33.75. 10. 506. 11. 8%. 12. 450.

11.5 Exercises

Name

Section   Date

Solve each of the following applications.

1. **Interest.** What interest will you pay on a $3400 loan for 1 year if the interest rate is 12%?

2. **Chemistry.** A chemist has 300 milliliters (mL) of solution that is 18% acid. How many milliliters of acid are in the solution?

3. **Payroll deductions.** Roberto has 26% of his pay withheld for deductions. If he earns $550 per week, what amount is withheld?

4. **Commission.** A real estate agent's commission rate is 6%. What will be the amount of the commission on the sale of an $85,000 home?

5. **Commission.** If a salesman is paid a $140 commission on the sale of a $2800 sailboat, what is his commission rate?

6. **Interest.** Ms. Jordan has been given a loan of $2500 for 1 year. If the interest charged is $275, what is the interest rate on the loan?

7. **Interest.** Joan was charged $18 interest for 1 month on a $1200 credit card balance. What was the monthly interest rate?

8. **Chemistry.** There is 117 grams (g) of acid in 900 g of a solution of acid and water. What percent of the solution is acid?

9. **Test scores.** On a test, Alice had 80% of the problems right. If she had 20 problems correct, how many questions were on the test?

10. **Sales tax.** A state sales tax rate is 3.5%. If the tax on a purchase is $7, what was the amount of the purchase?

11. **Loan amount.** Patty pays $525 interest for a 1-year loan at 10.5%. What was the amount of her loan?

12. **Commission.** A saleswoman is working on a 5% commission basis. If she wants to make $1800 in 1 month, how much must she sell?

ANSWERS

1. $408
2. 54 mL
3. $143
4. $5100
5. 5%
6. 11%
7. 1.5%
8. 13%
9. 25 questions
10. $200
11. $5000
12. $36,000

13. **Sales tax.** A state sales tax is levied at a rate of 6.4%. How much tax would one pay on a purchase of $260?

14. **Down payment.** Betty must make a $9\frac{1}{2}$% down payment on the purchase of a $2000 motorcycle. How much must she pay down?

15. **Commission.** If a house sells for $125,000 and the commission rate is $6\frac{1}{2}$%, how much will the salesperson make for the sale?

16. **Test scores.** Marla needs 70% on a final test to receive a C for a course. If the exam has 120 questions, how many questions must she answer correctly?

17. **Unemployment.** A study has shown that 102 of the 1200 people in the workforce of a small town are unemployed. What is the town's unemployment rate?

18. **Surveys.** A survey of 400 people found that 66 were left-handed. What percent of those surveyed were left-handed?

19. **Dropout rate.** Of 60 people who start a training program, 45 complete the course. What is the dropout rate?

20. **Manufacturing.** In a shipment of 250 parts, 40 are found to be defective. What percent of the parts are faulty?

21. **Surveys.** In a recent survey, 65% of those responding were in favor of a freeway improvement project. If 780 people were in favor of the project, how many people responded to the survey?

22. **Enrollments.** A college finds that 42% of the students taking a foreign language are enrolled in Spanish. If 1512 students are taking Spanish, how many foreign language students are there?

23. **Salary.** 22% of Samuel's monthly salary is deducted for withholding. If those deductions total $209, what is his salary?

24. **Budgets.** The Townsends budget 36% of their monthly income for food. If they spend $864 on food, what is their monthly income?

25. **Markup.** An appliance dealer marks up refrigerators 22% (based on cost). If the cost of one model was $600, what will its selling price be?

26. **Enrollments.** A school had 900 students at the start of a school year. If there is an enrollment increase of 7% by the beginning of the next year, what is the new enrollment?

27. **Land value.** A home lot purchased for $26,000 increased in value by 25% over 3 years. What was the lot's value at the end of the period?

28. **Depreciation.** New cars depreciate an average of 28% in their first year of use. What will a $9000 car be worth after 1 year?

29. **Enrollment.** A school's enrollment was up from 950 students in 1 year to 1064 students in the next. What was the rate of increase?

30. **Salary.** Under a new contract, the salary for a position increases from $11,000 to $11,935. What rate of increase does this represent?

31. **Markdown.** A stereo system is marked down from $450 to $382.50. What is the discount rate?

32. **Business.** The electricity costs of a business decrease from $12,000 one year to $10,920 the next. What is the rate of decrease?

33. **Price changes.** The price of a new van has increased $2030, which amounts to a 14% increase. What was the price of the van before the increase?

34. **Markdown.** A television set is marked down $75, to be placed on sale. If this is a 12.5% decrease from the original price, what was the selling price before the sale?

35. **Workforce.** A company had 66 fewer employees in July 1995 than in July 1994. If this represents a 5.5% decrease, how many employees did the company have in July 1995?

36. **Salary.** Carlotta received a monthly raise of $162.50. If this represented a 6.5% increase, what was her monthly salary before the raise?

37. **Stock.** Mr. Jackson buys stock for $15,000. At the end of 6 months, its value has decreased 7.5%. What is the stock worth at the end of the period?

38. **Population.** The population of a town increases 14% in 2 years. If the population was 6000 originally, what is the population after the increase?

26.	963
27.	$32,500
28.	$6480
29.	12%
30.	8.5%
31.	15%
32.	9%
33.	$14,500
34.	$600
35.	1200 employees
36.	$2500
37.	$13,875
38.	6840

39. $13.75

40. $337.50

41. $4494.40

42. $3434.70

43. $4630.50

44. $5955.08

45. 74.6%

46. 7550 thousand bbl

47. 17.22 million bbl

39. Markup. A store marks up merchandise 25% to allow for profit. If an item costs the store $11, what will its selling price be?

40. Payroll deductions. Tranh's pay is $450 per week. If deductions from his paycheck average 25%, what is the amount of his weekly paycheck (after deductions)?

Many percent problems involve calculating what is known as **compound interest.**

Suppose that you invest $1000 at 5% in a savings account for 1 year. For year 1, the interest is 5% of $1000, or 0.05 × $1000 = $50. At the end of year 1, you will have $1050 in the account.

$1000 $\xrightarrow{\text{At } 5\%}$ $1050
Start   Year 1

Now if you leave that amount in the account for a second year, the interest will be calculated on the original principal, $1000, plus the first year's interest, $50. This is called *compound interest.*
     For year 2, the interest is 5% of $1050, or 0.05 × $1050 = $52.50. At the end of year 2, you will have $1102.50 in the account.

$1000 $\xrightarrow{\text{At } 5\%}$ $1050 $\xrightarrow{\text{At } 5\%}$ $1102.50
Start   Year 1   Year 2

In Exercises 41 to 44, assume the interest is compounded annually (at the end of each year), and find the amount in an account with the given interest rate and principal.

41. $4000, 6%, 2 years   **42.** $3000, 7%, 2 years

43. $4000, 5%, 3 years   **44.** $5000, 6%, 3 years

Solve the following applications.

45. Automobiles. In 1990, there were an estimated 145.0 million passenger cars registered in the United States. The total number of vehicles registered in the United States for 1990 was estimated at 194.5 million. What percent of the vehicles registered were passenger cars?

46. Gasoline. Gasoline accounts for 85% of the motor fuel consumed in the United States every day. If 8882 thousand barrels (bbl) of motor fuel is consumed each day, how much gasoline is consumed each day in the United States?

47. Petroleum. In 1989, transportation accounted for 63% of U.S. petroleum consumption. If 10.85 million bbl of petroleum is used each day for transportation in the United States, what is the total daily petroleum consumption by all sources in the United States?

48. Pollution. Each year, 540 million metric tons (t) of carbon dioxide is added to the atmosphere by the United States. Burning gasoline and other transportation fuels is responsible for 35% of the carbon dioxide emissions in the United States. How much carbon dioxide is emitted each year by the burning of transportation fuels in the United States?

Answers

1. $408   **3.** $143   **5.** 5%   **7.** 1.5%   **9.** 25 questions   **11.** $5000
13. $16.64   **15.** $8125   **17.** 8.5%   **19.** 25%   **21.** 1200 people
23. $950   **25.** $732   **27.** $32,500   **29.** 12%   **31.** 15%
33. $14,500   **35.** 1200 employees   **37.** $13,875   **39.** $13.75
41. $4494.40   **43.** $4630.50   **45.** 74.6%   **47.** 17.22 million bbl

11.5 Supplementary Exercises

Name

Section   Date

Solve each of the following applications.

1. **Mixtures.** A can of mixed nuts contains 60% peanuts. If the can holds 32 ounces (oz), how many ounces of peanuts does it contain?

2. **Sales tax.** A sales tax rate is $5\frac{1}{2}$%. If Susan buys a television set for $450, how much sales tax must she pay?

3. **Interest.** You are charged $420 in interest on a $3000 loan for 1 year. What is the interest rate?

4. **Commission.** Mrs. Moore made a $450 commission on the sale of a $9000 pickup truck. What was her commission rate?

5. **Salary.** Cynthia makes a 5% commission on all her sales. She earned $1750 in commissions during 1 month. What were her gross sales for the month?

6. **Salary.** Nat places 6% of his salary in a payroll savings plan. If he saves $900 in 1 year, what is his annual salary?

ANSWERS

7. $133
8. 39 questions
9. 5.5%
10. 9%
11. 220 students
12. 2500 students
13. $12,900
14. 1917 students
15. 7%
16. 10.5%
17. $900
18. 70,000
19. $1332
20. 34.4 mi/gal

7. **Savings.** You are told that if you install storm windows, you will save 14% on your heating costs. If your annual cost for heating was $950 before the windows were installed, what will the annual savings be?

8. **Test scores.** On his 50-question biology examination, Bruce answered 78% correctly. How many questions did he answer correctly?

9. **Sales tax.** The sales tax on a $600 tape deck is $33. What is the sales tax rate?

10. **Utilities.** The utility costs for a business increased from $5000 to $5450 in 1 year. What was the rate of increase?

11. **Enrollment.** Of the students in a psychology class, 85% passed the midterm examination. If 187 students passed, how many students were in the class?

12. **Enrollment.** Students entering college are asked to take a mathematics placement exam. On the average, 35% of those taking the test place in beginning algebra. If 875 students were placed in beginning algebra this year, how many students took the examination?

13. **Salary.** Carrie earned $12,000 one year and then received a 7.5% raise. What is her new yearly salary?

14. **Enrollment.** A school's enrollment increases 6.5% in a given year. If the enrollment was 1800 students originally, how many students are there after the increase?

15. **Enrollment.** The enrollment of a 2-year college decreased from 4200 to 3906 in 1 year. What was the rate of decrease?

16. **Discounts.** A living room furniture set is discounted from $800 to $716 during a sale. What is the rate of discount?

17. **Salary.** Roger receives a raise of $112.50 per month. If this represents a 12.5% increase, what was his salary before the raise?

18. **Football.** A losing football team found its attendance down 6650 people per game. If this was a 9.5% decrease, what was the attendance before the decrease?

19. **Salary.** Miguel's salary is $1800 per month. If deductions average 26%, what is his take-home pay?

20. **Mileage.** A car was rated at 32 miles per gallon (mi/gal). A new model is advertised as having a 7.5% increase in mileage. What mileage should the new model have?

Using Your Calculator to Solve Percent Problems

In many everyday applications of percent, the computations required become quite lengthy, and so your calculator can be a great help. Let's look at some examples.

• Example 1

Solving a Problem Involving an Unknown Rate

In a test, 41 of 720 lightbulbs burn out before their advertised life of 700 hours (h). What percent of the bulbs fail to last the advertised life?

Solution We know the amount and base and want to find the percent (a rate). Let's use the percent proportion for the solution.

$$\frac{41}{720} = \frac{R}{100}$$

A points to 41, B points to 720

$$720R = 4100$$

$$\frac{720R}{720} = \frac{4100}{720}$$

Now use your calculator to divide

$4100 \div 720 = \boxed{5.6944444}$

5.7% of the lightbulbs fail.   We round the result to the nearest tenth.

● ● ● **CHECK YOURSELF 1**

Last month, 35 of the 475 emergency calls received by the local police department were false alarms. What percent of the calls were false alarms?

• Example 2

Solving a Problem Involving an Unknown Base

The price of a particular model of sofa has increased $48.20. If this represents an increase of 9.65%, what was the price before the increase?

Solution We want to find the base (the original price). Again, let's use the percent proportion for the solution.

$$A \searrow \frac{\$48.20}{B} = \frac{9.65}{100} \swarrow R$$

$$9.65B = 4820$$

$$\frac{9.65B}{9.65} = \frac{4820}{9.65}$$

Using the calculator gives

4820 ÷ 9.65 = 499.48187   Round to the nearest cent.

The original price was $499.48.

● ● ●   **CHECK YOURSELF 2**

The cost for medical insurance increased $136.40 last year. If this represents a 12.35% increase, what was the cost before the increase?

There is an alternative method for solving percent problems when increases or decreases are involved. Example 3 uses this second approach.

● Example 3

Solving a Problem Involving an Unknown Amount

A store marks up items 22.5% to allow for profit. If an item costs a store $36.40, what will the selling price be?

Let's diagram the problem:

Cost	Markup
100%	22.5%
$36.40	$?

Selling price
122.5%

Earlier we did a similar example in two steps, finding the markup and then adding that amount to the cost.

This method allows you to do the problem in one step.

This approach may lead to time-consuming hand calculations, but using a calculator reduces the amount of work involved.

Now the base is $36.40 and the rate is 122.5%, and we want to find the amount (the selling price).

$$\frac{A}{36.40} = \frac{122.5}{100}$$

so

$$A = \frac{122.5 \times 36.40}{100} = \$44.59$$

The selling price should be \$44.59.

● ● ● **CHECK YOURSELF 3**

An item costs \$75.40. If the markup is 36.2%, what is the selling price?

A similar approach will allow us to solve problems that involve a decrease in one step.

● **Example 4**

Solving a Problem Involving an Unknown Amount

Paul invests \$5250 in a piece of property. At the end of a 6-month period, the value has decreased 7.5%. What is the property worth at the end of the period?

Solution   Again, let's diagram the problem.

Original value
100%
or \$5250

Decrease   Ending value

7.5%   100% − 7.5% = 92.5%

So the amount (ending value) is found as

$$\frac{A}{5250} = \frac{92.5}{100}$$

$$A = \frac{92.5 \times \$5250}{100} = \$4856.25$$

The ending value is \$4856.25.

● ● ● **CHECK YOURSELF 4**

Tom buys a baseball card collection for \$750. After 1 year, the value has decreased 8.2%. What is the value of the collection after 1 year?

(margin note) Earlier we did a problem like this by finding the decrease and then subtracting from the original value. Again, using this method requires just one step.

● ● ● **CHECK YOURSELF ANSWERS**

1. 7.4%.   **2.** $1104.45.   **3.** $102.69.   **4.** $688.50.

Calculator Exercises

Name

Section   Date

A N S W E R S

1.	7.5%
2.	575
3.	28.6875
4.	21.3%
5.	555
6.	86.24
7.	11.2%
8.	16,720
9.	$153,625
10.	$53.27
11.	$117.60
12.	8.8%
13.	1387 students

Solve each of the following percent problems.

1. What percent is 648 of 8640?

2. 53.1875 is 9.25% of what number?

3. Find 7.65% of 375.

4. 17.4 is what percent (to the nearest tenth) of 81.5?

5. Find the base if 18.2% of the base is 101.01.

6. What is 3.52% of 2450?

7. What percent (to the nearest tenth) of 1625 is 182?

8. 22.5% of what number is 3762?

Solve each of the following applications.

9. **Sales.** What were Jamal's total sales in a given month if he earned a commission of $2458 at a commission rate of 1.6%?

10. **Payroll.** A retirement plan calls for a 3.18% deduction from your salary. What amount (to the nearest cent) will be deducted from your pay if your monthly salary is $1675?

11. **Salary.** You receive a 9.6% salary increase. If your salary was $1225 per month before the raise, how much will your raise be?

12. **Manufacturing.** In a shipment of 558 parts, 49 are found to be defective. What percent (to the nearest tenth of a percent) of the parts are faulty?

13. **Dropout rate.** Statistics show that an average of 42.4% of the students entering a 2-year program will complete their course work. If 588 students completed the program, how many students started?

14. $8500

15. 15.1%

16. $62.99

17. 20,319

18. $34.77

19. $86.19

20. $20,442.50

21. $1311.96

22. $5875.20

14. Interest. A time-deposit savings plan gives an interest rate of 6.42% on deposits. If the interest on an account for 1 year was $545.70, how much was deposited?

15. Taxes. The property taxes on a home increased from $832.10 to $957.70 in 1 year. What was the rate of increase (to the nearest tenth of a percent)?

16. Markdown. A dealer marks down the last year's model appliances 22.5% for a sale. If the regular price of an air conditioner was $279.95, how much will it be discounted (to the nearest cent)?

17. Population. The population of a town increases 4.2% in 1 year. If the original population was 19,500, what is the population after the increase?

18. Markup. A store marks up items 42.5% to allow for profit. If an item cost a store $24.40, what will its selling price be?

19. Markdown. An item that originally sold for $98.50 is marked down by 12.5% for a sale. Find its sale price (to the nearest cent).

20. Salary. Jerry earned $18,500 one year and then received a 10.5% raise. What is his new yearly salary?

21. Payroll deductions. Carolyn's salary is $1740 per month. If deductions average 24.6%, what is her take-home pay?

22. Investments. Yi Chen made a $6400 investment at the beginning of a year. By the end of the year, the value of the investment had decreased by 8.2%. What was its value at the end of the year?

ANSWERS

1. 7.5%   **3.** 28.6875   **5.** 555   **7.** 11.2%   **9.** $153,625
11. $117.60   **13.** 1387 students   **15.** 15.1%   **17.** 20,319   **19.** $86.19
21. $1311.96

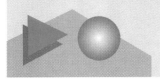

Self-Test
for Chapter 11

A N S W E R S

The purpose of the Self-Test is to help you check your progress and review for a chapter test in class. Allow yourself about 1 hour to take the test. When you are done, check your answers in the back of the book. If you missed any answers, be sure to go back and review the appropriate sections in the chapter and do the supplementary exercises provided there.

[11.1]   **1.** Use a percent to name the shaded portion of the diagram below.

In Exercises 2 and 3, write as fractions.

2. 7%   **3.** 72%

[11.2]   In Exercises 4 to 6, write as decimals.

4. 42%   **5.** 6%   **6.** 160%

In Exercises 7 to 10, write as percents.

7. 0.03   **8.** 0.042   **9.** $\dfrac{2}{5}$ **10.** $\dfrac{5}{8}$

[11.3]   In Exercises 11 to 13, identify the rate, base, and amount. *Do not solve* at this point.

11. 50 is 25% of 200   **12.** What is 8% of 500?

13. Purchase amount. A state sales tax rate is 6%. If the tax on a purchase is $30, what is the amount of the purchase?

[11.4]   In Exercises 14 to 21, solve the percent problems.

14. What is 4.5% of 250?   **15.** $33\dfrac{1}{3}$% of 1500 is what number?

1. 80%

2. $\dfrac{7}{100}$

3. $\dfrac{72}{100}$ or $\dfrac{18}{25}$

4. 0.42

5. 0.06

6. 1.6

7. 3%

8. 4.2%

9. 40%

10. 62.5%

11. (*A*) 50; (*R*%) 25%; (*B*) 200

12. (*A*) What is; (*R*%) 8%; (*B*) 500

13. (*R*%) 6%; (*A*) $30; (*B*) amount of purchase

14. 11.25

15. 500

661

16. Find 125% of 600.

17. What percent of 300 is 60?

18. 4.5 is what percent of 60?

19. 875 is what percent of 500?

20. 96 is 12% of what number?

21. 8.5% of what number is 25.5?

[11.5]   In Exercises 22 to 30, solve the applications.

22. Taxes. A state taxes sales at 6.2%. What tax will you pay on an item that costs $80?

23. Testing. You receive a grade of 75% on a test of 80 questions. How many questions did you have correct?

24. Markup. An item that costs a store $54 is marked up 30% (based on cost). Find its selling price.

25. Interests. Mrs. Sanford pays $300 in interest on a $2500 loan for 1 year. What is the interest rate for the loan?

26. Salary. Jovita's monthly salary is $2200. If the deductions for taxes from her monthly paycheck are $528, what percent of her salary goes for these deductions?

27. Markdown. A car is marked down $1128 from its original selling price of $9400. What is the discount rate?

28. Total sales. Sarah earns $540 in commissions in 1 month. If her commission rate is 3%, what were her total sales?

29. Enrollment. A community college has 480 more students in fall 1996 than in fall 1995. If this is a 7.5% increase, what was the fall 1995 enrollment?

30. Financing. Shawn arranges financing for his new car. The interest rate for the financing plan is 12%, and he will pay $1020 interest for 1 year. How much money did he borrow to finance the car?

Summary for Chapters 10–11

Ratios, Proportions, and Percents

$\frac{4}{7}$ can be thought of as "the ratio of 4 to 7."

$\frac{3}{5} = \frac{6}{10}$ is a proportion read "three is to five as six is to ten."

3 and 10 are the extremes.

5 and 6 are the means.

$5 \cdot 6 = 3 \cdot 10$

Ratio A means of comparing two numbers or quantities. A ratio can be written as a fraction.

Proportion A statement that two ratios are equal.

Extremes The first and fourth terms of a proportion.

Means The second and third terms of a proportion.

The Proportion Rule In a true proportion, the product of the means is equal to the product of the extremes.

Fractions and decimals are other ways of naming parts of a whole.

$21\% = \frac{21}{100} = 0.21$

Percent Another way of naming parts of a whole. Percent means per hundredths.

Solving Proportions for Unknown Values

To solve: $\frac{x}{5} = \frac{16}{20}$

$20x = 5 \cdot 16$

$20x = 80$

$\frac{20x}{20} = \frac{80}{20}$

$x = 4$

Check:

$\frac{4}{5} = \frac{16}{20}$

To Solve a Proportion

1. Set the product of the means equal to the product of the extremes.
2. Divide both terms of the resulting equation by the coefficient of the unknown.
3. Use the value found to replace the unknown in the original proportion. Multiply to check that the proportion is true.

Applying Proportions

To Solve a Problem by Using Proportions

1. Read the problem carefully to determine the given information.
2. Write the proportion necessary to solve the problem, using a letter to represent the unknown quantity. Be sure to include the units in writing the proportion.
3. Solve, answer the question of the original problem, and check the proportion as before.

Converting Between Fractions, Decimals, and Percents

$37\% = \frac{37}{100}$

$37\% = 0.37$

1. *To convert a percent to a fraction,* remove the percent symbol, and write the number over 100.
2. *To convert a percent to a decimal,* remove the percent symbol, and move the decimal point two places to the left.

663

$0.58 = 58\%$

$\dfrac{3}{5} = 0.60 = 60\%$

3. *To convert a decimal to a percent,* move the decimal point two places to the right, and attach the percent symbol.
4. *To convert a fraction to a percent,* write the decimal equivalent of the fraction, and then change that decimal to a percent.

Percent Problems

Every percent problem has the following three parts:

1. *The base.* This is the whole in the problem. It is the standard used for comparison. Label the base *B*.
2. *The amount.* This is the part of the whole being compared to the base. Label the amount *A*.
3. *The rate.* This is the ratio of the amount to the base. The rate is written as a percent. Label the rate *R*%.

45 is 30% of 150.

$\uparrow$ $\uparrow$ $\uparrow$

A   *R*%   *B*

$A = R \times B$

These quantities are related by the percent relationship, the equation

$$\text{Amount} = \text{rate} \times \text{base}$$

Solving Percent Problems

There are three types of percent problems.

- Finding the amount when the rate and the base are known
- Finding the rate when the amount and the base are known
- Finding the base when the rate and the amount are known

There are two methods for solving these problems.

What is 12% of 500?

Amount $= 0.12 \times 500$
$\qquad\quad = 60$

30 is what percent of 150?

Rate $= \dfrac{30}{150}$

$\qquad = 0.2$

$\qquad = 20\%$

48 is 30% of what number?

Method 1 Using the Percent Relationship

Finding an Unknown Amount
  1. Express the rate in decimal form.
  2. Multiply the rate by the base.

Finding an Unknown Rate
  1. Divide the amount by the base.
  2. Convert the decimal to a percent.

Finding an Unknown Base
  1. Convert the rate to a decimal.
  2. Divide the amount by the rate.

Base $= 48 \div 0.30 = 160$

Method 2 Using the Percent Proportion
The percent proportion is

$$\frac{A}{B} = \frac{R}{100}$$

To solve a percent problem using this proportion:
  1. Substitute the two known values into the proportion.
  2. Solve the proportion as before to find the unknown value.

What is 24% of 300?

$\dfrac{A}{300} = \dfrac{24}{100}$

$100A = 7200$

$\quad A = 72$

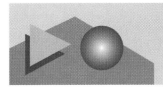

Summary Exercises

You should now be reviewing the material in Chapters 10 and 11 of the book. The following exercises will help in that process. Work all of the exercises carefully. Then check your answers in the back of the book. References are provided to the chapter and section for each exercise. If you made an error, go back and review the related material and do the supplementary exercises for that section.

A N S W E R S

1. $\dfrac{4}{17}$

2. $\dfrac{2}{3}$

3. $\dfrac{2}{5}$

4. $\dfrac{7}{36}$

5. $\dfrac{3}{4}$

[10.1]   In Exercises 1 to 5, write the following ratios in simplest form.

1. The ratio of 4 to 17

2. The ratio of 28 to 42

3. The ratio of 5 dimes to 5 quarters

4. The ratio of 7 in to 3 ft

5. The ratio of 72 h to 4 days

6. False

7. True

8. True

9. False

10. True

11. False

[10.2]   In Exercises 6 to 11, determine which of the statements are true proportions.

6. $\dfrac{3}{13} = \dfrac{7}{22}$

7. $\dfrac{8}{11} = \dfrac{24}{33}$

8. $\dfrac{9}{24} = \dfrac{12}{32}$

9. $\dfrac{7}{18} = \dfrac{35}{80}$

10. $\dfrac{5}{\frac{1}{6}} = \dfrac{120}{4}$

11. $\dfrac{0.8}{4} = \dfrac{12}{50}$

12. 2

13. 4

14. 4

15. 6

16. 16

17. 180

18. 30

19. 100

20. 0.5

[10.3]   In Exercises 12 to 20, solve for the unknown in each proportion.

12. $\dfrac{16}{24} = \dfrac{m}{3}$

13. $\dfrac{6}{a} = \dfrac{27}{18}$

14. $\dfrac{14}{35} = \dfrac{t}{10}$

15. $\dfrac{y}{22} = \dfrac{15}{55}$

16. $\dfrac{55}{88} = \dfrac{10}{p}$

17. $\dfrac{\frac{1}{2}}{18} = \dfrac{5}{w}$

18. $\dfrac{\frac{3}{2}}{9} = \dfrac{5}{a}$

19. $\dfrac{5}{x} = \dfrac{0.6}{12}$

20. $\dfrac{s}{2.5} = \dfrac{1.5}{7.5}$

21.	$67.50
22.	256 first-year students
23.	15 in.
24.	220 drives
25.	28 parts
26.	120 mi
27.	140 g
28.	960 ft²
29.	24 oz
30.	75%

[10.4] In Exercises 21 to 29, solve each application by using a proportion.

21. Ticket price. If 4 tickets to a civic theater performance cost $45, what will be the price for 6 tickets?

22. Enrollment. The ratio of first-year to second-year students at a school is 8 to 7. If there are 224 second-year students, how many first-year students are there?

23. Photo enlargement. A photograph that is 5 inches (in.) wide by 7 in. tall is to be enlarged so that the new height will be 21 in. What will be the width of the enlargement?

24. Worker output. Marcia assembles disk drives for a computer manufacturer. If she can assemble 11 drives in 2 hours (h), how many can she assemble in a workweek (40 h)?

25. Defective parts. A firm finds 14 defective parts in a shipment of 400. How many defective parts can be expected in a shipment of 800 parts?

26. Distance. The scale is $\frac{1}{4}$ in. = 10 miles (mi). How many miles apart are two towns that are 3 in. apart on the map?

27. Weight. A piece of tubing that is 16.5 centimeters (cm) long weighs 55 grams (g). What is the weight of a piece of the same tubing that is 42 cm long?

28. Area. If 1 quart (qt) of paint will cover 120 square feet (ft²), how many square feet will 2 gallons (gal) cover? (1 gal = 4 qt.)

29. Film developing. Instructions for a film developer are to use 6 ounces (oz) of powder to make 16 oz of the solution. How much powder is needed to make 2 qt of the developer? (1 qt = 32 oz.)

[11.1] In Exercise 30, use a percent to name the shaded portion of the diagram.

30.

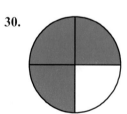

In Exercises 31 to 33, write the percent as a common fraction or a mixed number.

31. 2%   **32.** 20%   **33.** 150%

In Exercises 34 to 36, write the percents as decimals.

34. 75%   **35.** 13.5%   **36.** 225%

[11.2] In Exercises 37 to 41, write as percents.

37. 0.06   **38.** 0.375   **39.** 2.4

40. $\dfrac{2}{5}$ **41.** $2\dfrac{2}{3}$

[11.4] In Exercises 42 to 50, find the unknown.

42. 80 is 4% of what number?   **43.** 70 is what percent of 50?

44. 11% of 3000 is what number?   **45.** 24 is what percent of 192?

46. Find the base if 12.5% of the base is 625.   **47.** 90 is 120% of what number?

48. What is 9.5% of 700?   **49.** Find 150% of 50.

50. Find the base if 130% of the base is 780.

[11.5] In Exercises 51 to 60, solve the applications.

51. Commission. Joan works on a 4% commission basis. She sold $45,000 in merchandise during 1 month. What was the amount of her commission?

52. Discount rate. David buys a dishwasher that is marked down $77 from its original price of $350. What is the discount rate?

53. Chemistry. A chemist prepares a 400-milliliter (400-mL) acid-water solution. If the solution contains 30 mL of acid, what percent of the solution is acid?

ANSWERS

31. $\dfrac{1}{50}$

32. $\dfrac{1}{5}$

33. $1\dfrac{1}{2}$

34. 0.75

35. 0.135

36. 2.25

37. 6%

38. 37.5%

39. 240%

40. 40%

41. $266\dfrac{2}{3}\%$

42. 2000

43. 140%

44. 330

45. 12.5%

46. 5000

47. 75

48. 66.5

49. 75

50. 600

51. $1800

52. 22%

53. 7.5%

54. Price increase. The price of a new compact car has increased $459 over the previous year. If this amounts to a 4.5% increase, what was the price of the car before the increase?

55. Markdown. A store advertises, "Buy the red-tagged items at 25% off their listed price." If you buy a coat marked $136, what will you pay for the coat during the sale?

56. Salary. Tom has 6% of his salary deducted for a retirement plan. If that deduction is $78, what is his monthly salary?

57. Enrollment. A college finds that 35% of its science students take biology. If there are 252 biology students, how many science students are there altogether?

58. Increase rate. A company finds that its advertising costs increased from $72,000 to $76,680 in 1 year. What was the rate of increase?

59. Interest. A savings bank offers 5.25% on 1-year time deposits. If you place $3000 in an account, how much will you have at the end of the year?

60. Salary. Maria's company offers her a 9% pay raise. This will amount to a $99 per month increase in her salary. What is her monthly salary before and after the raise?

A N S W E R S

1. $\dfrac{2}{3}$

2. $\dfrac{3}{5}$

3. $\dfrac{2}{9}$

4. False

5. True

6. True

7. 21

8. 3

9. 60

10. 112.5

11. $66

12. 555 mi

13. 48 employees

This test is provided to help you in the process of reviewing Chapters 10 and 11. Answers are provided in the back of the book. If you missed any answers, be sure to go back and review the appropriate chapter sections.

In Exercises 1 to 3, write the ratios in simplest form.

1. The ratio of 16 to 24

2. The ratio of 6 dimes to 4 quarters

3. The ratio of 8 in. to 3 ft

In Exercises 4 to 6, determine which of the statements are true proportions.

4. $\dfrac{3}{10} = \dfrac{7}{23}$

5. $\dfrac{9}{22} = \dfrac{18}{44}$

6. $\dfrac{11}{18} = \dfrac{33}{54}$

In Exercises 7 to 10, solve for the unknown in each proportion.

7. $\dfrac{7}{y} = \dfrac{3}{9}$

8. $\dfrac{24}{40} = \dfrac{n}{5}$

9. $\dfrac{\frac{2}{3}}{8} = \dfrac{5}{a}$

10. $\dfrac{5}{m} = \dfrac{0.4}{9}$

In Exercises 11 to 17, solve each application by using a proportion.

11. Ticket costs. If 4 tickets to a theater cost $22, what will 12 tickets to the same performance cost?

12. Distance. Jeffrey drove 222 mi, using 6 gal of gasoline. At the same rate, how far can he travel on a full tank (15 gal) of gas?

13. Enrollment. A company finds that in one branch office, 6 of the 15 employees sign up for a new health plan. If the company has 120 employees overall, how many should be expected to sign up for the new plan?

14. Length. You are using a photocopy machine to reduce an advertisement that is 14 in. wide by 21 in. long. If the new width is to be 8 in., what will the new length be?

15. Distance. If the scale on a map is $\frac{1}{3}$ in. equals 10 mi, how far apart are two towns that are 4 in. apart on the map?

16. Salary. Diane worked 23.5 h on a part-time job and was paid $131.60. She is asked to work 25 h the next week at the same rate of pay. What salary will she receive?

17. Fertilizer production. Instructions for a powdered plant fertilizer call for 3 oz of powder to be mixed with water to form 16 oz of solution. How much of the fertilizer should be used to make 3 qt of solution? (1 qt = 32 oz.)

In Exercise 18, use a percent to name the shaded portion of the diagram.

18.

In Exercises 19 and 20, write the percent as a common fraction or a mixed number.

19. 45%

20. 175%

In Exercises 21 and 22, write the percents as decimals.

21. 55%

22. 17.5%

In Exercises 23 to 26, write as percents.

23. 0.125

24. 0.003

25. $\frac{4}{5}$

26. $\frac{5}{8}$

In Exercises 27 to 32, solve the percent problem.

27. 72 is 12% of what number?

28. Find 6.5% of 600.

29. 45.5 is what percent of 350?

30. $8\frac{1}{2}$% of 3000 is what number?

31. 120% of what number is 180?   **32.** What percent of 250 is 312.5?

In Exercises 33 to 40, solve the applications.

33. Interest rate. Sam takes out a $2000 loan for 1 year to pay for a remodeling project. If he will pay $250 in interest, what is the interest rate on the loan?

34. Commission sales. Jackie works on an 8% commission basis. If she wishes to earn $1400 in commissions in 1 month, how much must she sell during that period?

35. Course grade. Ann has calculated that she must score at least 68% in her last examination to receive a C in her chemistry course. She answered 42 of the 60 questions on the test correctly. Will she receive the C?

36. Enrollment. A college predicts that 66% of entering first-year students will take English composition. If 2500 first-year students are enrolled, how many composition students should be expected?

37. Gross pay. Martin averages 24% deductions from his paycheck for taxes and insurance. If in 1 month those deductions were $432, what was his pay before the deductions?

38. Home value. A home that was purchased for $95,000 increased in value by 16% over a 3-year period. What was its value at the end of that period?

39. Workforce size. The number of employees of a business increased from 440 at the end of one fiscal year to 473 at the end of the next. What was the rate of increase for that period?

40. Sales price. A stereo dealer decides to discount, or mark down, one model of video recorder for a sale. If the recorder's original price was $425 and it is to be discounted by 24%, find its sale price.

ANSWERS

31.	150
32.	125%
33.	12.5%
34.	$17,500
35.	She had 70%; yes
36.	1650 students
37.	$1800
38.	$110,200
39.	7.5%
40.	$323

ACROSS

1 Fraction tops

8 The GCF of two primes

9 One ____ One

10 Sick

11 Mix

12 Passenger

13 To the right of the tens place

15 Bill can do this to a bill

18 "And ____ ____ you my dear. . ."

21 The number ____

22 One of two dice

23 The person responsible for my happiness

24 Frozen water

25 The inverse of a fraction

DOWN

1 A kind of film

2 Two fractions can be this

3 Part of Pinnoccle

4 The bird chirps, the lion ____

5 "____ apply a day . . ."

6 Frequently

7 Fathers

14 Beethoven symphony

15 What discretion is the better part of

16 Sweet water

17 What you're writing on

19 Bank letters

20 Fishing gear

23 He's looking for an AWOL

MathWordPuzzle

Solution to MathWordPuzzle

12 GRAPHS

INTRODUCTION

A graphic artist might work for a sign shop, an advertising firm, or a magazine publisher. Since getting his degree, Donovan has worked for these three places, although many other jobs also require graphic arts.

Currently, Donovan does the graphic art for a "house organ," which is a magazine published by an employer for the employees only. Donovan says that the most difficult part of the job is producing graphs. Almost every issue contains several graphs containing information the employer wants the employees to see. Donovan must create a design that is pleasing to look at, uncluttered, and full of information. To do this, he must first make certain that he thoroughly understands the numbers to be graphed. ▬

© Stanley Rowin/The Picture Cube, Inc.

Statistics

In each of the following, locate the given values on a number line.

1. 7, 8, and 9.

2. $\frac{1}{3}$ and $\frac{6}{4}$

3. 3.25, 5.8, and 9.4

4. $\sqrt{21}$ and $\sqrt{35}$

Use the following graph to answer Questions 5–7.

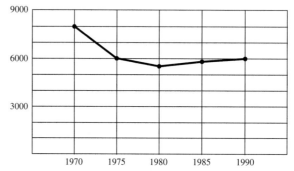

5. How many fewer family doctors were there in the United States in 1980 than in 1970?

6. What was the total change in family doctors between 1990 and 1970?

7. In what 5-year period was the decrease in family doctors the greatest? What was the decrease?

The following pie chart represents the portion of the $40,000,000 tourist industry that each country accounts for.

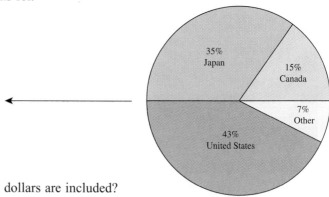

8. How many U.S. dollars are included?

9. How many Canadian and U.S. dollars are included?

10. How many non-U.S. dollars are included?

12.1 The Number Line

12.1 OBJECTIVES

1. Identify the location of positive integers on the number line.
2. Identify the location of fractions on the number line.
3. Identify the location of decimals on the number line.
4. Identify the location of square roots on the number line.

Fraction or decimal locations on a number line are always approximate.

Our work to this point has often had us comparing the size of two numbers. The easiest way to display the order of two or more numbers is on a number line. Pick any point on a number line; all numbers to the right are larger, and all numbers to the left are smaller. The space between the tick marks on a number line indicates all the numbers (fraction or decimal form) found between the integers. Let's first look at the location of fractions.

• Example 1

Locating Fractions on a Number Line

Locate $4\frac{1}{3}$ and $5\frac{7}{8}$ on the following number line.

We know that $4\frac{1}{3}$ is between 4 and 5 and that it is closer to 4. We also know that $5\frac{7}{8}$ is between 5 and 6, closer to 6. We will label those (approximate) locations with a point, and write the number under the point.

● ● ● CHECK YOURSELF 1

Locate $2\frac{2}{3}$ and $6\frac{1}{8}$ on the following number line.

Decimals are located in the same way as fractions are. Remember that these locations are only approximations.

• Example 2

Locating Decimals on a Number Line

Locate 3.23 and 5.88 on the following number line.

3.23 is between 3 and 4, closer to 3. 5.88 is between 5 and 6, closer to 6.

● ● ● CHECK YOURSELF 2

Locate 3.18 and 7.8 on the following number line.

In Example 3, we'll locate square roots on a number line.

• Example 3

Locating Square Roots on a Number Line

Locate $\sqrt{17}$ and $\sqrt{33}$ on the following number line.

$\sqrt{17}$ is slightly more than $\sqrt{16} = 4$, so it is located just to the right of 4. $\sqrt{33}$ is slightly less than $\sqrt{36} = 6$, so it is located just to the left of 6.

● ● ● CHECK YOURSELF 3

Locate $\sqrt{11}$ and $\sqrt{27}$ on the following number line.

● ● ● CHECK YOURSELF ANSWERS

1.

2.

3.

In each of the following, locate the given values on a number line.

1. 3, 4, and 6

2. 1, 5, and 7

3. 2, 9, and 11

4. 6, 8, and 10

5. 3, 10, and 12

6. 2, 6, and 10

7. $3\frac{2}{3}$ and $5\frac{5}{8}$

8. $8\frac{1}{5}$ and $6\frac{1}{7}$

9. $11\frac{1}{2}$ and $13\frac{3}{4}$

10. $8\frac{2}{7}$ and $9\frac{2}{15}$

11. $2\frac{2}{7}$ and $4\frac{1}{6}$

12. $5\frac{3}{8}$ and $6\frac{7}{9}$

13. 2.16 and 3.64

14. 4.15 and 7.89

15. 8.75 and 12.25

16. 9.65 and 5.35

17. 10.15 and 13.65

18. 9.75 and 6.42

19. $\sqrt{18}$ and $\sqrt{29}$

20. $\sqrt{40}$ and $\sqrt{53}$

21. $\sqrt{68}$ and $\sqrt{89}$

22. $\sqrt{105}$ and $\sqrt{98}$

1. See exercise
2. See exercise
3. See exercise
4. See exercise
5. See exercise
6. See exercise
7. See exercise
8. See exercise
9. See exercise
10. See exercise
11. See exercise
12. See exercise
13. See exercise
14. See exercise
15. See exercise
16. See exercise
17. See exercise
18. See exercise
19. See exercise
20. See exercise
21. See exercise
22. See exercise

Getting Ready for Section 12.2
[Section 3.7]

Find the mean of each set of numbers.

a. 1, 3, 5, 7, 9   **b.** 5, 8, 9, 11, 12   **c.** 5, 6, 10, 8, 9, 10

d. 21, 18, 16, 22, 24, 25   **e.** 8, 9, 1, 7, 7, 8, 9   **f.** 13, 15, 11, 17, 10, 10, 15

Answers

1. (number line, 0 to 10)

3. (number line, 0 to 10)

5. (number line, 0 to 10)

7. (number line, 0 to 8)

9. (number line, 0 to 15)

11. (number line, 0 to 10)

13. (number line, 0 to 10)

15. (number line, 0 to 13)

17. (number line, 0 to 15)

19. (number line, 0 to 10)

21. (number line, 0 to 10)

a. 5   **b.** 9   **c.** 8   **d.** 21

e. 7   **f.** 13

12.1   Supplementary Exercises

Name _____

Section _____   Date _____

In each of the following, locate the given values on a number line.

1. 4, 8, and 10

2. 3, 5, and 7

3. $5\frac{1}{3}$ and $6\frac{4}{9}$

4. $7\frac{1}{5}$ and $9\frac{2}{5}$

5. 3.68 and 8.56

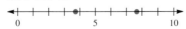

6. 4.17 and 9.65

7. $\sqrt{23}$ and $\sqrt{47}$

8. $\sqrt{12}$ and $\sqrt{69}$

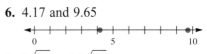
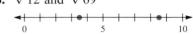

Line Graphs and Bar Graphs

12.2 OBJECTIVES

1. Read a line graph.
2. Read a bar graph.

Data are often displayed graphically. There are many types of graphs. In this section, we will discuss how to interpret line graphs, bar graphs, and pie charts.

A **graph** allows us to see how two different pieces of information are related. In our first example, we will look at a line graph. In a **line graph,** one of the pieces of information is often related to time (day, month, or year).

 Example 1

Reading a Line Graph

This graph represents the number of regular season games won by the Dallas Cowboys each year of the 1990s. Note that the information across the bottom indicates the year and the information along the side indicates the number of victories.

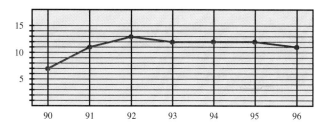

(*a*) How many games did they win in 1994?

We look across the bottom until we find 1994. We then look straight up until we see the line of the graph. Following across to the left, we see that they won 12 games in 1994.

(*b*) Find the mean number of games won by the Cowboys in the 1990s.

For each *x* (or dot) on the line, we look to the left to find how many victories it represents. We then add, using our definition of the mean, so

$$x = \frac{7 + 11 + 13 + 12 + 12 + 12 + 11}{7} = \frac{78}{7} = 11\frac{1}{7}$$

The mean number of games won is $11\frac{1}{6}$.

● ● ● ● **CHECK YOURSELF 1**

The following graph indicates the high temperatures in Baltimore, Maryland, for a week in September.

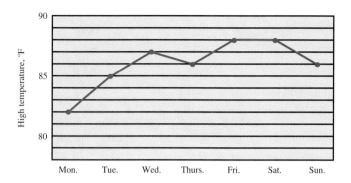

a. What was the high temperature on Friday?
b. Find the mean high temperature for that week.

Other kinds of graphs show the relationship of two sets of data. Perhaps the most common is the bar graph.

A bar graph is read in much the same manner as a line graph.

● Example 2

Reading a Bar Graph

The following bar graph represents the response to a 1995 Gallup poll that asked people what their favorite spectator sport was. In the graph, the information at the bottom describes the sport, and the information along the side describes the percentage of people surveyed. The height of the bar indicates the percentage of people who favor that particular sport.

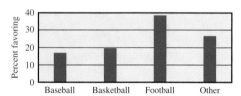

(*a*) Find the percentage of people for whom football is their favorite spectator sport.

As was the case with line graphs, we frequently have to estimate our answer when reading a bar graph. In this case, 38% would be a good estimate.

(*b*) Find the percentage of people for whom baseball is their favorite spectator sport.

Again, we can only estimate our answer. It appears to be approximately 16% of the people responding who favored baseball.

● ● ● **CHECK YOURSELF 2**

This bar graph represents the number of students who majored in each of five areas at Experimental Community College.

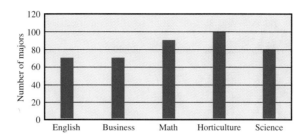

a. How many mathematics majors were there?
b. How many English majors were there?

Some bar graphs display additional information by using different colors or shading for different bars. With such graphs it is important to read the legend. The **legend** is the key that describes what each color or shade of the bar represents.

●Example 3

Reading the Legend of a Graph

The following bar graph represents the average student age at ECC.

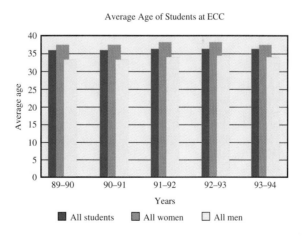

(*a*) What was the average age of female students in 93–94?

The legend tells us that the ages of all women are represented as the medium blue color. Looking at the height of the medium blue column for the year 93–94, we see the average age was about 38.

(*b*) Who tends to be older, male students or female students?

The medium blue bar is higher than the light blue bar in every year. Female students tend to be older than male students at ECC.

● ● ● **CHECK YOURSELF 3**

Use the graph in Example 3 to answer the following questions:

a. Did the average age of female students increase or decrease between 92–93 and 93–94?

b. What was the average age of male students in 91–92?

● ● ● **CHECK YOURSELF ANSWERS**

1. (a) 88°; **(b)** 86°.   **2. (a)** 90; **(b)** 70.   **3. (a)** It decreased; **(b)** 33 years.

Use the graph below, showing the yearly utility costs of a family, to solve Exercises 1 to 4.

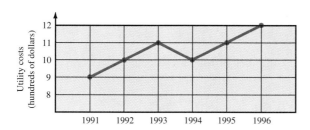

1. **$1000**
2. **$1050**
3. **$100**
4. **1991**
5. **3,000,000**
6. **1992**
7. **4,000,000**
8. **1993**

1. What was the cost in 1994?

2. What was the mean cost of utilities for this family in the 6 years from 1991 to 1996?

3. What was the decrease in the cost of utilities from 1993 to 1994?

4. In what year was the cost of utilities the smallest?

Use the graph below, showing automobile production in millions of cars from 1989 to 1995, to solve Exercises 5 to 8.

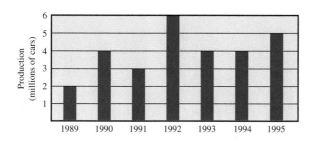

5. What was the production in 1991?

6. In what year did the greatest production occur?

7. Find the median number of cars produced in the 7 years.

8. In what year was the production decline the greatest compared to the previous year?

Use the graph below, showing the number of robberies in a town during the last 6 months of a year, to solve Exercises 9 to 13.

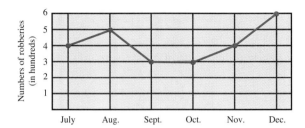

9. In which month did the greatest number of robberies occur?

10. How many robberies occurred in November?

11. Find the decrease in the number of robberies between August and September.

12. What was the mean number of robberies per month over the last 6 months?

13. What month showed the greatest increase in robberies over the previous month?

Getting Ready for Section 12.3
[Section 8.5]

Evaluate the following.

a. $12 \cdot 0.34$ **b.** $17 \cdot 0.02$ **c.** $183 \cdot 0.17$

d. $14{,}000 \cdot 0.03$ **e.** $6740 \cdot 0.25$ **f.** $11{,}500 \cdot 0.12$

Answers

1. $1000   **3.** $100   **5.** 3,000,000   **7.** 4,000,000   **9.** December
11. 200   **13.** December   **a.** 4.08   **b.** 0.34   **c.** 31.11   **d.** 420
e. 1685   **f.** 1380

A N S W E R S

a. 4.08

b. 0.34

c. 31.11

d. 420

e. 1685

f. 1380

Name

Section   Date

1. 2800

2. August 5

3. August 3

4. 4200 (on August 6)

Use the bar graph below, showing the attendance at a circus for 7 days in August, to solve Exercises 1 to 4.

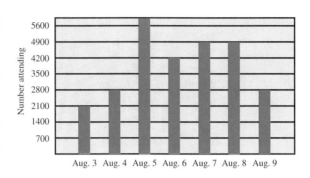

1. Find the attendance on August 4.

2. Which day had the greatest attendance?

3. Which day had the lowest attendance?

4. Find the median attendance over the 7 days.

12.3 Pie Charts

12.3 OBJECTIVES

1. Read a pie chart.
2. Interpret a pie chart.

When a graph represents how some unit is divided, a *pie chart* is frequently used.

As you might expect, a **pie chart** is a circle. Wedges (or sectors) are drawn in the circle to show how much of the whole each part makes up.

● Example 1

Reading a Pie Chart

This pie chart represents the results of a survey that asked students how they get to school most often.

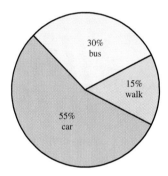

(*a*) What percentage of the students walk to school?

We see that 15% walk to school.

(*b*) What percentage of the students do not arrive by car?

Since 55% arrive by car, there are 100% − 55%, or 45%, who do not.

● ● ● CHECK YOURSELF 1

This pie chart represents the results of a survey that asked students whether they bought lunch, brought it, or skipped lunch altogether.

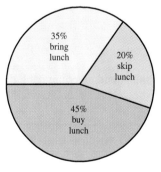

a. What percentage of the students skipped lunch?
b. What percentage of the students did not buy lunch?

If we know what the whole pie represents, we can also find out more about what each wedge represents. Example 2 illustrates this point.

•Example 2

Interpreting a Pie Chart

This pie chart shows how Sarah spent her $12,000 college scholarship.

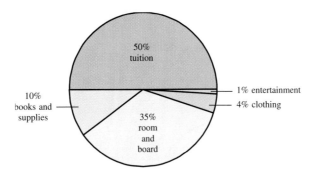

(*a*) How much did she spend on tuition?

50% of her $12,000 scholarship, or $6000.

(*b*) How much did she spend on clothing and entertainment?

Together, 5% of the money was spent on clothing and entertainment, and $0.05 \times 12,000 = 600$. Therefore, $600 was spent on clothing and entertainment.

● ● ● CHECK YOURSELF 2

This pie chart shows how Rebecca spends an average school day.

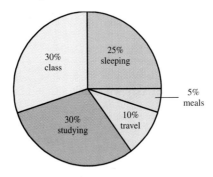

a. How many hours does she spend sleeping each day?
b. How many hours does she spend altogether studying and in class?

● ● ● CHECK YOURSELF ANSWERS

1. (a) 20%; **(b)** 55%.   **2. (a)** 6 h; **(b)** 14.4 h.

A N S W E R S

1. $270,000

2. $60,000

3. $90,000

4. $120,000

5. $60,000

a. Smallest: 2; largest: 9

b. Smallest: 8; largest: 17

c. Smallest: 17; largest: 27

d. Smallest: 15; largest: 39

e. Smallest: 1; largest: 17

f. Smallest: 7; largest: 23

The following pie chart shows the budget for a local company. The total budget is $600,000.

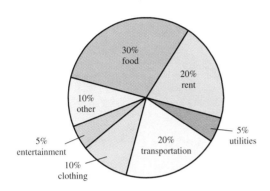

Find the amount budgeted in each of the following categories.

1. Production

2. Taxes

3. Research

4. Operating expenses

5. Miscellaneous

 **Getting Ready for Section 12.4
[Section 3.7]**

Find the largest and smallest number in each of the following data sets.

a. 5, 8, 2, 7, 9

b. 11, 17, 10, 8, 9

c. 27, 26, 21, 18, 19, 17, 23

d. 31, 15, 36, 32, 38, 39, 31, 15

e. 5, 9, 8, 2, 3, 9, 1, 1, 1, 11, 17, 4, 7, 6, 6

f. 15, 18, 12, 7, 23, 7, 8, 9, 23, 11, 16, 16

A N S W E R S

1. $270,000   **3.** $90,000   **5.** $60,000   **a.** Smallest: 2; largest: 9
b. Smallest: 8; largest: 17   **c.** Smallest: 17; largest: 27   **d.** Smallest: 15;
largest: 39   **e.** Smallest: 1; largest: 17   **f.** Smallest: 7; largest: 23

Name

Section   Date

The following pie chart shows the distribution of a person's total yearly income of $24,000.

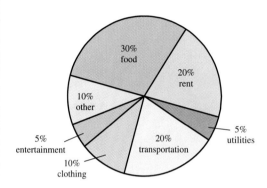

Find the amount budgeted for each category.

1. Food

2. Rent

3. Utilities

4. Transportation

5. Clothing

6. Entertainment

Creating Bar Graphs

12.4 OBJECTIVES

1. Appropriately scale the horizontal axis.
2. Appropriately scale the vertical axis.
3. Draw a bar graph.

In Section 12.2, we discussed bar graphs. In that section, you were asked to read some bar graphs. Now, you will have the opportunity to create some bar graphs.

The first part of creating bar graphs from a data set is to determine the scaling for the horizontal and vertical axes. The first example deals with scaling the axes.

● Example 1

Scaling the Horizontal and Vertical Axes

The following data represent the number of children that each president of the United States had.

0, 5, 6, 0, 2, 4, 0, 4, 10, 15, 0, 6, 2, 3, 0, 4, 5, 4, 8, 5, 3, 5, 3, 2, 6, 3, 3, 0, 2, 2, 5, 1, 1, 2, 2, 2, 4, 4, 4, 6, 1

We want to set up a bar graph showing the relationship between the number of children the presidents had (0 to 15) and the number of presidents who had a given number of children. How should we set up our axes and what scale should we use?

Our horizontal axis will be the number of children, and the scale will go from 0 (the smallest number of children any president had) to 15 (the largest number of children any president had), in intervals of 1. How do we determine the scale for the vertical axis? We need to make the vertical axis go as high as the greatest number of presidents who had two children, since more presidents had two children than any other number. Since there were eight presidents who had two children, the vertical axis needs to go to 8:

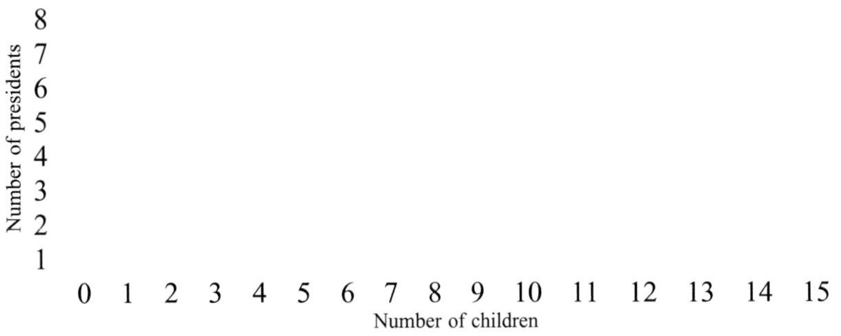

(graph with vertical axis "Number of presidents" labeled 1–8, horizontal axis "Number of children" labeled 0–15)

● ● ● CHECK YOURSELF 1

The following data represents the number of daughters that each president of the United States had.

0, 2, 5, 0, 2, 1, 0, 0, 4, 5, 0, 5, 1, 0, 0, 0, 2, 1, 1, 1, 1, 3, 1, 2, 1, 1, 3, 0, 0, 0, 1, 1, 0, 1, 2, 2, 1, 1, 1, 2, 1

How should the axes be scaled?

Now that we have scaled the axes, we can construct the graph from Example 1.

• Example 2

Creating a Bar Graph

Finish the bar graph for the number of children each president had.

We count the number of presidents that had no children, which is 6, and draw a bar at 0 on the horizontal axis that goes up to the number 6 on the vertical axis. We do the same for every possible number of children through the maximum (15).

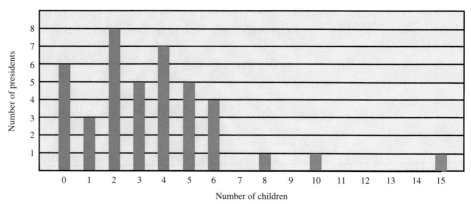

● ● ● CHECK YOURSELF 2

Finish the bar graph for the number of daughters each president had.

● ● ● CHECK YOURSELF ANSWERS

1. The horizontal axis should go from 0 to 5. The vertical axis should go from 0 to 16.

2.

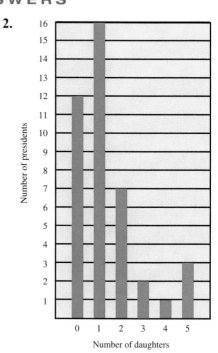

Name

Section   Date

1. The following data represent the number of children that the members of the faculty had.

0, 4, 1, 2, 2, 1, 3, 4, 0, 2, 1, 3, 2, 4, 3, 5, 0, 2, 1, 3, 2, 3, 1, 2, 5, 0, 2, 1, 2, 1

How should the axes be scaled? Draw a bar graph that represents the data.

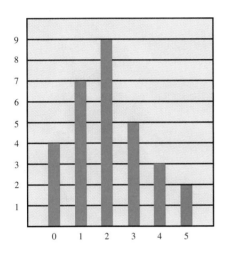

2. The following are grades obtained in a recent statistics quiz given to a class of 30 students.

9, 8, 7, 8, 5, 7, 9, 6, 9, 8, 10, 9, 5, 8, 7, 9, 8, 10, 8, 6, 9, 7, 8, 8, 9, 7, 6, 8, 8, 10

How should the axes be scaled? Draw a bar graph to represent the data.

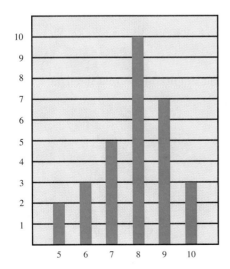

Answers

1. Horizontal axis: from 25 to 30; vertical axis: from 1 to 8

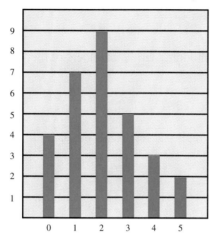

Name

Section   Date

1. The following data represent the number of sales calls made per week for each of the last 26 weeks.

28, 26, 27, 29, 25, 27, 26, 28, 29, 27, 28, 28, 29, 30, 30, 26, 27, 28, 29, 27, 25, 28, 26, 28, 27, 28

How should the axes be scaled? Draw a bar graph to represent the data.

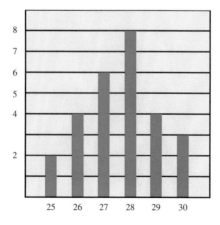

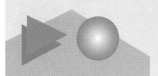

Self-Test for Chapter 12

A N S W E R S

1. See exercise

2. See exercise

3. See exercise

4. See exercise

5. December

6. August and September

7. 16,000

8. 30,000

9. 10,000

In each of the following, locate the given values on a number line.

1. 5, 8, 10

2. 5, 9, 3

3. $3\frac{1}{2}$, $5\frac{3}{4}$, $1\frac{1}{2}$

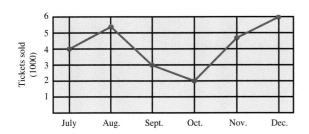

4. 5.20, 6.75, and 2.4

The graph below shows ticket sales for the last 6 months of the year.

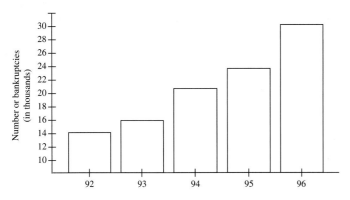

5. What month had the greatest number of ticket sales?

6. Between what two months did the greatest decrease in ticket sales occur?

The bar graph below represents the number of bankruptcy filings during the past 5 years.

7. How many people filed for bankruptcy in 1993?

8. How many people filed for bankruptcy in 1996?

9. What was the increase in filings from 1994 to 1996?

10. What was the increase in filings from 1992 to 1996?

11. Which year had the greatest increase in filings?

The following pie chart represents the way a company ships its goods.

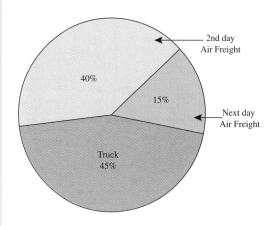

12. What percentage was shipped by next day air freight?

13. What percentage was shipped by truck?

14. What percentage was shipped by 2nd day air freight?

15. What percentage was not shipped by air freight?

CHAPTER 13 MEASUREMENT

INTRODUCTION

Every reporter must be prepared to deal with the interpretation of numerical information. Whether it is the details of a plane crash or the results of a new popularity poll, it is impossible to report on the findings if one cannot interpret them. But no reporter spends as much time working with numbers as one in the business and sports sections. Rebecca has worked in both areas.

From the time she worked on the school paper in high school, Rebecca wanted to become an editor for a city paper. When she got her first assignment as a reporter, the editor told her that she should try reporting for every section of the paper if she wanted to achieve her goal. So far she has reported for the life, business, and sports sections. Rebecca is enjoying covering sports, but she is ready to move on toward her coveted editorial position.

© MONKMEYER/Conklin

Name

Section   Date

Measurement

This pretest will point out any difficulties you may be having with measurement. Do all the problems. Then check your answers with those in the back of the book.

1. 9 ft = _____ in.

2. Simplify: 5 min 90s

3. Add: 4 gal 5 qt and 8 gal 2 qt

4. 5 km = _____ m

5. 7 m = _____ cm

6. 8000 mm = _____ m

7. 6000 g = _____ kg

8. 15 kg = _____ lb

9. 2000 mL = _____ L

10. 8 L = _____ mL

13.1 The Units of the English System

13.1 OBJECTIVES

1. Convert between two English units of measure.
2. Use a unit ratio to convert between English units of measure.

Don't let the name mislead you. The English system is no longer used in England. England has converted to the metric system, which we will look at in Sections 13.3 to 13.5.

Many arithmetic problems involve **units of measure.** When we measure an object, we give it a number and some unit. For instance, we might say a board is 6 feet long, a container holds 4 quarts, or a package weighs 5 pounds. Feet, quarts, and pounds are the units of measure.

The system you are probably most familiar with is called the **English system of measurement.** This system is used in the United States and a few other countries.

The following table lists the units of measurement you should be familiar with.

English Units of Measure and Equivalents

Length		Weight	
1 foot (ft)	= 12 inches (in.)	1 pound (lb)	= 16 ounces (oz)
1 yard (yd)	= 3 ft	1 ton	= 2000 lb
1 mile (mi)	= 5280 ft		
Volume		**Time**	
1 pint (pt)	= 16 fluid ounces (fl oz)	1 minute (min)	= 60 seconds (s)
1 quart (qt)	= 2 pt	1 hour (h)	= 60 min
1 gallon (gal)	= 4 qt	1 day	= 24 h
		1 week	= 7 days

You may want to use the equivalencies shown in the table to change from one unit to another. Let's look at one approach.

Converting Units in the English System

To change from one unit to another, replace the unit of measure with the appropriate equivalent measure and multiply.

• Example 1

Converting Within the English System

We write 5 ft as 5(1 ft) and then change 1 ft to 12 in.

$$5 \text{ ft} = 5(1 \text{ ft}) = 5(12 \text{ in.}) = 60 \text{ in.} \qquad \text{Replace 1 ft with 12 in.}$$

$$3 \text{ lb} = 3(1 \text{ lb}) = 3(16 \text{ oz}) = 48 \text{ oz} \qquad \text{Replace 1 lb with 16 oz.}$$

$$6 \text{ gal} = 6(1 \text{ gal}) = 6(4 \text{ qt}) = 24 \text{ qt} \qquad \text{Replace 1 gal with 4 qt.}$$

$$48 \text{ in.} = 48(1 \text{ in.}) = 48\left(\frac{1}{12} \text{ ft}\right) = 4 \text{ ft} \qquad \text{Since 12 in.} = 1 \text{ ft, 1 in.} = \frac{1}{12} \text{ ft.}$$

$$180 \text{ min} = 180(1 \text{ min}) = 180\left(\frac{1}{60} \text{ h}\right) = 3 \text{ h} \qquad \text{Since 60 min} = 1 \text{ h, 1 min} = \frac{1}{60} \text{ h.}$$

• • • **CHECK YOURSELF 1**

Complete each of the following statements.

a. 4 ft = _____ in.   **b.** 12 qt = _____ pt
c. 48 fl oz = _____ pt   **d.** 240 s = _____ min

Here is another idea that may help you convert units. You can use a *unit ratio* to convert from one unit to another. A *unit ratio* is a fraction whose value is 1.

> **Using Unit Ratios**
>
> To decide which unit ratio to use, just choose one with the unit you *want* in the numerator (inches in Example 12) and the unit you *want to remove* in the denominator (feet in Example 12).

• Example 2

Using the Unit Ratio to Convert

Convert 5 ft to inches.

To convert from feet to inches, you can multiply by the ratio 12 in./1 ft. So, to convert 5 ft to inches, write

$\dfrac{12 \text{ in.}}{1 \text{ ft}}$ is a *unit ratio*. It can be reduced to 1.

$$5 \text{ ft} = 5 \text{ ft} \left(\dfrac{12 \text{ in.}}{1 \text{ ft}} \right)$$

We are multiplying by 1, and so the value of the expression is not changed.

$$= 60 \text{ in.}$$

Note that we can divide out units just as we do numbers.

• • • **CHECK YOURSELF 2**

Use a unit ratio to complete each of the following statements.

a. 240 min = _____ h   **b.** 64 qt = _____ gal

Historically, units were associated with various things. A foot was the length of a foot, of course. The yard was the distance from the end of a nose to the fingertips of an outstretched arm. Objects were weighed by comparing them with grains of barley.

You have now had a chance to use two different methods for converting from one unit of measurement to another. Use whichever approach seems easiest for you.

From our work so far, it should be clear that one big disadvantage of the English system is that the relationships between units are all different. One foot is 12 in., 1 lb is 16 oz, and so on. We will see in Sections 13.3 to 13.5 that this problem doesn't exist in the metric system.

• • • **CHECK YOURSELF ANSWERS**

1. (a) 48; **(b)** 24; **(c)** 3; **(d)** 4.   **2. (a)** $240 \text{ min} \left(\dfrac{1 \text{ h}}{60 \text{ min}} \right) = 4 \text{ h}$; **(b)** 16.

Complete the following statements.

1. 8 ft = _____ in.

2. 9 gal = _____ qt

3. 3 lb = _____ oz

4. 300 s = _____ min

5. 360 min = _____ h

6. 7 lb = _____ oz

7. 4 days = _____ h

8. 5 pt = _____ fl oz

9. 16 qt = _____ gal

10. 6 h = _____ min

11. 10,000 lb = _____ tons

12. 11 min = _____ s

13. 30 pt = _____ qt

14. 5 mi = _____ ft

15. 64 oz = _____ lb

16. 64 fl oz = _____ pt

17. 7 yd = _____ ft

18. 540 min = _____ h

19. 39 ft = _____ yd

20. 24 qt = _____ gal

21. 6 h = _____ min

22. 15 ft = _____ in.

23. 80 fl oz = _____ pt

24. 5 yd = _____ ft

25. 2

26. 8

27. 28

28. 9

29. 36,960

30. 12

31. 480

32. 36

33. 15

34. 3

35. 32

36. 168

37. 435

38. 21.5

39. 3.5

40. 1.25

41. 3.75

42. $3\frac{2}{3}$

43. 24.8

44. 56.64

45. 18 billion lb

46. 40 billion lb

25. 120 s = _____ min

26. 192 h = _____ days

27. 7 gal = _____ qt

28. 18 pt = _____ qt

29. 7 mi = _____ ft

30. 192 oz = _____ lb

31. 8 min = _____ s

32. 18 qt = _____ pt

33. 360 h = _____ days

34. 6000 lb = _____ tons

35. 16 qt = _____ pt

36. 7 days = _____ h

37. $7\frac{1}{4}$ h = _____ min

38. 43 pt = _____ qt

39. 56 oz = _____ lb

40. 20 fl oz = _____ pt

41. 225 s = _____ min

42. 44 in. = _____ ft

43. 1.55 lb = _____ oz

44. 4.72 ft = _____ in.

Solve the following applications.

45. Environment. In 1989, the United States emitted approximately 9 million tons of suspended particulates into the atmosphere. How many pounds of (suspended) particulates did the United States emit in 1989?

46. Environment. In 1989, the United States emitted approximately 20 million tons of volatile organic compounds into the atmosphere. How many pounds of volatile organic compounds did the United States emit in 1989?

ANSWERS

47. **(a)** John is traveling at a speed of 60 mi/h. What is his speed in ft/s?

(b) Use the information in part (a) to develop a method to convert any speed from mi/h to ft/s.

48. Refer to several sources and write a brief history of how the units that are currently used in the English system of measurement originated. Discuss some units that were previously used but are no longer in use today.

49. A unit of measurement used in surveying is the **chain.** There are 80 chains in a mile. If you measured the distance from your home to school, how many chains would you have traveled?

(a) 88 ft/s;
47. (b)

48.

49.

a. 231

b. 315

c. 189

d. 19

e. 1815

f. 119

Getting Ready for Section 13.2 [Sections 1.4, 1.7, 2.3, 3.3]

Evaluate the following.

a. $146 + 58 + 27$ **b.** $215 + 37 + 63$ **c.** $238 - 49$

d. $55 - 36$ **e.** 363×5 **f.** $\dfrac{476}{4}$

Answers

1. 96   **3.** 48   **5.** 6   **7.** 96   **9.** 4   **11.** 5   **13.** 15   **15.** 4
17. 21   **19.** 13   **21.** 360   **23.** 5   **25.** 2   **27.** 28   **29.** 36,960
31. 480   **33.** 15   **35.** 32   **37.** 435   **39.** 3.5   **41.** 3.75   **43.** 24.8
45. 18 billion lb   **47.**   **49.**   **a.** 231   **b.** 315   **c.** 189

d. 19   **e.** 1815   **f.** 119

© 1998 McGraw-Hill Companies

703

Name

Section   Date

A N S W E R S

1. 132

2. 12

3. 24

4. 8

5. 64

6. 7

7. 3

8. 3

9. 5

10. 96

11. 26,400

12. 4

13. 6

14. 1260

15. 168

16. 4

17. 8

18. 420

Complete each of the following statements.

1. 11 ft = _____ in.

2. 4 yd = _____ ft

3. 6 gal = _____ qt

4. 480 min = _____ h

5. 4 lb = _____ oz

6. 84 in. = _____ ft

7. 180 s = _____ min

8. 72 h = _____ days

9. 80 fl oz = _____ pt

10. 6 pt = _____ fl oz

11. 5 mi = _____ ft

12. 64 oz = _____ lb

13. 24 qt = _____ gal

14. 21 min = _____ s

15. 7 days = _____ h

16. 8000 lb = _____ tons

17. 16 pt = _____ qt

18. 7 h = _____ min

 **Denominate Numbers**

13.2 OBJECTIVES

1. Simplify denominate numbers.
2. Perform operations with denominate numbers.

Numbers used to give measurements are called **denominate numbers.** A denominate number is a number with a unit or name attached.

$$\left.\begin{array}{l} 3 \text{ ft} \\ 4 \text{ yd} \\ 5 \text{ lb} \\ 6 \text{ qt} \end{array}\right\}$$ These are all denominate numbers.

Numbers without units attached are called **abstract numbers.**

A denominate number may involve two or more different units. We regularly combine feet and inches, pounds and ounces, and so on. The measures 5 ft 10 in and 6 lb 7 oz are examples. In combining units, you should let the larger unit represent *as much as is possible:* The denominate number should always be simplified. For example, 7 ft 3 in. is in simplest form, while 4 ft 18 in. is *not* in simplest form; 18 in. can be written as a combination of feet and inches.

Our first example shows the steps of simplifying a denominate number.

● Example 1

Simplifying Denominate Numbers

(*a*) Simplify 4 ft 18 in.

18 in. is larger than 1 ft and can be simplified.

Write 18 in. as 1 ft 6 in. since 12 in. is 1 ft.

$$4 \text{ ft } 18 \text{ in.} = \underbrace{4 \text{ ft} + \overbrace{1 \text{ ft}}^{18 \text{ in.}} + 6 \text{ in.}}$$
$$= 5 \text{ ft } 6 \text{ in.}$$

(*b*) Simplify 5 h 75 min.   Write 75 min as 1 h 15 min since 1 h is 60 min.

$$5 \text{ h } 75 \text{ min} = \underbrace{5 \text{ h} + \overbrace{1 \text{ h}}^{75 \text{ min}} + 15 \text{ min}}$$
$$= 6 \text{ h } 15 \text{ min}$$

● ● ● **CHECK YOURSELF 1**

a. Simplify 5 lb 24 oz.   **b.** Simplify 7 ft 20 in.

Denominate numbers with the same units are called *like numbers.* We can always add or subtract denominate numbers according to the following rule.

Adding Denominate Numbers

STEP 1 Arrange the numbers so that the like units are in the same vertical column.
STEP 2 Add in each column.
STEP 3 Simplify if necessary.

Example 2 illustrates this rule for adding denominate numbers.

●Example 2

Adding Denominate Numbers

Add 5 ft 4 in., 6 ft 7 in., and 7 ft 9 in.

The columns here represent inches and feet.

$$
\begin{array}{r}
5 \text{ ft} \quad 4 \text{ in.} \\
6 \text{ ft} \quad 7 \text{ in.} \\
+ \ 7 \text{ ft} \quad 9 \text{ in.} \\
\hline
18 \text{ ft} \ 20 \text{ in.} \\
= 19 \text{ ft} \quad 8 \text{ in.}
\end{array}
$$

Arrange in a vertical column.

Add in each column.

Be sure to simplify the results.

Simplify as before.

● ● ● CHECK YOURSELF 2

Add 3 h 15 min, 5 h 50 min, and 2 h 40 min.

To subtract denominate numbers, we have a similar rule.

Subtracting Denominate Numbers

STEP 1 Arrange the numbers so that the like units are in the same vertical column.
STEP 2 Subtract in each column. You may have to borrow from the larger unit at this point.
STEP 3 Simplify if necessary.

Consider the following example of subtracting denominate numbers.

●Example 3

Subtracting Denominate Numbers

Subtract 3 lb 6 oz from 8 lb 13 oz.

8 lb 13 oz   Arrange vertically.

−3 lb 6 oz

5 lb 7 oz   Subtract in each column.

● ● ● CHECK YOURSELF 3

Subtract 5 ft 9 in. from 10 ft 11 in.

As step 2 points out, subtracting denominate numbers may involve borrowing.

● Example 4

Subtracting Denominate Numbers

Subtract 5 ft 8 in. from 9 ft 3 in.

Borrowing with denominate numbers is not the same as in the place-value system where we always borrowed 10.

9 ft 3 in.

−5 ft 8 in.   Do you see the problem? We cannot subtract in the inches column.

To complete the subtraction, we borrow 1 ft and rename. The "borrowed" number will depend on the units involved.

~~9 ft~~ ~~3 in.~~   9 ft becomes 8 ft 12 in. Combine the 12 in. with the original 3 in.

8 ft 15 in.

−5 ft 8 in.

3 ft 7 in.   We can now subtract.

● ● ● CHECK YOURSELF 4

Subtract 3 lb 9 oz from 8 lb 5 oz.

Certain types of problems involve multiplying or dividing denominate numbers by abstract numbers, that is, numbers without a unit of measure attached. The following rule is used.

Multiplying or Dividing by Abstract Numbers

STEP 1   Multiply or divide each part of the denominate number by the abstract number.

STEP 2   Simplify if necessary.

Examples 5 and 6 illustrate this procedure.

• Example 5

Multiplying Denominate Numbers

(*a*) Multiply 4 × 5 in.

4 × 5 in. = 20 in. or 1 ft 8 in.

(*b*) Multiply 3 × (2 ft 7 in.).

Multiply each part of the denominate number by 3.

$$\begin{array}{r} 2 \text{ ft} \quad 7 \text{ in.} \\ \times \qquad\quad 3 \\ \hline 6 \text{ ft} \; 21 \text{ in.} \end{array}$$

Simplify. The product is 7 ft 9 in.

● ● ● CHECK YOURSELF 5

Multiply 5 lb 8 oz by 4.

Division is illustrated in the next example.

• Example 6

Dividing Denominate Numbers

Divide 8 lb 12 oz by 4.

Divide each part of the denominate number by 4.

$$\frac{8 \text{ lb } 12 \text{ oz}}{4} = 2 \text{ lb } 3 \text{ oz}$$

● ● ● CHECK YOURSELF 6

Divide 9 ft 6 in. by 3.

● ● ● CHECK YOURSELF ANSWERS

1. (a) 6 lb 8 oz; **(b)** 8 ft 8 in.   **2.** 11 h 45 min.   **3.** 5 ft 2 in.
4. Rename 8 lb 5 oz as 7 lb 21 oz. Then subtract for the result, 4 lb 12 oz.
5. 20 lb 32 oz, or 22 lb.   **6.** 3 ft 2 in.

Simplify.

1. 4 ft 18 in.

2. 6 lb 20 oz

3. 7qt 5 pt

4. 7 yd 50 in.

5. 5 gal 9 qt

6. 3 min 110 s

7. 9 min 75 s

8. 9 h 80 min

Add.

9. 8 lb 7 oz
     + 6 lb 15 oz

10. 9 ft 7 in.
      + 3 ft 10 in.

11. 3 h 20 min
       4 h 25 min
     + 5 h 35 min

12. 5 yd 2 ft
       4 yd
     + 6 yd 1 ft

13. 4 lb 7 oz, 3 lb 11 oz, and 5 lb 8 oz

14. 7 ft 8 in., 8 ft 5 in., and 9 ft 7 in.

Subtract.

15. 9 lb 15 oz
      − 5 lb 8 oz

16. 7 ft 11 in.
      − 4 ft 3 in.

17. 6 h 30 min
      − 3 h 50 min

18. 7 gal 3 qt
      − 1 gal 3 qt

1. 5 ft 6 in.

2. 7 lb 4 oz

3. 9 qt 1 pt

4. 8 yd 14 in.

5. 7 gal 1 qt

6. 4 min 50 s

7. 10 min 15 s

8. 10 h 20 min

9. 15 lb 6 oz

10. 13 ft 5 in.

11. 13 h 20 min

12. 16 yd

13. 13 lb 10 oz

14. 25 ft 8 in.

15. 4 lb 7 oz

16. 3 ft 8 in.

17. 2 h 40 min

18. 6 gal

19. 2 yd 2 ft

20. 4 h 55 min

21. 3 lb 4 oz

22. 3 ft 4 in.

23. 13 ft 3 in.

24. 21 min 40 s

25. 2 ft 3 in.

26. 4 lb 5 oz

27. 4 min 7 s

28. 5 h 8 min

29. 25 ft 6 in.

30. 13 h 30 min

31. 3 ft 2 in.

32. 1 lb 13 oz

33. Yes, 8 in. will remain

19. Subtract 2 yd 2 ft from 5 yd 1 ft.

20. Subtract 2 h 30 min from 7 h 25 min.

Multiply.

21. 4×13 oz

22. 4×10 in.

23. $3 \times (4 \text{ ft } 5 \text{ in.})$

24. $5 \times (4 \text{ min } 20 \text{ s})$

Divide.

25. $\dfrac{4 \text{ ft } 6 \text{ in.}}{2}$

26. $\dfrac{12 \text{ lb } 15 \text{ oz}}{3}$

27. $\dfrac{16 \text{ min } 28 \text{ s}}{4}$

28. $\dfrac{25 \text{ h } 40 \text{ min}}{5}$

Solve each of the following applications.

29. Construction. A railing for a deck requires pieces of cedar 4 ft 8 in., 11 ft 7 in., and 9 ft 3 in. long. What is the total length of material that is needed?

30. Working hours. Ted worked 3 h 45 min on Monday, 5 h 30 min on Wednesday, and 4 h 15 min on Friday. How many hours did he work during the week?

31. Sewing. A pattern requires a 2-ft 10-in. length of fabric. If a 2-yd length is used, what length remains?

32. Cooking. You use 2 lb 8 oz of hamburger from a package that weighs 4 lb 5 oz. How much is left over?

33. Framing. A picture frame is to be 2 ft 6 in. long and 1 ft 8 in. wide. A 9-ft piece of molding is available for the frame. Will this be enough for the frame?

34. **Plumbing.** A plumber needs two pieces of plastic pipe that are 6 ft 9 in. long and 1 piece that is 2 ft 11 in. long. He has a 16-ft piece of pipe. Is this enough for the job?

35. **Photography.** Mark uses 1 pt 9 fl oz and then 2 pt 10 fl oz from a container of film developer that holds 3 qt. How much of the developer remains?

36. **Weight limits.** Some flights limit passengers to 44 lb of checked-in luggage. Susan checks three pieces, weighing 20 lb 5 oz, 7 lb 8 oz, and 15 lb 7 oz. By how much was she under or over the limit?

37. **Total weight.** Six packages weighing 2 lb 9 oz each are to be mailed. What is the total weight of the packages?

38. **Construction.** A bookshelf requires four boards 3 ft 8 in. long and two boards 2 ft 10 in. long. How much lumber will be needed for the bookshelf?

39. **Unit pricing.** You can buy three 12-oz cans of peanuts for $3 or one large can containing 2 lb 8 oz for the same price. Which is the better buy?

40. **Travel.** Rich, Susan, and Marc agree to share the driving on a 12-h trip. Rich has driven for 4 h 45 min, and Susan has driven for 3 h 30 min. How long must Marc drive to complete the trip?

Solve each of the following.

41. 6 gal 2 qt

42. 11 weeks 6 days 1 h

43. 3 yd 1 ft 7 in.

44. 2 gal 1 qt 1 pt

45. 6 weeks 1 day 13 h 20 min

46. 17 gal 14 oz

47.

48. 229.5 gal

41. 2 gal 3 qt 1 pt
    +3 gal 2 qt 1 pt

42. 7 weeks 3 days 15 hours
    +3 weeks 9 days 10 hours

43. 13 yd 15 ft 10 in.
    − 9 yd 16 ft 15 in.

44. 8 gal 3 qt 2 pt
    −5 gal 5 qt 3 pt

45. 2 weeks 7 days 18 h 40 min
    ×   2

46. 4 gal 5 qt 3 pt 10 oz
    ×   2

47. The average person takes about 17 breaths per minute. How many breaths have you taken in your lifetime?

48. The average breath takes in about $1\frac{1}{2}$ pints of air. The air is about 20% oxygen. Of the oxygen that we breath, about 25% of it makes its way into our blood stream. How much oxygen does the average person take in every day?

49. 26 yr 4 mo 23 days

50. 810 gallons

51.

52. 0.008 h or 30.4 s

a. 360

b. 24,500

c. 38,500

d. 7.2

e. 1.56

f. 0.55

---

49. The average number of times a human heart beats per minute is about 72. At what age has a person's heart beat 1 billion times?

50. Each time a human heart beats, it carries about 1 ounce of blood. How many gallons of blood does the average human heart carry in a day?

51. What is your age in seconds? (Don't forget about the leap years!)

52. A greyhound can run at an average rate of 37 mi/h for $\frac{5}{16}$ of a mile. How long does it take the greyhound to run $\frac{5}{16}$ of a mile?

Getting Ready for Section 13.3
[Sections 8.8, 9.3]

Evaluate the following.

a. 36×10 **b.** 245×100 **c.** 38.5×1000

d. $72 \div 10$ **e.** $156 \div 100$ **f.** $550 \div 1000$

Answers

1. 5 ft 6 in.   **3.** 9 qt 1 pt   **5.** 7 gal 1 qt   **7.** 10 min 15 s   **9.** 15 lb 6 oz
11. 13 h 20 min   **13.** 13 lb 10 oz   **15.** 4 lb 7 oz   **17.** 2 h 40 min
19. 2 yd 2 ft   **21.** 3 lb 4 oz   **23.** 13 ft 3 in.   **25.** 2 ft 3 in.
27. 4 min 7 s   **29.** 25 ft 6 in.   **31.** 3 ft 2 in.   **33.** Yes, 8 in. will remain.
35. 1 pt 13 fl oz   **37.** 15 lb 6 oz   **39.** The three small cans contain 36 oz, or 2 lb 4 oz. The larger can is the better buy.   **41.** 6 gal 2 qt   **43.** 3 yd 1 ft 7 in.
45. 6 weeks 1 day 13 h 20 min   **47.**   **49.** 26 yr 4 mo 23 days

51.   **a.** 360   **b.** 24,500   **c.** 38,500   **d.** 7.2

e. 1.56   **f.** 0.55

13.2 Supplementary Exercises

Name _____

Section _____ Date _____

ANSWERS

Simplify.

1. 3 ft 23 in.

2. 4 lb 20 oz

3. 7 min 80 s

4. 8 h 150 min

Add.

5.   3 lb 9 oz
      + 5 lb 10 oz

6.   5 h 20 min
       3 h 40 min
     + 2 h 20 min

7. 4 yd, 5 yd, 2 ft, 2 ft, and 6 yd 1 ft

Subtract.

8.   7 ft 11 in.
      − 2 ft 4 in.

9.   3 h 30 min
      − 1 h 50 min

10. 8 lb 14 oz from 10 lb 5 oz

Multiply.

11. 3×20 oz

12. $3 \times (1 \text{ h } 25 \text{ min})$

Divide

13. $\dfrac{10 \text{ lb } 12 \text{ oz}}{2}$

14. $\dfrac{15 \text{ ft } 6 \text{ in.}}{3}$

1. 4 ft 11 in.

2. 5 lb 4 oz

3. 8 min 20 s

4. 10 h 30 min

5. 9 lb 3 oz

6. 11 h 20 min

7. 16 yd 2 ft

8. 5 ft 7 in.

9. 1 h 40 min

10. 1 lb 7 oz

11. 3 lb 12 oz

12. 4 h 15 min

13. 5 lb 6 oz

14. 5 ft 2 in.

15. **Candle making.** Mike wants to fill four candle molds with hot wax. The molds require 10 oz, 1 lb 8 oz, 1 lb 14 oz, and 2 lb of wax. How much wax will he need for the candles?

16. **Working hours.** John worked 6 h 15 min, 8 h, 5 h 50 min, 7 h 30 min, and 6 h during 1 week. What were the total hours worked?

17. **Construction.** A bookcase requires two boards of length 3 ft 10 in., two of length 2 ft 6 in., and one of length 3 ft 8 in. Will a 16-ft board give enough material?

18. **Construction.** A room requires two pieces of floor molding 12 ft 8 in. long, one piece 6 ft 5 in. long, and one piece 10 ft long. Will 42 ft of molding be enough for the job?

19. **Running.** Jackie finished a long-distance race in 1 h 55 min 30 s. Jorge's time for the race was 2 h 15 min 40 s. How much better was Jackie's time?

20. **Gardening.** You use 1 pt 10 fl oz of liquid fertilizer from a full container that holds 3 pt 8 fl oz. How much of the liquid remains?

Metric Units of Length

13.3 OBJECTIVES

1. Know the meanings of metrix prefices.
2. Estimate metric units of length.
3. Convert metric units of length.

Even in the United States, the metric system is used in science, medicine, the automotive industry, the food industry, and many other areas.

The basic unit of length in the metric system is also spelled *metre* (the British spelling).

In the metric system, you don't have to worry about things like 12 in. to 1 foot, 5280 ft to 1 mile, and all that.

The meter is one of the basic units of the International System of Units (abbreviated SI). This is a standardization of the metric system agreed to by scientists in 1960.

There is a standard pattern of abbreviation in the metric system. We will introduce the abbreviation for each term as we go along. The abbreviation for meter is m (no period!).

In Sections 13.1 and 13.2 we studied the English system of measurement, which is used in the United States and a few other countries. Our work will now concentrate on the **metric system,** used throughout the rest of the world.

The metric system is based on one unit of length, the **meter (m).** In the eighteenth century the meter was defined to be one ten-millionth of the distance from the north pole to the equator. Today the meter is scientifically defined in terms of a wavelength in the spectrum of krypton-86 gas.

One big advantage of the metric system is that you can convert from one unit to another by simply multiplying or dividing by powers of 10. This advantage and the need for uniformity throughout the world have led to legislation that will promote the use of the metric system in the United States.

Let's see how the metric system works. We will start with measures of length and compare a basic English unit, the yard, with the meter.

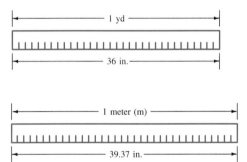

As you can see, the meter is just slightly longer than the yard. It is used for measuring the same things you might measure in feet or yards. Look at Example 1 to get a feel for the size of the meter.

● Example 1

Estimating Metric Length

A room might be 6 meters (6 m) long.

A building lot could be 30 m wide.

A fence is 2 m tall.

● ● ● CHECK YOURSELF 1

Try to estimate the following lengths in meters.

a. A traffic lane is _____ m wide.
b. A small car is _____ m long.
c. You are _____ m tall.

For other units of length, the meter is multiplied or divided by powers of 10. One commonly used unit is the **centimeter (cm).**

Comparing Centimeters (cm) to Meters (m)

$$1 \text{ centimeter (cm)} = \frac{1}{100} \text{ meter (m)}$$

The drawing below relates the centimeter and the meter:

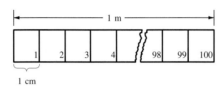

There are 100 cm in 1 m.

Just to give you an idea of the size of the centimeter, it is about the width of your little finger. There are about $2\frac{1}{2}$ cm to 1 in., and the unit is used to measure small objects. Look at Example 2 to get a feel for the length of a centimeter.

•Example 2

Estimating Metric Length

A small paperback book is 10 cm wide.

A playing card is 8 cm long.

A ballpoint pen is 16 cm long.

● ● ● CHECK YOURSELF 2

Try to estimate each of the following. Then use a metric ruler to check your guess.

a. This page is _____ cm long.
b. A dollar bill is _____ cm long.
c. The seat of the chair you are on is _____ cm from the floor.

To measure *very* small things, the **millimeter (mm)** is used. To give you an idea of its size, the millimeter is about the thickness of a new dime.

Comparing Millimeters (mm) to Meters (m)

$$1 \text{ millimeter (mm)} = \frac{1}{1000} \text{ m}$$

The diagram below will help you see the relationships of the three units we have looked at.

Note that there are 10 mm to 1 cm.

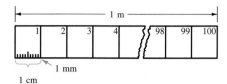

To get used to the millimeter, consider Example 3.

● Example 3

Estimating Metric Length

Standard camera film is 35 mm wide.

A small paper clip is 5 mm wide.

Some cigarettes are 100 mm long.

● ● ● CHECK YOURSELF 3

Try to estimate each of the following. Then use a metric ruler to check your guess.

a. Your pencil is _____ mm wide.
b. The tabletop you are working on is _____ mm thick.

The **kilometer (km)** is used to measure long distances. The kilometer is about six-tenths of a mile.

Comparing Kilometers (km) to Meters (m)

1 kilometer (km) = 1000 m

The prefix "kilo" means 1000. You are already familiar with this. For instance, 1 kilowatt (kW) = 1000 watts (W).

Example 4 shows how to get used to the kilometer.

● Example 4

Estimating Metric Length

The distance from New York to Boston is 325 km.

A popular distance for road races is 10 km.

Now that you have seen the four commonly used units of length in the metric system, you can review with the following Check Yourself exercise.

● ● ●   **CHECK YOURSELF 4**

Choose the most reasonable measure in each of the following statements.

a. The width of a doorway: 50 mm, 1 m, or 50 cm.
b. The length of your pencil: 20 m, 20 mm, or 20 cm.
c. The distance from your house to school: 500 km, 5 km, or 50 m.
d. The height of a basketball center: 2.2 m, 22 m, or 22 cm.
e. The width of a matchbook: 30 cm, 30 mm, or 3 mm.

Of course, this is easy. All we need to do is move the decimal point to the right or left the required number of places. Again, that's the big advantage of the metric system.

Remember: The *smaller* the unit, the *more* units it takes, so *multiply.*

As we said earlier, to convert units of measure within the metric system, all we have to do is multiply or divide by the appropriate power of 10.

Converting Metric Measurements to Smaller Units

To convert to a *smaller* unit of measure, we *multiply* by a power of 10, moving the decimal point *to the right*.

● **Example 5**

Converting Metric Length

5.2 m = 520 cm	Multiply by 100 to convert from meters to centimeters.
8 km = 8000 m	Multiply by 1000.
6.5 m = 6500 mm	Multiply by 1000.
2.5 cm = 25 mm	Multiply by 10.

● ● ●   **CHECK YOURSELF 5**

Complete the following. Remember, you don't need to do any calculation. Just move the decimal point the appropriate number of places, and write the answer.

a. 3 km = _____ m   **b.** 4.5 m = _____ cm
c. 1.2 m = _____ mm   **d.** 6.5 cm = _____ mm

Remember: The *larger* the unit, the *fewer* units it takes, so *divide.*

Converting Metric Measurements to Larger Units

To convert to a *larger* unit of measure, we *divide* by a power of 10, moving the decimal point *to the left*.

•Example 6

Converting Metric Length

43 mm = 4.3 cm   Divide by 10.

3000 m = 3 km   Divide by 1000.

450 cm = 4.5 m   Divide by 100.

● ● ● **CHECK YOURSELF 6**

Complete the following statements.

a. 750 cm = _____ m   **b.** 5000 m = _____ km

c. 78 mm = _____ cm   **d.** 3500 mm = _____ m

We have introduced all the commonly used units of linear measure in the metric system. There are other prefixes that can be used to form other linear measures. The The prefix "deci" means $\frac{1}{10}$, "deka" means times 10, and "hecto" means times 100. Their use is illustrated in the following table.

Using Metric Prefixes

1 *milli*meter (mm) = $\frac{1}{1000}$ m

1 *centi*meter (cm) = $\frac{1}{100}$ m

1 *deci*meter (dm) = $\frac{1}{10}$ m

1 meter (m)

1 *deka*meter (dam) = 10 m

1 *hecto*meter (hm) = 100 m

1 *kilo*meter (km) = 1000 m

You may find the following chart helpful when converting between metric units.

To convert to smaller units ⟶

km	hm	dam	m	dm	cm	mm
1000 m	100 m	10 m	1 m	0.1 m	0.01 m	0.001 m

⟵ To convert to larger units

To convert between metric units, just move the decimal point to the left or right the number of places indicated by the chart.

• Example 7

Converting Between Metric Lengths

(*a*) 800 cm = ? m

To convert from centimeters to meters, you can see from the chart that you must move the decimal point *two places to the left.*

800 cm = 8,00 m = 8 m

(*b*) 500 m = ? km

To convert from meters to kilometers, move the decimal point *three places to the left.*

500 m = ,500 km = 0.5 km

(*c*) 6 m = ? mm

To convert from meters to millimeters, move the decimal point *three places to the right.*

6 m = 6000, mm

● ● ● CHECK YOURSELF 7

Complete each statement.

a. 800 cm = _____ m   **b.** 370 mm = _____ m
c. 4500 m = _____ km

● ● ● CHECK YOURSELF ANSWERS

1. (a) About 3 m; **(b)** perhaps 5 m; **(c)** You are probably between 1.5 and 2 m tall.
2. (a) About 28 cm; **(b)** almost 16 cm; **(c)** about 45 cm.
3. (a) About 8 mm; **(b)** probably between 25 and 30 mm.
4. (a) 1 m; **(b)** 20 cm; **(c)** 5 km; **(d)** 2.2 m; **(e)** 30 mm.
5. (a) 3000 m; **(b)** 450 cm; **(c)** 1200 mm; **(d)** 65 mm.
6. (a) 7.5 m; **(b)** 5 km; **(c)** 7.8 cm; **(d)** 3.5 m.   **7. (a)** 8 m; **(b)** 0.37 m; **(c)** 4.5 km.

Choose the most reasonable measure.

1. The height of a ceiling
 (*a*) 25 m
 (*b*) 2.5 m
 (*c*) 25 cm

2. The diameter of a quarter
 (*a*) 24 mm
 (*b*) 2.4 mm
 (*c*) 24 cm

3. The height of a kitchen counter

 (*a*) 9 m
 (*b*) 9 cm
 (*c*) 90 cm

4. The diagonal measure of a television screen
 (*a*) 50 mm
 (*b*) 50 cm
 (*c*) 5 m

5. The height of a two-story building
 (*a*) 7 m
 (*b*) 70 m
 (*c*) 70 cm

6. An hour's drive on a freeway
 (*a*) 9 km
 (*b*) 90 m
 (*c*) 90 km

7. The width of a roll of cellophane tape

 (*a*) 1.27 mm
 (*b*) 12.7 mm
 (*c*) 12.7 cm

8. The width of a sheet of typing paper
 (*a*) 21.6 cm
 (*b*) 21.6 mm
 (*c*) 2.16 cm

9. The thickness of window glass
 (*a*) 5 mm
 (*b*) 5 cm
 (*c*) 50 mm

10. The height of a refrigerator
 (*a*) 16 m
 (*b*) 16 cm
 (*c*) 160 cm

11. The length of a ballpoint pen

 (*a*) 16 mm
 (*b*) 16 m
 (*c*) 16 cm

12. The width of a hand-held calculator key
 (*a*) 1.2 mm
 (*b*) 12 mm
 (*c*) 12 cm

Complete each statement, using a metric unit of length.

13. A playing card is 6 _____ wide.

14. The diameter of a penny is 19 _____ .

1.	(*b*)
2.	(*a*)
3.	(*c*)
4.	(*b*)
5.	(*a*)
6.	(*c*)
7.	(*b*)
8.	(*a*)
9.	(*a*)
10.	(*c*)
11.	(*c*)
12.	(*b*)
13.	cm
14.	mm

15. m

16. cm

17. m

18. km

19. mm

20. km

21. m

22. mm

23. km

24. cm

25. 3

26. 1.5

27. 800

28. 7.7

29. 25,000,000

30. 5

31. 250

32. 0.15

33. 7

34. 900

35. 80

36. 450

37. 5000

38. 4

39. 5000

40. 7000

15. A doorway is 2 _____ high.

16. A table knife is 22 _____ long.

17. A basketball court is 28 _____ long.

18. A commercial jet flies 800 _____ per hour.

19. The width of a nail file is 12 _____ .

20. The distance from New York to Washington, D.C., is 360 _____ .

21. A recreation room is 6 _____ long.

22. A ruler is 22 _____ wide.

23. A long-distance run is 35 _____ .

24. A paperback book is 11 _____ wide.

Complete each statement.

25. 3000 mm = _____ m

26. 150 cm = _____ m

27. 8 m = _____ cm

28. 77 mm = _____ cm

29. 250 km = _____ cm

30. 500 cm = _____ m

31. 25 cm = _____ mm

32. 150 mm = _____ m

33. 7000 m = _____ km

34. 9 m = _____ cm

35. 8 cm = _____ mm

36. 45 cm = _____ mm

37. 5 km = _____ m

38. 4000 m = _____ km

39. 5 m = _____ mm

40. 7 km = _____ m

Use a metric ruler to measure the necessary dimensions, and complete the statements.

41. The perimeter of the parallelogram is _____ cm.

42. The perimeter of the triangle is _____ mm.

43. The perimeter of the rectangle below is _____ cm.

44. Its area is _____ cm².

45. The perimeter of the square below is _____ mm.

46. The area of the square in problem 45 is _____ mm².

A N S W E R S

41. 12

42. 110

43. 14

44. 12

45. 100

46. 625

ANSWERS

47. 94.2

48. 706.5

49.

a. 3.3

b. 0.7

c. 0.92

d. 2.1

e. 30.6

f. 238.5

---

47. The circumference of the circle below is _____ mm.

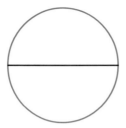

48. The area of the circle in Exercise 47 is _____ mm^2.

49. **(a)** Determine the world record speed for both men and women in meters per second (m/s) for the following events: 100-, 400-, 1500-, and 5000-m run. The record times can be found at www./hkkk.fi/~niininen/athl.html

(b) Rank all the speeds obtained in order from fastest to slowest.

50. What units in the metric system would you use to measure each of the following quantities?

(a) Distance from Los Angeles to New York

(b) Your waist measurement

(c) Width of a hair

(d) Your height

● Getting Ready for Section 13.4 [Section 10.3]

Solve the following proportions.

a. $\dfrac{x}{3} = \dfrac{11}{10}$

b. $\dfrac{10}{x} = \dfrac{100}{7}$

c. $\dfrac{23}{x} = \dfrac{100}{4}$

d. $\dfrac{3}{10} = \dfrac{x}{7}$

e. $\dfrac{85}{100} = \dfrac{x}{36}$

f. $\dfrac{10}{45} = \dfrac{53}{x}$

Answers

1. (*b*)   **3.** (*c*)   **5.** (*a*)   **7.** (*b*)   **9.** (*a*)   **11.** (*c*)   **13.** cm
15. m   **17.** m   **19.** mm   **21.** m   **23.** km   **25.** 3   **27.** 800
29. 25,000,000   **31.** 250   **33.** 7   **35.** 80   **37.** 5000   **39.** 5000
41. 12   **43.** 14   **45.** 100   **47.** 94.2   **49.**   **a.** 3.3   **b.** 0.7
c. 0.92   **d.** 2.1   **e.** 30.6   **f.** 238.5

13.3 Supplementary Exercises

Name

Section   Date

Choose the most reasonable measure.

1. A marathon race
(*a*) 40 km
(*b*) 400 km
(*c*) 400 m

2. The distance around your wrist
(*a*) 15 mm
(*b*) 15 cm
(*c*) 1.5 m

3. The diameter of a penny
(*a*) 19 cm
(*b*) 1.9 mm
(*c*) 19 mm

4. The width of a portable television screen
(*a*) 28 mm
(*b*) 28 cm
(*c*) 2.8 m

5. The height of a doorway
(*a*) 200 mm
(*b*) 20 m
(*c*) 2 m

6. The length of your car key
(*a*) 60 mm
(*b*) 60 cm
(*c*) 6 m

Complete each statement, using a metric unit of length.

7. A matchbook is 39 _____ wide.

8. The distance from San Francisco to Los Angeles is 618 _____.

9. A 1-lb coffee can has a diameter of 10 _____.

10. A fence is 2 _____ high.

11. A living room is 5 _____ long.

12. A pencil is 19 _____ long.

Complete each statement.

13. 2 km = _____ m

14. 3 cm = _____ mm

15. 25 mm = _____ cm

16. 8 m = _____ mm

17. 6 cm = _____ m

18. 8 m = _____ km

1.	(*a*)
2.	(*b*)
3.	(*c*)
4.	(*b*)
5.	(*c*)
6.	(*a*)
7.	mm
8.	km
9.	cm
10.	m
11.	m
12.	cm
13.	2000
14.	30
15.	2.5
16.	8000
17.	0.6
18.	0.008

19. 3000 mm = _____ m   **20.** 2 m = _____ cm

Use a metric ruler to measure the necessary dimensions, and complete the statements.

21. The perimeter of this parallelogram is _____ cm.

22. The perimeter of this rectangle is _____ mm.

23. The area of the rectangle in Exercise 22 is _____ mm^2.

24. The circumference of this circle is _____ cm.

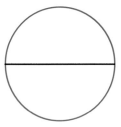

Metric Units of Weight

13.4

1. Estimate metric units of weight.
2. Convert metric units of weight.

Technically, the gram is a unit of *mass* rather than weight. Weight is the force of gravity on an object. Thus, astronauts *weigh less* on the moon than on earth even though their masses are unchanged. For common use on earth, the terms "mass" and "weight" are still used interchangeably.

The basic unit of weight in the metric system is a very small unit called the **gram.** Think of a paper clip. It weighs roughly 1 gram (g). About 28 g will make 1 oz in the English system. The gram is most often used to measure items that are fairly light. For heavier items, a more convenient unit of weight is the **kilogram (km).** From the prefix "kilo" you should know that a kilogram is equal to 1000 grams.

Comparing Kilograms (kg) to Grams (g)
1 kilogram (kg) = 1000 grams (g) (A kilogram is just a bit more than 2 lb.)

• Example 1

Estimating Metric Weight

The weight of a box of breakfast cereal is 320 g.

A woman might weigh 50 kg.

The weight of a nickel is 5 g.

 CHECK YOURSELF 1

Choose the most reasonable measure.

a. A penny: 30 g, 3 g, or 3 kg.   **b.** A bar of soap: 120 g, 12 g, or 1.2 kg.
c. A car: 5000 kg, 1000 kg, or 5000 g.

Another metric unit of weight that you will encounter is the **milligram.**

The prefix "milli" means one thousandth.

Comparing Milligrams (mg) to Grams (g)
1 milligram (mg) = $\dfrac{1}{1000}$ g

A milligram is an extremely small unit. However, it is used, for example, in medicine for measuring drug amounts. Thus, a pill might contain 300 mg of aspirin.

Just as with units of length, converting metric units of weight is simply a matter of moving the decimal point. The following chart will help.

To convert to smaller units ⟶

*kg, g, and mg are the units in common use.

kg*	hg	dag	g*	dg	cg	mg*
1000 g	100 g	10 g	1 g	0.1 g	0.01 g	0.001 g

⟵ To convert to larger units

![Example 2]

Converting Metric Weight

Complete the following statements.

We are converting to a *smaller* unit.

(*a*) 7 kg = ? g

7 kg = 7000 g = 7000 g

Move the decimal point *three places to the right* (to multiply by 1000).

We are converting to a *larger* unit.

(*b*) 5000 mg = ? g

5000 mg = 5 000 g = 5 g

Move the decimal point *three places to the left* (to divide by 1000).

● ● ● **CHECK YOURSELF 2**

a. 3000 g = _____ kg

b. 500 cg = _____ g

For very heavy objects, the *metric ton* is used.

1 metric ton = 1000 kg

We will use this measure in Example 3.

![Example 3]

Converting to Metric Tons

Complete the following

(*a*) 7500 kg = ? metric tons

7500 kg = 7.500 metric tons

Move the decimal point three places to the left to divide by 1000.

(*b*) 12.25 metric tons = ? kg

12.25 metric tons = 12250 kg

Move the decimal point three places to the right to multiply by 1000.

● ● ● **CHECK YOURSELF 3**

a. 13400 kg = ? metric tons

b. 0.76 metric tons = ? kg

● ● ● **CHECK YOURSELF ANSWERS**

1. (a) 3 g; **(b)** 120 g; **(c)** 1000 kg.   **2. (a)** 3; **(b)** 5.

3. (a) 13.4 metric tons; **(b)** 760 kg.

Choose the most reasonable measure of weight.

1. A nickel
(*a*) 5 kg
(*b*) 5 g
(*c*) 50 g

2. A portable television set
(*a*) 8 g
(*b*) 8 kg
(*c*) 80 kg

3. A flashlight battery
(*a*) 8 g
(*b*) 8 kg
(*c*) 80 g

4. A 10-year-old boy
(*a*) 30 kg
(*b*) 3 kg
(*c*) 300 g

5. A Volkswagen Rabbit
(*a*) 100 kg
(*b*) 1000 kg
(*c*) 1000 g

6. A 10-lb bag of flour
(*a*) 45 kg
(*b*) 4.5 kg
(*c*) 45 g

7. A dinner fork
(*a*) 50 g
(*b*) 5 g
(*c*) 5 kg

8. A can of spices
(*a*) 3 g
(*b*) 300 g
(*c*) 30 g

9. A slice of bread
(*a*) 2 g
(*b*) 20 g
(*c*) 2 kg

10. A house paintbrush
(*a*) 120 g
(*b*) 12 kg
(*c*) 12 g

11. A sugar cube
(*a*) 2 mg
(*b*) 20 g
(*c*) 2 g

12. A salt shaker
(*a*) 10 g
(*b*) 100 g
(*c*) 1 g

Complete each statement, using a metric unit of weight.

13. A marshmallow weighs 5 _____ .

14. A toaster weighs 2 _____ .

15. 1 _____ is $\frac{1}{1000}$ g.

1. (*b*)

2. (*b*)

3. (*c*)

4. (*a*)

5. (*b*)

6. (*b*)

7. (*a*)

8. (*c*)

9. (*b*)

10. (*a*)

11. (*c*)

12. (*b*)

13. g

14. kg

15. mg

16. A bag of peanuts weighs 100 _____.

17. An electric razor weighs 250 _____.

18. A soup spoon weighs 50 _____.

19. A heavyweight boxer weighs 98 _____.

20. A vitamin C tablet weighs 500 _____.

21. A cigarette lighter weighs 30 _____.

22. A clock radio weighs 1.5 _____.

23. A household broom weighs 300 _____.

24. A 60-watt light bulb weighs 25 _____.

Complete each statement.

25. 8 kg = _____ g

26. 5000 mg = _____ g

27. 9500 kg = _____ metric tons

28. 3 kg = _____ g

29. 1.45 metric tons = _____ kg

30. 12,500 kg = _____ metric tons

31. 3 g = _____ mg

32. 2000 g = _____ kg

Solve the following applications.

33. Emissions. The United States emitted 61.4 million metric tons (t) of carbon monoxide (CO) into the atmosphere in 1987. One metric ton equals 1000 kg. How many kilograms of CO were emitted to the atmosphere in the United States during 1987?

34. Emissions. The United States emitted 19.5 million t of nitrogen oxides (NO) into the atmosphere in 1987. One metric ton (1 t) equals 1000 kg. How many kilograms of NO were emitted to the atmosphere in the United States during 1987?

35. Mass (weight) and volume are connected in the metric system. The weight of water in a cube 1 cm on a side is 1g. Does such a relationship exist in the English system of measurement? If so, what is it? If not, why not?

35.

a. True

b. False

c. True

d. True

e. True

f. False

● Getting Ready for Section 13.5 [Section 10.2]

Which of the following are true proportions?

a. $\dfrac{28}{7} = \dfrac{8}{2}$ **b.** $\dfrac{14}{9} = \dfrac{7}{11}$ **c.** $\dfrac{73}{4} = \dfrac{146}{8}$

d. $\dfrac{30}{24} = \dfrac{5}{4}$ **e.** $\dfrac{56}{6} = \dfrac{28}{3}$ **f.** $\dfrac{111}{9} = \dfrac{17}{6}$

Answers

1. (*b*)   **3.** (*c*)   **5.** (*b*)   **7.** (*a*)   **9.** (*b*)   **11.** (*c*)   **13.** g   **15.** mg
17. g   **19.** kg   **21.** g   **23.** g   **25.** 8000   **27.** 9.5   **29.** 14,500
31. 3000   **33.** 61.4 billion kg   **35.**   **a.** True   **b.** False

c. True   **d.** True   **e.** True   **f.** False

Name _____

Section _____ Date _____

Choose the most reasonable measure of weight.

1. A quarter
   (a) 6 g
   (b) 6 kg
   (c) 60 g

2. A tube of toothpaste
   (a) 20 kg
   (b) 200 g
   (c) 20 g

3. A refrigerator
   (a) 120 kg
   (b) 1200 kg
   (c) 12 kg

4. A paperback book
   (a) 1.2 kg
   (b) 120 g
   (c) 12 g

5. A package of cigarettes
   (a) 2.6 g
   (b) 26 g
   (c) 2.6 kg

6. A bicycle
   (a) 0.6 kg
   (b) 60 kg
   (c) 6 kg

Complete each statement, using a metric unit of weight.

7. A loaf of bread weighs 500 _____ .

8. A compact car weighs 900 _____ .

9. A television set weighs 25 _____ .

10. 1 centigram (cg) is $\dfrac{1}{100}$ _____ .

11. A table knife weighs 70 _____ .

12. A football player's weight might be listed as 115 _____ .

Complete each statement.

13. 5 kg = _____ g

14. 2000 g = _____ kg

15. 5 metric tons = _____ kg

16. 2000 mg = _____ g

Metric Units of Volume

13.5 OBJECTIVES

1. Estimate metric units of volume.
2. Convert metric units of volume.

The liter is related to the meter. It is defined as the volume of a cube 10 cm on each edge, so

$1\ L = 1000\ cm^3$.

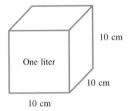

One liter

10 cm

10 cm

10 cm

This unit of volume is also spelled *litre* (the British spelling).

In measuring volume or capacity in the English system, we use the liquid measures of ounces, pints, quarts, and gallons.

In the metric system, the basic unit of volume is the **liter (L).** The liter is just slightly more than the quart and would be used for soft drinks (many are already sold here by the liter), milk, oil, gasoline, and so on.

The metric unit that is used to measure smaller volumes is the **milliliter (mL).** From the prefix we know that it is one thousandth of a liter.

Comparing Liters (L) to Milliliters (mL)

1 liter (L) = 1000 milliliters (mL)

The milliliter is the volume of a cube 1 cm on each edge. So 1 mL is equal to 1 cm^3. These units can be used interchangeably.

Note: Scientists give measurements of volume in terms of cubic centimeters (cm^3). The liter is *not* an SI unit.

Example 1 will help you get used to the metric units of volume.

●Example 1

Estimating Metric Volume

A teaspoon is about 5 mL or 5 cm^3.

A 6-oz cup of coffee is about 180 mL.

A quart of milk is 946 mL (just less than 1 L).

A gallon is just less than 4 L.

Now try these exercises.

● ● ● CHECK YOURSELF 1

Choose the most reasonable measure.

a. A can of soup: 3 L, 30 mL, or 300 mL.
b. A pint of cream: 4.73 L, 473 mL, or 47.3 mL.
c. A home-heating oil tank: 100 L, 1000 L, or 1000 mL.
d. A tablespoon: 150 mL, 1.5 L, or 15 mL.

Converting metric units of volume is again just a matter of moving the decimal point. A chart similar to the ones you saw earlier may be helpful.

*L, cL, and mL are the most commonly used units. We have shown the other units simply to indicate that the prefixes and abbreviations are used in a consistent fashion.

To convert to smaller units ───────────→

kL	hL	daL	L*	dL	cL*	mL*
1000 L	100 L	10 L	1 L	0.1 L	0.01 L	0.001 L

←─────────── To convert to larger units

• Example 2

Converting Metric Volume

Complete the following statements.

(*a*) 4 L = ? mL

We are converting to a *smaller* unit.

From the chart above, we see that we should move the decimal point three places to the right (to multiply by 1000).

4 L = 4000 mL = 4000 mL

(*b*) 3500 mL = ? L

We are converting to a *larger* unit.

Move the decimal point three places to the left (to divide by 1000).

3500 mL = 3 500 L = 3.5 L

(*c*) 30 cL = ? mL

We are converting to a *smaller* unit.

Move the decimal point one place to the right (to multiply by 10).

30 cL = 30.0 mL = 300 mL

● ● ● CHECK YOURSELF 2

Complete the following statements.

a. 5 L = _____ mL   **b.** 7500 mL = _____ L
c. 550 mL = _____ cL

● ● ● CHECK YOURSELF ANSWERS

1. (a) 300 mL; **(b)** 473 mL; **(c)** 1000 L; **(d)** 15 mL.
2. (a) 5000; **(b)** 7.5; **(c)** 55.

13.5 Exercises

Name

Section   Date

Choose the most reasonable measure of volume.

1. A bottle of wine
  (*a*) 75 mL
  (*b*) 7.5 L
  (*c*) 750 mL

2. A gallon of gasoline
  (*a*) 400 mL
  (*b*) 4 L
  (*c*) 40 L

3. A bottle of perfume
  (*a*) 15 mL
  (*b*) 150 mL
  (*c*) 1.5 L

4. A can of frozen orange juice
  (*a*) 1.5 L
  (*b*) 150 mL
  (*c*) 15 mL

5. A hot-water heater
  (*a*) 200 mL
  (*b*) 50 L
  (*c*) 200 L

6. An oil drum
  (*a*) 220 L
  (*b*) 220 mL
  (*c*) 22 L

7. A bottle of ink
  (*a*) 60 cm^3
  (*b*) 6 cm^3
  (*c*) 600 cm^3

8. A cup of tea
  (*a*) 18 ml
  (*b*) 180 mL
  (*c*) 18 L

9. A jar of mustard
  (*a*) 150 mL
  (*b*) 15 L
  (*c*) 15 mL

10. A bottle of aftershave lotion
  (*a*) 50 mL
  (*b*) 5 L
  (*c*) 5 mL

11. A cream pitcher
  (*a*) 12 mL
  (*b*) 120 mL
  (*c*) 1.2 L

12. One tablespoon
  (*a*) 1.5 mL
  (*b*) 1.5 L
  (*c*) 15 mL

Complete each statement, using a metric unit of volume.

13. A can of tomato soup is 300 _____.

14. 1 _____ is $\dfrac{1}{100}$ L.

15. A saucepan holds 1.5 _____.

ANSWERS
1. (*c*)
2. (*b*)
3. (*a*)
4. (*b*)
5. (*c*)
6. (*a*)
7. (*a*)
8. (*b*)
9. (*a*)
10. (*a*)
11. (*b*)
12. (*c*)
13. mL
14. cL
15. L

16. A thermos bottle contains 500 _____ of liquid.

17. A coffee pot holds 720 _____.

18. A garbage can will hold 120 _____.

19. A car's engine capacity is 2000 cm^3. It is advertised as a 2.0 _____ model.

20. A bottle of vanilla extract contains 60 _____.

21. 1 _____ is $\frac{1}{10}$ cL.

22. A can of soft drink is 35 _____.

23. A garden sprinkler delivers 8 _____ of water per minute.

24. 1 kL is 1000 _____.

Complete each statement.

25. 7 L = _____ mL

26. 4000 cm^3 = _____ L

27. 4 hL = _____ L

28. 7 L = _____ cL

29. 8000 mL = _____ L

30. 12 L = _____ mL

31. 5 L = _____ cm^3

32. 2 L = _____ cL

33. 75 cL = _____ mL

34. 5 kL = _____ L

35. 5 L = _____ cL

36. 400 mL = _____ cL

37. (a) Determine how many liters of gasoline your car will hold.

(b) Using current prices, determine what a liter of gasoline should cost to make it competitive.

(c) How much would it cost to fill your car?

38. Do the following doses of medicine seem reasonable or unreasonable?

(a) Take 5 L of Koapectate every morning.

(b) Soak your feet in 5 L of epson salt bath every evening.

(c) Inject yourself with $\frac{3}{4}$ L of insulin every day.

Answers

1. (c)   **3.** (a)   **5.** (c)   **7.** (a)   **9.** (a)   **11.** (b)   **13.** mL (or cm^3)

15. L   **17.** mL   **19.** L   **21.** mL   **23.** L   **25.** 7000   **27.** 0.04

29. 8   **31.** 5000   **33.** 750   **35.** 500   **37.**

Name _____

Section _____ Date _____

ANSWERS

1. (a)	
2. (a)	
3. (c)	
4. (a)	
5. (b)	
6. (b)	
7. mL	
8. L	
9. L	
10. mL	
11. mL	
12. L	
13. 5000	
14. 6	
15. 9000	
16. 0.01	

Choose the most reasonable measure of volume.

1. A bottle of cough syrup
(a) 100 cm^3
(b) 10 cm^3
(c) 1 L

2. A watering can
(a) 10 L
(b) 100 L
(c) 100 mL

3. The gas tank of your car
(a) 500 mL
(b) 5 L
(c) 50 L

4. A bottle of eye drops
(a) 18 cm^3
(b) 180 cm^3
(c) 1.8 L

5. A can of soft drink
(a) 3.5 L
(b) 350 mL
(c) 35 mL

6. A punch bowl
(a) 200 L
(b) 20 L
(c) 200 mL

Complete each statement, using a metric unit of volume.

7. A glass of beer holds 250 _____ .

8. The crankcase of an automobile takes 5.5 _____ of oil.

9. A pressure cooker holds 20 _____ .

10. The correct dosage for a cough medicine is 40 _____ .

11. A bottle of iodine holds 20 _____ .

12. A large mixing bowl holds 6 _____ .

Complete each statement.

13. 5 L = _____ mL

14. 6000 cm^3 = _____ L

15. 9 L = _____ cm^3

16. 10 mL = _____ L

Using Your Calculator to Convert Between the English and Metric Systems

Occasionally, it is necessary to convert between the English and metric systems. When this is necessary, it is best to have a calculator and a conversion table. In this section, we provide conversion tables for length, weight, and volume. We begin by looking at length.

The following table will help you to convert from the metric to the English system for measuring length.

Converting Between Metric and English Units (Length)

1 m = 39.37 in.
1 cm = 0.394 in.
1 km = 0.62 mi

• Example 1

Converting Between Metric and English Units of Length

Complete the following statements.

There are all kinds of ways to convert between the systems. You can just multiply or divide by the appropriate factor. We have chosen to use the unit ratio idea introduced in the last chapter.

(*a*) 3 m = ? in.

$$3 \text{ m} = 3 \text{ m} \left(\frac{39.37 \text{ in.}}{1 \text{ m}} \right) = 118.11 \text{ in.}$$

This is a unit ratio equal to 1.

(*b*) 25 km = ? mi

$$25 \text{ km} = 25 \text{ km} \left(\frac{0.62 \text{ mi}}{1 \text{ km}} \right) = 15.5 \text{ mi}$$

● ● ● **CHECK YOURSELF 1**

Complete the following statements.

a. 5 cm = _____ in.   **b.** 40 km = _____ mi

Note: Guidebooks for U.S. travelers give this tip for a quick conversion from kilometers to miles: Multiply the number of kilometers by 6 and drop the last digit. The result will be the approximate number of miles. For example, a speed limit sign reads 90 km/h.

What this does is multiply the number of kilometers by 0.6.

$90 \times 6 = 540$ ↖ Drop this digit.

Multiply by 6 and drop the last digit, in this case 0.

90 km/h is approximately 54 mi/h.

You may also want to be able to convert units of the English system to those of the metric system. You can use the following conversion factors.

Converting Between English and Metric Units (Length)

1 in. = 2.54 cm

1 yd = 0.914 m

1 mi = 1.6 km

• Example 2

Converting Between English and Metric Units of Length

Complete the following statements.

(*a*) 5 in. = ? cm

$$5 \text{ in.} = 5 \text{ in.} \left(\frac{2.54 \text{ cm}}{1 \text{ in.}} \right) \qquad \frac{2.54 \text{ cm}}{1 \text{ in.}} \text{ is a unit ratio equal to 1.}$$

$$= 12.7 \text{ cm}$$

(*b*) 12 mi = ? km

$$12 \text{ mi} = 12 \text{ mi} \left(\frac{1.6 \text{ km}}{1 \text{ mi}} \right) = 19.2 \text{ km}$$

(*c*) 5 yd = ? m

$$5 \text{ yd} = 5 \text{ yd} \left(\frac{0.914 \text{ m}}{1 \text{ yd}} \right) = 4.57 \text{ m}$$

● ● ● **CHECK YOURSELF 2**

Complete the following statements.

a. 8 in. = _____ cm

b. 20 mi = _____ km

To convert between units of weight in the two systems, use the following.

Converting Between English and Metric Units (Weight)

1 kg = 2.2 lb

1 lb = 0.45 kg

1 oz = 28 g

• Example 3

Converting Between English and Metric Units of Weight

(*a*) A roast beef weighs 3 kg. What is its weight in pounds?

$$3 \text{ kg} = 3 \text{ kg} \left(\frac{2.2 \text{ lb}}{1 \text{ kg}} \right) = 6.6 \text{ lb}$$

(*b*) A jar of spices weighs $1\frac{1}{2}$ oz. What is its weight in grams?

$$1\frac{1}{2} \text{ oz} = 1.5 \text{ oz} = 1.5 \text{ oz} \left(\frac{28 \text{ g}}{1 \text{ oz}} \right) = 42 \text{ g}$$

(*c*) A package weighs 5 lb. What is its weight in kilograms?

$$5 \text{ lb} = 5 \text{ lb} \left(\frac{0.45 \text{ kg}}{1 \text{ lb}} \right) = 2.25 \text{ kg}$$

● ● ●   **CHECK YOURSELF 3**

a. A radio weighs 4 kg. What is its weight in pounds?

b. A bag of peanuts weighs $3\frac{1}{2}$ oz. What is its weight in grams?

c. If your cat weighs 8 lb, what is her weight in kilograms?

If you need to convert between units of volume in the English and metric systems, you can use the following table.

Converting Between Metric and English Units (Volume)

1 L = 1.06 qt

1 qt = 0.95 L

1 fluid ounce (fl oz) = 30 mL

$= 30 \text{ cm}^3$

Examples 4 and 5 show how to use these conversion factors.

● Example 4

Converting Between English and Metric Units of Volume

A bottle's volume is 3 L. What is its capacity in quarts?

$$3 \text{ L} = 3 \cancel{L} \left(\frac{1.06 \text{ qt}}{1 \cancel{L}} \right) = 3.18 \text{ qt}$$ Note that $\dfrac{1.06 \text{ qt}}{1 \text{ L}}$ is a unit ratio equal to 1.

● ● ● **CHECK YOURSELF 4**

A soft drink is bottled in a 2-L container. What does it contain in quarts?

● Example 5

Converting Between English and Metric Units of Volume

A gasoline tank holds 16 gal. How many liters will it hold?

$$16 \text{ gal} = 16 \cancel{\text{gal}} \left(\frac{4 \text{ qt}}{1 \cancel{\text{gal}}} \right) = 64 \text{ qt}$$ First we convert to quarts.

$$= 64 \cancel{\text{qt}} \left(\frac{0.95 \text{ L}}{1 \cancel{\text{qt}}} \right) = 60.8 \text{ L}$$ Then we multiply by the unit ratio.

● ● ● **CHECK YOURSELF 5**

A hot-water tank will hold 40 gal. What is its volume in liters?

● ● ● **CHECK YOURSELF ANSWERS**

1. (a) 1.97 in.; (b) 24.8 mi.   **2.** (a) 20.32 cm; (b) 32 km.
3. (a) 8.8 lb; (b) 98 g; (c) 3.6 kg.   **4.** 2.12 qt.   **5.** 152 L.

Calculator Exercises

A N S W E R S

Complete each statement.

1. 250 km = _____ mi

2. 9 cm = _____ in.

3. 150 mi = _____ km

4. 9 yd = _____ m

5. 2.6 m = _____ in.

6. 72 in. = _____ cm

7. 6 lb = _____ kg

8. 8 oz = _____ g

9. 0.25 kg = _____ oz

10. 5 lb = _____ g

11. 4 qt = _____ L

12. 7 L = _____ qt

13. 8 fl oz = _____ mL

14. 15.9 gal = _____ L

15. 760 mL = _____ qt

16. 15 L = _____ gal

Solve the following applications.

17. Football. A football team's fullback weighs 250 lb. How many kilograms does he weigh?

18. Travel. Samantha's speedometer reads in kilometers per hour. If the legal speed limit is 55 mi/h, how fast can she drive?

1.	155
2.	3.546
3.	240
4.	8.226
5.	102.4
6.	183
7.	2.7
8.	224
9.	8.8
10.	2270
11.	3.8
12.	7.42
13.	240
14.	60.42
15.	0.806
16.	3.975
17.	114
18.	88.57 km/h

Answers

1. 155   **3.** 240   **5.** 102.4   **7.** 2.7   **9.** 8.8   **11.** 3.8   **13.** 240
15. 0.806   **17.** 114   Due to rounding, the answer to Exercise 17 will be 113 if students multiply by 0.45, and it will be 114 if they divide by 2.2.

743

ACROSS

1 6000 milliliters
6 1760 yards
7 paradise
8 LV + LV
9 extra terrestrial
10 out of whack
13 tome
14 seven thousandths g

DOWN

1 600 centimeters
2 top
3 ____ de France
4 ____ dm in one m
5 sixty thousand g
10 presidential nickname
11 didn't lose
12 read only memory

MathWork Puzzle

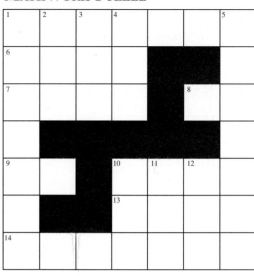

Solution To MathWorkPuzzle

Self-Test
for Chapter 13

A N S W E R S

1. 96

2. 6

3. 48

4. 6

5. 6 ft 9 in.

6. 7 min 30 s

7. 11 ft 5 in.

8. 2 lb 9 oz

9. 15 h 20 min

10. 4 lb 6 oz

11. $1050

12. (b)

13. (b)

14. mm

The purpose of the Self-Test is to help you check your progress and review for a chapter test in class. Allow yourself about 1 hour to take the test. When you are done, check your answers in the back of the book. If you missed any answers, be sure to go back and review the appropriate sections in the chapter and do the supplementary exercises provided there.

[13.1]   In Exercises 1 to 4, complete the statements.

1. 8 ft = _____ in.

2. 360 min = _____ h

3. 3 pt = _____ fl oz

4. 96 oz = _____ lb

In Exercises 5 and 6, simplify.

5. 5 ft 21 in.

6. 6 min 90 s

[13.2]   In Exercises 7 to 10, do the indicated operations.

7.   7 ft 9 in.
      +3 ft 8 in.

8.   7 lb 3 oz
      −4 lb 10 oz

9. 4 × (3 h 50 min)

10. $\dfrac{12 \text{ lb } 18 \text{ oz}}{3}$

11. Total cost. The Martins are fencing in a rectangular yard that is 110 ft long by 40 ft wide. If the fencing costs $3.50 per linear foot, what will be the total cost of the fencing?

[13.3]   In Exercises 12 and 13, choose the reasonable measure.

12. The width of your hand
   (a) 50 cm
   (b) 10 cm
   (c) 1 m

13. The speed limit on a freeway
   (a) 9 km/h
   (b) 90 km/h
   (c) 90 m/h

Use a metric unit of length to complete the statement.

14. The width of your pencil is about 6 _____.

15.	3000
16.	5000
17.	8
18.	3.1
19.	*(c)*
20.	*(b)*
21.	kg
22.	3000
23.	1.35
24.	140
25.	7000
26.	*(c)*
27.	*(b)*
28.	mL
29.	2000
30.	3
31.	4.75
32.	240

In Exercises 15 to 18, complete the statements.

15. 3 km = _____ m

16. 5 m = _____ mm

17. 800 cm = _____ m

18. 5 km = _____ mi

[13.4] In Exercises 19 and 20, choose the most reasonable measure.

19. A package of gum
    (a) 3 g
    (b) 3 kg
    (c) 30 g

20. A football player
    (a) 12 kg
    (b) 120 kg
    (c) 120 g

21. A living room sofa weighs 80 _____.

In Exercises 22 to 25, complete the statements.

22. 3 kg = _____ g

23. 3 lb = _____ kg

24. 5 oz = _____ g

25. 7 g = _____ mg

[13.5] In Exercises 26 and 27, choose the most reasonable measure.

26. The gas tank of your car
    (a) 600 L
    (b) 6 L
    (c) 60 L

27. A small can of tomato juice
    (a) 4 L
    (b) 400 mL
    (c) 40 mL

Use a metric unit of volume to complete the statement.

28. A drinking glass holds 250 _____.

In Exercises 29 to 32, complete the statements.

29. 2 L = _____ mL

30. 300 cL = _____ L

31. 5 qt = _____ L

32. 8 fl oz = _____ mL

CHAPTER

14

AN INTRODUCTION TO ALGEBRA

INTRODUCTION

Many people think that meteorologists are the people who report the weather on television. Although a few of these people are meteorologists, most are better classified as reporters or television personalities.

Danielle works for the U.S. Weather Service. Her degree in meteorology was preceded by a bachelor degree in physics. She enjoys the inexact science of weather forecasting, but she specializes in the study (and prediction) of tornadoes. ——————————————————

An Introduction to Algebra

1. 8

2. −9

3. (a) 7

   (b) −12

4. (a) −22

   (b) 17

5. (a) −36

   (b) 54

6. (a) 3

   (b) −5

7. 30

8. −4

9. 4

10. 3

This pretest will point out any difficulties you may be having with signed numbers. Do all the problems. Then check your answers with those in the back of the book.

1. The absolute value of -8 is _____ .

2. The opposite of 9 is _____ .

3. (a) $12 + (-5) =$

   (b) $-9 + (-3) =$

4. (a) $-7 - 15 =$

   (b) $14 - (-3) =$

5. (a) $(-3)(12) =$

   (b) $(-6)(-9) =$

6. (a) $\dfrac{-24}{-8} =$

   (b) $40 \div (-8) =$

7. Evaluate $5xy$ if $x = 3$ and $y = 2$.

8. Evaluate $\dfrac{5a - 2b}{b - a}$ if $a = 4$ and $b = -2$.

9. Solve for x: $7x = 28$

10. Solve for x: $2x + 4 = 10$

The Integers

749

14.1 OBJECTIVES

1. Use signed numbers.
2. Find the opposite of a number.
3. Find the absolute value of a number.

Up until now you have been working with two groups of numbers, whole numbers and fractions or decimals. There are many applications in arithmetic that cannot be solved with these numbers.

Suppose you want to represent a temperature of 10° below zero, a debt of $50, or an altitude 100 ft below sea level. These situations involve a new set of numbers called **negative numbers.**

Recall that, in Chapter 1, we represented the whole numbers by points on a number line.

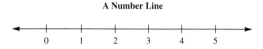

To form a number line, choose a convenient unit of measure and mark off equally spaced points from the origin (the point corresponding to 0).

We now want to extend the number line by marking off units to the *left* of 0.

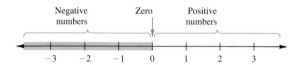

Numbers to the right of 0 on the number line are called **positive numbers.** Numbers to the left of 0 are called **negative numbers.** Zero is neither positive nor negative.

Since −3 is to the left of 0, it is a negative number. Read −3 as "negative three."

To indicate a negative number, we use a minus sign (−) in front of the number. Positive numbers may be written with a plus sign (+) or with no sign at all, in which case the number is understood to be positive.

• Example 1

Identifying Signed Numbers

+6 is a positive number.

−9 is a negative number.

5 is a positive number.

0 is neither positive nor negative.

If no sign appears, a number (other than 0) is positive.

● ● ● CHECK YOURSELF 1

Label each of the following as positive, negative, or neither.

a. +3   **b.** 7   **c.** −5   **d.** 0

Together, the sets of positive and negative numbers are called **signed numbers.** An important idea in our work with signed numbers is the *opposite* of a number. Every number has an opposite.

Opposite of a Number

The **opposite** of a number corresponds to a point the same distance from 0 as the given number, but in the opposite direction.

• Example 2

Writing the Opposite of a Signed Number

The opposite of a *positive* number is *negative*.

(*a*) The opposite of 5 is −5.

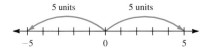

Both numbers are located 5 units from 0.

The opposite of a *negative* number is *positive*.

(*b*) The opposite of −3 is 3.

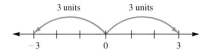

Both numbers correspond to points that are 3 units from 0.

● ● ●   **CHECK YOURSELF 2**

a. What is the opposite of 8?   **b.** What is the opposite of −9?

Place a minus sign in front of the number.

Again place a minus sign in front of the number.

We write the opposite of 5 as −5. You can now think of −5 in two ways: as negative 5 and as the opposite of 5.

Using the same idea, we can write the opposite of a negative number. The opposite of −3 is −(−3). Since we know from looking at the number line that the opposite of −3 is 3, this means that

$$-(-3) = 3$$

So the opposite of a negative number must be positive.

Let's summarize our results:

The Opposite of a Signed Number

1. The opposite of a positive number is negative.
2. The opposite of a negative number is positive.
3. The opposite of 0 is 0.

We also need to define the *absolute value,* or magnitude, of a signed number.

Absolute Value

The **absolute value** of a signed number is the distance (on the number line) between the number and 0.

• Example 3

Finding the Absolute Value

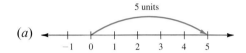

(*a*)

The absolute value of 5 is 5.   5 is 5 units from 0.

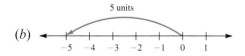

(*b*)

The absolute value of −5 is 5.   −5 is also 5 units from 0.

We usually write the absolute value of a number by placing vertical bars before and after the number. For Example 3 we can write

|5| is read "the absolute value of 5."

$$|5| = 5 \qquad \text{and} \qquad |-5| = 5$$

● ● ● **CHECK YOURSELF 3**

Complete the following statements.

a. The absolute value of 9 is _____.
b. The absolute value of −12 is _____.
c. $|-6| =$ **d.** $|15| =$

All the numbers we looked at in this section were natural numbers, 0, or the negatives of natural numbers. These numbers make up the set of integers.

The Set of Integers

The **set of integers** consists of the natural numbers, their negatives, and 0.

• Example 4

Identifying Integers

3, −5, and 0 are integers.   −5 is the negative of the natural number 5.

$\dfrac{3}{4}$ and 0.5 are *not* integers.   Common and decimal fractions do *not* belong to the set of integers.

● ● ●   **CHECK YOURSELF 4**

Which of the following are integers?

$$7, 0, \frac{4}{7}, -5, 0.2$$

Of course, there are also many negative numbers that are not integers. For instance, negative fractions or decimals are not integers.

● Example 5

Identifying Negative Numbers

$-\frac{2}{3}$, -0.3, and -3.5 are negative numbers.

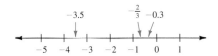

However, $-\frac{2}{3}$, -0.3, and -3.5 are *not* integers. The set of integers includes only the

natural numbers, 0, and the negatives of the natural numbers.

● ● ●   **CHECK YOURSELF 5**

Which of the following are negative integers?

$$-3, -\frac{4}{5}, 2.3, -2.3, -7, 0$$

● ● ●   **CHECK YOURSELF ANSWERS**

1. (a) positive; **(b)** positive; **(c)** negative; **(d)** neither.   **2. (a)** -8; **(b)** 9.
3. (a) 9; **(b)** 12; **(c)** 6; **(d)** 15.   **4.** 7, 0, -5.   **5.** -3 and -7.

14.1 Exercises

Indicate whether the following statements are true or false.

1. The opposite of 7 is −7.

2. The opposite of −10 is −10.

3. −9 is an integer.

4. 5 is an integer.

5. The opposite of −11 is 11.

6. The absolute value of 8 is 8.

7. $|-6| = -6$

8. $-(-30) = -30$

9. −12 is not an integer.

10. The opposite of −18 is 18.

11. $|7| = -7$

12. The absolute value of −9 is −9.

13. $-(-8) = 8$

14. $\dfrac{2}{3}$ is not an integer.

15. $|-20| = 20$

16. The absolute value of −3 is 3.

17. $\dfrac{3}{5}$ is an integer.

18. 0.8 is an integer.

19. 0.15 is not an integer.

20. $|-9| = -9$

21. $\dfrac{5}{7}$ is not an integer.

22. 0.23 is not an integer.

23. $-(-7) = -7$

24. The opposite of 15 is −15.

ANSWERS	
1.	True
2.	False
3.	True
4.	True
5.	True
6.	True
7.	False
8.	False
9.	False
10.	True
11.	False
12.	False
13.	True
14.	True
15.	True
16.	True
17.	False
18.	False
19.	True
20.	False
21.	True
22.	True
23.	False
24.	True

Complete each of the following statements.

25. The absolute value of −10 is _____.   **26.** −(−16) = _____.

27. $|-20|$ = _____.   **28.** The absolute value of −12 is __.

29. The absolute value of 7 is _____.   **30.** The opposite of −9 is _____.

31. The opposite of 30 is _____.   **32.** $|-15|$ = _____.

33. −(−6) = _____.   **34.** The absolute value of 0 is _____.

35. $|50|$ = _____.   **36.** The opposite of 18 is _____.

Complete each of the following statements, using the symbol <, =, or >.

37. −2 _____ −3   **38.** −10 _____ −5

39. −20 _____ −10   **40.** −15 _____ −14

41. $|3|$ _____ 3   **42.** $|-5|$ _____ −5

43. −4 _____ $|-4|$ **44.** 7 _____ $|7|$

45. (a) Every number has an opposite. The opposite of 5 is −5. In English, a similar situation exists for words. For example, the opposite of "regular" is "irregular." Write the opposite of these words:

irredeemable, uncomfortable, uninteresting, uninformed, irrelevant, immoral.

(b) Note that the idea of an opposite is usually expressed by a prefix such as "un" or "in." What other prefixes can be used to negate or change the meaning of a word to its opposite? List four words using these prefixes, and use the words in a sentence.

46. (a) What is the difference between positive integers and nonnegative integers?

(b) What is the difference between negative and nonpositive integers?

A N S W E R S

a. $7\frac{1}{2}$

b. $1\frac{1}{4}$

c. $6\frac{1}{5}$

d. $1\frac{1}{3}$

e. $2\frac{5}{6}$

f. $11\frac{1}{2}$

Getting Ready for Section 14.2
[Section 7.5]

Perform the indicated operations.

a. $2\frac{1}{4} + 5\frac{1}{4}$

b. $3\frac{7}{8} - 2\frac{5}{8}$

c. $2\frac{7}{10} + 3\frac{5}{10}$

d. $4\frac{1}{6} - 2\frac{5}{6}$

e. $5\frac{5}{12} - 2\frac{7}{12}$

f. $7\frac{5}{8} + 3\frac{7}{8}$

Answers

1. True **3.** True **5.** True **7.** False **9.** False **11.** False

13. True **15.** True **17.** False **19.** True **21.** True **23.** False

25. 10 **27.** 20 **29.** 7 **31.** -30 **33.** 6 **35.** 50 **37.** $-2 > -3$

39. $-20 < -10$ **41.** $|3| = 3$ **43.** $-4 < |-4|$ **45.** 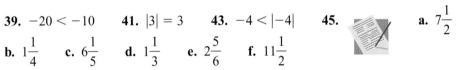 **a.** $7\frac{1}{2}$

b. $1\frac{1}{4}$ **c.** $6\frac{1}{5}$ **d.** $1\frac{1}{3}$ **e.** $2\frac{5}{6}$ **f.** $11\frac{1}{2}$

A N S W E R S

Indicate whether the following statements are true or false.

ANSWERS	
1. True	
2. True	
3. True	
4. False	
5. False	
6. True	
7. True	
8. True	
9. False	
10. True	
11. False	
12. False	
13. 12	
14. 15	
15. 8	
16. 15	
17. $^-20$	
18. 18	

1. The opposite of 9 is -9.   **2.** The absolute value of -5 is 5.

3. The opposite of -10 is 10.   **4.** The opposite of -7 is -7.

5. The absolute value of -18 is -18.   **6.** -3 is an integer.

7. $-(-7) = 7$ **8.** $|-8| = 8$

9. $|50| = -50$ **10.** 0.8 is not an integer.

11. 0.5 is an integer.   **12.** $\dfrac{3}{4}$ is an integer.

Complete each of the following statements.

13. The absolute value of -12 is _____.   **14.** $|-15| =$ _____.

15. The opposite of -8 is _____.   **16.** The absolute value of 15 is ____.

17. The opposite of 20 is _____.   **18.** $-(-18) =$ _____.

Adding and Subtracting Signed Numbers

14.2 OBJECTIVES

1. Add signed numbers.
2. Subtract signed numbers.

Remember that in Chapter 1, we used the number line to picture addition. Using that approach, we add positive numbers by moving to the *right* on the number line.

To use the number line in adding signed numbers, we will move *to the right* for positive numbers, but *to the left* for negative numbers.

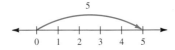

Finding the Sum of Two Signed Numbers

To avoid confusion, we always use parentheses when two signs are together. So write $5 + (-2)$, *not* $5 + -2$.

To find the sum $5 + (-2)$, first move 5 units *to the right* of 0:

Then count 2 units *to the left,* to add negative 2.

We see that

$$5 + (-2) = 3$$

● ● ● **CHECK YOURSELF 1**

Use the number line to find the sum

$$8 + (-3)$$

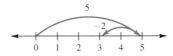

Finding the Sum of Two Signed Numbers

Find the sum $-2 + 5$.

Since the first addend is *negative,* we start by moving left.

First move 2 units *to the left* of 0. Then, to add 5, move 5 units back *to the right*. We see that

$$-2 + 5 = 3$$

757

●●● **CHECK YOURSELF 2**

Find the sum.

$-7 + 9$

● Example 3

Finding the Sum of Two Signed Numbers

Find the sum $4 + (-5)$.

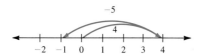

Move 4 units *to the right* of 0. Then move 5 units back *to the left* to add negative 5.

$4 + (-5) = -1$

●●● **CHECK YOURSELF 3**

Find the sum.

$3 + (-7)$

We can also use the number line to picture addition when two negative numbers are involved. Example 4 illustrates this approach.

● Example 4

Finding the Sum of Two Signed Numbers

Find the sum $-2 + (-3)$.

Move 2 units *to the left* of 0. Then move 3 more units *to the left* to add negative 3. We see that

$-2 + (-3) = -5$

●●● **CHECK YOURSELF 4**

Find the sum.

$-7 + (-5)$

You may have noticed some patterns in the previous examples. These patterns will let you do much of the addition mentally. Look at the following rule.

This means that the sum of two positive numbers is positive and the sum of two negative numbers is negative.

To Add Signed Numbers

1. If two numbers have the same sign, add their absolute values. Give the sum the sign of the original numbers.

● Example 5

Finding the Sum of Two Signed Numbers

Find the sums.

(*a*) $5 + 2 = 7$ The sum of two positive numbers is positive.

(*b*) $-2 + (-6) = -8$ Add the absolute values ($2 + 6 = 8$). Give the sum the sign of the original numbers.

● ● ●   CHECK YOURSELF 5

Find the sums.

a. $6 + 7$ **b.** $-8 + (-7)$

You have no doubt also noticed that when you add a positive and a negative number, sometimes the answer is positive and sometimes it is negative. This depends on which number has the larger absolute value (the distance from 0). This leads us to the second part of our addition rule.

To Add Signed Numbers

2. If two numbers have different signs, subtract the smaller absolute value from the larger. Give the sum the sign of the number with the larger absolute value.

● Example 6

Finding the Sum of Two Signed Numbers

Find the sum.

$6 + (-2) = 4$ The numbers have different signs. Subtract the absolute values ($6 - 2 = 4$). The result is positive, since 6 has the larger absolute value.

● ● ● **CHECK YOURSELF 6**

Find the sums.

a. $8 + (-3)$ **b.** $12 + (-10)$

● Example 7

Finding the Sum of Two Signed Numbers

Find the sums.

(*a*) $4 + (-7) = -3$ Subtract the absolute values ($7 - 4 = 3$). The result is negative, since -7 has the larger absolute value.

In general, the sum of a number and its opposite is 0.

(*b*) $8 + (-8) = 0$ Subtract the absolute values ($8 - 8 = 0$). Don't worry about the sign, since 0 is neither positive nor negative.

● ● ● **CHECK YOURSELF 7**

Find the sums.

a. $3 + (-8)$ **b.** $-9 + 15$ **c.** $9 + (-9)$

Earlier we listed some properties of addition. To extend these, what about the order in which we add two numbers? Let's look at this question in Example 8.

● Example 8

Finding the Sum of Two Signed Numbers

Find the sums.

$-2 + 7 = 7 + (-2) = 5$

$-3 + (-4) = -4 + (-3) = -7$

Note that in both cases the order in which we add the numbers does not affect the sum. This leads us to the following property of signed numbers.

The Commutative Property of Addition

The *order* in which we add two numbers does not change the sum. Addition is **commutative.** In symbols, for any numbers a and b,

$a + b = b + a$

● ● ●   **CHECK YOURSELF 8**

Show that $-8 + 2 = 2 + (-8)$.

What if we want to add more than two numbers? Another property of addition will be helpful. Look at Example 9.

● Example 9

Finding the Sum of Three Signed Numbers

Find the sum $2 + (-3) + (-4)$. First,

$$\underline{2 + (-3)} + (-4) \qquad \text{Add the first two numbers.}$$
$$= \quad -1 \quad + (-4) \qquad \text{Then add the third to that sum.}$$
$$= \quad -5$$

Now let's try a second approach.

$$2 + \underline{[(-3) + (-4)]} \qquad \text{This time, add the second and}$$
$$\qquad\qquad\qquad\qquad\qquad \text{third numbers.}$$
$$= \quad 2 + \quad (-7) \qquad \text{Then add the first number to that sum.}$$
$$= -5$$

Do you see that it makes no difference which way we group numbers in addition? The final sum is not changed. We can state the following.

> **The Associative Property of Addition**
>
> The way we *group* numbers does not change the sum. Addition is **associative.** In symbols, for any numbers a, b, and c,
> $(a + b) + c = a + (b + c)$

● ● ●   **CHECK YOURSELF 9**

Show that $-2 + (-3 + 5) = [-2 + (-3)] + 5$.

Let's look at a special property of 0.

> **The Additive Identity**
>
> The sum of any number and 0 is just that number. Because of this special property, 0 is called the **additive identity.** In symbols, for any number a,
> $a + 0 = a$

• Example 10

Finding Sums That Involve Addition of Zero

(a) $-8 + 0 = -8$

(b) $0 + (-20) = -20$

● ● ● **CHECK YOURSELF 10**

Find the sum. $-6 + 0$

We saw in Section 14.1 that every number has an opposite. The opposite of a number is also called its **additive inverse.** The additive inverse of a is $-a$.

• Example 11

Identifying the Additive Inverse

We can also call -6 the *opposite* of 6.

The additive inverse of 6 is -6.
The additive inverse of -8 is $-(-8)$, or 8.

● ● ● **CHECK YOURSELF 11**

Find the additive inverse for each number.

a. 8   **b.** -9

Use the following rule to add opposite numbers.

> **The Additive Inverse**
>
> The sum of any number and its additive inverse is 0. In symbols, for any number a,
>
> $a + (-a) = 0$

• Example 12

Finding the Sum of Two Additive Inverses

Find the sums.

(a) $6 + (-6) = 0$

(b) $-8 + 8 = 0$

● ● ● **CHECK YOURSELF 12**

Find the sum.

$9 + (-9)$

So far we have looked only at addition of integers. The process is the same if we want to add other negative numbers.

● **Example 13**

Finding the Sum of Two Signed Mixed Numbers

Find the sums.

(a) $3\frac{3}{4} + \left(-2\frac{1}{4}\right) = 1\frac{1}{2}$

Subtract the absolute values $\left(3\frac{3}{4} - 2\frac{1}{4} = 1\frac{2}{4} = 1\frac{1}{2}\right)$. The sum is positive since $3\frac{3}{4}$ has the larger absolute value.

(b) $-0.5 + (-0.2) = -0.7$

Add the absolute values $(0.5 + 0.2 = 0.7)$. The sum is negative.

● ● ● **CHECK YOURSELF 13**

Find the sums.

a. $-2\frac{1}{2} + \left(-3\frac{1}{2}\right)$

b. $5.3 + (-4.3)$

We can also add more than two signed numbers by repeated use of the addition rule. Our final example illustrates.

● **Example 14**

Finding the Sum of Two Signed Decimal Numbers

Note that by the associative property, the sum would be the same if the numbers were grouped differently for the addition. Try adding the second and third numbers as the first step.

Find the sum.

$\underbrace{2.5 + (-1.5)}_{} + (-0.5)$ Add the first two numbers.

$= \quad 1 \quad + (-0.5)$ Now add the third.

$= \quad 0.5$

● ● ● **CHECK YOURSELF 14**

Find the sum.

$$-3.5 + (-1.5) + (-2.5)$$

Now we turn our attention to the subtraction of signed numbers. In Chapter 1, subtraction was called the *inverse* operation to addition. This is extremely useful now because it means that any subtraction problem can be written as a problem in addition. Let's see how it works with the following rule.

> **To Subtract Signed Numbers**
>
> To subtract signed numbers, add the first number and the *opposite* of the number being subtracted. In symbols, by definition
>
> $a - b = a + (-b)$

To find the difference $a - b$, we add a and the opposite of b.

Example 15 illustrates this property.

● **Example 15**

Finding the Difference of Two Signed Numbers

(*a*) Subtract $5 - 3$.

$$5 - 3 = 5 + (-3) = 2$$ To subtract 3, we can add the opposite of 3.

The opposite of 3

(*b*) Subtract $2 - 5$.

$$2 - 5 = 2 + (-5) = -3$$

The opposite of 5

● ● ● **CHECK YOURSELF 15**

Find each difference, using the definition of subtraction.

a. $8 - 3$ **b.** $7 - 9$

Let's look at two more examples that involve two negative numbers.

• Example 16

Finding the Difference of Two Signed Numbers

By the definition, add the opposite of 4, -4, to the value -3.

(*a*) Subtract $-3 - 4$.

$$-3 - 4 = -3 + (-4) = -7 \qquad \text{-4 is the opposite of 4.}$$

(*b*) Subtract $-10 - 15$.

$$-10 - 15 = -10 + (-15) = -25 \qquad \text{-15 is the opposite of 15.}$$

● ● ● **CHECK YOURSELF 16**

Find each difference.

a. $-5 - 9$ **b.** $-12 - 6$

Now let's see how the definition is applied in subtracting a negative number.

• Example 17

Finding the Difference of Two Signed Numbers

Find each difference.

By the definition of subtraction, we add the opposite of -3. Remember, the opposite of -3 is 3.

(*a*) $5 - (-3) = 5 + 3 = 8$ 3 is the opposite of -3.

(*b*) $7 - (-8) = 7 + 8 = 15$ 8 is the opposite of -8.

(*c*) $-9 - (-5) = -9 + 5 = -4$ 5 is the opposite of -5.

● ● ● **CHECK YOURSELF 17**

Find each difference.

a. $7 - (-5)$ **b.** $5 - (-9)$ **c.** $-10 - (-8)$

Signed numbers other than integers are subtracted in exactly the same way, as our final examples illustrate.

●Example 18

Finding the Difference of Two Signed Decimals

Subtract.

$$3.5 - (-2.25) = 3.5 + 2.25 = 5.75$$

● ● ● **CHECK YOURSELF 18**

Find the difference.

$$-4.3 - (-7.3)$$

●Example 19

Finding the Difference of Two Signed Mixed Numbers

Subtract.

$$-2\frac{3}{4} - \left(-1\frac{1}{4}\right) = -2\frac{3}{4} + 1\frac{1}{4} = -1\frac{1}{2}$$

● ● ● **CHECK YOURSELF 19**

Find the difference.

$$5\frac{7}{8} - \left(-2\frac{3}{8}\right)$$

● ● ● **CHECK YOURSELF ANSWERS**

1. 5.   **2.** 2.   **3.** −4.   **4.** −12.   **5. (a)** 13; **(b)** −15.

6. (a) 5; **(b)** 2.   **7. (a)** −5; **(b)** 6; **(c)** 0.   **8.** −6 = −6.   **9.** 0 = 0.

10. −6.   **11. (a)** −8; **(b)** 9.   **12.** 0.   **13. (a)** −6; **(b)** 1.   **14.** −7.5.

15. (a) 5; **(b)** −2.   **16. (a)** −14; **(b)** −18.   **17. (a)** 12; **(b)** 14; **(c)** −2.

18. 3.   **19.** $8\frac{1}{4}$.

Add.

1. $-6 + (-5)$

2. $3 + 9$

3. $8 + (-4)$

4. $-6 + (-7)$

5. $4 + (-6)$

6. $9 + (-2)$

7. $7 + 9$

8. $-5 + 9$

9. $(-11) + 5$

10. $5 + (-8)$

11. $-8 + (-7)$

12. $8 + (-7)$

13. $(-16) + 15$

14. $(-12) + 10$

15. $-8 + 0$

16. $7 + (-7)$

17. $-9 + 10$

18. $-6 + 8$

19. $-4 + 4$

20. $5 + (-20)$

21. $7 + (-13)$

22. $0 + (-10)$

23. $-8 + 5$

24. $-7 + 3$

25. $6 + (-6)$

26. $-9 + 9$

	ANSWERS
1.	−11
2.	12
3.	4
4.	−13
5.	−2
6.	7
7.	16
8.	4
9.	−6
10.	−3
11.	−15
12.	1
13.	−1
14.	−2
15.	−8
16.	0
17.	1
18.	2
19.	0
20.	−15
21.	−6
22.	−10
23.	−3
24.	−4
25.	0
26.	0

27. -16	
28. -25	
29. $2\frac{1}{4}$	
30. -0.5	
31. 3.4	
32. $-9\frac{2}{5}$	
33. $2\frac{7}{8}$	
34. -0.9	
35. -1.8	
36. $-\frac{1}{2}$	
37. -8	
38. -5	
39. -12	
40. 2	
41. -7	
42. 0	
43. -6	
44. -10	
45. -2	
46. 2	
47. 6	
48. -5	
49. -11	
50. -21	
51. -20	
52. -6	
53. -40	
54. -40	

27. $(-10) + (-6)$

28. $-18 + (-7)$

29. $5\frac{3}{8} + \left(-3\frac{1}{8}\right)$

30. $(0.7) + (-1.2)$

31. $-3.8 + 7.2$

32. $-4\frac{1}{10} + \left(-5\frac{3}{10}\right)$

33. $-4\frac{9}{16} + 7\frac{7}{16}$

34. $-3.8 + 2.9$

35. $-1.5 + (-0.3)$

36. $2\frac{1}{2} + (-3)$

37. $4 + (-7) + (-5)$

38. $-7 + 8 + (-6)$

39. $-2 + (-6) + (-4)$

40. $12 + (-6) + (-4)$

41. $-3 + (-7) + 5 + (-2)$

42. $7 + (-8) + (-9) + 10$

43. $-7 + (-3) + (-4) + 8$

44. $-8 + (-5) + (-4) + 7$

Subtract.

45. $5 - 7$

46. $7 - 5$

47. $9 - 3$

48. $4 - 9$

49. $-8 - 3$

50. $-15 - 6$

51. $-12 - 8$

52. $9 - 15$

53. $-22 - 18$

54. $-25 - 15$

55. $3 - (-2)$

56. $4 - (-2)$

57. $-2 - (-3)$

58. $-9 - (-6)$

59. $-5 - (-5)$

60. $7 - (-9)$

61. $10 - (-5)$

62. $-8 - (-8)$

63. $38 - (-12)$

64. $50 - (-25)$

65. $-15 - (-25)$

66. $-20 - (-30)$

67. $-25 - (-15)$

68. $-30 - (-20)$

69. $-0.5 - 1.5$

70. $0.25 - 0.75$

71. $3.5 - (-2.5)$

72. $-3.2 - (-1.2)$

73. $5 - \left(-3\frac{1}{2}\right)$

74. $-2 - 1\frac{1}{2}$

75. $-2\frac{1}{4} - \left(-3\frac{3}{4}\right)$

76. $3\frac{1}{4} - \left(-1\frac{3}{4}\right)$

77. $-7 - (-5) - 6$

78. $-5 - (-8) - 10$

79. $-10 - 8 - (-7)$

80. $3 - (-9) - 10$

	ANSWERS
55.	5
56.	6
57.	1
58.	-3
59.	0
60.	16
61.	15
62.	0
63.	50
64.	75
65.	10
66.	10
67.	-10
68.	-10
69.	-2
70.	-0.5
71.	6
72.	-2
73.	$8\frac{1}{2}$
74.	$-3\frac{1}{2}$
75.	$1\frac{1}{2}$
76.	5
77.	-8
78.	-7
79.	-11
80.	2

Solve the following applications.

81. Temperature. The temperature in Chicago dropped from 22°F at 4 PM to −11°F at midnight. What was the drop in temperature?

82. Banking. Charley's checking account had $225 deposited at the beginning of 1 month. After he wrote checks for the month, the account was $65 *overdrawn*. What amount of checks did he write during the month?

83. Elevator stops. Micki entered the elevator on the 34th floor. From that point the elevator went up 12 floors, down 27 floors, down 6 floors, and up 15 floors before she got off. On what floor did she get off the elevator?

84. Submarines. A submarine dives to a depth of 500 ft below the ocean's surface. It then dives another 217 ft before climbing 140 ft. What is the depth of the submarine?

85. Military vehicles. A helicopter is 600 ft above sea level and a submarine directly below it is 325 ft below sea level. How far apart are they?

86. Bank balance. Tom has received an overdraft notice from the bank telling him that his account is overdrawn by $142. How much must he deposit in order to have $625 in his account?

87. Change in temperature. At 9:00 AM, Jose had a temperature of 99.8°. It rose another 2.5° before falling 3.7° by 1:00 PM. What was his temperature at 1:00 PM?

88. Bank balance. Aaron had $769 in his bank account on June 1. He deposited $125 and $986 during the month and wrote checks for $235, $529, and $712 during June. What was his balance at the end of the month?

89. Complete the following problems: "4 − (−9) is the same as _____." Write an application problem that might be answered using this subtraction.

90. Explain the difference between these two phrases:
"A number less than 7" and "a number subtracted from 7"
Use both algebra and English to explain the meaning of these phrases. Write some other ways of expressing subtraction in English.

Getting Ready for Section 14.3
[Section 6.3]

Multiply.

a. $\dfrac{2}{3} \cdot \dfrac{6}{7}$

b. $\dfrac{7}{8} \cdot \dfrac{4}{21}$

c. $\dfrac{9}{10} \cdot \dfrac{5}{18}$

d. $\dfrac{5}{12} \cdot \dfrac{9}{20}$

e. $\dfrac{10}{21} \cdot \dfrac{14}{15}$

f. $\dfrac{21}{25} \cdot \dfrac{5}{28}$

Answers

1. -11 **3.** 4 **5.** -2 **7.** 16 **9.** -6 **11.** -15 **13.** -1

15. -8 **17.** 1 **19.** 0 **21.** -6 **23.** -3 **25.** 0 **27.** -16

29. $2\dfrac{1}{4}$ **31.** 3.4 **33.** $2\dfrac{7}{8}$ **35.** -1.8 **37.** -8 **39.** -12

41. -7 **43.** -6 **45.** -2 **47.** 6 **49.** -11 **51.** -20 **53.** -40

55. 5 **57.** 1 **59.** 0 **61.** 15 **63.** 50 **65.** 10 **67.** -10

69. -2 **71.** 6 **73.** $8\dfrac{1}{2}$ **75.** $1\dfrac{1}{2}$ **77.** -8 **79.** -11 **81.** $33°F$

83. 28th floor **85.** 925 ft **87.** 98.6° **89.**   **a.** $\dfrac{4}{7}$ **b.** $\dfrac{1}{6}$ **c.** $\dfrac{1}{4}$

d. $\dfrac{3}{16}$ **e.** $\dfrac{4}{9}$ **f.** $\dfrac{3}{20}$

ANSWERS

a. $\dfrac{4}{7}$

b. $\dfrac{1}{6}$

c. $\dfrac{1}{4}$

d. $\dfrac{3}{16}$

e. $\dfrac{4}{9}$

f. $\dfrac{3}{20}$

14.2 Supplementary Exercises

Name

Section   Date

Add.

1. $6 + (-3)$

2. $11 + (-7)$

3. $7 + (-7)$

4. $7 + (-2)$

5. $8 + (-9)$

6. $5 + (-9)$

7. $-6 + (-3)$

8. $-12 + 12$

9. $-8 + 8$

ANSWERS

1. 3

2. 4

3. 0

4. 5

5. -1

6. -4

7. -9

8. 0

9. 0

10. -14

11. -3

12. -10

13. -7

14. 4

15. $1\frac{1}{2}$

16. -2

17. -5

18. $-1\frac{1}{8}$

19. -6

20. -6

21. 0

22. -4

23. -3

24. -5

25. -14

26. -18

27. -10

28. -16

29. 11

30. 8

31. 3

32. 3

33. 60

34. 0

35. -6

36. -2

37. 5

38. -1

10. $-5 + (-9)$ **11.** $-10 + 7$ **12.** $0 + (-10)$

13. $-7 + 0$ **14.** $9 + (-5)$ **15.** $3\frac{5}{8} + \left(-2\frac{1}{8}\right)$

16. $-2.7 + 0.7$ **17.** $-3.5 + (-1.5)$ **18.** $-2\frac{3}{16} + 1\frac{1}{16}$

19. $5 + (-8) + (-3)$ **20.** $-8 + 7 + (-5)$ **21.** $-8 + (-7) + 10 + 5$

22. $(-6) + (-3) + (-5) + 10$

Subtract.

23. $6 - 9$ **24.** $10 - 15$ **25.** $-9 - 5$

26. $-11 - 7$ **27.** $50 - 60$ **28.** $32 - 48$

29. $8 - (-3)$ **30.** $5 - (-3)$ **31.** $-6 - (-9)$

32. $-4 - (-7)$ **33.** $45 - (-15)$ **34.** $-3 - (-3)$

35. $-4.2 - 1.8$ **36.** $-2.8 - (-0.8)$ **37.** $3\frac{1}{2} - \left(-1\frac{1}{2}\right)$

38. $-2\frac{1}{4} - \left(-1\frac{1}{4}\right)$

Using Your Calculator to Add and Subtract Signed Numbers

Your scientific (or graphing) calculator has a key that makes a number negative. This key is different from the "minus" key. The negative key is found on the bottom row of the calculator. It is marked either $\boxed{+/-}$ or $\boxed{(\ -\ )}$. With a scientific calculator, this key is pressed *after* the number you wish to make negative is entered. All of the instructions in this section will assume that you have a scientific calculator.

• Example 1

Entering a Negative Number into the Calculator

Enter -24 in your calculator.

(*a*) -24

24 $\boxed{+/-}$

The 12 goes between positive and negative in the display. The final display is 12, because there are an even number of negative signs in front of the 12.

(*b*) $-(-(-(-12)))$

12 $\boxed{+/-}$ $\boxed{+/-}$ $\boxed{+/-}$ $\boxed{+/-}$

● ● ● **CHECK YOURSELF 1**

Enter each number into your calculator.

a. -36 **b.** $-(-(-6))$

• Example 2

Adding Signed Numbers

Find the sum for each pair of signed numbers.

(*a*) $256 + (-297)$

256 $\boxed{+}$ 297 $\boxed{+/-}$ $\boxed{=}$ $\boxed{-41}$

(*b*) $-312 + (-569)$

312 $\boxed{+/-}$ $\boxed{+}$ 569 $\boxed{+/-}$ $\boxed{=}$ $\boxed{-881}$

● ● ● **CHECK YOURSELF 2**

Find the sum for each pair of signed numbers.

a. $-368 + 547$ **b.** $-596 + (-834)$

• Example 3

Subtracting Signed Numbers

Find the difference for $-356 - (-469)$.

356 $\boxed{+/-}$ $\boxed{-}$ 469 $\boxed{+/-}$ $\boxed{=}$ 113 $\boxed{}$

● ● ● CHECK YOURSELF 3

Find the differences

a. $349 - (-49)$ **b.** $-294 - (-137)$

● ● ● CHECK YOURSELF ANSWERS

1. (a) -36; **(b)** -6.   **2. (a)** 179; **(b)** -1430.   **3. (a)** 398; **(b)** -157.

Calculator Exercises

A N S W E R S

1.	130
2.	292
3.	555
4.	130
5.	-917
6.	-1477
7.	-780
8.	-779
9.	72
10.	342
11.	690
12.	535

Using your calculator, perform the following operations.

1. $345 + (-215)$ **2.** $415 + (-123)$ **3.** $679 + (-124)$

4. $345 + (-215)$ **5.** $-789 + (-128)$ **6.** $-910 + (-567)$

7. $-349 + (-431)$ **8.** $-412 + (-367)$ **9.** $47 - (-25)$

10. $123 - (-219)$ **11.** $234 - (-456)$ **12.** $412 - (-123)$

Answers

1. 130   **3.** 555   **5.** -917 **7.** -780 **9.** 72   **11.** 690

Multiplying and Dividing Signed Numbers

14.3 OBJECTIVES

1. Multiply signed numbers.
2. Divide signed numbers.

In Section 14.2, we looked at addition involving signed numbers. When we first considered multiplication, we thought of it as repeated addition. Let's see what our work with addition tells us about the multiplication of signed numbers. We can interpret multiplication as repeated addition to find a product.

$$3 \cdot 4 = 4 + 4 + 4 = 12$$

We can use the same idea to find products involving negative numbers.

● Example 1

Finding the Product of Two Signed Numbers

Multiply.

Note that we use parentheses ( ) to indicate multiplication when negative numbers are involved.

(a) $(3)(-4) = (-4) + (-4) + (-4) = -12$
(b) $(4)(-5) = (-5) + (-5) + (-5) + (-5) = -20$

● ● ● CHECK YOURSELF 1

Find the product by writing as repeated addition.

$4(-3)$

Looking at the products we found by repeated addition in Example 1 should suggest our first rule for multiplying signed numbers.

To Multiply Signed Numbers

1. The product of two numbers with different signs is negative.

The rule is easy to use. To multiply two numbers with different signs, just multiply their absolute values and attach a minus sign to the product.

● Example 2

Finding the Product of Two Signed Numbers

Find each product.
$(5)(-6) = -30$

$$(10)(-12) = -120$$

$$(-7)(9) = -63$$

The product must have two decimal places.

$$(1.5)(-0.3) = -0.45$$

The product is negative. You can simplify as before in finding the product.

$$\left(-\frac{5}{8}\right)\left(\frac{4}{15}\right) = -\left(\frac{\overset{1}{\cancel{5}}}{\underset{2}{\cancel{8}}} \times \frac{\overset{1}{\cancel{4}}}{\underset{3}{\cancel{15}}}\right)$$

$$= -\frac{1}{6}$$

● ● ● **CHECK YOURSELF 2**

Find each product.

a. $(15)(-5)$　　　　**b.** $(-0.8)(0.2)$　　　　**c.** $\left(-\frac{2}{3}\right)\left(\frac{6}{7}\right)$

The product of two negative numbers is harder to visualize. The following pattern may help you see how we can determine the sign of the product.

Decreasing by 1

$$\begin{array}{l}(3)(-2) = -6 \\ (2)(-2) = -4 \\ (1)(-2) = -2 \\ (0)(-2) = 0 \\ (-1)(-2) = 2 \\ (-2)(-2) = 4\end{array}$$

Do you see that the product is *increasing* by 2 each time?

We already know that the product of two positive numbers is positive.

This suggests that the product of two negative numbers is positive, and this is in fact the case. To extend our multiplication rule, we have the following.

To Multiply Signed Numbers

2. The product of two numbers with the same sign is positive.

● Example 3

Finding the Product of Two Signed Numbers

Find each product.

$$8 \cdot 7 = 56$$

$$(-9)(-6) = 54$$

$$(-0.5)(-2) = 1$$

Since the numbers have the same sign, the product is positive.

CHECK YOURSELF 3

Find each product.

a. $(5)(7)$ **b.** $(-8)(-6)$ **c.** $(9)(-6)$ **d.** $(-1.5)(-4)$

CAUTION

Be Careful! $(-8)(-6)$ tells you to multiply. The parentheses are *next to* one another. The expression $-8 - 6$ tells you to subtract. The numbers are *separated* by the operation sign.

To multiply more than two signed numbers, just apply the multiplication rule repeatedly.

● Example 4

Finding the Product of a Set of Signed Numbers

Multiply.

$$\begin{aligned}
&\quad (5)(-7)(-3)(-2) \\
&= (-35)(-3)(-2) \\
&= (105)(-2) \\
&= \quad -210
\end{aligned}$$

$(5)(-7) = -35$

$(-35)(-3) = 105$

● ● ● **CHECK YOURSELF 4**

Find the product.

$(-4)(3)(-2)(5)$

We saw in Section 14.2 that the commutative and associative properties for addition could be extended to signed numbers. The same is true for multiplication. What about the order in which we multiply? Look at the following examples.

● Example 5

Using the Commutative Property of Multiplication

Find the products.

$(-5)(7) = (7)(-5) = -35$

$(-6)(-7) = (-7)(-6) = 42$

The order in which we multiply does not affect the product. This gives us the following rule.

The centered dot represents multiplication. This could have been written as

$a \times b = b \times a$

The Commutative Property of Multiplication

The order in which we multiply does not change the product. Multiplication is *commutative*. In symbols, for any *a* and *b*,
$a \cdot b = b \cdot a$

● ● ● **CHECK YOURSELF 5**

Show that $(-8)(-5) = (-5)(-8)$.

What about the way we group numbers in multiplication? Look at Example 6.

●Example 6

Using the Associative Property

Multiply.

The symbols [ ] are called *brackets* and are used to group numbers in the same way as parentheses.

$$[(3)(-7)](-2) \qquad \text{or} \qquad (3)[(-7)(-2)]$$
$$= (-21)(-2) \qquad\qquad\quad = (3)(14)$$
$$= 42 \qquad\qquad\qquad\qquad = 42$$

We group the first two numbers on the left and the second two numbers on the right. Note that the product is the same in either case.

The Associative Property of Multiplication

The way we *group* the numbers does not change the product. Multiplication is *associative*. In symbols, for any *a*, *b*, and *c*,
$(a \cdot b) \cdot c = a \cdot (b \cdot c)$

● ● ● **CHECK YOURSELF 6**

Show that $[(2)(-6)](-3) = (2)[(-6)(-3)]$.

Two numbers, 0 and 1, have special properties in multiplication.

The Multiplicative Identity

The product of 1 and any number is that number. We call 1 the *multiplicative identity*. In symbols, for any *a*,
$a \cdot 1 = a$

• Example 7

Multiplying Signed Numbers by 1

Find the products.

$(-8)(1) = -8$
$(1)(-15) = -15$

● ● ● **CHECK YOURSELF 7**

Find the product.

$(-10)(1)$

What about multiplication by 0? As we saw earlier:

Multiplying by Zero

The product of 0 and any number is 0. In symbols, for any a,
$a \cdot 0 = 0$

• Example 8

Multiplying Signed Numbers by Zero

Find the products.

$(-9)(0) = 0$
$(0)(-23) = 0$

● ● ● **CHECK YOURSELF 8**

Find the product.

$(0)(-12)$

A detailed explanation of why the product of two negative numbers must be positive concludes our discussion of multiplying signed numbers.

The Product of Two Negative Numbers

The following argument shows why the product of two negative numbers must be positive.

From our earlier work, we know that a number added to its opposite is 0.	$5 + (-5) = 0$
Multiply both sides of the statement by -3:	$(-3)[5 + (-5)] = (-3)(0)$
A number multiplied by 0 is 0, so on the right we have 0.	$(-3)[5 + (-5)] = 0$
We can now use the distributive property on the left.	$(-3)(5) + (-3)(-5) = 0$
Since we know that $(-3)(5) = -15$, the statement becomes	$-15 + (-3)(-5) = 0$

We now have a statement of the form $-15 + \square = 0$. This asks, "What number must we add to -15 to get 0, where $\square$ is the value of $(-3)(-5)$?" The answer is, of course, 15. This means that

$(-3)(-5) = 15$ The product must be positive.

It doesn't matter what numbers we use in the argument. The product of two negative numbers will always be positive.

In Chapter 3, we said that multiplication and division are related operations. So every division problem can be stated as an equivalent multiplication problem.

We used this earlier to check our division!

$8 \div 4 = 2$ Since $8 = 4 \cdot 2$

$\dfrac{12}{3} = 4$ Since $12 = 3 \cdot 4$

Since the operations are related, the rule of signs for multiplication is also true for division.

To Divide Signed Numbers

1. If two numbers have the same sign, the quotient is positive.
2. If two numbers have different signs, the quotient is negative.

• Example 9

Dividing Two Signed Numbers

Divide.

The numbers 20 and -5 have different signs, and so the quotient is negative.

$20 \div (-5) = -4$ Since $20 = (-5)(-4)$

• • • CHECK YOURSELF 9

Write the multiplication statement that is equivalent to

$$36 \div (-4) = -9$$

• Example 10

Dividing Two Signed Numbers

Divide.

The two numbers have the same sign, and so the quotient is positive.

$$\frac{-20}{-5} = 4 \qquad \text{Since } -20 = (-5)(4)$$

• • • CHECK YOURSELF 10

Find each quotient.

a. $\dfrac{48}{-6}$

b. $(-50) \div (-5)$

As you would expect, division with fractions or decimals uses the same rule for signs. Example 11 illustrates this concept.

• Example 11

Dividing Two Signed Numbers

First note that the quotient is positive. Then invert the divisor and multiply.

$$\left(-\frac{3}{5}\right) \div \left(-\frac{9}{20}\right) = \frac{\overset{1}{\cancel{3}}}{\underset{1}{\cancel{5}}} \cdot \frac{\overset{4}{\cancel{20}}}{\underset{3}{\cancel{9}}} = \frac{4}{3} = 1\frac{1}{3}$$

• • • CHECK YOURSELF 11

Find each quotient.

a. $-\dfrac{5}{8} \div \dfrac{3}{4}$

b. $-4.2 \div (-0.6)$

Be very careful when 0 is involved in a division problem. Remember that 0 divided by any nonzero number is just 0. However, division *by* 0 is not allowed.

●Example 12

Dividing Signed Numbers When Zero Is Involved

$$0 \div 7 = 0 \qquad\qquad \frac{0}{-4} = 0$$

A statement like $-9 \div 0$ has no meaning. There is no answer to the problem. Just write "undefined."

$-9 \div 0$ is undefined. $\qquad \dfrac{-5}{0}$ is undefined.

CHECK YOURSELF 12

Find the quotient if possible.

a. $\dfrac{0}{-7}$

b. $\dfrac{-12}{0}$

CHECK YOURSELF ANSWERS

1. $(-3) + (-3) + (-3) + (-3) = -12$. **2. (a)** -75; **(b)** -0.16; **(c)** $-\dfrac{4}{7}$.

3. (a) 35; **(b)** 48; **(c)** -54; **(d)** 6. **4.** 120. **5.** $40 = 40$. **6.** $36 = 36$.

7. -10. **8.** 0. **9.** $36 = (-4)(-9)$. **10. (a)** -8; **(b)** 10.

11. (a) $-\dfrac{5}{6}$; **(b)** 7. **12. (a)** 0; **(b)** undefined.

Multiply.

1. $7 \cdot 8$

2. $(6)(-12)$

3. $(4)(-3)$

4. $15 \cdot 5$

5. $(-8)(9)$

6. $(-8)(3)$

7. $(-7)(-6)$

8. $(-12)(-2)$

9. $(-10)(0)$

10. $(10)(-10)$

11. $(-8)(-8)$

12. $(0)(-50)$

13. $(20)(-4)$

14. $(-25)(-8)$

15. $(-9)(-12)$

16. $(-9)(-9)$

17. $(-20)(1)$

18. $(1)(-30)$

19. $(-40)(5)$

20. $(-25)(5)$

21. $(1.8)(-0.2)$

22. $(-2.4)(-0.5)$

23. $\left(-\dfrac{7}{10}\right)\left(-\dfrac{5}{14}\right)$

24. $\left(-\dfrac{3}{8}\right)\left(\dfrac{4}{9}\right)$

25. $(-0.5)(-1.2)$

26. $(2.5)(-0.3)$

ANSWERS

1. 56

2. -72

3. -12

4. 75

5. -72

6. -24

7. 42

8. 24

9. 0

10. -100

11. 64

12. 0

13. -80

14. 200

15. 108

16. 81

17. -20

18. -30

19. -200

20. -125

21. -0.36

22. 1.2

23. $\dfrac{1}{4}$

24. $\dfrac{1}{6}$

25. 0.6

26. -0.75

27. $-\dfrac{1}{6}$

28. $\dfrac{2}{3}$

29. 9

30. -8

31. -9

32. -8

33. -64

34. 81

35. 16

36. -25

37. 120

38. 60

39. -80

40. -70

41. -150

42. 300

43. 144

44. 240

45. -5

46. 5

47. 6

48. 10

49. -10

50. -6

51. 8

52. -7

53. -4

54. -5

27. $\left(\dfrac{5}{8}\right)\left(-\dfrac{4}{15}\right)$

28. $\left(-\dfrac{8}{21}\right)\left(-\dfrac{7}{4}\right)$

29. $(-3)^2$

30. $(-2)^3$

31. -3^2

32. -2^3

33. $(-4)^3$

34. $(-3)^4$

35. $(-2)^4$

36. -5^2

37. $(-5)(3)(-8)$

38. $(4)(-3)(-5)$

39. $(-2)(-8)(-5)$

40. $(-7)(-5)(-2)$

41. $(2)(-5)(-3)(-5)$

42. $(-2)(-5)(-5)(-6)$

43. $(-4)(-3)(-6)(-2)$

44. $(-8)(3)(-2)(5)$

Divide.

45. $15 \div (-3)$

46. $\dfrac{35}{7}$

47. $\dfrac{48}{8}$

48. $-20 \div (-2)$

49. $\dfrac{-50}{5}$

50. $-36 \div 6$

51. $\dfrac{-24}{-3}$

52. $\dfrac{42}{-6}$

53. $-20 \div 5$

54. $-45 \div 9$

55. $-72 \div 8$

56. $\dfrac{-60}{-12}$

57. $\dfrac{60}{-15}$

58. $70 \div (-10)$

59. $18 \div (-1)$

60. $\dfrac{-250}{-25}$

61. $\dfrac{0}{-9}$

62. $\dfrac{-12}{0}$

63. $-144 \div (-12)$

64. $\dfrac{0}{-10}$

65. $-7 \div 0$

66. $\dfrac{-25}{1}$

67. $\dfrac{-150}{6}$

68. $\dfrac{-80}{-16}$

69. $-4.5 \div (-0.9)$

70. $-\dfrac{2}{3} \div \dfrac{4}{9}$

71. $-\dfrac{7}{9} \div \left(-\dfrac{14}{3}\right)$

72. $(-0.8) \div (-0.4)$

73. $\dfrac{7}{10} \div \left(-\dfrac{14}{25}\right)$

74. $\dfrac{0.75}{-0.5}$

75. $\dfrac{-7.5}{1.5}$

76. $-\dfrac{5}{8} \div \left(-\dfrac{5}{16}\right)$

ANSWERS

55. -9

56. 5

57. -4

58. -7

59. -18

60. 10

61. 0

62. Undefined

63. 12

64. 0

65. Undefined

66. -25

67. -25

68. 5

69. 5

70. $-\dfrac{3}{2}$ or $-1\dfrac{1}{2}$

71. $\dfrac{1}{6}$

72. 2

73. $-\dfrac{5}{4}$ or $-1\dfrac{1}{4}$

74. -1.5

75. -5

76. 2

77. -2

78. -4

79. -2

80. 5

81. 4

82. -3

To evaluate an expression involving a fraction (indicating division), we evaluate the numerator and then the denominator. We then divide the numerator by the denominator as the last step. Using this approach, find the value of each of the following expressions.

77. $\dfrac{5 - 15}{2 + 3}$

78. $\dfrac{4 - (-8)}{2 - 5}$

79. $\dfrac{-6 + 18}{-2 - 4}$

80. $\dfrac{-4 - 21}{3 - 8}$

81. $\dfrac{(5)(-12)}{(-3)(5)}$

82. $\dfrac{(-8)(-3)}{(2)(-4)}$

Answers

1. 56 **3.** -12 **5.** -72 **7.** 42 **9.** 0 **11.** 64 **13.** -80

15. 108 **17.** -20 **19.** -200 **21.** -0.36 **23.** $\dfrac{1}{4}$ **25.** 0.6

27. $-\dfrac{1}{6}$ **29.** $(-3)^2 = (-3)(-3) = 9$ **31.** -9 **33.** -64 **35.** 16

37. 120 **39.** -80 **41.** -150 **43.** 144 **45.** -5 **47.** 6

49. -10 **51.** 8 **53.** -4 **55.** -9 **57.** -4 **59.** -18 **61.** 0

63. 12 **65.** Undefined   **67.** -25 **69.** 5 **71.** $\dfrac{1}{6}$ **73.** $-\dfrac{5}{4}$ or $-1\dfrac{1}{4}$

75. -5 **77.** -2 **79.** -2 **81.** 4

Name _____

Section _____ Date _____

Multiply.

1. $8 \cdot 9$

2. $(7)(-9)$

3. $(-6)(7)$

4. $(-12)(-7)$

5. $(10)(-8)$

6. $8 \cdot 6$

7. $(-11)(1)$

8. $(-10)(7)$

9. $(-15)(-7)$

10. $(-5)(0)$

11. $(1.8)(-0.5)$

12. $\left(-\dfrac{5}{6}\right)\left(\dfrac{9}{20}\right)$

13. $\left(-\dfrac{5}{8}\right)\left(-\dfrac{4}{9}\right)$

14. $(-0.25)(12)$

15. $(-5)^2$

16. -7^2

17. $(-3)^3$

18. $(-2)^4$

19. $(-3)(-7)(-2)$

20. $(-4)(3)(-6)$

21. $(-5)(3)(-2)(-2)$

22. $(5)(-3)(4)(-2)$

1. 72

2. -63

3. -42

4. 84

5. -80

6. 48

7. -11

8. -70

9. 105

10. 0

11. -0.9

12. $-\dfrac{3}{8}$

13. $\dfrac{5}{18}$

14. -3

15. 25

16. -49

17. -27

18. 16

19. -42

20. 72

21. -60

22. 120

Divide.

23. $12 \div (-4)$

24. $\dfrac{-45}{9}$

25. $\dfrac{24}{3}$

26. $-60 \div (-5)$

27. $-72 \div 8$

28. $\dfrac{81}{9}$

29. $-42 \div (-7)$

30. $\dfrac{-12}{0}$

31. $\dfrac{0}{-3}$

32. $48 \div (-6)$

33. $\dfrac{-8}{-1}$

34. $-48 \div (-4)$

35. $-2.5 \div 0.5$

36. $-\dfrac{5}{8} \div \left(-\dfrac{3}{4}\right)$

37. $\dfrac{3}{10} \div \left(-\dfrac{4}{5}\right)$

38. $(-4.5) \div (-0.5)$

Using Your Calculator to Multiply and Divide Signed Numbers

Finding the product of two signed numbers using a calculator is relatively straightforward.

• Example 1

Multiplying Signed Numbers

Find the product. $457 \times (-734)$

457 $\boxed{\times}$ 734 $\boxed{+/-}$ $\boxed{=}$ $\boxed{-335438}$

● ● ● **CHECK YOURSELF 1**

Find the products.

a. $36 \times (-91)$ **b.** $-12 \times (-284)$

Finding the quotient of signed numbers is also straightforward.

• Example 2

Dividing Signed Numbers

Find the quotient. $-376 \div 16$

376 $\boxed{+/-}$ $\boxed{\div}$ 16 $\boxed{=}$ $\boxed{-23.5}$

● ● ● **CHECK YOURSELF 2**

Find the quotient. $-7865 \div -242$

We can also use the calculator to raise a signed number to a power.

• Example 3

Raising a Number to a Power

Evaluate.

$(-3)^6$

3 $\boxed{+/-}$ $\boxed{y^x}$ 6 $\boxed{=}$ $\boxed{729}$

● ● ● **CHECK YOURSELF 3**

Evaluate.

$(-2)^9$

● ● ● **CHECK YOURSELF ANSWERS**

1. (a) -3276; **(b)** 3408.   **2.** 32.5.   **3.** -512.

Calculator Exercises

Name

Section   Date

ANSWERS

1. -525

2. -675

3. -936

4. -1736

5. 952

6. 1349

7. -13.8

8. -4.57

9. -15.67

10. -4.46

11. 16.28

12. 3.05

13. 2

14. 3.68

15. -1024

16. 625

Use your calculator to multiply and divide the following.

1. $25 \times (-21)$ **2.** $15 \times (-45)$

3. $78 \times (-12)$ **4.** $(-56) \times 31$

5. $(-34) \times (-28)$ **6.** $(-71) \times (-19)$

7. $345 \div (-25)$ **8.** $128 \div (-28)$

9. $(-564) \div 36$ **10.** $(-232) \div 52$

11. $(-456) \div (-28)$ **12.** $(-128) \div (-42)$

13. $(-28) \div (-14)$ **14.** $(-456) \div (-124)$

15. $(-4)^5$ **16.** $(-5)^4$

Answers

1. -525 **3.** -936 **5.** 952   **7.** -13.8 **9.** -15.67 **11.** 16.28
13. 2   **15.** -1024

14.4 Evaluating Algebraic Expressions

Remember that a letter used to represent a number is called a *variable*.

The variables are *b* and *h*.

A, B, and R are all variables.

The next step in getting ready for algebra involves using **algebraic,** or **literal expressions.** You have already seen examples of these expressions, which use letters to represent numbers. Some algebraic expressions that we have used are

$b \cdot h$ The area of a rectangle

$\dfrac{A}{B} = \dfrac{R}{100}$ The percent proportion

Often you will want to find the value of an expression. You can do this when you know the number values for each of the letters or variables. The process is called **evaluating the expression.**

To evaluate an expression, just replace each variable or letter with its number value. Then do the necessary operations.

Note: In algebra, writing letters together, such as *xy*, means multiplication, or "*x* times *y*." You must supply the "times" sign ($\cdot$) with numbers, because $2 \cdot 3$ is entirely different from 23.

Of course, 2×3 and (2)(3) also mean "2 times 3."

● Example 1

Evaluating an Algebraic Expression

Suppose that $x = 2$ and $y = 3$. To evaluate *xy*, replace *x* with 2 and *y* with 3. Then multiply.

$xy = 2 \cdot 3 = 6$

● ● ●   **CHECK YOURSELF 1**

If $x = 4$ and $y = 5$, evaluate *xy*.

● Example 2

Evaluating an Algebraic Expression

If $x = 3$ and $y = 7$, evaluate $4xy$.

$4xy = 4 \cdot 3 \cdot 7 = 84$ Replace *x* with 3 and *y* with 7. Then multiply.

● ● ●   **CHECK YOURSELF 2**

If $x = 5$ and $y = 8$, evaluate $5xy$.

You may want to return to Section 3.5 and review the order of operations before going on.

Often, evaluating an expression will involve applying the rules for the order of operations. These rules were introduced in Section 3.5 and are restated here for reference.

The Order of Operations

STEP 1   Do any operations inside the parentheses.
STEP 2   Evaluate any powers.
STEP 3   Do all multiplication and division in order from left to right.
STEP 4   Do all addition and subtraction in order from left to right.

● Example 3

Evaluating an Algebraic Expression

If $x = 3$ and $y = 8$, evaluate $3x + 2y$.

Remember: Do any multiplication (or division) *before* any addition (or subtraction).

$$
\begin{aligned}
3x + 2y &= 3 \cdot 3 + 2 \cdot 8 \qquad &\text{Replace } x \text{ with 3 and } y \text{ with 8.} \\
&= 9 + 16 \qquad &\text{Multiply, } then \text{ add.} \\
&= 25
\end{aligned}
$$

● ● ●   **CHECK YOURSELF 3**

If $x = 4$ and $y = 5$, evaluate $5x - 3y$

● Example 4

Evaluating an Algebraic Expression That Involves Powers

Remember: If the expression involves a square, a cube, or some other power, find that value first. Then proceed as before.

Evaluate $5x^2$ when $x = 2$.

$5x^2 = 5 \cdot 2^2 = 5 \cdot 4 = 20$ Evaluate the power, *then* multiply.

● ● ●   **CHECK YOURSELF 4**

Evaluate $3x^2$ when $x = 4$.

If the variables of an expression have negative values, you should be especially careful with the rule for signs. Consider Example 5.

● Example 5

Evaluating an Algebraic Expression That Involves Negative Numbers

Evaluate $5x - 3y$ if $x = 8$ and $y = -3$.

$$5x - 3y = 5 \cdot 8 - (3)(-3)$$

Multiply first. Be sure to use the rule of signs for multiplication. Now subtract.

$$= 40 - (-9)$$
$$= 40 + 9 = 49$$

● ● ● **CHECK YOURSELF 5**

If $x = 5$ and $y = -4$, evaluate $4x - 7y$.

Example 6 shows a negative number raised to a power.

● Example 6

Evaluating an Algebraic Expression That Involves Negative Numbers

Find the value of $3x^2$ if $x = -2$.

$$3x^2 = (3)(-2)^2$$

Evaluate the power first:

$$= 3 \cdot 4$$

Here $(-2)^2 = (-2)(-2) = 4$

$$= 12$$

● ● ● **CHECK YOURSELF 6**

Find the value of $2x^2 - 5y$ when $x = -2$ and $y = 4$.

Expressions involving parentheses can be evaluated easily if you remember our earlier rule: *Evaluate the expression inside the parentheses first.* Then continue to evaluate the expression as before.

● Example 7

Evaluating an Algebraic Expression That Involves Parentheses

Evaluate $3(x + y)$ when $x = 7$ and $y = 4$.

$$3(x + y) = 3(7 + 4)$$
$$= 3(11)$$
$$= 33$$

Replace *x* with 7 and *y* with 4. Evaluate the expression inside the parentheses as the first step. Then multiply.

● ● ● **CHECK YOURSELF 7**

Evaluate $2(x + y)$ when $x = 3$ and $y = 4$.

Many expressions in algebra involve fractions. To evaluate a fraction, start by evaluating the numerator. Then evaluate the denominator. As the *last step,* divide the numerator by the denominator.

● **Example 8**

Evaluating an Algebraic Expression That Involves a Fraction

Evaluate $\dfrac{2ab}{c}$ if $a = 5$, $b = -3$, and $c = 10$.

$$\frac{2ab}{c} = \frac{(2)(5)(-3)}{10} = \frac{-30}{10}$$

Evaluate the numerator. Be careful to apply the rules for signs in multiplication. Then divide the numerator by the denominator.

$$= -3$$

● ● ● **CHECK YOURSELF 8**

Evaluate $\dfrac{5xy}{z}$ if $x = -4$, $y = -3$, and $z = 12$.

● ● ● **CHECK YOURSELF ANSWERS**

1. 20.   **2.** $5xy = 5 \cdot 5 \cdot 8 = 200$.   **3.** $5x - 3y = 5 \cdot 4 - 3 \cdot 5 = 5$.
4. 48.   **5.** $4x - 7y = 4 \cdot 5 - (7)(-4)$ **6.** $2x^2 - 5y = 2(-2)^2 - 5 \cdot 4$
$\qquad\qquad\qquad\quad = 20 + 28 \qquad\qquad\qquad\qquad = 2 \cdot 4 - 5 \cdot 4 = -12$.
$\qquad\qquad\qquad\quad = 48$.
7. $2(x + y) = 2(3 + 4)$ **8.** $\dfrac{5xy}{z} = \dfrac{(5)(-4)(-3)}{12} = 5$.
$\qquad\qquad\quad = 2(7)$
$\qquad\qquad\quad = 14$

Evaluate each of the following expressions if $x = 5$, $y = 4$, $w = -2$, and $z = -3$.

1. $6x$

2. $5y$

3. $-2y$

4. $-4w$

5. $5xy$

6. $3yw$

7. $6wz$

8. $-3xz$

9. $2x + 3y$

10. $5x + 2y$

11. $3w + 4z$

12. $2x + 5z$

13. $3x - 4w$

14. $4y - 2z$

15. $5w - 2z$

16. $3y - 4w$

17. x^2

18. z^2

19. w^3

20. y^3

21. $2y^2$

22. $3w^2$

23. $x^2 + 2z^2$

24. $y^2 - 3w^2$

Evaluate each of the following expressions if $x = -2$, $y = 4$, $s = -3$, and $t = 6$.

25. $2(y + t)$

26. $3(x + y)$

1.	30
2.	20
3.	-8
4.	8
5.	100
6.	-24
7.	36
8.	45
9.	22
10.	33
11.	-18
12.	-5
13.	23
14.	22
15.	-4
16.	20
17.	25
18.	9
19.	-8
20.	64
21.	32
22.	12
23.	43
24.	4
25.	20
26.	6

27. $3(s - y)$ **28.** $5(x + s)$

29. $4(s + t)$ **30.** $6(t - s)$

31. $5(y + 3s)$ **32.** $6(2x + 3s)$

33. $x(2y + t)$ **34.** $s(3y + 2t)$

35. $2(y^2 + t^2)$ **36.** $4(x^2 + y^2)$

37. $3(2y^2 + s^2)$ **38.** $5(t^2 - 2x^2)$

39. $\dfrac{3xy}{t}$ **40.** $\dfrac{2yt}{3x}$

41. $\dfrac{5s + t}{s}$ **42.** $\dfrac{3t - x}{2y + s}$

43. $\dfrac{3x^2 - 4s}{3y - t}$ **44.** $\dfrac{2s^2 + 3t}{s - t}$

There are many real-world applications of our work in evaluating expressions. One such application is in the use of formulas. The following are just a few examples.

45. Distance. A physics formula for distance is $d = rt$. Find d if $r = 55$, $t = 4$.

46. Interest. A formula from business for interest is $I = Prt$. Find I if $P = 3000$, $r = 0.06$, and $t = 3$.

47. Height. A formula from physics for the height of an object is $h = 64t - 16t^2$. Find h if $t = 2$.

48. Area. A formula from geometry for the area of a trapezoid is $A = \dfrac{1}{2}h(a + b)$. Find A if $h = 8$, $a = 10$, and $b = 16$.

49. Account amount. A formula from business for the amount in an account is $A = P(1 + rt)$. Find A if $P = 5000$, $r = 0.08$, and $t = 2$.

50. Sum. A formula from mathematics for a sum is $S = \dfrac{n}{2}[a + (n-1)d]$. Find S if

$a = 5$, $d = 3$, and $n = 20$.

51. Evaluate $(x + y)^2$, $x^2 + y^2$, and $x^2 + 2xy + y^2$ for $x = 2$ and $y = 4$. What can you conclude about the three expressions?

52. Describe the difference between the expressions $3x^5$ and $(3x)^5$. Illustrate the difference for $x = -2$.

50. 620

51.

52.

a. 0

b. 3

c. 0

d. -4

e. 0

f. 6

● Getting Ready for Section 14.5
[Section 14.2]

Add.

a. $5 + (-5)$ **b.** $-2 + 5$

c. $-7 + 7$ **d.** $-8 + 4$

e. $9 + (-9)$ **f.** $10 + (-4)$

Answers

1. 30   **3.** -8 **5.** 100   **7.** 36   **9.** $2x + 3y = 2 \cdot 5 + 3 \cdot 4 = 10 + 12 = 22$

11. -18 **13.** 23   **15.** -4 **17.** 25   **19.** -8 **21.** 32

23. $x^2 + 2z^2 = 5^2 + 2(-3)^2 = 25 + 2 \cdot 9 = 25 + 18 = 43$

25. 20   **27.** -21 **29.** 12   **31.** -25

33. $x(2y + t) = (-2)(2 \cdot 4 + 6) = (-2)(8 + 6) = (-2)(14) = -28$

35. 104   **37.** 123   **39.** $\dfrac{3xy}{t} = \dfrac{(3)(-2)(4)}{6} = \dfrac{-24}{6} = -4$

41. 3   **43.** 4   **45.** 220   **47.** 64   **49.** 5800   **51.**

a. 0   **b.** 3   **c.** 0   **d.** -4 **e.** 0   **f.** 6

Name _____

Section _____ Date _____

ANSWERS

Evaluate each of the following expressions if $x = 4$, $y = 3$, $z = -3$, and $w = -2$.

1. $5x$	**2.** $3z$
3. $4xy$	**4.** $5yz$
5. $4x + 3w$	**6.** $6w - 2z$
7. $3z - 4w$	**8.** z^2
9. $3y^2$	**10.** $2z^3$
11. $x^2 + 2z^2$	**12.** $y^2 - 2w^2$

Evaluate each of the following expressions if $a = -3$, $b = 5$, $c = -2$, and $d = 4$.

13. $3(b + d)$	**14.** $2(b - c)$
15. $2(a + c)$	**16.** $5(c - d)$
17. $a(2b + c)$	**18.** $d(b^2 - c^2)$
19. $\dfrac{2ab}{c}$	**20.** $\dfrac{3bd}{5c}$
21. $\dfrac{2b - c}{a + b}$	**22.** $\dfrac{3b - a}{2d + c}$

1. 20
2. −9
3. 48
4. −45
5. 10
6. −6
7. −1
8. 9
9. 27
10. −54
11. 34
12. 1
13. 27
14. 14
15. −10
16. −30
17. −24
18. 84
19. 15
20. −6
21. 6
22. 3

▷14.5 Solving Equations

15.2 OBJECTIVES

1. Recognize a solution to an equation.
2. Find a solution for an equation by using addition and multiplication rules.

In the last section we learned how to evaluate algebraic expressions by replacing the letters or variables in the expression with numerical values. Sometimes algebraic expressions are used in equations. An **equation** is a statement that two expressions are equal.

$$\underbrace{3x - 1}_{} = \underbrace{2x + 5}_{} \quad \text{is an equation.}$$

Left side   Equals sign   Right side

This equation says that "$3x - 1$ is equal to $2x + 5$." As you can see, an equals sign (=) separates the two sides of the equation.

One of the most important ideas in algebra is finding the solution of an equation. An equation may be either true or false depending on the value given to the variable.

We sometimes say that the solution *satisfies* the equation.

Solution for an Equation

A **solution** for an equation is a numerical value for the variable or letter that makes the equation a true statement.

This means that the sides of the equation must be equal when the letter is replaced by the number. This is illustrated in the following example.

● Example 1

Recognizing a Solution to an Equation

Is 6 a solution for the following equation?

$$3x - 1 = 2x + 5$$

To find out, replace x with 6.

$$3 \cdot 6 - 1 \overset{?}{=} 2 \cdot 6 + 5$$
$$18 - 1 \overset{?}{=} 12 + 5$$
$$17 = 17$$

6 satisfies the equation.

Since the two sides are equal, 6 is a solution.

● ● ●

CHECK YOURSELF 1

Is 5 a solution for the following equation?

$$3x - 3 = 2x + 2$$

799

• Example 2

Recognizing a Solution to an Equation

(*a*) Is 2 a solution for the following equation?

$$2x + 5 = x + 3$$

Replace x with 2.

$$2 \cdot 2 + 5 \stackrel{?}{=} 2 + 3$$
$$4 + 5 \stackrel{?}{=} 2 + 3$$
$$9 \neq 5 \qquad \text{This is read "nine is not equal to five."}$$

2 does not satisfy the equation.

Since the sides are not equal, 2 is *not* a solution for the equation.

(*b*) Is -2 a solution? This time replace x with -2.

$$(2)(-2) + 5 \stackrel{?}{=} -2 + 3$$
$$-4 + 5 \stackrel{?}{=} -2 + 3$$
$$1 = 1 \qquad \text{(True)}$$

-2 satisfies the equation.

Since the sides now have the same value, -2 is a solution.

CHECK YOURSELF 2

For the equation $3x + 1 = 2x - 3$,

a. Is 2 a solution?   **2.** Is -4 a solution?

Note: You may be wondering whether an equation can have more than one solution. It certainly can.

The equation $x^2 = 4$ has two solutions, 2 and -2, because

$$2^2 = 4 \qquad \text{and} \qquad (-2)^2 = 4$$

However, in this section we will be looking at **linear equations in one variable.** These are equations in one variable or letter (we have used x) that appears to the first power. No other powers (x^2, x^3, etc.) can appear. These linear equations in one variable will have at most *one solution.*

• Example 3

Recognizing a Solution to an Equation

$x + 5 = 8$ is a linear equation in one variable, x. The solution for the equation is 3, since $3 + 5 = 8$.

You can easily find the solution for an equation like the one in Example 3 by guessing the answer to the question, "What plus 5 is 8?"

• • •   CHECK YOURSELF 3

Find the solution for the equation $x + 2 = 9$.

___

There is nothing wrong with guessing the answer when you can. It is just that in more difficult equations, guessing is a lot harder, and we need methods to handle these cases.

For more complicated equations, you will need something more than guesswork. So let's start to develop a set of rules that will allow you to solve any linear equation. First we will need a definition.

Definition of Equivalent Equations

Equations that have exactly the same solution are called **equivalent equations**.

Check for yourself that 3 satisfies each equation.

For example, $3x + 2 = 11$, $3x = 9$, and $x = 3$ are all equivalent equations because they all have the same solution, 3.

We say that a linear equation is *solved* when the variable is alone on the left side of the equation and only a number appears on the right.

An equation is *solved* when we write it as an equivalent equation of the form

$x = \square$ where $\square$ is some number

Here is the first rule we will need for solving equations:

Solving Equations

THE ADDITION RULE   Adding or subtracting the same number on each side of an equation gives an equivalent equation.

Remember: An equation is a statement that the two sides are equal. Adding or subtracting the same thing from both sides maintains the equality or balance.

Let's work through our first example using this rule.

• Example 4

Using the Addition Rule to Find the Solution to an Equation

Solve $x - 3 = 9$.

$$
\begin{array}{rl}
x - 3 = & 9 \\
\underline{+\,3 \quad +\ 3} & \\
x + 0 = & 12 \\
x = & 12
\end{array}
$$

Adding 3 "undoes" the subtraction and removes everything but the variable x from the left side.

Since $x + 0 = x$, you do not need to write the 0.

Remember: $-3 + 3 = 0$.

Since 12 is a solution for the equivalent equation $x = 12$, it is a solution for our original equation. We have found the solution 12 for the equation $x - 3 = 9$. It is always a good idea to check your work. Replace x with 12 in the original equation.

$$12 - 3 \overset{?}{=} 9$$
$$9 = 9$$

This is a true statement, so 12 is a solution for the original equation.

● ● ● **CHECK YOURSELF 4**

Solve and check.

$$x - 5 = 8$$

Let's look at a slightly different example. This time we will need to subtract the same number from both sides of the equation.

● Example 5

Using the Addition Rule to Find the Solution to an Equation

Solve and check $x + 5 = 9$.

$$
\begin{array}{rr}
x + 5 = & 9 \\
- 5 & -5 \\
\hline
x \quad = & 4
\end{array}
$$
Here 5 is *added* to the variable on the left side. *Subtract* 5 from both sides to undo the addition.

To check:

$$4 + 5 \overset{?}{=} 9$$
$$9 = 9$$

● ● ● **CHECK YOURSELF 5**

Solve and check.

a. $x + 8 = 12$ **b.** $x + 6 = -3$

Let's look at a different type of equation. What if we know that $5x = 15$? Adding or subtracting numbers won't help. We need a second rule for solving equations.

> ### Solving Equations
>
> **THE MULTIPLICATION RULE** Multiplying or dividing both sides of an equation by the same nonzero number gives an equivalent equation.

Again, as long as you do the *same* thing to *both* sides of the equation, the balance is maintained.

Let's see how this second rule is used to solve equations.

● Example 6

Using the Multiplication Rule to Find the Solution to an Equation

Solve and check $5x = 15$.

Solution On the left, x is multiplied by 5. We can use division to undo that multiplication.

$$\frac{5x}{5} = \frac{15}{5}$$ Use the multiplication rule to divide both sides by 5.

Note: $\dfrac{\overset{1}{5x}}{\underset{1}{5}} = 1x = x$

$$\frac{\overset{1}{5x}}{\underset{1}{5}} = 3$$ We divide numerator and denominator by 5, leaving x alone on the left side. The right side simplifies to 3.

$$x = 3$$

To check, replace x with 3 in the original equation.

$$5 \cdot 3 \overset{?}{=} 15$$
$$15 = 15 \qquad \text{True, so 3 is the solution.}$$

● ● ● CHECK YOURSELF 6

Solve and check.

$8x = 32$

● Example 7

Using the Multiplication Rule to Find the Solution to an Equation

Solve and check $-5x = 30$.

$$\frac{\cancel{-5}x}{\cancel{-5}} = \frac{30}{-5}$$ Here we must divide both sides by -5 to leave x alone on the left side.

$$x = -6$$

To check, we replace x with -6.

$$-5x = 30$$
$$(-5)(-6) \overset{?}{=} 30$$
$$30 = 30 \qquad \text{True, so } -6 \text{ is the solution.}$$

● ● ● **CHECK YOURSELF 7**

Solve and check.

$-7x = 35$

Now let's look at an example that will require multiplying both sides of the equation by the same number.

●Example 8

Using the Multiplication Rule to Find the Solution to an Equation

Solve and check $\dfrac{x}{3} = 5$.

Solution Here the variable is *divided* by 3. We can use multiplication to undo the division. Use the multiplication rule to multiply both sides by 3.

$3\left(\dfrac{x}{3}\right) = 3 \cdot 5$ Multiply both sides by 3.

$3\left(\dfrac{x}{3}\right) = 15$ We divide numerator and denominator by 3 on the left, leaving x alone on that side.

$x = 15$

To check, replace x with 15 in the original equation.

$\dfrac{x}{3} = 5$

$\dfrac{15}{3} \stackrel{?}{=} 5$

$5 = 5$ True, so 15 is the solution.

● ● ● **CHECK YOURSELF 8**

Solve and check.

$\dfrac{x}{4} = 8$

In all the equations we have looked at so far, either the addition or the multiplication rule was used in finding the solution. Sometimes we will need both rules in solving an equation.

•Example 9

Using the Addition and Multiplication Rules

Solve and check $2x + 3 = 7$.

Solution   In this equation, the variable x is multiplied by 2. Then 3 is added. The two operations mean that we will need both of our rules in order to solve for x. Start with the addition rule. We subtract 3 to undo the addition on the left.

$$\begin{aligned} 2x + 3 &= 7 \\ -3 &\quad -3 \\ \hline 2x &= 4 \end{aligned}$$

Now use the multiplication rule. We divide both sides by 2 to undo the multiplication.

$$\frac{2x}{2} = \frac{4}{2}$$
$$x = 2$$

To check, replace x with 2 in the original equation.

$$2x + 3 = 7$$
$$2 \cdot 2 + 3 \stackrel{?}{=} 7$$
$$4 + 3 \stackrel{?}{=} 7$$
$$7 = 7 \quad \text{True, so 2 is the solution.}$$

● ● ● **CHECK YOURSELF 9**

Solve and check.

$$3x + 5 = 14$$

As our examples have shown, we always start with the addition rule to undo any addition or subtraction. Then we apply the multiplication rule to undo any multiplication or division. Here is a summary of those steps.

Solving Equations

Step 1 Add or subtract the same number on each side of the equation until the variable is on one side of the equation.

Step 2 Multiply or divide both sides of the equation by the same nonzero number so that the variable is left by itself on one side of the equation.

Step 3 Check your solution in the original equation.

•Example 10

Using the Addition and Multiplication Rules

Solve and check $3x - 5 = 10$.

$$3x - 5 = 10$$
$$\underline{+\ 5 \qquad +5}$$
$$3x \quad = 15$$

Add 5 to both sides to undo the subtraction.

$$\frac{3x}{3} = \frac{15}{3}$$

Now divide both sides by 3.

$$x = 5$$

To check, replace x with 5.

Note: We always return to the *original equation* for our check.

$$3x \quad - \quad 5 = 10$$
$$3 \cdot 5 - 5 \overset{?}{=} 10$$
$$15 - 5 \overset{?}{=} 10$$
$$10 = 10 \qquad \text{True, so 5 is the solution.}$$

● ● ● **CHECK YOURSELF 10**

Solve and check.

$$4x - 7 = 5$$

The variable may appear in any position in an equation. The process of solving the equations remains much the same. Just apply the rules carefully.

•Example 11

Using the Addition and Multiplication Rules

Solve and check $3 - 2x = 9$.

$$3 - 2x = \quad 9$$
$$\underline{-3 \qquad\qquad -3}$$
$$-2x = \quad 6$$

Subtract 3 from both sides.

$$\frac{-2x}{-2} = \frac{6}{-2}$$

Now divide both sides by -2 to leave x alone on the left side of the equation.

$$x = -3$$

To check, replace x with -3.

$$3 - 2x = 9$$
$$3 - 2(-3) \overset{?}{=} 9 \qquad \text{Multiply first. Be careful with the rules for signs.}$$
$$3 + 6 \overset{?}{=} 9$$
$$9 = 9 \qquad \text{True, so } -3 \text{ is the solution.}$$

●●● **CHECK YOURSELF 11**

Solve and check.

$10 - 3x = 1$

You may also have to combine multiplication with addition or subtraction to solve an equation. Our final example illustrates.

●Example 12

Using the Addition and Multiplication Rules

Solve and check $\dfrac{x}{3} - 5 = 2$.

$$\dfrac{x}{3} - 5 = 2$$

$$\underline{\quad +5 \quad +5 \quad} \qquad \text{Add 5 to both sides.}$$

$$\dfrac{x}{3} \qquad = 7$$

$$3\left(\dfrac{x}{3}\right) = 3 \cdot 7 \qquad \text{Now multiply both sides by 3.}$$

$$x = 21$$

To check, replace x with 21.

$$\dfrac{x}{3} - 5 = 2$$

$$\dfrac{21}{3} - 5 \stackrel{?}{=} 2$$

$$7 - 5 \stackrel{?}{=} 2$$

$$2 = 2 \qquad \text{True, so 21 is the solution.}$$

●●● **CHECK YOURSELF 12**

Solve and check.

$\dfrac{x}{4} + 3 = 7$

● ● ● **CHECK YOURSELF ANSWERS**

1. Yes.   **2. (a)** 2 is not a solution; **(b)** −4 is a solution.   **3.** $x = 7$.

4. 13.   **5. (a)** 4; **(b)** −9.   **6.** 4.   **7.** −5.   **8.** 32.   **9.** 3.

10.
$$\begin{aligned} 4x - 7 &= 5 \\ \underline{+\,7} &\ \underline{+7} \\ 4x &= 12 \\ \frac{4x}{4} &= \frac{12}{4} \\ x &= 3 \end{aligned}$$
Check: $4 \cdot 3 - 7 \stackrel{?}{=} 5$
$12 - 7 \stackrel{?}{=} 5$
$5 = 5.$

11.
$$\begin{aligned} 10 - 3x &= 1 \\ \underline{-10} &\ \underline{-10} \\ -3x &= -9 \\ \frac{-3x}{-3} &= \frac{-9}{-3} \\ x &= 3. \end{aligned}$$

12. 16.

Solve and check.

1. $x - 3 = 4$ **2.** $x + 7 = 12$

3. $x + 4 = 10$ **4.** $x - 5 = 3$

5. $x + 5 = -5$ **6.** $x - 7 = -3$

7. $x - 6 = -5$ **8.** $x + 4 = -2$

9. $x + 8 = -2$ **10.** $x - 5 = -7$

11. $7x = 28$ **12.** $9x = 45$

13. $-10x = -30$ **14.** $-8x = 72$

15. $6x = -42$ **16.** $3x = -36$

17. $-5x = 30$ **18.** $-7x = -35$

19. $\dfrac{x}{4} = 6$ **20.** $\dfrac{x}{8} = 4$

21. $\dfrac{x}{5} = -10$ **22.** $\dfrac{x}{9} = -4$

23. $2x + 5 = 9$ **24.** $3x - 4 = 5$

25. $4x - 5 = 7$ **26.** $5x + 7 = 12$

1. 7
2. 5
3. 6
4. 8
5. −10
6. 4
7. 1
8. −6
9. −10
10. −2
11. 4
12. 5
13. 3
14. −9
15. −7
16. −12
17. −6
18. 5
19. 24
20. 32
21. −50
22. −36
23. 2
24. 3
25. 3
26. 1

27. 9

28. −5

29. −2

30. −2

31. −4

32. −3

33. 7

34. 5

35. 3

36. −2

37. 24

38. 12

39. −15

40. −42

41.

42.

27. $3x - 10 = 17$

28. $4x - 1 = -21$

29. $4x - 3 = -11$

30. $6x + 5 = -7$

31. $5x + 6 = -14$

32. $3x - 2 = -11$

33. $5 - 3x = -16$

34. $3 - 4x = -17$

35. $6 - 5x = -9$

36. $7 - 2x = 11$

37. $\dfrac{x}{3} - 5 = 3$

38. $\dfrac{x}{4} + 4 = 7$

39. $\dfrac{x}{5} + 7 = 4$

40. $\dfrac{x}{6} - 5 = -12$

41. Suppose $x = 8$ is a solution to the equation $x - a = -13$. Find the value of a.

42. The definition of a linear equation $ax + b = 0$ states that a and b are real numbers, with $a \neq 0$. Explain what happens if $a = 0$? What type of equation would result?

Answers

1. 7   **3.** 6   **5.** -10 **7.** $x - 6 = -5$ Check: $1 - 6 = -5$ **9.** -10

$$\begin{array}{rcl} +6 & & +6 \\ \hline x & = & 1 \end{array}$$

$-5 = -5$

11. 4   **13.** 3   **15.** $\dfrac{6x}{6} = \dfrac{-42}{6}$; $x = -7$; check: $(6)(-7) \stackrel{?}{=} -42$; $-42 = -42$

17. -6 **19.** $\dfrac{x}{4} = 6$; $4\left(\dfrac{x}{4}\right) = 4 \cdot 6$; $x = 24$; check: $\dfrac{24}{4} \stackrel{?}{=} 6$; $6 = 6$

21. -50 **23.** 2   **25.** 3   **27.** 9

29. $4x - 3 = -11$ $\dfrac{4x}{4} = \dfrac{-8}{4}$; $x = -2$; check: $4(-2) - 3 \stackrel{?}{=} -11$ **31.** -4

$$\begin{array}{rcl} +3 & = & +3 \\ \hline 4x & = & -8 \end{array}$$

$-11 = -11$

33. 7   **35.** 3   **37.** 24   **39.** -15 **41.**

Name

Section   Date

A N S W E R S

Solve and check.

1. 11

2. −2

3. 6

4. −8

5. −4

6. 8

7. −5

8. 10

9. −7

10. 20

11. −36

12. 3

13. 8

14. 4

15. −5

16. −7

17. 2

18. −5

19. −12

20. 30

1. $x - 5 = 6$

2. $x + 6 = 4$

3. $x + 3 = 9$

4. $x + 5 = -3$

5. $x - 4 = -8$

6. $9x = 72$

7. $-8x = 40$

8. $-5x = -50$

9. $-8x = 56$

10. $\dfrac{x}{4} = 5$

11. $\dfrac{x}{9} = -4$

12. $2x + 3 = 9$

13. $4x - 5 = 27$

14. $5x - 4 = 16$

15. $3x - 5 = -20$

16. $3x - 4 = -25$

17. $4 - 3x = -2$

18. $9 - 2x = 19$

19. $\dfrac{x}{4} + 8 = 5$

20. $\dfrac{x}{3} - 4 = 6$

Self-Test
for Chapter 14

A N S W E R S

The purpose of the Self-Test is to help you check your progress and review for a chapter test in class. Allow yourself about 1 hour to take the test. When you are done, check your answers in the back of the book. If you missed any answers, be sure to go back and review the appropriate sections in the chapter and do the supplementary exercises provided there.

[14.1]   In Exercises 1 to 4, complete the statements.

1. The absolute value of -9 is_____.   **2.** $-(-8) =$ _____.

3. $|-7| =$ _____.   **4.** The opposite of -10 is _____.

[14.2]   In Exercises 5 to 12, add.

5. $9 + (-3)$ **6.** $8 + (-10)$

7. $-7 + (-5)$ **8.** $-7 + 7$

9. $-9 + 0$

In Exercises 10 to 14, subtract.

10. $-3 + (-7) + 12$ **11.** $8 - 11$

12. $-3 - 7$ **13.** $7 - (-3)$

14. $-5 - (-9)$

[14.3]   In Exercises 15 to 20, multiply.

15. $(-7)(4)$ **16.** $(-9)(-6)$

17. $(-2.5)(-0.3)$ **18.** $(-12)(0)$

19. $(2)(-3)(-8)$ **20.** $(-6)^2$

In Exercises 21 to 26, divide.

21. $20 \div (-5)$ **22.** $\dfrac{-28}{-4}$

23. $\dfrac{-50}{10}$ **24.** $(-40) \div (10)$

25. $0 \div (-7)$ **26.** $12 \div 0$

813

#	Answer
1.	9
2.	8
3.	7
4.	10
5.	6
6.	−2
7.	−12
8.	0
9.	−9
10.	2
11.	−3
12.	−10
13.	10
14.	4
15.	−28
16.	54
17.	.75
18.	0
19.	48
20.	36
21.	−4
22.	7
23.	−5
24.	−4
25.	0
26.	Undefined

27. 20

28. −11

29. 27

30. −28

31. −4

32. 2

33. 8

34. −3

35. 6

36. 56

37. −30

38. −9

39. 6

40. 2

[14.4] In Exercises 27 to 32, evaluate the expressions
if $x = 5$, $y = 2$, $z - 3$, and $w = -4$.

27. $2xy$ **28.** $2x + 7z$

29. $3z^2$ **30.** $4(x + 3w)$

31. $\dfrac{2w}{y}$ **32.** $\dfrac{2x - w}{2y - z}$

[14.5] In Exercises 33 to 40, solve and check each equation.

33. $x - 5 = 3$ **34.** $x - 4 - -7$

35. $8x = 48$ **36.** $\dfrac{x}{7} = 8$

37. $\dfrac{x}{-6} - 5$ **38.** $3x - 5 \div 32$

39. $15 - 2x = 3$ **40.** $-25 + 19 - 3x$

814

The Number Line

A **number line** shows the relative position of all numbers, including integers, fractions, and decimals.

Graphs

A **graph** relates two different pieces of information. Three basic types of graphs are line graphs, bar charts, and pie charts.

Line graph In line graphs, one of the axes is usually related to time.

Bar chart Bar charts are graphs that usually show only one category of information.

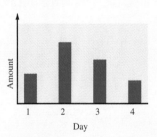

Pie chart Pie charts are graphs that show the component parts of a whole.

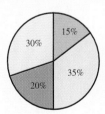

The English System of Measurement

The system of measurement in common use in the United States.

Denominate Numbers

5 ft and 10 m are denominate numbers.

7 is an abstract number.

A number with a unit of measure attached. Numbers without units attached are called *abstract numbers*.

To add 4 ft 7 in. and 5 ft 10 in.:

```
   4 ft  7 in.
 +5 ft 10 in.
   9 ft 17 in.
or 10 ft  5 in.
```

To Add Like Denominate Numbers

1. Arrange the numbers so that the like units are in the same column.
2. Add in each column.
3. Simplify if necessary.

To subtract:
```
  5 h 20 min
 -3 h 40 min
```

Borrow and rename:
```
  4 h 80 min
 -3 h 40 min
  1 h 40 min
```

To Subtract Like Denominate Numbers

1. Arrange the numbers so that the like units are in the same column.
2. Subtract in each column. You may have to borrow from the larger unit at this point.
3. Simplify if necessary.

$2 \times (3 \text{ yd } 2 \text{ ft}) = 6 \text{ yd } 4 \text{ ft}$, or 7 yd 1 ft

To Multiply or Divide Denominate Numbers by Abstract Numbers

1. Multiply or divide each part of the denominate number by the abstract number.
2. Simplify if necessary.

The Metric System of Measurement

The system of measurement used throughout the rest of the world.

Basic Metric Units

Length meter (m)
Mass or Weight gram (g)
Volume liter (L)

Basic Metric Prefixes

*These are the most commonly used and should be memorized.

*milli** means $\dfrac{1}{1000}$ *kilo** means 1000

*centi** means $\dfrac{1}{100}$ *hecto* means 100

deci means $\dfrac{1}{10}$ *deka* means 10

Converting Metric Units

You can use the following chart.

To convert to smaller units ⟶

km 1000 m	hm 100 m	dam 10 m	m 1 m	dm 0.1 m	cm 0.01 m	mm 0.001 m

⟵ To convert to larger units

To convert between metric units, just move the decimal point the same number of places to the left or right as indicated by the chart.

500 cm = ? m

To convert from centimeters to meters, move the decimal point two places to the *left*.

5 00. cm = 5 m

3 L = ? mL

To convert from liters to milliliters, move the decimal point three places to the *right*.

3 L = 3.000 mL = 3000 mL

Conversions between units of volume (liters) or units of weight (grams) work in exactly the same fashion.

To convert to smaller units ⟶

kL 1000 L	hL 100 L	daL 10 L	L* 1 L	dL .1 L	cL* .01 L	mL* .001 L

⟵ To convert to larger units

To convert to smaller units ⟶

kg 1000 g	hg 100 g	dag 10 g	g* 1 g	dg .1 g	cg* .01 g	mg* .001 g

⟵ To convert to larger units

Conversions Between the Systems

Converting Between Metric and English Units

1 meter (m)   = 39.37 inches (in.)
1 centimeter (cm) = 0.394 in.
1 kilometer (km) = 0.62 mile (mi)
1 kilogram (kg)   = 2.2 pounds (lb)
1 liter (L)   = 1.06 quarts (qt)

Converting Between English and Metric Units

1 inch (in.)   = 2.54 centimeters (cm)
1 yard (yd)   = 0.914 meter (m)
1 mile (mi)   = 1.6 kilometers (km)
1 quart (qt)   = 0.95 liter (L)
1 fluid ounce (fl oz) = 30 milliliters (mL)
                   = 30 cubic centimeters (cm^3)
1 pound (lb)   = 0.45 kilogram (kg)
1 ounce (oz)   = 28 grams (g)

English Units of Measure and Equivalents

Length
1 foot (ft)   = 12 inches (in.)
1 yard (yd)   = 3 ft
1 mile (mi)   = 5280 ft

Weight
1 pound (lb)   = 16 ounces (oz)
1 ton   = 2000 lb

Volume
1 pint (pt)   = 16 fluid ounces
                    (fl oz)
1 quart (qt)   = 2 pt
1 gallon (gal) = 4 qt

Time
1 minute (min) = 60 seconds (s)
1 hour (h)   = 60 min
1 day   = 24 h
1 week   = 7 days

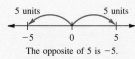

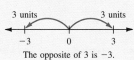

The opposite of 5 is −5.

The opposite of 3 is −3.

Signed Numbers—The Terms

Positive Numbers Numbers used to name points to the right of 0 on the number line.

Negative Numbers Numbers used to name points to the left of 0 on the number line.

Signed Numbers A set containing both positive and negative numbers.

Opposites Two numbers are opposites if the points name the same distance from 0 on the number line, but in opposite directions.

The opposite of a positive number is negative.

The opposite of a negative number is positive.

0 is its own opposite.

The integers are
$\{\ldots, -3, -2, -1, 0, 1, 2, 3, \ldots\}$
The absolute value of a number a is written $|a|$.

$|7| = 7$ $|-8| = 8$

Integers The set consisting of the natural numbers, their opposites, and 0.

Absolute Value The distance on the number line between the point named by a number and 0.

The absolute value of a number is always positive or 0.

Operations on Signed Numbers

To Add Signed Numbers

$5 + 8 = 13$

$-3 + (-7) = -10$

$5 + (-3) = 2$

$7 + (-9) = -2$

1. If two numbers have the same sign, add their absolute values. Give the sum the sign of the original numbers.
2. If two numbers have different signs, subtract the smaller absolute value from the larger. Give the sum the sign of the number with the larger absolute value.

To Subtract Signed Numbers

$4 - (-2) = 4 + 2 = 6$

The opposite of −2.

To subtract signed numbers, add the first number and the opposite of the number being subtracted.

To Multiply Signed Numbers

To multiply signed numbers, multiply the absolute values of the numbers. Then attach a sign to the product according to the following rules:

1. If the numbers have the same sign, the product is positive.
2. If the numbers have different signs, the product is negative.

To Divide Signed Numbers

To divide signed numbers, divide the absolute values of the numbers. Then attach a sign to the quotient according to the following rules:

1. If the numbers have the same sign, the quotient is positive.
2. If the numbers have different signs, the quotient is negative.

$5 \cdot 7 = 35$

$(-4)(-6) = 24$

$(8)(-7) = -56$

$\dfrac{-8}{-2} = 4$

$27 \div (-3) = -9$

$\dfrac{-16}{8} = -2$

$2x + 3y$ is an algebraic expression. The variables are x and y.

Algebraic Expressions

Algebraic Expressions An expression that contains numbers and letters (called *variables*).

Term A number or the product of a number and one or more variables.

Evaluating Algebraic Expressions

To evaluate an algebraic expression:

1. Replace each variable or letter with its number value.
2. Do the necessary arithmetic, following the rules for the order of operations.

Evaluate $2x + 3y$ if $x = 5$ and $y = -2$.

$2x + 3y$
$= 2 \cdot 5 + (3)(-2)$
$= 10 - 6 = 4$

Combining Like Terms

To combine like terms:

1. Add or subtract the coefficients (the numbers multiplying the variables).
2. Attach the common variable.

$5x + 2x = 7x$
$\quad\quad 5 + 2$

$8a - 5a = 3a$
$\quad\quad 8 - 5$

$5x - 1 = 3x + 7$ is an algebraic equation.

4 is a solution for the equation above because

$5 \cdot 4 - 1 \overset{?}{=} 3 \cdot 4 + 7$
$20 - 1 \overset{?}{=} 12 + 7$
$19 = 19$

Algebraic Equations

Algebraic Equation A statement that two expressions are equal. An equation may be either true or false.

Solution A numerical value for the variable or letter that makes the equation a true statement.

$2x + 1 = x + 5$

$x + 1 = 5$

and

$x = 4$

are equivalent equations. They all have the solution 4.

Equivalent Equations Equations that have exactly the same solutions.

Rules for Writing Equivalent Equations

$$\begin{array}{r} 3x + 5 = 17 \\ -5 \quad -5 \\ \hline 3x \quad = 12 \end{array}$$

$$\frac{3x}{3} = \frac{12}{3}$$

$$x = 4$$

1. Adding or subtracting the same term on each side of an equation gives an equivalent equation.
2. Multiplying or dividing both sides of an equation by the same nonzero number gives an equivalent equation.

Solving Equations

Solve and check:

$7x - 2 = 3x + 10$

$$\begin{array}{r} 7x - 2 = \quad 3x + 10 \\ -3x \quad\quad -3x \\ \hline 4x - 2 = \quad\quad 10 \\ +2 \quad\quad\quad + 2 \\ \hline 4x \quad = \quad\quad 12 \end{array}$$

$$\frac{4x}{4} = \frac{12}{4}$$

$$x = 3$$

To check:

$$7x - 2 = 3x + 10$$

$$7 \cdot 3 - 2 \stackrel{?}{=} 3 \cdot 3 + 10$$

$$19 = 19$$

1. Add or subtract the same term on each side of the equation until all terms involving the variable are on one side of the equation and the numbers are on the other.
2. Multiply or divide both sides of the equation by the same nonzero number so that the variable is left by itself on one side of the equation.
3. Check your solution in the original equation.

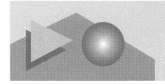

Summary Exercises

A N S W E R S

1. 250,000

2. 5000

3. 84

4. 128

5. 240

6. 20

You should now be reviewing the material in Chapters 12–14 of the book. The following exercises will help in that process. Work all the exercises carefully. Then check your answers in the back of the book. References are provided there to the chapter and section for each problem. If you made an error, go back and review the related material and do the supplementary exercises for that section.

[12.2]   **1.** According to the following line graph, how many more personal computers were sold in 1989 than in 1986?

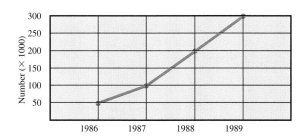

2. According to the following bar graph, how many more students were enrolled in the university in 1989 than in 1980?

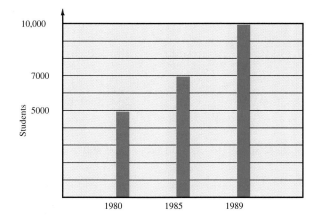

[13.1]   In Exercises 3 to 6, complete each statement.

3. 7 ft = _____ in.   **4.** 8 lb = _____ oz

5. 4 min = _____ s   **6.** 5 gal = _____ qt

© 1998 McGraw-Hill Companies

821

7. 8 ft 10 in.

8. 9 lb 4 oz

9. 12 ft 6 in.

10. 2 lb 9 oz

11. 10 h 30 min

12. 2 min 9 s

13. 14 ft 8 in.

14. The three smaller bottles

15. 5000

16. 3000

17. 300

18. 4

19. 3

20. 2

21. 600

22. 132

23. 2.4

24. 562

25. 5.6

26. 2

27. *a*

28. *a*

[13.2]   In Exercises 7 and 8, simplify.

7. 7 ft 22 in.   **8.** 8 lb 20 oz

In Exercises 9 to 12, do the indicated operations.

9.   5 ft 8 in.   **10.**   5 lb 8 oz
       +6 ft 10 in.   −2 lb 10 oz

11. 3 × (3 h 30 min)   **12.** $\dfrac{10 \text{ min } 45 \text{ s}}{5}$

In Exercises 13 and 14, solve the applications.

13. A plan for a bookcase requires three pieces of lumber 2 ft 8 in. long and two pieces 3 ft 4 in. long. What is the total length of material that is needed?

14. You can buy three bottles of dishwashing liquid, each containing 1 pt 6 fl oz, on sale for $2.40. For the same price you can buy a large container holding 2 qt. Which is the better buy?

[13.3–13.5]   In Exercises 15 to 20, complete each statement.

15. 5 km = _____ m   **16.** 3 L = _____ mL

17. 30 cL = _____ mL   **18.** 400 cm = _____ m

19. 3000 cm^3 = _____ L   **20.** 2000 mg = _____ g

In Exercises 21 to 26, complete each statement.

21. 0.6 L = _____ cm^3 **22.** 1.32 m = _____ cm

23. 0.24 cL = _____ mL   **24.** 0.562 km = _____ m

25. 0.0056 kg = _____ g   **26.** 0.002 m = _____ mm

In Exercises 27 to 28, choose the most reasonable metric measure.

27. The width of a fence gate: (*a*) 1.5 m   (*b*) 15 cm   (*c*) 15 m

28. The volume of a home aquarium: (*a*) 40 L   (*b*) 400 L   (*c*) 400 mL

[14.1] In Exercises 29 to 32, complete the statements.

29. The absolute value of 12 is _____.

30. The opposite of -8 is _____.

31. $|-3| = $ _____.

32. $-(-20) = $ _____.

[14.2] In Exercises 33 to 38, add.

33. $15 + (-7)$

34. $4 + (-9)$

35. $-8 + (-3)$

36. $2\frac{1}{2} + (-2)$

37. $-3.75 + (-1.25)$

38. $5 + (-6) + (-3)$

In Exercises 39 to 44, subtract.

39. $15 - 20$

40. $-10 - 5$

41. $2 - (-3)$

42. $-7 - (-3)$

43. $5\frac{3}{4} - \left(-\frac{3}{4}\right)$

44. $-2.75 - 1.25$

[14.3] In Exercises 45 to 50, multiply.

45. $(-12)(-3)$

46. $(-10)(8)$

47. $(-0.5)(3)$

48. $\left(-\frac{3}{8}\right)\left(-\frac{4}{5}\right)$

49. $(-4)^2$

50. $(-2)(7)(-3)$

29. 12

30. 8

31. 3

32. 20

33. 8

34. -5

35. -11

36. $\frac{1}{2}$

37. -5

38. -4

39. -5

40. -15

41. 5

42. -4

43. $6\frac{1}{2}$

44. -4

45. 36

46. -80

47. -1.5

48. $\frac{3}{10}$

49. 16

50. 42

In Exercises 51 to 56, divide.

51. $48 \div (-12)$ **52.** $\dfrac{-33}{-3}$

53. $-9 \div 0$ **54.** $0.75 \div (-10)$

55. $-\dfrac{7}{9} \div \left(-\dfrac{2}{3}\right)$ **56.** $0 \div (-12)$

[14.4] In Exercises 57 to 62, evaluate the expressions if $w = 3$, $x = 2$, $y = -4$, and $z = -8$.

57. $3xy$ **58.** $3x - 4y$

59. $x^2 + z^2$ **60.** $3(x + 2y)$

61. $\dfrac{2x}{y}$ **62.** $\dfrac{y - 3z}{2y + w}$

[14.5] In Exercises 63 to 68, solve each equation and check your solution.

63. $x - 3 = 5$ **64.** $x + 7 = -2$

65. $-9x = 81$ **66.** $3x - 7 = 8$

67. $5 - 8x = 11$ **68.** $\dfrac{x}{4} - 7 = 3$

A N S W E R S

1. 96

2. 96

3. 9 ft 8 in.

4. 11 lb 5 oz

5. 1 min 35 s

6. 8000

7. 3

8. 5

9. 250

10. 1995 and 1996

11. −20

12. 7

13. 12

14. 20

This test is provided to help you in the process of reviewing Chapters 12 –14. Answers are provided in the back. If you missed any answers, be sure to go back and review the appropriate chapter sections.

1. 8 ft = _____ in.

2. 6 lb = _____ oz

Simplify the quantity.

3. 8 ft 20 in.

In Exercises 4 and 5, do the indicated operations.

4.   7 lb   9 oz
      +3 lb 12 oz

5.   4 min 10 s
      −2 min 35 s

In Exercises 6 to 9, complete each statement.

6. 8 km = _____ m

7. 3000 mg = _____ g

8. 500 cm = _____ m

9. 25 cL = _____ mL

10. According to the line graph, between what two years was the increase in benefits the greatest?

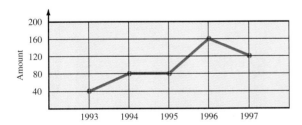

In Exercises 11 to 14, complete the statements.

11. The opposite of 20 is _____.

12. The absolute value of −7 is _____.

13. − (−12) =

14. |−20| =

In Exercises 15 to 24, add.

15. $-12 + (-6)$

16. $-7 + 7$

17. $3\frac{2}{3} + \left(-2\frac{1}{3}\right)$

18. $9 + (-4) + (-7)$

19. $-8 - (-8)$

20. $3.25 - (-1.75)$

21. $(-8)(-12)$

22. $(-6)(15)$

23. $0.75 \div (-0.3)$

24. $-2.5 \div 0$

In Exercises 25 to 28, evaluate the expressions if $a = 5$, $b = -3$, $c = 4$, and $d = -2$.

25. $6ad$

26. $3b^2$

27. $3(c - 2d)$

28. $\dfrac{2a - 7d}{a - b}$

In Exercises 29 and 30, solve each equation and check your solution.

29. $x - 7 = -15$

30. $5x - 8 = 27$

Answers to Pretests, Even-Numbered Section Exercises, Calculator Exercises, Self-Tests, Summary Exercises, Cumulative Tests

Pretest for Chapter 1

1. One hundred seven thousand, nine hundred forty-five (Sec. 1.1)
2. Associative property (Sec. 1.2) **3.** 758 (Sec. 1.3) **4.** 39,662 (Sec. 1.4) **5.** 57,800 (Sec. 1.5) **6.** False (Sec. 1.5) **7.** 323 (Sec. 1.6)
8. 4844 (Sec. 1.7) **9.** 93 (Sec. 1.7) **10.** 14 ft (Sec. 1.8)

Exercises 1.1

2. $(6 \times 100) + (3 \times 10) + 7$ **4.** $(2 \times 10,000) + (7 \times 100) + (2 \times 10) + 1$ **6.** Thousands **8.** Hundreds **10.** Hundred thousands
12. Hundred millions **14.** Twenty-one thousand, eight hundred twelve
16. One hundred three thousand, nine hundred **18.** 350,359
20. 4,230,000,000 **22.** Twenty-two thousand, two hundred twenty-two
24. 46,789 **26.**   **28.**   **30.**

Exercises 1.2

2. addend; addend; sum **4.** 7 **6.** 14 **8.** 7 **10.** 16 **12.** 14
14. 8 **16.** 9 **18.** 13 **20.** 13 **22.** 12 **24.** 12 **26.** 12 **28.** 16
30. 11 **32.** 15 **34.** 15 **36.** 15 **38.** 15 **40.** 7 **42.** 5 **44.** 15
46. 23 **48.** Associative property of addition **50.** Commutative property of addition **52.** Additive identity property **54.** Associative property of addition **56.**

Exercises 1.3

2. 18 **4.** 95 **6.** 687 **8.** 769 **10.** 5898 **12.** 7986 **14.** 9208
16. 47,728 **18.** 87 **20.** 2798 **22.** 589 **24.** 2889 **26.** 29
28. 168 **30.** 2475 **32.** 668 **34.** 569 **36.** 357 mi **38.** $8695
40. 99 **42.** 999 people **44.** $1999 **46.**

Exercises 1.4

2. 72 **4.** 153 **6.** 154 **8.** 450 **10.** 1355 **12.** 779 **14.** 1224
16. 13,222 **18.** 5281 **20.** 6320 **22.** 112,330 **24.** 119,655
26. 414 **28.** 316 **30.** 3787 **32.** 42,022 **34.** 32, 38, 44, 50
36. 41, 49, 57, 65 **38.** 5660 points **40.** $3088 **42.** 261 mi
44. Department Totals: Office, $95,052; Production, $253,859; Sales, $105,635; Warehouse, $41,523; **Monthly Total:** Oct., $168,016; Nov., $168,971; Dec., $159,082 **46.**

Calculator Exercises

2. 846 **4.** 1299 **6.** 40,134 **8.** 1125 **10.** 45,694 **12.** 2,598,960

Exercises 1.5

2. 70 **4.** 580 **6.** 700 **8.** 6700 **10.** 4000 **12.** 40,000 **14.** 39,000
16. 600,000 **18.** 930,000 **20.** Estimate: 280 Sum: 278
22. Estimate: 330 Sum: 326 **24.** Estimate: 4600 Sum: 4614
26. Estimate: 7500 Sum: 7503 **28.** Estimate: 10,000 Sum: 9925
30. Estimate: 35,000 Sum: 35,255 **32.** < **34.** > **36.** >
38. $27 **40.** $15 **42.**

Exercises 1.6

2. 7 minuend, 5 subtrahend, 2 difference: $2 + 5 = 7$ **4.** 43 **6.** 54
8. 52 **10.** 273 **12.** 320 **14.** 2034 **16.** 2610 **18.** 56,572

20. 32,202 **22.** 35 **24.** 140 **26.** 110 **28.** $704 **30.** $210
32. 332 votes **34.** $423 **36.**

Exercises 1.7

2. 37 **4.** 17 **6.** 435 **8.** 384 **10.** 275 **12.** 177 **14.** 337
16. 2874 **18.** 1957 **20.** 2775 **22.** 3458 **24.** 1651 **26.** 27,838
28. 24,800 **30.** 13,355 **32.**

Monthly Income	$1620
House payment	343
Balance	1277
Car payment	183
Balance	1094
Food	312
Balance	782
Clothing	89
Amount remaining	693

34. Incorrect **36.** Correct **38.** 174 students **40.** $28
42. 43 points **44.** 7595 mi **46.** 90, 73, 56, 39 **48.** 34
50.

4	3	8
9	5	1
2	7	6

52.

7	12	1	14
2	13	8	11
16	3	10	5
9	6	15	4

54.

Calculator Exercises

2. 187 **4.** 7685 **6.** 178,750 **8.** 402 **10.** 61,712 mi^2

Exercises 1.8

2. 16 in. **4.** 32 ft **6.** 23 yd **8.** 26 in. **10.** 14 in **12.** 100 ft

Self-Test for Chapter 1

1. Ten thousands (Sec. 1.1) **2.** Twenty-three thousand, five hundred forty-three (Sec. 1.1) **3.** 408,520,000 (Sec. 1.1) **4.** Commutative property of addition (Sec. 1.2) **5.** Associative property of addition (Sec. 1.2) **6.** Additive identity property (Sec. 1.2) **7.** 565 (Sec. 1.3) **8.** 4558 (Sec. 1.3) **9.** 7173 (Sec. 1.4) **10.** 1918 (Sec. 1.4)
11. 2731 (Sec. 1.4) **12.** 13,103 (Sec. 1.4) **13.** 25,979 (Sec. 1.4)
14. 21,696 (Sec. 1.4) **15.** 55,978 (Sec. 1.4) **16.** 8550, 8500, 9000 (Sec. 1.5) **17.** 2970, 3000, 3000 (Sec. 1.5) **18.** Incorrect (Sec. 1.5)
19. Correct (Sec. 1.5) **20.** < (Sec. 1.5) **21.** > (Sec. 1.5) **22.** 235 (Sec. 1.6) **23.** 12,220 (Sec. 1.6) **24.** 429 (Sec. 1.7) **25.** 3239 (Sec. 1.7) **26.** 30,770 (Sec. 1.7) **27.** 40,555 (Sec. 1.7) **28.** 72 lb (Sec. 1.7) **29.** 16 in. (Sec. 1.8) **30.** 12 in. (Sec. 1.8)

Pretest for Chapter 2

1. factors, product (Sec. 2.1) **2.** 1, 2, 3, 6, 11, 22, 33, 66 (Sec. 2.1)
3. Associative property (Sec. 2.1) **4.** 21,080 (Sec. 2.2) **5.** 19,992 (Sec. 2.3) **6.** 56,100 (Sec. 2.4) **7.** (a) 26 (b) 56 (Sec. 2.5)
8. $151 (Sec. 2.5) **9.** 192 (Sec. 2.6) **10.** (a) 10 yd^2 (b) 12 ft^3 (Sec. 2.7)

Exercises 2.1

2. 20 **4.** Factors, product **6.** 1, 2, 3, 4, 6, 9, 12, 18, 36
8. Multiples **10.** 28 **12.** 45 **14.** 36 **16.** 7 **18.** 0 **20.** 32
22. 56 **24.** 35 **26.** 63 **28.** 42 **30.** 48 **32.** 81 **34.** 49 **36.** 25
38. 21 **40.** 0 **42.** 56 **44.** 0 **46.** Multiplicative identity property
48. Commutative property of multiplication **50.** Multiplication
property of zero **52.** Distributive property of multiplication
54. Associative property of multiplication **56.** Multiplicative identity
property **58.** Commutative property of multiplication
60. Multiplication property of zero

Exercises 2.2

2. 96 **4.** 265 **6.** 8127 **8.** 4445 **10.** 35,126 **12.** 62,037
14. 183,144 **16.** 155,075 **18.** 2754 **20.** 8806 **22.** 1636
24. 117,260 **26.** 972 cards **28.** 825 cal **30.** 45 pages **32.** 84 seats
34.

Exercises 2.3

2. 2842 **4.** 5100 **6.** 19,293 **8.** 342,300 **10.** 27,900 **12.** 228,762
14. 203,740 **16.** 666,333 **18.** 1,254,047 **20.** 6,451,794
22. 18,816 **24.** 163,812 **26.** 432 pictures **28.** 2400 labels
30. $24,612 **32.** 25,024 ft **34.** 12,650 aspirin **36.** 1204 signatures
38. 1075 papers **40.** 2636 bushels **42.**

Exercises 2.4

2. 670 **4.** 73,000 **6.** 345,600 **8.** 420,000 **10.** 2320 **12.** 816,300
14. 112,220 **16.** 1,340,900 **18.** 16,890 **20.** 428,400 **22.** 335,000
24. 848,000 **26.** 900 **28.** 4200 **30.** 80,000 **32.** 350,000
34. 1500 students **36.** $20,000 **38.**

Exercises 2.5

2. 16 **4.** 1 **6.** 39 **8.** 23 **10.** 0 **12.** 58 **14.** 72 **16.** 28
18. 12 **20.** 40 **22.** 98 **24.** 98 **26.** $5340 **28.** 172 people

Calculator Exercises

2. 2450 **4.** 233,700 **6.** 61,017,734 **8.** 2880 **10.** 12,956,736
12. 675,990 **14.** 29 **16.** 3 **18.** 3 **20.** 88 **22.** 88
24. $ 8 $ 15
     16   31
     32   63
     64   127
    128   255
    256   511
    512   1023

Exercises 2.6

2. 8 **4.** 25 **6.** 243 **8.** 256 **10.** 1 **12.** 7 **14.** 100
16. 10,000,000 **18.** 512 **20.** 36 **22.** 49 **24.** 48 **26.** 144
28. 28 **30.** 21 **32.** 69 **34.** No **36.** Yes **38.** Yes
40.

Exercises 2.7

2. 18 in.^{2} **4.** 16 ft^2 **6.** 30 in.^{2} **8.** 9 in.^{2} **10.** 700 ft^2 **12.** 74 ft^2
14. 234 ft^2 **16.** 48 yd^2 **18.** 128 in.^{3} **20.** 27 yd^3 **22.** 30 in.^{3}

24. 6 squares **26.** No, the area is 440 ft^2. **28.** $260 **30.** 720 ft^2
32. 128 ft^3 **34.** 324 ft^3

Self-Test for Chapter 2

1. 1, 2, 3, 4, 6, 8, 12, 16, 24, 48 (Sec. 2.1) **2.** 6, 12, 18, 24, 30
(Sec. 2.1) **3.** Commutative property of multiplication (Sec. 2.1)
4. Multiplicative identity property (Sec. 2.1) **5.** Multiplication
property of 0 (Sec. 2.1) **6.** Associative property of multiplication
(Sec. 2.1) **7.** Distributive property of multiplication over addition
(Sec. 2.1) **8.** 174 (Sec. 2.2) **9.** 1911 (Sec. 2.2) **10.** 4984
(Sec. 2.3) **11.** 55,414 (Sec. 2.3) **12.** 252 people (Sec. 2.3)
13. $308,750 (Sec. 2.3) **14.** 900 uniforms (Sec. 2.3) **15.** 4000
books (Sec. 2.3) **16.** 2,453,016 (Sec. 2.3) **17.** 53,000 (Sec. 2.4)
18. 226,800 (Sec. 2.4) **19.** 321,840 (Sec. 2.4) **20.** 150,000 (Sec. 2.4)
21. 920,000 (Sec. 2.4) **22.** 17 (Sec. 2.5) **23.** 22 (Sec. 2.5)
24. 1 (Sec. 2.5) **25.** 75 (Sec. 2.5) **26.** 75 (Sec. 2.5) **27.** $2224
(Sec. 2.5) **28.** 3769 mi (Sec. 2.5) **29.** 625 (Sec. 2.6)
30. 1 (Sec. 2.6) **31.** 108 (Sec. 2.6) **32.** 9 (Sec. 2.6) **33.** 12 in^2
(Sec. 2.7) **34.** 36 ft^2 (Sec. 2.7) **35.** 24 ft^3 (Sec. 2.7) **36.** 81 yd^3
(Sec. 2.7) **37.** $1200 (Sec. 2.7)

Pretest for Chapter 3

1. Divisor: 8; dividend: 61; quotient: 7; remainder 5 (Sec. 3.1)
2. 7 (Sec. 3.2) **3.** Undefined (Sec. 3.2) **4.** 1204 r3 (Sec. 3.3)
5. 270 r1 (Sec. 3.3) **6.** 81 r131 (Sec. 3.4) **7.** 1041 r3 (Sec. 3.4)
8. 7 (Sec. 3.6) **9.** $47 (Sec. 3.6) **10.** 277 mi (Sec. 3.7)

Exercises 3.1

2. Quotient, divisor, dividend **4.** 5 **6.** $35 = 5 \times 7$
8. $7 \times 6 + 4 = 46$ **10.** 6 **12.** 6 **14.** 7 **16.** 8 **18.** 7 **20.** 7
22. 5 r3 **24.** 5 r5 **26.** 6 r2 **28.** 4 r4 **30.** 8 r3 **32.** 9 r2
34. $9 **36.** 8 cars **38.** 6 printers **40.**

Exercises 3.2

2. 9 **4.** 1 **6.** 10 **8.** Undefined **10.** 8 **12.** Undefined **14.** 0
16. Undefined **18.** 1 **20.** 0 **22.**

Exercises 3.3

2. 8 r6 **4.** 58 **6.** 49 r3 **8.** 987 r3 **10.** 391 r8 **12.** 8657 r2
14. 22,184 r4 **16.** 15 r3 **18.** 13 r1 **20.** 243 **22.** 318 **24.** 131 r1
26. 488 r5 **28.** 406 r3 **30.** 1274 r5 **32.** 8573 r1 **34.** 3128 r3
36. 6 **38.** 58 **40.** 4 cards **42.** 32 sections

Exercises 3.4

2. 21 r2 **4.** 7 r36 **6.** 6 r48 **8.** 14 r1 **10.** 65 r35 **12.** 345 r20
14. 164 r37 **16.** 207 r26 **18.** 4 r837 **20.** 36 r68 **22.** 20 r110
24. 52 r159 **26.** 301 r120 **28.** 706 r126 **30.** 43 mi/gal **32.** $16
34. $24 **36.** $229 **38.** 62 mi/h **40.** 579 cars **42.** $91
44. 106 shares **34.**

Exercises 3.5

2. 25 **4.** 6 **6.** 25 **8.** 100 **10.** 67 **12.** 40 **14.** 25
16. 10 houses **18.** 20 boxes **20.** 15 shirts

Exercises 3.6

2. 17 **4.** 6 **6.** 4 **8.** 24 **10.** 12 **12.** 4 **14.** 128 **16.** 12
18. 96 **20.** 4 **22.** 4 **24.** 4

Calculator Exercises
2. 78 **4.** 647 **6.** 2456 **8.** 16 **10.** 16 **12.** 25 **14.** 10 **16.** 3

Exercises 3.7
2. 16 **4.** 58 **6.** 36 **8.** 11 **10.** 23 **12.** 15 **14.** 16 **16.** 69
18. 169 mi **20.** 211 students **22.** 68 points **24.** 1996 by $2
26. 396 kWh **28.** 268 kWh **30.** 176

Self-Test for Chapter 3
1. Divisor, dividend; quotient, remainder (Sec. 3.1) **2.** 7 (Sec. 3.1)
3. 6 (Sec. 3.1) **4.** 8 r5 (Sec. 3.1) **5.** 9 r4 (Sec. 3.1) **6.** 8 (Sec. 3.2)
7. 1 (Sec. 3.2) **8.** 0 (Sec. 3.2) **9.** Undefined (Sec. 3.2) **10.** 123
(Sec. 3.3) **11.** 492 r6 (Sec. 3.3) **12.** 3041 r2 (Sec. 3.3) **13.** 6 r23
(Sec. 3.4) **14.** 14 r41 (Sec. 3.4) **15.** 76 r7 (Sec. 3.4) **16.** 24 r191
(Sec. 3.4) **17.** 22 r21 (Sec. 3.4) **18.** 209 r145 (Sec. 3.4) **19.** S223
(Sec. 3.4) **20.** 35 mi/gal (Sec. 3.4) **21.** $248 (Sec. 3.4) **22.** $28
(Sec. 3.4) **23.** 6 (Sec. 3.5) **24.** 20 (Sec. 3.5) **25.** 5 (Sec. 3.6)
26. 7 (Sec. 3.6) **27.** 3 (Sec. 3.6) **28.** 16 (Sec. 3.6) **29.** 20 (Sec.
3.6) **30.** 3 (Sec. 3.6) **31.** 15 (Sec. 3.7) **32.** 8 (Sec. 3.7) **33.** 204
riders (Sec. 3.7) **34.** 93 points (Sec. 3.7)

Pretest for Chapter 4
1. 1, 2, 3, 6, 7, 14, 21, 42 (Sec. 4.1) **2.** Prime: 2, 3, 7, 17, 23;
composite: 6, 9, 13, 21 (Sec. 4.1) **3.** 2 and 3 (Sec. 4.1) **4.** 2, 3, and
5 (Sec. 4.1) **5.** $2 \times 2 \times 3 \times 5$ (Sec. 4.2) **6.** $2 \times 5 \times 5 \times 7$ (Sec. 4.2)
7. 4 (Sec. 4.3) **8.** 6 (Sec. 4.3) **9.** 150 (Sec. 4.4) **10.** 144 (Sec. 4.4)

Exercises 4.1
2. 1, 2, 3, 6 **4.** 1, 2, 3, 4, 6, 12 **6.** 1, 3, 7, 21 **8.** 1, 2, 4, 8, 16, 32
10. 1, 2, 3, 6, 11, 22, 33, 66 **12.** 1, 37 **14.** 15, 49, 55, 87, 91, 105
16. 59, 61, 67, 71, 73 **18.** 45, 72, 378, 570, 585, 4530 **20.** 260,
570, 4530, 8300 **22.**

Exercises 4.2
2. 2×11 **4.** 5×7 **6.** $2 \times 3 \times 7$ **8.** 2×47 **10.** $2 \times 3 \times 3 \times 5$
12. $2 \times 2 \times 5 \times 5$ **14.** $2 \times 2 \times 2 \times 11$ **16.** $2 \times 2 \times 2 \times 2 \times 5 \times 5$
18. $2 \times 2 \times 3 \times 11$ **20.** $2 \times 3 \times 5 \times 11$ **22.** $2 \times 2 \times 5 \times 5 \times 5$
24. $2 \times 3 \times 3 \times 5 \times 13$ **26.** 3, 5 **28.** 4, 7 **30.**

Exercises 4.3
2. 3 **4.** 2 **6.** 11 **8.** 14 **10.** 1 **12.** 12 **14.** 18 **16.** 35 **18.** 18
20. 15 **22.** 16 **24.** 36 **26.**

Exercises 4.4
2. 15 **4.** 18 **6.** 36 **8.** 60 **10.** 30 **12.** 84 **14.** 50 **16.** 70
18. 72 **20.** 420 **22.** 72 **24.** 60 **26.** 84 **28.** 40 **30.** 120
32. 420 **34.**

Self-Test for Chapter 4
1. 1, 2, 3, 6, 9, 18 (Sec. 4.1) **2.** 1, 2, 3, 6, 7, 14, 21, 42 (Sec. 4.1)
3. 1, 17 (Sec. 4.1) **4.** 13, 29, 37 (Sec. 4.1) **5.** 21, 51, 91 (Sec. 4.1)
6. 41, 43, 47, 53, 59 (Sec. 4.1) **7.** 2, 3, and 5 (Sec. 4.1) **8.** None
(Sec. 4.1) **9.** 2 and 3 (Sec. 4.1) **10.** 3 and 5 (Sec. 4.1)

11. $2 \times 3 \times 7$ (Sec. 4.2) **12.** $2^3 \times 3^2$ (Sec. 4.2) **13.** $2 \times 3 \times 5 \times 7$
(Sec. 4.2) **14.** $2^3 \times 3^2 \times 11$ (Sec. 4.2) **15.** 5 (Sec. 4.3)
16. 6 (Sec. 4.3) **17.** 8 (Sec. 4.3) **18.** 1 (Sec. 4.3) **19.** 10 (Sec. 4.3)
20. 4 (Sec. 4.3) **21.** 14 (Sec. 4.3) **22.** 22 (Sec. 4.3) **23.** 35
(Sec. 4.4) **24.** 18 (Sec. 4.4) **25.** 60 (Sec. 4.4) **26.** 90 (Sec. 4.4)
27. 20 (Sec. 4.4) **28.** 24 (Sec. 4.4) **29.** 168 (Sec. 4.4) **30.** 180
(Sec. 4.4)

Summary Exercises for Chapters 1–4
1. Hundreds **2.** Hundred thousands (Sec. 1.1) **3.** Twenty-seven
thousand, four hundred twenty-eight **4.** Two hundred thousand, three
hundred five (Sec. 1.1) **5.** 37,583 **6.** 300,400 (Sec. 1.1)
7. Commutative property of addition **8.** Associative property of
addition (Sec. 1.2) **9.** 1389 **10.** 24,552 **11.** 14,722 **12.** 969
people (Sec. 1.4) **13.** 7000 **14.** 16,000 **15.** 550,000 (Sec. 1.5)
16. < **17.** > (Sec. 1.5) **18.** 4478 **19.** 18,800 **20.** 1763
21. $536 (Sec. 1.7) **22.** 18 ft (Sec. 1.8) **23.** factor **24.** multiple
25. Commutative property of multiplication **26.** Distributive property
of multiplication over addition (Sec. 2.1) **27.** 1856 **28.** 1075
29. 154,602 (Sec. 2.3) **30.** 42,657 (Sec. 2.3) **31.** $630 (Sec.
2.3) **32.** 30,960 (Sec. 2.4) **33.** 600,000 (Sec. 2.4) **34.** 28 **35.** 36
36. 36 (Sec. 2.5) **37.** $28,380 (Sec. 2.5) **38.** 40 **39.** 1000 (Sec.
2.6) **40.** 18 in.² **41.** 144 in.³ (Sec. 2.7) **42.** 0 **43.** Undefined
(Sec. 3.2) **44.** 308 r5 **45.** 55 r12 **46.** 28 mi/gal (Sec. 3.4)
47. 10 **48.** 21 (Sec. 3.5) **49.** 4 (Sec. 3.6) **50.** 88 (Sec. 3.7)
51. 1, 2, 4, 13, 26, 52 **52.** 1, 41 (Sec. 4.1) **53.** Prime: 2, 5, 7, 11,
17, 23, 43; Composite: 14, 21, 27, 39 (Sec. 4.1) **54.** 2 and 5
55. None (Sec. 4.1) **56.** $2^4 \times 3$ **57.** $2^2 \times 3 \times 5 \times 7$ (Sec. 4.2)
58. 5 **59.** 1 (Sec. 4.3) **60.** 120 (Sec. 4.4)

Cumulative Test for Chapters 1–4
1. Hundred thousands **2.** Three hundred two thousand, five hundred
twenty-five **3.** 2,430,000 **4.** Commutative property **5.** Additive
identity **6.** Associative property **7.** 966 **8.** 23,351 **9.** 5900
10. 950,000 **11.** 7700 **12.** > **13.** < **14.** 3861 **15.** 17,465
16. 905 **17.** $7579 **18.** Associative property of multiplication
19. Multiplicative identity **20.** Distributive property of multiplication
over addition **21.** 378,214 **22.** 686,000 **23.** 7695 **24.** 600,000
25. $1008 **26.** 67 r43 **27.** 103 r176 **28.** 38 **29.** 56 **30.** 36
31. 8 **32.** $58 **33.** 85 **34.** Prime: 5, 13, 17, 31; composite: 9, 22,
27, 45 **35.** 2 and 3 **36.** $2^3 \times 3 \times 11$ **37.** 12 **38.** 8 **39.** 60
40. 90

Pretest for Chapter 5
1. $\dfrac{5}{7}$ ← Numerator, ← Denominator **2.** Proper: $\dfrac{5}{6}, \dfrac{3}{8}, \dfrac{20}{21}, \dfrac{5}{11}$;
improper: $\dfrac{8}{7}, \dfrac{13}{9}, \dfrac{15}{8}, \dfrac{9}{9}, \dfrac{16}{5}$; mixed numbers: $2\dfrac{3}{5}, 7\dfrac{2}{9}, 3\dfrac{2}{7}$ (Sec. 5.2)
3. $9\dfrac{1}{4}$ (Sec. 5.3) **4.** $\dfrac{46}{7}$ (Sec. 5.3) **5.** Yes (Sec. 5.4) **6.** $\dfrac{3}{5}$
(Sec. 5.4) **7.** 14 (Sec. 5.5) **8.** $\dfrac{3}{5}$ (Sec. 5.5) **9.** $\dfrac{9}{12}, \dfrac{10}{12}$ (Sec. 5.5)
10. $\dfrac{27}{72}, \dfrac{40}{72},$ and $\dfrac{42}{72}$ (Sec. 5.5)

Exercises 5.1
2. Numerator: 5, denominator: 12 **4.** Numerator: 9, denominator: 14
6. $\dfrac{2}{3}$ **8.** $\dfrac{3}{7}$ **10.** $\dfrac{5}{8}$ **12.** $\dfrac{8}{8}$ **14.** $\dfrac{4}{4}$ **16.** $\dfrac{4}{7}$ **18.** $\dfrac{5}{9}$ **20.** $\dfrac{2}{5}$
22. Hamburgers: $\dfrac{5}{9}$; not: $\dfrac{4}{9}$ **24.** $4 \div 5$ **26.**

Exercises 5.2
2. Improper 4. Proper 6. Mixed number 8. Improper
10. Mixed number 12. Improper 14. Proper 16. Improper
18. $1\frac{2}{3}$ 20. $2\frac{1}{4}$ 22. $2\frac{2}{3}$ 24. $1\frac{2}{7}$ 26. $3\frac{3}{8}$ 28. $4\frac{1}{6}$ 30. $8\frac{2}{7}$
32. $12\frac{7}{12}$ 34. 20 36. 8 38. $\frac{17}{6}$ 40. $\frac{37}{8}$ 42. $\frac{7}{1}$ 44. $\frac{20}{9}$
46. $\frac{73}{10}$ 48. $\frac{67}{5}$ 50. $\frac{601}{4}$ 52. $\frac{1003}{4}$ 54. \$4.75 56. $9\frac{1}{2}$ gal
58. (a) $\frac{1}{2}$; (b) $\frac{3}{6} = \frac{1}{2}$

Exercises 5.3
2. Yes 4. No 6. No 8. Yes 10. Yes 12. No

Exercises 5.4
2. $\frac{4}{5}$ 4. $\frac{3}{10}$ 6. $\frac{4}{5}$ 8. $\frac{7}{8}$ 10. $\frac{2}{5}$ 12. $\frac{3}{8}$ 14. $\frac{3}{5}$ 16. $\frac{1}{3}$ 18. $\frac{2}{3}$
20. $\frac{8}{11}$ 22. $\frac{3}{7}$ 24. $\frac{7}{8}$ 26. $\frac{21}{32}$ 28. $\frac{42}{55}$ 30. $\frac{1}{10}$ 32. $\frac{1}{4}$
34. $\frac{3}{10}$ 36. $\frac{9}{10}$ 38. $\frac{13}{25}$ 40.

Calculator Exercises
2. $\frac{11}{12}$ 4. $\frac{7}{11}$ 6. $\frac{3}{5}$ 8. $\frac{8}{11}$

Exercises 5.5
2. 7 4. 8 6. 40 8. 60 10. 45 12. 15 14. 110 16. 45
18. $\frac{4}{9}, \frac{5}{11}$ 20. $\frac{8}{9}, \frac{9}{10}$ 22. $\frac{5}{18}, \frac{1}{3}, \frac{7}{12}$ 24. $\frac{13}{32}, \frac{9}{16}, \frac{5}{8}$ 26. <
28. < 30. > 32. < 34. $\frac{25}{30}, \frac{24}{30}$ 36. $\frac{15}{42}, \frac{16}{42}$ 38. $\frac{6}{30}, \frac{10}{30}, \frac{5}{30}$
40. $\frac{75}{120}, \frac{36}{120}, \frac{70}{120}$ 42. $\frac{3}{16}$ in. 44. $\frac{3}{8}$ in. 46.

Self-Test for Chapter 5
1. $\frac{5}{6}$ ← Numerator / ← Denominator (Sec. 5.1) 2. $\frac{5}{8}$ ← Numerator / ← Denominator (Sec. 5.1)
3. $\frac{3}{5}$ ← Numerator / ← Denominator (Sec. 5.1) 4. Proper: $\frac{10}{11}, \frac{1}{2}$; Improper: $\frac{9}{5}, \frac{7}{7}$,
$\frac{8}{1}$; Mixed number: $2\frac{3}{5}$ (Sec. 5.2) 5. $4\frac{1}{4}$ (Sec. 5.2) 6. $4\frac{1}{4}$ (Sec. 5.2)
7. $9\frac{1}{4}$ (Sec. 5.2) 8. 3 (Sec. 5.2) 9. 15 (Sec. 5.2) 10. $\frac{37}{7}$ (Sec. 5.2) 11. $\frac{35}{8}$ (Sec. 5.2) 12. $\frac{74}{9}$ (Sec. 5.2) 13. Yes (Sec. 5.3)
14. Yes (Sec. 5.3) 15. No (Sec. 5.3) 16. $\frac{7}{9}$ (Sec. 5.4) 17. $\frac{3}{7}$
(Sec. 5.4) 18. $\frac{8}{23}$ (Sec. 5.4) 19. 28 (Sec. 5.5) 20. 42 (Sec. 5.5)
21. 105 (Sec. 5.5) 22. $\frac{4}{7}, \frac{5}{8}$ (Sec. 5.5) 23. $\frac{5}{14}, \frac{8}{21}$ (Sec. 5.5)
24. $\frac{4}{15}, \frac{1}{3}, \frac{2}{5}$ (Sec. 5.5) 25. > (Sec. 5.5) 26. < (Sec. 5.5)
27. $\frac{8}{20}, \frac{15}{20}$ (Sec. 5.5) 28. $\frac{20}{72}, \frac{21}{72}$ (Sec. 5.5) 29. $\frac{9}{24}, \frac{6}{24}, \frac{20}{24}$
(Sec. 5.5) 30. $\frac{27}{60}, \frac{44}{60}, \frac{35}{60}$ (Sec. 5.5)

Pretest for Chapter 6
1. $\frac{15}{28}$ (Sec. 6.1) 2. $4\frac{2}{3}$ (Sec. 6.2) 3. $4\frac{1}{5}$ (Sec. 6.2) 4. $\frac{2}{3}$ (Sec. 6.3)
5. $12\frac{3}{5}$ (Sec. 6.3) 6. $\frac{4}{5}$ (Sec. 6.4) 7. $\frac{2}{21}$ (Sec. 6.4) 8. $\frac{3}{4}$ (Sec. 6.4)
9. $20\frac{5}{8}$ in.² (Sec. 6.4) 10. 14 (Sec. 6.4)

Exercises 6.1
2. $\frac{10}{63}$ 4. $1\frac{1}{15}$ 6. $\frac{8}{11}$ 8. $\frac{10}{33}$ 10. $\frac{7}{15}$ 12. $\frac{2}{15}$ 14. $3\frac{1}{5}$
16. $\frac{7}{15}$ 18. $\frac{1}{2}$

Exercises 6.2
2. $\frac{5}{9}$ 4. $1\frac{2}{5}$ 6. $1\frac{3}{5}$ 8. $1\frac{3}{10}$ 10. $5\frac{5}{6}$ 12. $4\frac{4}{21}$ 14. $4\frac{1}{2}$
16. $3\frac{8}{9}$ 18. $7\frac{1}{2}$ 20. 14 22. $16\frac{2}{3}$ 24. $6\frac{2}{3}$ 26. 95 28. $40\frac{1}{3}$
30. 21 32. $3\frac{3}{4}$ cups 34. $12\frac{1}{12}$ yd 36.

Calculator Exercises
2. $\frac{1}{6}$ 4. $\frac{8}{35}$ 6. $\frac{2}{7}$ 8. $\frac{6}{7}$ 10. $\frac{5}{24}$ 12. $\frac{6}{77}$ 14. $\frac{63}{170}$

Exercises 6.3
2. $\frac{5}{21}$ 4. $\frac{22}{75}$ 6. $\frac{4}{9}$ 8. $1\frac{1}{44}$ 10. $1\frac{3}{5}$ 12. $1\frac{1}{3}$ 14. 9 16. 30
18. $7\frac{1}{2}$ 20. $6\frac{1}{2}$ 22. 22 24. $\frac{1}{6}$ 26. $1\frac{2}{3}$ 28. 14 30. $\frac{3}{4}$ 32. 16
34. $1\frac{1}{14}$ 36. $1\frac{3}{5}$ 38. 10 40. 24 42. 2 44. \$67.50 46. \$700,
\$1050 48. $\frac{7}{15}$ 50. 91 mi 52. $8\frac{1}{3}$ acres 54. $\frac{9}{13}$ yd² 56. 27 ft³
58. $2\frac{1}{4}$ in.²

Exercises 6.4
2. $1\frac{1}{5}$ 4. $\frac{5}{6}$ 6. $\frac{55}{72}$ 8. $1\frac{7}{33}$ 10. $\frac{2}{3}$ 12. $\frac{5}{12}$ 14. 63 16. $\frac{1}{8}$
18. $\frac{1}{12}$ 20. $6\frac{2}{3}$ 22. $2\frac{2}{5}$ 24. $\frac{2}{3}$ 26. $\frac{3}{14}$ 28. $\frac{1}{4}$ 30. $3\frac{3}{10}$
32. $\frac{1}{15}$ 34. $1\frac{1}{4}$ 36. $2\frac{2}{9}$ 38. $2\frac{4}{5}$ 40. 24 bowls 42. \$16,000
44. 44 books 46. 26 shirts 48. 64 lots 50. $\frac{1}{2}$ mi²
52.

Calculator Exercises
2. $\frac{2}{3}$ 4. $\frac{4}{3}$ or $1\frac{1}{3}$ 6. $\frac{1}{9}$ 8. $\frac{3}{2}$ or $1\frac{1}{2}$ 10. $\frac{1}{14}$ 12. $\frac{4}{3}$ or $1\frac{1}{3}$

Self-Test for Chapter 6
1. $\frac{10}{21}$ (Sec. 6.1) 2. $\frac{9}{16}$ (Sec. 6.1) 3. $\frac{1}{6}$ (Sec. 6.1) 4. $2\frac{1}{7}$ (Sec. 6.1)
5. $3\frac{3}{7}$ (Sec. 6.1) 6. $9\frac{1}{5}$ (Sec. 6.1) 7. $\frac{2}{3}$ (Sec. 6.1) 8. $\frac{8}{15}$ (Sec. 6.1)

9. $\frac{4}{15}$ (Sec. 6.1)   **10.** 4 (Sec. 6.1)   **11.** $6\frac{1}{9}$ (Sec. 6.1)   **12.** $7\frac{1}{5}$

(Sec. 6.1)   **13.** $4\frac{10}{11}$ (Sec. 6.1)   **14.** $\frac{1}{2}$ (Sec. 6.1)   **15.** $\frac{7}{12}$ yd (Sec. 6.3)

16. $1.32 (Sec. 6.3)   **17.** 20 yd² (Sec. 6.3)   **18.** 190 mi (Sec. 6.3)

19. $1\frac{1}{7}$ (Sec. 6.4)   **20.** $\frac{5}{8}$ (Sec. 6.4)   **21.** $\frac{2}{27}$ (Sec. 6.4)   **22.** 16

(Sec. 6.4)   **23.** $3\frac{3}{10}$ (Sec. 6.4)   **24.** $\frac{2}{3}$ (Sec. 6.4)   **25.** $1\frac{3}{11}$ (Sec. 6.4)

26. $2\frac{2}{3}$ (Sec. 6.4)   **27.** 47 homes (Sec. 6.4)   **28.** 48 mi/h (Sec. 6.4)

29. 64 sheets (Sec. 6.4)   **30.** 48 books (Sec. 6.4)

Pretest for Chapter 7

1. $\frac{5}{7}$ (Sec. 7.1)   **2.** 48 (Sec. 7.2)   **3.** $\frac{17}{30}$ (Sec. 7.3)   **4.** $2\frac{1}{8}$ (Sec. 7.3)

5. $\frac{4}{15}$ (Sec. 7.4)   **6.** $\frac{5}{72}$ (Sec. 7.4)   **7.** $6\frac{17}{24}$ (Sec. 7.5)   **8.** $1\frac{1}{36}$

(Sec. 7.5)   **9.** $18\frac{1}{4}$ yd² (Sec. 7.5)   **10.** $4\frac{3}{8}$ points (Sec. 7.5)

Exercises 7.1

2. $\frac{5}{7}$ **4.** $\frac{9}{16}$ **6.** $\frac{1}{2}$ **8.** $\frac{1}{2}$ **10.** 1   **12.** $\frac{4}{5}$ **14.** $1\frac{2}{5}$ **16.** $1\frac{1}{2}$

18. $1\frac{1}{3}$ **20.** $\frac{7}{10}$ **22.** $1\frac{7}{12}$ **24.** $1\frac{7}{20}$ **26.** $\frac{4}{5}$ of a dollar   **28.** $\frac{2}{3}$ h

30. $\frac{8}{3}$ in. or $2\frac{2}{3}$ in.

Exercises 7.2

2. 15   **4.** 12   **6.** 30   **8.** 120   **10.** 60   **12.** 72   **14.** 144   **16.** 420

18. 12   **20.** 66   **22.** 48   **24.** 360

Exercises 7.3

2. $\frac{14}{15}$ **4.** $\frac{7}{18}$ **6.** $\frac{9}{10}$ **8.** $\frac{3}{10}$ **10.** $\frac{23}{40}$ **12.** $\frac{27}{40}$ **14.** $\frac{43}{60}$ **16.** $\frac{5}{12}$

18. $\frac{47}{75}$ **20.** $1\frac{19}{150}$ **22.** $\frac{3}{4}$ **24.** $1\frac{7}{24}$ **26.** $1\frac{11}{20}$ **28.** $\frac{7}{8}$ lb

30. $\frac{17}{30}, \frac{13}{30}$ **32.** $1\frac{11}{12}$ mi   **34.** $\frac{5}{8}$

Calculator Exercises

2. $\frac{7}{4}$ or $1\frac{3}{4}$ **4.** $\frac{97}{66}$ or $\frac{31}{66}$ **6.** $\frac{37}{56}$ **8.** $\frac{13}{16}$ **10.** $\frac{47}{40}$ or $1\frac{7}{40}$

12. $\frac{13}{15}$ **14.** $\frac{77}{72}$ or $1\frac{5}{72}$

Exercises 7.4

2. $\frac{3}{7}$ **4.** $\frac{2}{5}$ **6.** $\frac{1}{15}$ **8.** $\frac{1}{3}$ **10.** $\frac{11}{18}$ **12.** $\frac{23}{42}$ **14.** $\frac{1}{10}$ **16.** $\frac{19}{60}$

18. $\frac{13}{60}$ **20.** $\frac{1}{12}$ **22.** $\frac{1}{90}$ **24.** $\frac{11}{48}$ **26.** $\frac{1}{5}$ **28.** $\frac{5}{16}$ in.   **30.** $\frac{1}{16}$ lb

32. $\frac{7}{16}$ **34.** Yes—$\frac{1}{3}$ gal remains   **36.** $\frac{7}{12}$ **38.** $24\frac{3}{4}$ in. or $2\frac{1}{6}$ ft

Exercises 7.5

2. $11\frac{2}{3}$ **4.** 7   **6.** $10\frac{1}{3}$ **8.** $3\frac{5}{12}$ **10.** $11\frac{11}{36}$ **12.** $7\frac{1}{40}$ **14.** $10\frac{19}{20}$

16. $16\frac{1}{18}$ **18.** $12\frac{11}{24}$ **20.** $2\frac{2}{3}$ **22.** $3\frac{2}{7}$ **24.** $4\frac{19}{30}$ **26.** $3\frac{41}{60}$

28. $6\frac{17}{21}$ **30.** $2\frac{1}{3}$ **32.** $3\frac{5}{9}$ **34.** 3   **36.** $1\frac{17}{30}$ **38.** $6\frac{5}{8}$ lb

40. $11\frac{1}{12}$ mi   **42.** $5\frac{3}{8}$ in.   **44.** $5\frac{1}{8}$ in.   **46.** $\frac{7}{8}$ lb   **48.** $13\frac{3}{8}$ yd

50. $\frac{3}{4}$ in.   **52.** 4 in.   **54.** $2\frac{5}{12}$ mi   **56.** $3\frac{3}{8}$ mi   **58.** $\frac{3}{8}$

Calculator Exercises

2. $14\frac{5}{6}$ **4.** $7\frac{1}{14}$ **6.** $20\frac{19}{24}$ **8.** $180\frac{113}{135}$ **10.** $3\frac{3}{22}$ **12.** $7\frac{2}{15}$

14. $32\frac{91}{180}$ **16.** $12\frac{1}{15}$

Self-Test for Chapter 7

1. $\frac{9}{10}$ (Sec. 7.1)   **2.** $\frac{2}{3}$ (Sec. 7.1)   **3.** $1\frac{1}{3}$ (Sec. 7.1)   **4.** 60 (Sec. 7.2)

5. 36 (Sec. 7.2)   **6.** $\frac{4}{5}$ (Sec. 7.3)   **7.** $\frac{25}{42}$ (Sec. 7.3)   **8.** $\frac{19}{24}$ (Sec. 7.3)

9. $1\frac{11}{60}$ (Sec. 7.3)   **10.** $1\frac{23}{40}$ (Sec. 7.3)   **11.** $\frac{7}{12}$ (Sec. 7.3)   **12.** $1\frac{5}{12}$

cups (Sec. 7.3)   **13.** $\frac{1}{3}$ (Sec. 7.4)   **14.** $\frac{1}{9}$ (Sec. 7.4)   **15.** $\frac{23}{30}$ (Sec. 7.4)

16. $\frac{1}{4}$ h (Sec. 7.4)   **17.** $7\frac{7}{10}$ (Sec. 7.5)   **18.** $10\frac{1}{4}$ (Sec. 7.5)   **19.** $7\frac{11}{12}$

(Sec. 7.5)   **20.** $12\frac{3}{40}$ (Sec. 7.5)   **21.** $1\frac{3}{4}$ (Sec. 7.5)   **22.** $1\frac{11}{18}$ (Sec. 7.5)

23. $3\frac{23}{24}$ (Sec. 7.5)   **24.** $1\frac{8}{15}$ (Sec. 7.5)   **25.** $9\frac{1}{7}$ (Sec. 7.5)

26. $13\frac{11}{20}$ (Sec. 7.5)   **27.** $5\frac{3}{4}$ h (Sec. 7.5)   **28.** $24\frac{3}{4}$ in. (Sec. 7.5)

29. $4\frac{3}{4}$ in. (Sec. 7.5)   **30.** $30\frac{5}{6}$ yd (Sec. 7.5)

Summary Exercises for Chapters 5–7

1. Fraction $\frac{3}{8}$, numerator 3, denominator 8   **2.** Fraction $\frac{5}{6}$ numerator 5,

denominator 6 (Sec. 5.1)   **3.** Proper: $\frac{2}{3}, \frac{7}{10}$ improper: $\frac{5}{4}, \frac{45}{8}, \frac{7}{7}, \frac{9}{1}$,

$\frac{12}{5}$ mixed numbers: $2\frac{3}{7}, 3\frac{4}{5}, 5\frac{2}{9}$ (Sec. 5.2)   **4.** $6\frac{5}{6}$ **5.** 4   **6.** $\frac{61}{8}$

7. $\frac{43}{10}$ (Sec. 5.3)   **8.** No   **9.** Yes (Sec. 5.4)   **10.** $\frac{2}{3}$ **11.** $\frac{3}{5}$ **12.** $\frac{7}{9}$

13. $\frac{16}{21}$ (Sec. 5.5)   **14.** 15   **15.** 32 (Sec. 5.6)   **16.** $\frac{7}{12}, \frac{5}{8}$ **17.** $\frac{7}{10}$,

$\frac{4}{5}, \frac{5}{6}$ **18.** >   **19.** =   **20.** <   **21.** $\frac{4}{24}, \frac{21}{24}$ **22.** $\frac{36}{120}, \frac{75}{120}, \frac{70}{120}$

(Sec. 5.6)   **23.** $\frac{1}{9}$ **24.** $\frac{1}{6}$ (Sec. 6.1)   **25.** $1\frac{1}{2}$ **26.** $2\frac{1}{8}$ (Sec. 6.1)

27. $9\frac{3}{5}$ **28.** $11\frac{1}{3}$ **29.** 8 (Sec. 6.3)   **30.** $\frac{2}{3}$ **31.** $\frac{5}{6}$ **32.** $\frac{3}{16}$

(Sec. 6.4)   **33.** $1\frac{1}{2}$ **34.** $\frac{3}{7}$ (Sec. 6.4)   **35.** 220 mi   **36.** $204

37. 52 mi/h   **38.** 48 lots (Sec. 6.4)   **39.** $\frac{7}{9}$ (Sec. 7.1)   **40.** $1\frac{3}{5}$

(Sec. 7.1)   **41.** $1\frac{4}{9}$ **42.** $\frac{31}{36}$ (Sec. 7.2)   **43.** $1\frac{41}{60}$ (Sec. 7.3)   **44.** $\frac{1}{4}$

45. $\frac{7}{18}$ **46.** $\frac{7}{60}$ **47.** $\frac{7}{54}$ **48.** $\frac{35}{72}$ (Sec. 7.4)   **49.** $10\frac{2}{7}$ **50.** $9\frac{37}{60}$

51. $9\frac{17}{24}$ **52.** $4\frac{1}{3}$ **53.** $6\frac{1}{24}$ **54.** $2\frac{19}{24}$ **55.** $4\frac{7}{10}$ (Sec. 7.5)

56. $69\frac{1}{16}$ in.   **57.** $19\frac{9}{16}$ in.   **58.** $1\frac{3}{8}$ in.   **59.** $53\frac{9}{16}$ in.   **60.** Yes.

$\frac{5}{12}$ yd (Sec. 7.5)

Cumulative Test for Chapters 5–7

1. Fraction: $\frac{5}{8}$ numerator: 5 denominator: 8   **2.** Proper: $\frac{7}{12}, \frac{3}{7}$

improper: $\frac{10}{8}, \frac{9}{9}, \frac{7}{1}$ mixed number: $3\frac{1}{5}, 2\frac{2}{3}$ **3.** $2\frac{4}{5}$ **4.** 4   **5.** $\frac{13}{3}$

6. $\frac{63}{8}$ **7.** Yes   **8.** No   **9.** $\frac{2}{3}$ **10.** $\frac{3}{8}$ **11.** $\frac{6}{11}, \frac{5}{9}$ **12.** $\frac{8}{15}, \frac{3}{5}, \frac{7}{10}$

13. $\frac{15}{24}, \frac{14}{24}$ **14.** $\frac{24}{36}, \frac{20}{36}, \frac{27}{36}$ **15.** $\frac{8}{27}$ **16.** $\frac{4}{15}$ **17.** $5\frac{2}{5}$ **18.** $22\frac{2}{3}$

19. $\frac{3}{4}$ **20.** $1\frac{1}{3}$ **21.** $4\frac{1}{2}$ **22.** $\frac{5}{6}$ **23.** $1\frac{1}{2}$ **24.** $540   **25.** 88 sheets

26. $\frac{4}{5}$ **27.** $\frac{61}{75}$ **28.** $1\frac{31}{40}$ **29.** $\frac{1}{2}$ **30.** $\frac{5}{36}$ **31.** $\frac{5}{9}$ **32.** $6\frac{2}{7}$

33. $8\frac{1}{24}$ **34.** $4\frac{5}{9}$ **35.** $4\frac{1}{24}$ **36.** $3\frac{5}{8}$ **37.** $3\frac{13}{24}$ **38.** $14\frac{19}{30}$ h

39. $\frac{5}{8}$ in.   **40.** $1\frac{11}{12}$ h   **41.**

Pretest for Chapter 8

1. Thousandths (Sec. 8.1)   **2.** 2,371; two and three hundred seventy-one thousandths (Sec. 8.1)   **3.** (a) 14.28 (b) 63.29 (Sec. 8.3)
4. 2.375 (Sec. 8.4)   **5.** $2.36 (Sec. 8.4)   **6.** (a) 0.86037 (b) 536.2
(Sec. 8.5)   **7.** 2.36 (Sec. 8.2)   **8.** $11.95 (Sec. 8.5)   **9.** 25.12 yd
(Sec. 8.6)   **10.** (a) 153.86 in² (b) 7.5 in² (Sec. 8.6) (c) 6 ft²

Exercises 8.1

2. Tenths   **4.** Hundred thousandths   **6.** 0.371   **8.** 3.5   **10.** 7.0431
12. Three hundred seventy-one thousandths   **14.** Two hundred fifty-one ten thousandths   **16.** Twenty-three and fifty-six thousandths

18. 0.0253   **20.** 12.245   **22.** $\frac{765}{100,000}$ **24.** $4\frac{171}{10,000}$ **26.** <

28. >   **30.** <   **32.** =   **34.**

Exercises 8.2

2. 6.79   **4.** 5.8   **6.** 2.358   **8.** 1.5   **10.** 0.8536   **12.** 52.873
14. 12.547   **16.** 503.82   **18.** 56.3583   **20.** 56.36

Exercises 8.3

2. 3.22   **4.** 3.735   **6.** 28.29   **8.** 31.135   **10.** 23.675   **12.** 25.7717
14. 2.209   **16.** 46.86   **18.** 42.15   **20.** 280.101   **22.** 1.718
24. 67.28   **26.** 8.2 mi   **28.** 5.725 in.   **30.** $173.15   **32.** $241.24
34. 76.84 yds   **36.** 11.535 mi   **38.** $50   **40.** $1008   **42.**

Exercises 8.4

2. 3.03   **4.** 28.78   **6.** 18.497   **8.** 2.63   **10.** 2.99   **12.** 36.85
14. 6.15   **16.** 32.375   **18.** 26.125   **20.** 6.88   **22.** 6.18   **24.** 7.62
26. 0.0575 in.   **28.** $15.42   **30.** $25.98   **32.** 0.89 in. below normal
34. $256.71, $238.71, $152.93, $403.38, $202.14   **36.** $1110.24,
$554.47, $681.24, $627.35   **38.**
40.

2.4	8.4	7.2
10.8	6	1.2
4.8	3.6	9.6

Calculator Exercises

2. 21.876   **4.** 12,807.13   **6.** 5.175   **8.** 10.385   **10.** 2.1925
12. $77.64 overdrawn   **14.** $974.46

Exercises 8.5

2. 27.95   **4.** 42.32   **6.** 1984.5   **8.** 4.275   **10.** 16.697   **12.** 0.5304
14. 3.6666   **16.** 21.576   **18.** 0.665   **20.** 0.046368   **22.** 204.16
24. 0.01918   **26.** $1483.80   **28.** $252.45   **30.** $18.85   **32.** $337.60
34. $261.30   **36.** 18.6 gal   **38.** $11,634.24   **40.** $24

Calculator Exercises

2. 0.1067   **4.** 755.811   **6.** 196.66995   **8.** 0.000512   **10.** 0.271441
12. $224.64   **14.** $182.25

Exercises 8.6

2. 15.7 ft   **4.** 23.6 ft   **6.** 22 ft   **8.** 24 in.   **10.** 35.7 in.   **12.** 113 ft²
14. 201 ft²   **16.** $7\frac{1}{14}$ in.²   **18.** $9.42   **20.** $58.88   **22.** 127.2 ft²
24. 44.1 ft²   **26.** 21.5 in.²

Exercises 8.7

2. 32 in.²   **4.** 35 in.²   **6.** 33 ft²   **8.** 84.5 yd²   **10.** 48 yd²
12. 2.5 acres   **14.** $360   **16.** $900   **18.**   **20.**
22.

Exercises 8.8

2. 89.5   **4.** 24.1   **6.** 580   **8.** 0.25   **10.** 9500   **12.** 23,420   **14.** 360
16. 5800   **18.** 530 cm   **20.** $178.00

Calculator Exercises

2. 4128   **4.** 8,163,000   **6.** 52,340,000   **8.** 412,340   **10.** 61,356

Self-Test for Chapter 8

1. 0.431 (Sec. 8.1)   **2.** 5.13 (Sec. 8.1)   **3.** Four hundred thirty-one thousandths (Sec. 8.1)   **4.** Five and thirteen hundredths (Sec. 8.1)
5. > (Sec. 8.1)   **6.** < (Sec. 8.1)   **7.** 2.6 (Sec. 8.2)   **8.** 23.34 (Sec. 8.2)
9. 2.208 (Sec. 8.3)   **10.** 3.521 (Sec. 8.3)   **11.** 40.764 (Sec. 8.3)
12. 10.805 (Sec. 8.3)   **13.** 50.8 gal (Sec. 8.3)   **14.** 10.15 in. (Sec. 8.3)
15. 2.58 (Sec. 8.4)   **16.** 4.875 (Sec. 8.4)   **17.** 6.515 (Sec. 8.4)
18. 937.3 mi (Sec. 8.4)   **19.** $279.57 (Sec. 8.4)   **20.** 1.75 in. (Sec. 8.4)
21. 21.46 (Sec. 8.5)   **22.** 1.5718 (Sec. 8.5)   **23.** 0.0094 (Sec. 8.5)
24. 37.41 cm² (Sec. 8.5)   **25.** $202.10 (Sec. 8.5)   **26.** $470 (Sec. 8.5)
27. 41.7 in. (Sec. 8.6)   **28.** 27.0 ft (Sec. 8.6)   **29.** 139.32 in.²
(Sec. 8.7)   **30.** 8.325 ft² (Sec. 8.7)   **31.** 5.4 (Sec. 8.8)   **32.** 84,320
(Sec. 8.8)

Pretest for Chapter 9

1. 4.25 (Sec. 9.1)   **2.** 1.67 (Sec. 9.1)   **3.** $5.68 (Sec. 9.1)   **4.** (a) 3.42
(b) 2.435 (Sec. 9.2)   **5.** $7.60 (Sec. 9.2)   **6.** 0.0534 (Sec. 9.3)

7. (a) 0.375 (b) 0.29 (Sec. 9.4)   **8.** $\frac{39}{50}$ (Sec. 9.5)   **9.** $8^2 + 15^2 = 289$

and $17^2 = 289$ (Sec. 9.6)   **10.** 10 (Sec. 9.6)

Exercises 9.1

2. 5.49   **4.** 0.92   **6.** 0.345   **8.** 4.384   **10.** 4.6   **12.** 3.26
14. 0.345   **16.** 2.816   **18.** 0.66   **20.** 0.093   **22.** 0.047   **24.** 0.09
26. $24.58   **28.** $0.80 or 80¢   **30.** $15.78   **32.** $36.72   **34.** 48.1
36. 17.7

Exercises 9.2
2. 13.55 **4.** 4.6 **6.** 0.565 **8.** 21.4 **10.** 6.215 **12.** 2.45
14. 0.215 **16.** 12.8 **18.** 2.76 **20.** 0.254 **22.** 2.36 **24.** 0.018
26. 2.37 **28.** 0.295 **30.** $9.42 **32.** 640 nails **34.** 15.5 mi/gal
36. 16.4 ft³ **38.** 4.65 in. **40.** Mean: $260.75 **42.** Mean: 111.67
44. **46.**

Calculator Exercises
2. 58.5 **4.** 62.5 **6.** 23.48 **8.** 4.9 **10.** 1.59 **12.** 2.835
14. 230 lots **16.** 31.4 mi/gal

Exercises 9.3
2. 0.51 **4.** 0.03817 **6.** 0.0841 **8.** 0.0072 **10.** 0.0036
12. 0.00573 **14.** $235 **16.** 0.75 L **18.** 76¢

Exercises 9.4
2. 0.8 **4.** 0.3 **6.** 0.125 **8.** 0.55 **10.** 0.4375 **12.** 0.53125
14. 0.58 **16.** $0.0\overline{5}$ **18.** $0.\overline{27}$ **20.** 7.75

Exercises 9.5
2. $\frac{3}{10}$ **4.** $\frac{3}{5}$ **6.** $\frac{97}{100}$ **8.** $\frac{379}{1000}$ **10.** $\frac{3}{4}$ **12.** $\frac{13}{20}$ **14.** $\frac{29}{250}$
16. $\frac{9}{40}$ **18.** $\frac{23}{40}$ **20.** $\frac{67}{1000}$ **22.** $\frac{17}{400}$ **24.** $5\frac{7}{10}$ **26.** $3\frac{31}{100}$
28. $15\frac{7}{20}$ **30.** <

Calculator Exercises
2. 0.6875 **4.** 0.29 **6.** 0.147 **8.** $0.\overline{63}$ **10.** 3.8 **12.** 8.1875
14. $\frac{55}{100}=\frac{11}{20}$ **16.** $\frac{1}{10}$ **18.** $\frac{125}{1000}=\frac{1}{8}$

Exercises 9.6
2. 11 **4.** 14 **6.** y **8.** No **10.** Yes **12.** Yes **14.** 13 **16.** 24
18. c **20.** c

Calculator Exercises
2. 12 **4.** 32 **6.** 28 **8.** 73 **10.** 58 **12.** 5.6 **14.** 6.5 **16.** 15.8

Self-Test for Chapter 9
1. 2.75 (Sec. 9.1) **2.** 2.385 (Sec. 9.1) **3.** 0.46 (Sec. 9.1) **4.** 0.145
(Sec. 9.1) **5.** $23.28 (Sec. 9.1) **6.** 0.65 in. (Sec. 9.1) **7.** 6.7 (Sec.
9.2) **8.** 3.225 (Sec. 9.2) **9.** 2.84 (Sec. 9.2) **10.** 5.53 (Sec. 9.2)
11. 2.02 (Sec. 9.2) **12.** 0.541 (Sec. 9.2) **13.** 30.3 mi/gal (Sec. 9.2)
14. 29 shirts (Sec. 9.2) **15.** 45 lots (Sec. 9.2) **16.** 3.857 (Sec. 9.3)
17. 0.02847 (Sec. 9.3) **18.** 0.003795 (Sec. 9.3) **19.** $5.37 (Sec. 9.3)
20. 0.828 L (Sec. 9.3) **21.** 0.875 (Sec. 9.4) **22.** 2.5625 (Sec. 9.4)
23. $0.\overline{63}$ (Sec. 9.4) **24.** $\frac{29}{100}$ (Sec. 9.4) **25.** $\frac{14}{25}$ (Sec. 9.4) **26.** $\frac{313}{400}$
(Sec. 9.4) **27.** $\frac{49}{100}$ (Sec. 9.5) **28.** $\frac{3}{8}$ (Sec. 9.5) **29.** $7\frac{1}{5}$ (Sec. 9.5)
30. $23\frac{39}{50}$ (Sec. 9.5) **31.** 13 (Sec. 9.6) **32.** 16 (Sec. 9.6) **33.** 20
(Sec. 9.6) **34.** 10 (Sec. 9.6) **35.** 24 (Sec. 9.6)

Summary Exercises for Chapters 8–9
1. Hundredths **2.** Ten thousandths (Sec. 8.1) **3.** 0.37 **4.** 0.0307
(Sec. 8.1) **5.** Seventy-one thousandths (Sec. 8.1) **6.** Twelve and
thirty-nine hundredths **7.** 4.5 **8.** 400.037 (Sec. 8.1) **9.** >

10. = **11.** < **12.** > (Sec. 8.1) **13.** 5.84 **14.** 9.572 **15.** 4.876
(Sec. 8.2) **16.** 45.94 cm² (Sec. 8.2) **17.** 3.47 **18.** 18.852
19. 37.728 **20.** 20.533 (Sec. 8.3) **21.** 22.2 mi **22.** 6.15 cm
(Sec. 8.3) **23.** 23.46 **24.** 4.245 **25.** 1.075 **26.** 6.62 (Sec.
8.4) **27.** $61.75 **28.** $18.93 (Sec. 8.4) **29.** 16.416 **30.** 0.000261
31. 69.44 **32.** 0.0012275 (Sec. 8.5) **33.** $271.15 **34.** $287.50
35. $152.10 (Sec. 8.5) **36.** 37.7 ft **37.** 22.9 in. **38.** 314 ft² (Sec.
8.6) **39.** 750 ft² **40.** 400 in.² (Sec. 8.7) **41.** 180 ft², 20 yd²
42. $787.50 (Sec. 8.7) **43.** 52 **44.** 450 (Sec. 8.7) **45.** $5742 (Sec.
8.7) **46.** 0.385 **47.** 4.65 **48.** 0.322 (Sec. 9.1) **49.** $23.45
50. 39.3 mi/gal (Sec. 9.1) **51.** 2.66 **52.** 8.45 **53.** 1.3 **54.** 0.089
(Sec. 9.2) **55.** 54 lots **56.** 29.8 mi/gal (Sec. 9.2) **57.** 0.76
58. 0.0807 **59.** 0.0457 (Sec. 9.3) **60.** $7.09 (Sec. 9.3) **61.** 0.4375
62. 0.429 **63.** $0.2\overline{6}$ **64.** 3.75 (Sec. 9.4) **65.** $\frac{21}{100}$ **66.** $\frac{21}{250}$
67. $5\frac{7}{25}$ (Sec. 9.5) **68.** 18 **69.** 28 **70.** 55 (Sec. 9.6)

Cumulative Test for Chapters 8–9
1. Ten thousandths **2.** 0.049 **3.** Two and fifty-three hundredths
4. 12.017 **5.** < **6.** > **7.** 16.64 **8.** 47.253 **9.** 12.803
10. 50.2 gal **11.** 10.54 **12.** 24.375 **13.** 3.888 **14.** $3.06
15. 17.437 **16.** 0.02793 **17.** 1.4575 **18.** 7.525 in.² **19.** 735
20. 12,570 **21.** $543 **22.** 0.598 **23.** 23.57 **24.** 10.05 ft
25. 0.465 **26.** 2.35 **27.** 0.051 **28.** 2.55 **29.** 2.385 **30.** 7.35
31. 0.067 **32.** 32 lots **33.** 0.004983 **34.** 0.00523 **35.** $573.40
36. 0.4375 **37.** 0.429 **38.** $0.\overline{63}$ **39.** $\frac{9}{125}$ **40.** $4\frac{11}{25}$

Pretest for Chapter 10
1. $\frac{7}{10}$ (Sec. 10.1) **2.** $\frac{4}{3}$ (Sec. 10.1) **3.** Yes (Sec. 10.2) **4.** No (Sec.
10.2) **5.** 10 (Sec. 10.3) **6.** 15 (Sec. 10.3) **7.** 12 (Sec. 10.3)
8. $6.30 (Sec. 10.3) **9.** 32 (Sec. 10.3) **10.** 11 gal (Sec. 10.3)

Exercises 10.1
2. $\frac{5}{4}$ **4.** $\frac{5}{12}$ **6.** $\frac{4}{3}$ **8.** $\frac{5}{8}$ **10.** $\frac{23}{36}$ **12.** $\frac{10}{9}$ **14.** $\frac{2}{3}$ **16.** $\frac{5}{6}$
18. $\frac{7}{48}$ **20.** $\frac{2}{9}$ **22.** $\frac{1}{3}$ **24.** $\frac{14}{15}$ **26.** $\frac{7\text{ cups}}{4\text{ loaves}}$ **28.** $\frac{\$53}{8\text{ h}}$ **30.** $\frac{3}{5}$
32. $\frac{36\text{ mi}}{1\text{ gal}}$ **34.** $\frac{29}{98}$ **36.** $\frac{1}{8}$ **38.** $\frac{3}{50}$ **40.** $\frac{4}{3}$ **42.** $\frac{5}{3}$

Exercises 10.2
2. Means: 4, 6; extremes: 3, 8 **4.** Means: x, 20; extremes: 5,
24 **6.** Means: 8, 15; extremes: 3, 40 **8.** Means: 7, n; extremes: 4,
28 **10.** True **12.** True **14.** False **16.** True **18.** False **20.** True
22. False **24.** True **26.** True **28.** True **30.** False **32.** True
34. True **36.** False **38.** False **40.** True **42.** True
44.

Exercises 10.3
2. 2 **4.** 6 **6.** 4 **8.** 25 **10.** 30 **12.** 3 **14.** 5 **16.** 14 **18.** 28
20. 75 **22.** 27 **24.** 15 **26.** 3 **28.** 22 **30.** 30 **32.** 3 **34.** 100
36. 36 **38.** 100 **40.** 2 **42.** $1.44 **44.** $11.16 **46.** $3.00
48. 2500 women **50.** 110 chairs **52.** $2100 **54.** 240 mi
56. $13.75 **58.** $6000 **60.** 9 lbs **62.** 10 ft **64.** 14 ft **66.** 1 in.
68. 24 ft **70.** $60 **72.** 20 km **74.** 900 ft² **76.** 90 rolls
78. 48,000 people **80.** 15 **82.** 9 **84.**

Calculator Exercises

2. 27.5 **4.** 15.22 **6.** 23.24 **8.** $263.77 **10.** 11 parts
12. 63.20 items **14.** 1.24 in. **16.** (*a*) **18.** (*b*) **20.** (*c*) **22.** (*d*)

Self-Test for Chapter 10

1. $\frac{7}{19}$ (Sec. 10.1) **2.** $\frac{5}{3}$ (Sec. 10.1) **3.** $\frac{2}{3}$ (Sec. 10.1) **4.** $\frac{1}{12}$ (Sec. 10.1) **5.** $\frac{26}{33}, \frac{26}{7}$ (Sec. 10.1) **6.** Means: 13, 18; extremes: 6, 39 (Sec. 10.2) **7.** Means: *a*, 15; extremes: 5, 21 (Sec. 10.2) **8.** True (Sec. 10.2) **9.** False (Sec. 10.2) **10.** True (Sec. 10.2) **11.** False (Sec. 10.2) **12.** 2 (Sec. 10.3) **13.** 12 (Sec. 10.3) **14.** 25 (Sec. 10.3) **15.** 45 (Sec. 10.3) **16.** 20 (Sec. 10.3) **17.** 18 (Sec. 10.3) **18.** 3 (Sec. 10.3) **19.** 48 (Sec. 10.3) **20.** 6 (Sec. 10.3) **21.** 16 (Sec. 10.3) **22.** $2.28 (Sec. 10.3) **23.** 644 points (Sec. 10.3) **24.** 576 mi (Sec. 10.3) **25.** 3000 no votes (Sec. 10.3) **26.** 420 mi (Sec. 10.3) **27.** 10 in. (Sec. 10.3) **28.** 12 ft (Sec. 10.3) **29.** 600 mufflers (Sec. 10.3) **30.** 24 tsp (Sec. 10.3)

Pretest for Chapter 11

1. $\frac{7}{100}$ (Sec. 11.1) **2.** 0.23 (Sec. 11.1) **3.** 3.5% or $3\frac{1}{2}$% (Sec. 11.2) **4.** 80% (Sec. 11.2) **5.** 63 (Sec. 11.3) **6.** 9% (Sec. 11.3) **7.** 600 (Sec. 11.4) **8.** $560 (Sec. 11.5) **9.** 5% (Sec. 11.5) **10.** $1200 (Sec. 11.5)

Exercises 11.1

2. 60% **4.** 40% **6.** 5% **8.** 29% **10.** 45% **12.** 36% **14.** 85% **16.** 62% **18.** $\frac{17}{100}$ **20.** $\frac{1}{5}$ **22.** $\frac{12}{25}$ **24.** $\frac{13}{25}$ **26.** $\frac{7}{20}$ **28.** $\frac{12}{25}$ **30.** $1\frac{2}{5}$ **32.** $1\frac{1}{3}$ **34.** 0.7 **36.** 0.75 **38.** 0.27 **40.** 0.07 **42.** 2.5 **44.** 1.6 **46.** 0.105 **48.** 0.035 **50.** 0.005 **52.** 0.0825 **54.** $\frac{21}{25}$

56. *a*, 4; *b*, 3; *c*, 6; *d*, 5; *e*, 1; *f*, 2

Exercises 11.2

2. 9% **4.** 13% **6.** 63% **8.** 45% **10.** 30% **12.** 60% **14.** 250% **16.** 500% **18.** 9.5% or $9\frac{1}{2}$% **20.** 8.5% or $8\frac{1}{2}$% **22.** 0.8% or $\frac{4}{5}$% **24.** 0.1% or $\frac{1}{10}$% **26.** 80% **28.** 50% **30.** 75% **32.** 87.5% or $87\frac{1}{2}$% **34.** 120% **36.** $66\frac{2}{3}$% **38.** 18.75% or $18\frac{3}{4}$% **40.** 45.5% **42.** 56.6%

Exercises 11.3

2. (*A*) 150, (*R*) 20%, (*B*) 750 **4.** (*A*) 200, (*R*) 40%, (*B*) 500 **6.** (*A*) 80, (*R*) unknown, (*B*) 400 **8.** (*A*) 30, (*R*) unknown, (*B*) 150 **10.** (*A*) Unknown, (*R*) 60%, (*B*) 250 **12.** (*A*) 150, (*R*) 75%, (*B*) unknown **14.** (*A*) $209, (*R*) 22%, (*B*) unknown **16.** (*A*) Unknown, (*R*) 80%, (*B*) 16 **18.** (*A*) 75, (*R*) unknown, (*B*) 750 **20.** (*A*) Unknown, (*R*) 6.5%, (*B*) $5000 **22.**

Exercises 11.4

2. 80 **4.** 480 **6.** 90 **8.** 6% **10.** 7% **12.** 14% **14.** 600 **16.** 1200 **18.** 350 **20.** 690 **22.** 3200 **24.** 130% **26.** 120% **28.** 250 **30.** 900 **32.** 66 **34.** 17.5 **36.** 551 **38.** $33\frac{1}{3}$%

40. 9.75% **42.** 8.5% **44.** 400 **46.** 5000 **48.** 350 **50.** 750 **52.** 120 **54.** 6000 **56.**

Exercises 11.5

2. 54 mL **4.** $5100 **6.** 11% **8.** 13% **10.** $200 **12.** $36,000 **14.** $190 **16.** 84 questions **18.** 16.5% **20.** 16% **22.** 3600 students **24.** $2400 **26.** 963 **28.** $6480 **30.** 8.5% **32.** 9% **34.** $600 **36.** $2500 **38.** 6840 **40.** $337.50 **42.** $3434.70 **44.** $5955.08 **46.** 7550 thousand bbl **48.** 189 million

Calculator Exercises

2. 575 **4.** 21.3% **6.** 86.24 **8.** 16,720 **10.** $53.27 **12.** 8.8% **14.** $8500 **16.** $62.99 **18.** $34.77 **20.** $20,442.50 **22.** $5875.20

Self-Test for Chapter 11

1. 80% (Sec. 11.1) **2.** $\frac{7}{100}$ (Sec. 11.1) **3.** $\frac{72}{100}$ or $\frac{18}{25}$ (Sec. 11.1) **4.** 0.42 (Sec. 11.2) **5.** 0.06 (Sec. 11.2) **6.** 1.6 (Sec. 11.2) **7.** 3% (Sec. 11.2) **8.** 4.2% (Sec. 11.2) **9.** 40% (Sec. 11.2) **10.** 62.5% (Sec. 11.2) **11.** (*A*) 50; (*R*%) 25%; (*B*) 200 (Sec. 11.3) **12.** (*A*) What is: (*R*%) 8%; (*B*) 500 (Sec. 11.3) **13.** (*R*%) 6%; (*A*) $30; (*B*) amount of purchase (Sec. 11.3) **14.** 11.25 (Sec. 11.4) **15.** 500 (Sec. 11.4) **16.** 750 (Sec. 11.4) **17.** 20% (Sec. 11.4) **18.** 7.5% (Sec. 11.4) **19.** 175% (Sec. 11.4) **20.** 800 (Sec. 11.4) **21.** 300 (Sec. 11.4) **22.** $4.96 (Sec. 11.5) **23.** 60 questions (Sec. 11.5) **24.** $70.20 (Sec. 11.5) **25.** 12% (Sec. 11.5) **26.** 24% (Sec. 11.5) **27.** 12% (Sec. 11.5) **28.** $18,000 (Sec. 11.5) **29.** 6400 students (Sec. 11.5) **30.** $8500 (Sec. 11.5)

Summary Exercises for Chapters 10–11

1. $\frac{4}{17}$ **2.** $\frac{2}{3}$ **3.** $\frac{2}{5}$ **4.** $\frac{7}{36}$ **5.** $\frac{3}{4}$ (Sec. 10.1) **6.** False **7.** True **8.** True **9.** False **10.** True **11.** False (Sec. 10.2) **12.** 2 **13.** 4 **14.** 4 **15.** 6 **16.** 16 **17.** 180 **18.** 30 **19.** 100 **20.** 0.5 (Sec. 10.3) **21.** $67.50 **22.** 256 first-year students **23.** 15 in. **24.** 220 drives **25.** 28 parts **26.** 120 mi **27.** 140 g **28.** 960 ft^2 **29.** 24 oz (Sec. 10.4) **30.** 75% **31.** $\frac{1}{50}$ **32.** $\frac{1}{5}$ **33.** $1\frac{1}{2}$ **34.** 0.75 **35.** 0.135 **36.** 2.25 (Sec. 11.1) **37.** 6% **38.** 37.5% **39.** 240% **40.** 40% **41.** $266\frac{2}{3}$% (Sec. 11.2) **42.** 2000 **43.** 140% **44.** 330 **45.** 12.5% **46.** 5000 **47.** 75 **48.** 66.5 **49.** 75 **50.** 600 (Sec. 11.4) **51.** $1800 **52.** 22% **53.** 7.5% **54.** $10,200 **55.** $102 **56.** $1300 **57.** 720 students **58.** 6.5% **59.** $3157.50 **60.** $1100 before; $1199 after (Sec. 11.5)

Cumulative Test for Chapters 10–11

1. $\frac{2}{3}$ **2.** $\frac{3}{5}$ **3.** $\frac{2}{9}$ **4.** False **5.** True **6.** True **7.** 21 **8.** 3 **9.** 60 **10.** 112.5 **11.** $66 **12.** 555 mi **13.** 48 employees **14.** 12 in. **15.** 120 mi **16.** $140 **17.** 18 oz **18.** 70% **19.** $\frac{9}{20}$ **20.** $1\frac{3}{4}$ **21.** 0.55 **22.** 0.175 **23.** 12.5% **24.** 0.3% or $\frac{3}{10}$% **25.** 80% **26.** 62.5% **27.** 600 **28.** 39 **29.** 13% **30.** 255 **31.** 150 **32.** 125% **33.** 12.5% **34.** $17,500 **35.** She had 70%, yes **36.** 1650 students **37.** $1800 **38.** $110,200 **39.** 7.5% **40.** $323

Pretest for Chapter 12

1.  (Sec. 12.1)

2. (Sec. 12.1)

3. (Sec. 12.1)

4. (Sec. 12.1)

5. Approximately 2500 (Sec. 12.2) **6.** −2000 (Sec. 12.2)
7. 1970–1975; 2000 (Sec. 12.2) **8.** $17,200,000 (Sec. 12.3)
9. $23,200,000 (Sec. 12.3) **10.** $22,800,000 (Sec. 12.3)

Exercises 12.1

2.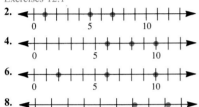

4.

6.

8.

10.

12.

14.

16.

18.

20.

22.

Exercises 12.2
2. $1050 **4.** 1991 **6.** 1992 **8.** 1993 **10.** 400
12. 417 (rounded)

Exercises 12.3
2. $60,000 **4.** $120,000

Exercises 12.4
2.
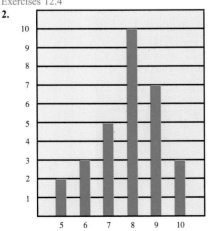

1. 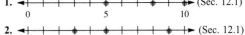 (Sec. 12.1)

2. (Sec. 12.1)

3. (Sec. 12.1)

4. (Sec. 12.1)

5. December (Sec. 12.2) **6.** August and September (Sec. 12.2)
7. 16,000 (Sec. 12.2) **8.** 30,000 (Sec. 12.2) **9.** 10,000 (Sec. 12.2)
10. 16,000 (Sec. 12.2) **11.** 1995–1996 (Sec. 12.2) **12.** 15% (Sec. 12.3) **13.** 45% (Sec. 12.3) **14.** 40% (Sec. 12.3) **15.** 45% (Sec. 12.3)

Pretest for Chapter 13
1. 108 (Sec. 13.1) **2.** 6 min 30 seconds (Sec. 13.1) **3.** 13 gal 3 qt (Sec. 13.2) **4.** 5000 (Sec. 13.3) **5.** 700 (Sec. 13.3) **6.** 8 (Sec. 13.3)
7. 6 (Sec. 13.4) **8.** 33 (Sec. 13.4) **9.** 2 (Sec. 13.5) **10.** 8000 (Sec. 13.5)

Exercises 13.1
2. 36 **4.** 5 **6.** 112 **8.** 80 **10.** 360 **12.** 660 **14.** 26,400 **16.** 4
18. 9 **20.** 6 **22.** 180 **24.** 15 **26.** 8 **28.** 9 **30.** 12 **32.** 36
34. 3 **36.** 168 **38.** 21.5 **40.** 1.25 **42.** $3\frac{2}{3}$ **44.** 56.64
46. 40 billion lb **48.**

Exercises 13.2
2. 7 lb 4 oz **4.** 8 yd 14 in. **6.** 4 min 50 s **8.** 10 h 20 min
10. 13 ft 5 in. **12.** 16 yd **14.** 25 ft 8 in. **16.** 3 ft 8 in. **18.** 6 gal
20. 4 h 55 min **22.** 3 ft 4 in. **24.** 21 min 40 s **26.** 4 lb 5 oz
28. 5 h 8 min **30.** 13 h 30 min **32.** 1 lb 13 oz **34.** No; the length needed is 16 ft 5 in. **36.** Under, 12 oz **38.** 20 ft 4 in. **40.** 3 h 45 min **42.** 11 weeks 6 days 1 h **44.** 2 gal 1 qt 1 pt **46.** 17 gal 14 oz
48. 229.5 gal **50.** 810 gal **52.** 304 s

Exercises 13.3
2. (*a*) **4.** (*b*) **6.** (*c*) **8.** (*a*) **10.** (*c*) **12.** (*b*) **14.** mm **16.** cm
18. km **20.** km **22.** mm **24.** cm **26.** 1.5 **28.** 7.7 **30.** 5
32. 0.15 **34.** 900 **36.** 450 **38.** 4 **40.** 7000 **42.** 110 **44.** 12
46. 625 **48.** 706.5

Exercises 13.4
2. (*b*) **4.** (*a*) **6.** (*b*) **8.** (*c*) **10.** (*a*) **12.** (*b*) **14.** kg **16.** g
18. g **20.** mg **22.** kg **24.** g **26.** 5 **28.** 3000 **30.** 12.5 **32.** 2
34. 19.5 billion kg

Exercises 13.5
2. (*b*) **4.** (*b*) **6.** (*a*) **8.** (*b*) **10.** (*a*) **12.** (*c*) **14.** cL **16.** mL
18. L **20.** mL **22.** cL **24.** L **26.** 4 **28.** 700 **30.** 12,000
32. 200 **34.** 5000 **36.** 40 **38.**

Calculator Exercises
2. 3.546 **4.** 8.226 **6.** 183 **8.** 224 **10.** 2270 **12.** 7.42
14. 60.42 **16.** 3.975 **18.** 88.57 km/h

Self-Test for Chapter 13
1. 96 (Sec. 13.1) **2.** 6 (Sec. 13.1) **3.** 48 (Sec. 13.1) **4.** 6 (Sec.

13.1) **5.** 6 ft 9 in. (Sec. 13.1) **6.** 7 min 30 s (Sec. 13.1) **7.** 11 ft 5 in. (Sec. 13.2) **8.** 2 lb 9 oz (Sec. 13.2) **9.** 15 h 20 min (Sec. 13.2) **10.** 4 lb 6 oz (Sec. 13.2) **11.** $1050 (Sec. 13.2) **12.** (*b*) (Sec. 13.3) **13.** (*b*) (Sec. 13.3) **14.** mm (Sec. 13.3) **15.** 3000 (Sec. 13.3) **16.** 5000 (Sec. 13.3) **17.** 8 (Sec. 13.3) **18.** 3.1 (Sec. 13.3) **19.** (*c*) (Sec. 13.4) **20.** (*b*) (Sec. 13.4) **21.** kg (Sec. 13.4) **22.** 3000 (Sec. 13.4) **23.** 1.35 (Sec. 13.4) **24.** 140 (Sec. 13.4) **25.** 7000 (Sec. 13.4) **26.** (*c*) (Sec. 13.5) **27.** (*b*) (Sec. 13.5) **28.** mL (Sec. 13.5) **29.** 2000 (Sec. 13.5) **30.** 3 (Sec. 13.5) **31.** 4.75 (Sec. 13.5) **32.** 240 (Sec. 13.5)

Pretest for Chapter 14

1. 8 (Sec. 14.1) **2.** −9 (Sec. 14.1) **3.** (a) 7; (b) −12 (Sec. 14.2) **4.** (a) −22; (b) 17 (Sec. 14.2) **5.** (a) −36; (b) 54 (Sec. 14.3) **6.** (a) 3; (b) −5 (Sec. 14.3) **7.** 30 **8.** −4 **9.** 4 **10.** 3

Exercises 14.1

2. False **4.** True **6.** True **8.** False **10.** True **12.** False **14.** True **16.** True **18.** False **20.** False **22.** True **24.** True **26.** 16 **28.** 12 **30.** 9 **32.** 15 **34.** 0 **36.** −18 **38.** < **40.** < **42.** > **44.** = **46.**

Exercises 14.2

2. 12 **4.** −13 **6.** 7 **8.** 4 **10.** −3 **12.** 1 **14.** −2 **16.** 0 **18.** 2 **20.** −15 **22.** −10 **24.** −4 **26.** 0 **28.** −25 **30.** −0.5 **32.** $-9\frac{2}{5}$ **34.** −0.9 **36.** $-\frac{1}{2}$ **38.** −5 **40.** 2 **42.** 0 **44.** −10 **46.** 2 **48.** −5 **50.** −21 **52.** −6 **54.** −40 **56.** 6 **58.** −3 **60.** 16 **62.** 0 **64.** 75 **66.** 10 **68.** −10 **70.** −0.5 **72.** −2 **74.** $-3\frac{1}{2}$ **76.** 5 **78.** −7 **80.** 2 **82.** $290 **84.** 577 ft below sea level **86.** $767 **88.** $404 **90.**

Calculator Exercises

2. 292 **4.** 130 **6.** −1477 **8.** −779 **10.** 342 **12.** 535

Exercises 14.3

2. −72 **4.** 75 **6.** −24 **8.** 24 **10.** −100 **12.** 0 **14.** 200 **16.** 81 **18.** −30 **20.** −125 **22.** 1.2 **24.** $-\frac{1}{6}$ **26.** −0.75 **28.** $\frac{2}{3}$ **30.** −8 **32.** −8 **34.** 81 **36.** −25 **38.** 60 **40.** −70 **42.** 300 **44.** 240 **46.** 5 **48.** 10 **50.** −6 **52.** −7 **54.** −5 **56.** 5 **58.** −7 **60.** 10 **62.** Undefined **64.** 0 **66.** −25 **68.** 5 **70.** $-\frac{3}{2}$ or $-1\frac{1}{2}$ **72.** 2 **74.** −1.5 **76.** 2 **78.** −4 **80.** 5 **82.** −3

Calculator Exercises

2. −675 **4.** −1736 **6.** 1349 **8.** −4.57 **10.** −4.46 **12.** 3.05 **14.** 3.68 **16.** 625

Exercises 14.4

2. 20 **4.** 8 **6.** −24 **8.** 45 **10.** 33 **12.** −5 **14.** 22 **16.** 20 **18.** 9 **20.** 64 **22.** 12 **24.** 4 **26.** 6 **28.** −25 **30.** 54 **32.** −78 **34.** −72 **36.** 80 **38.** 140 **40.** −8 **42.** 4 **44.** −4 **46.** 540 **48.** 104 **50.** 620 **52.**

Exercises 14.5

2. 5 **4.** 8 **6.** 4 **8.** −6 **10.** −2 **12.** 5 **14.** −9 **16.** −12 **18.** 5 **20.** 32 **22.** −36 **24.** 3 **26.** 1 **28.** −5 **30.** −2 **32.** −3 **34.** 5 **36.** −2 **38.** 12 **40.** −42 **42.**

Self-Test for Chapter 14

1. 9 (Sec. 14.1) **2.** 8 (Sec. 14.1) **3.** 7 (Sec. 14.1) **4.** 10 (Sec. 14.1) **5.** 6 (Sec. 14.2) **6.** −2 (Sec. 14.2) **7.** −12 (Sec. 14.2) **8.** 0 (Sec. 14.2) **9.** −9 (Sec. 14.2) **10.** 2 (Sec. 14.2) **11.** −3 (Sec. 14.2) **12.** −10 (Sec. 14.2) **13.** 10 (Sec. 14.2) **14.** 4 (Sec. 14.2) **15.** −28 (Sec. 14.3) **16.** 54 (Sec. 14.3) **17.** 75 (Sec. 14.3) **18.** 0 (Sec. 14.3) **19.** 48 (Sec. 14.3) **20.** 36 (Sec. 14.3) **21.** −4 (Sec. 14.3) **22.** 7 (Sec. 14.3) **23.** −5 (Sec. 14.3) **24.** −4 (Sec. 14.3) **25.** 0 (Sec. 14.3) **26.** Undefined (Sec. 14.4) **27.** 20 (Sec. 14.4) **28.** −11 (Sec. 14.4) **29.** 27 (Sec. 14.4) **30.** −28 (Sec. 14.4) **31.** −4 (Sec. 14.4) **32.** 2 (Sec. 14.4) **33.** 8 (Sec. 14.5) **34.** −3 (Sec. 14.5) **35.** 6 (Sec. 14.5) **36.** 56 (Sec. 14.5) **37.** −30 (Sec. 14.5) **38.** −9 (Sec. 14.5) **39.** 6 (Sec. 14.5) **40.** 2 (Sec. 14.5)

Summary Exercises for Chapters 12–14

1. 250,000 **2.** 5000 (Sec. 12.2) **3.** 84 **4.** 128 **5.** 240 **6.** 20 (Sec. 12.1) **7.** 8 ft 10 in. **8.** 9 lb 4 oz (Sec. 13.2) **9.** 12 ft 6 in. **10.** 2 lb 9 oz **11.** 10 h 30 min **12.** 2 min 9 s (Sec. 13.2) **13.** 14 ft 8 in. **14.** The three smaller bottles (Sec. 13.2) **15.** 5000 **16.** 3000 **17.** 300 **18.** 4 **19.** 3 **20.** 2 (Sec. 13.5) **21.** 600 **22.** 132 **23.** 2.4 **24.** 562 **25.** 5.6 **26.** 2 (Sec. 13.5) **27.** *a* **28.** *a* (Sec. 13.5) **29.** 12 **30.** 8 **31.** 3 **32.** 20 **33.** 8 **34.** −5 **35.** −11 **36.** $\frac{1}{2}$ **37.** −5 **38.** −4 **39.** −5 **40.** −15 **41.** 5 **42.** −4 **43.** $6\frac{1}{2}$ **44.** −4 **45.** 36 **46.** −80 **47.** −1.5 **48.** $\frac{3}{10}$ **49.** 16 **50.** 42 **51.** −4 **52.** 11 **53.** Undefined **54.** −0.075 **55.** $\frac{7}{6}$ or $1\frac{1}{6}$ **56.** 0 **57.** −24 **58.** 22 **59.** 68 **60.** −18 **61.** −1 **62.** −4 **63.** 8 **64.** −9 **65.** −9 **66.** 5 **67.** $-\frac{3}{4}$ **68.** 40

Cumulative Test for Chapters 12–14

1. 96 **2.** 96 **3.** 9 ft 8 in. **4.** 11 lb 5 oz **5.** 1 min 35 s **6.** 8000 **7.** 3 **8.** 5 **9.** 250 **10.** 1995 and 1996 **11.** −20 **12.** 7 **13.** 12 **14.** 20 **15.** −18 **16.** 0 **17.** $1\frac{1}{3}$ **18.** −2 **19.** 0 **20.** 5 **21.** 96 **22.** −60 **23.** −2.5 **24.** Undefined **25.** −60 **26.** 27 **27.** 24 **28.** 3 **29.** −8 **30.** 7

Index